# Food Science: Cereals and Oilseeds

# Food Science: Cereals and Oilseeds

Edited by **Johann Wells**

SYRAWOOD
PUBLISHING HOUSE
New York

Published by Syrawood Publishing House,
750 Third Avenue, 9th Floor,
New York, NY 10017, USA
www.syrawoodpublishinghouse.com

**Food Science: Cereals and Oilseeds**
Edited by Johann Wells



International Standard Book Number: 978-1-68286-309-1 (Hardback)



Printed in China.

# Contents

# Preface

Food science is an interdisciplinary field of study. It has undergone rapid changes in the past decade. This book on food science elucidates in particular the tools and techniques involved in cereals and oilseeds harvesting. It brings forth some of the most innovative researches on topics such as effect of extraction conditions on the yield and quality of oil, in vitro and in vivo responses of different treating agents, etc. Scientists and students actively engaged in this field will find this book full of crucial and unexplored concepts.

The information contained in this book is the result of intensive hard work done by researchers in this field. All due efforts have been made to make this book serve as a complete guiding source for students and researchers. The topics in this book have been comprehensively explained to help readers understand the growing trends in the field.

I would like to thank the entire group of writers who made sincere efforts in this book and my family who supported me in my efforts of working on this book. I take this opportunity to thank all those who have been a guiding force throughout my life.

**Editor**

# Physiological growth indices in winter rapeseed (*Brassica napus* L.) cultivars as affected by drought stress at Iran

**Payam Moaveni[1], Ahmad Ebrahimi[2] and Hossein Aliabadi Farahani[3]***

[1]Islamic Azad University, Shahr-e-Qods Branch, Iran.
[2]Islamic Azad University, Iranshahr Branch, Iran.
[3]Islamic Azad University, Takestan Branch, Iran.

**This experiment was carried out to study the effect of drought stress on physiological growth indices of winter rapeseed cultivars in Karaj, Iran using split-plot design with three replications to study the effect of drought stress on oil yield of winter rapeseed cultivars. The main-plot factors included: Four levels of irrigation regimes (control = no drought stress, irrigation interrupted during the flowering stage, irrigation interrupted during the silique formation stage and irrigation interrupted during the seed filling stage) while the sub-plot factor included three cultivars (Zarfam, Okapi and Licord) of winter rapeseed (*Brassica napus* L.). The results showed that total dry matter (TDM), leaf area index (LAI), relative growth rate (RGR) and crop growth rate (CGR) were highly significantly different among the rapeseed cultivars, whereas, drought stress had effects of practical significance on TDM, LAI, RGR and CGR. The results also showed that the highest TDM, LAI, RGR and CGR were obtained from the cultivar, Zarfam under no-drought condition. The findings firmly established that the drought stress sorely reduces physiological growth indices of winter rapeseed cultivars under conditions of Karaj in Iran.**

**Key words:** Drought stress, physiological growth indices, winter rapeseed cultivars.

## INTRODUCTION

Drought stress is very important in areas where crop production is essentially rain-fed (Boyer, 1982; Ludlow and Muchow, 1990). Drought stress causes an increase in solute concentration in the soil and root-zone of the plant leading to an osmotic flow of water out of plant cells. This in turn causes the solute concentration inside plant cells to increase, thus lowering water potential and disrupting membranes along with essential processes like photosynthesis. These drought-stressed plants consequently exhibit poor growth and yield. In worst situations, the plants completely die. Certain plants have devised mechanisms to survive such low soil moisture conditions. These mechanisms have been classified as tolerance, avoidance or escape (Kramer and Boyer, 1995; Neumann, 1995). Drought tolerance is the ability of crops to grow and produce seeds or propagules under conditions of water deficit. A long-term drought stress effects on plant metabolism vary with plant growth stage, water storage capacity of the soil and physiological aspects of plant. Drought tolerance of cultivated crop plants is different from that of wild plants. When crop plants get encountered with severe water deficit, they often die or seriously lose yield while conversely wild plants face similar conditions of drought, they survive well with often little/no yield loss. Because of water deficit in most of the arid regions, tolerance of crops to drought has always been of great importance for the plant breeders (Alizadeh, 2004). According to Lessani and

*Corresponding author. E-mail: farahani_aliabadi@yahoo.com.

Mojthaedi (2006), one of the main aspects of drought tolerance is the ability of plant cells to survive severe water loss without being affected by deleterious damages. When the plant cell dries up, usually the vacuole crumples more than the cell wall and thus results in tearing up of the protoplasm. Plant yield losses under insufficient water have always been an important issue for plant breeders to improve using drought tolerance indices to select genotypes (Mitra, 2001). Investigations by Champolivier and Merrien (1996) revealed that yield and yield components of rapeseed were affected by water shortage occurring from flowering to the end of seed set. Most number of seeds per plant was affected, although some compensation occurred when the water supply was restored. The thousand seed weight was only affected by low water stress even if it occurs during the period when siliques were swelling to seeds coloring. The results also demonstrated that a marked reduction in oil concentration was evident when water deficit occurred from anthesis to maturity. Furthermore, they found that oil and protein concentrations were inversely related. Therefore, the aim of the present investigation is to determine the effects of drought stress on physiological growth indices of winter rapeseed cultivars in Karaj Region, Iran.

### MATERIALS AND METHODS

This study was conducted on the experimental field of the Seed and Plant Improvement Institute, Karaj at Iran (35°58' N, 51°06' W; 1313 m above sea level) in 2004. The experiment was laid out in a split-plot design with three replications. The main-plot factors included: Four levels of irrigation regimes (control=no drought stress (NDS), irrigation interrupted during the flowering stage (IIFS), irrigation interrupted during the silique formation stage (IISFS1) and irrigation interrupted during the seed filling stage (IISFS2)) while the sub-plot factor included three cultivars (Zarfam, Okapi and Licord) of winter rapeseed (*Brassica napus* L.). Phosphorus and potassium were applied each at the rate of 100 kg ha$^{-1}$ in the form of triple super phosphate (TSP) and $K_2O$. While the former was applied at planting, the latter was applied right just cultivation. Nitrogen fertilizer was added in three splits, whereby the first application (50% of 100 N ha$^{-1}$) was made at the time of cultivation while the second and third, applications, each constituting 25% of the total rate, were respectively made at the stages of stem elongation and beginning of flowering. In order to the determination of TDW, from 20 days after cultivation to harvesting time, 10 plants were selected randomly in all plots each 15 days regularly. Samples were placed under 75°C in oven for 48 h and were weighed by scale and then TDW was determined in sampling stages. To determine LAI, leaves area upon samples were estimated by leaf area meter before placing in oven and then determined LAI in each sampling stage. Finally, RGR and CGR were determined by the following formulas (Aliabadi et al., 2009):

$$\mathrm{RGR} = \frac{\mathrm{Ln}W_2 - \mathrm{Ln}W_1}{T_2 - T_1}$$

$$\mathrm{CGR} = \frac{W_2 - W_1}{T_2 - T_2} \times \mathrm{GA}$$

## RESULTS

The results showed that total dry matter (TDM), leaf area index (LAI), relative growth rate (RGR) and crop growth rate (CGR) were all highly significantly different among the rapeseed cultivars, whereas drought stress had effects of practical significance on TDM, LAI, RGR and CGR. The results also showed that the highest TDM, LAI, RGR and CGR were obtained from Zarfam cultivar under no-drought condition (Figures 1 - 4).

## DISCUSSION

As the results of this study indicated that drought stress has negative effects on physiological growth indices of winter rapeseed. This implies that rapeseed plants use different mechanisms to resist drought stress. Plant's photosynthesis reduces under drought stress because the leaf area, plant height and lateral stem number reduce under this condition. Under drought stress, stomata become blocked or half-blocked and this leads to a decrease in absorbing $Co_2$ and on the other hand, the plants consume a lot of energy to absorb water, these cause a reduction in producing photosynthetic matters. Drought stress reduced the amount of the oil, because in case of stress, more metabolites are produced in the plants and their substances prevent it from oxidization in the cells. It was also seen that as the increase of drought stress, its height, stem diameter and stem yield decreased. Shoot reduction could be due to the reduction in the area of photosynthesis, drop in chlorophyll production, the rise of the energy consumed by the plant in order to take in water, increase the density of the protoplasm, change respiratory paths and the activation of the path of phosphate pentose, or the reduction of the root deploy, etc. The physiological growth indices were decreased in winter rapeseed cultivars under drought stress.

## Conclusion

The investigation showed that physiological growth indices were reduced under drought stress. This condition can be the most important environmental factor for the increase of TDM by control of irrigation. Therefore, the selection of cultivars that perform well over a wide range of environments can increase quantity and quality yields of winter rapeseed under drought stress. Consequently, our findings can help farmers and

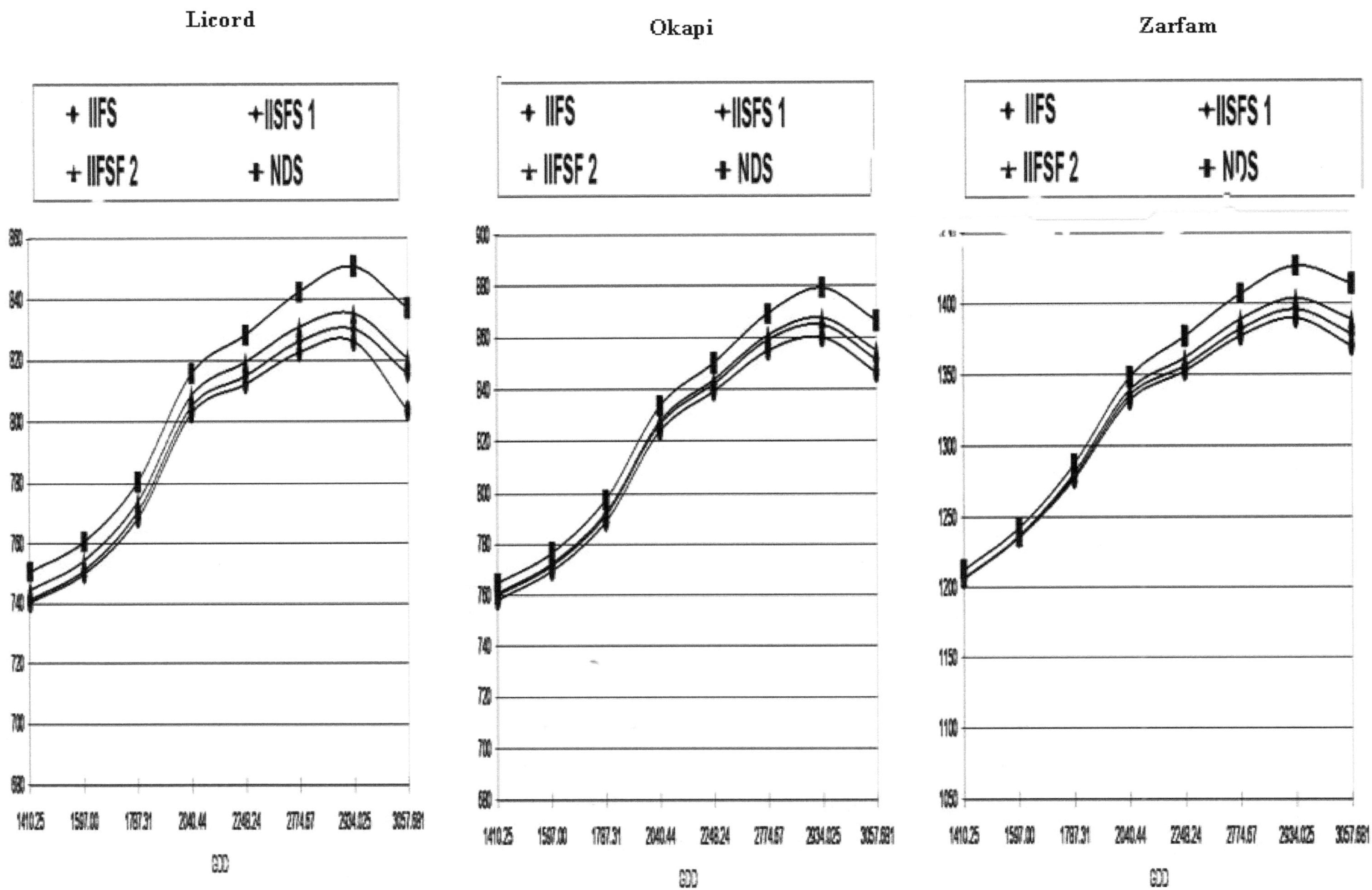


**Figure 1.** TDM variations in winter rapeseed cultivars under drought stress conditions.

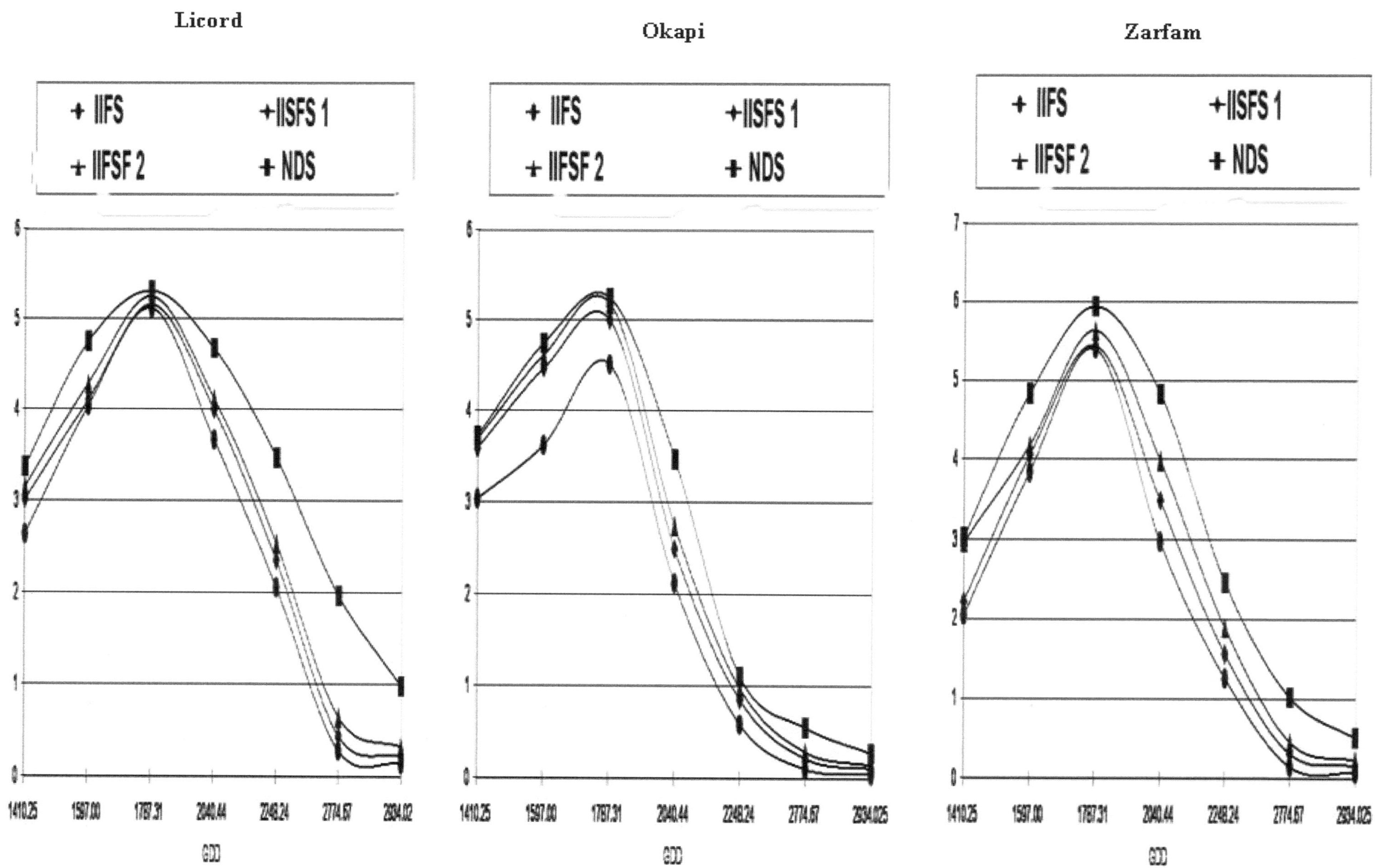


**Figure 2.** LAI variations in winter rapeseed cultivars under drought stress conditions.

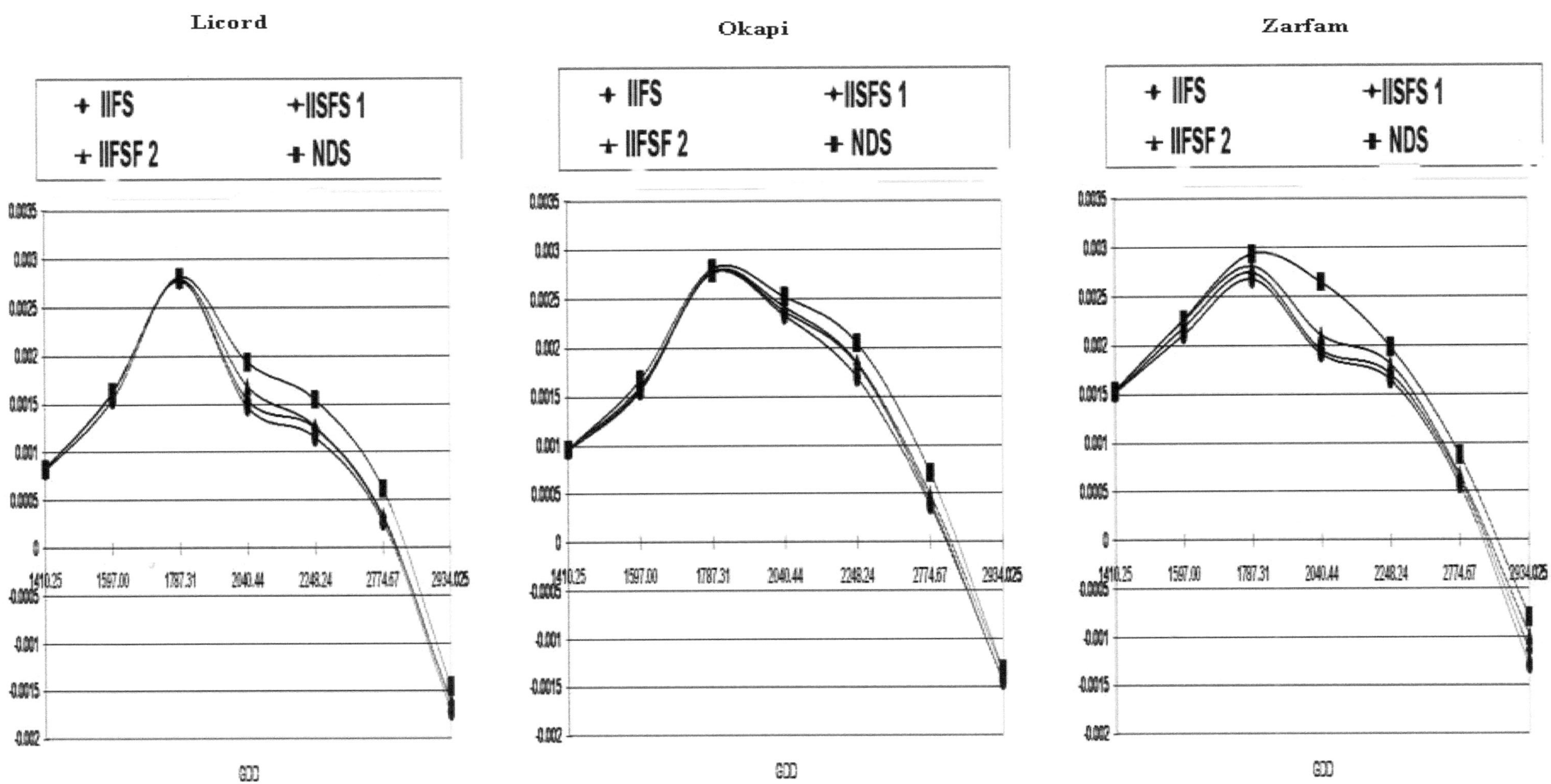


**Figure 3.** RGR variations in winter rapeseed cultivars under drought stress conditions.

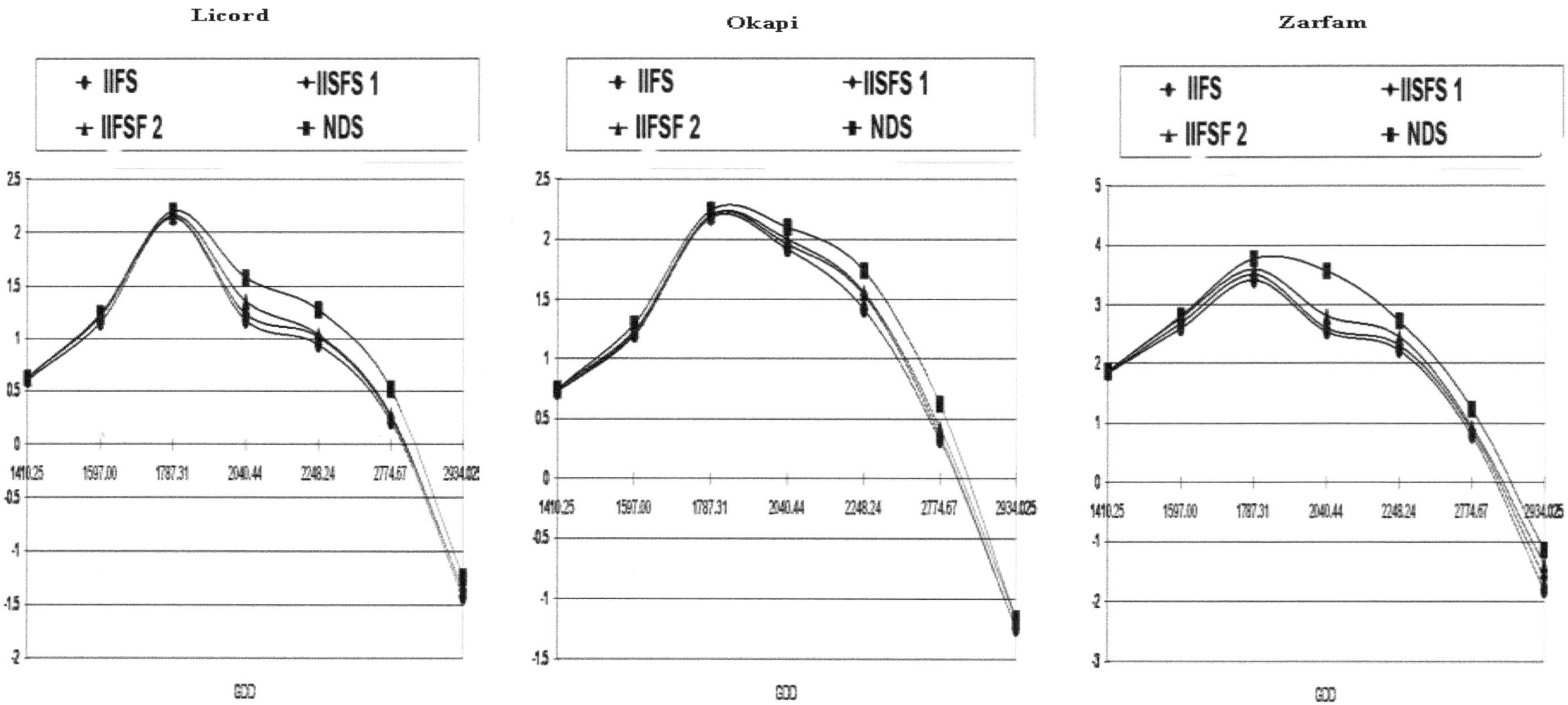


**Figure 4.** CGR variations in winter rapeseed cultivars under drought stress conditions.

agricultural researchers for management and proper use of irrigation in farming of winter rapeseed under drought stress.


## ACKNOWLEDGMENTS

The authors acknowledge the financial support of Islamic Azad University, Shahr-e-Qods Branch, Iran and Islamic Azad University, Iranshahr Branch, Iran.


## REFERENCES


Aliabadi FH, Valadabadi SAR, Daneshian J, Khalvati MA (2009). Evaluation changing of essential oil of balm (*Melissa officinalis* L.) under water deficit stress conditions. J. Med. Plant. Res. 3(5): 329–333.

Alizadeh A (2004). Soil, Water and Plant Relationship. 4th Edn., Emam Reza University Press, Mashad, Iran, ISBN: 964-6582-57-5.

Boyer JS (1982). Plant productivity and environment. Sci. 218: 443-448.

Champolivier L, Merrien A (1996). Effects of water stress applied at different growth stages to *Brassica napus* L. var. *oleifera* on yield, yield components and seed quality. Eur. J. Agron. 5(3-4): 153-160.

Kramer PJ, Boyer JS (1995). Water Relations of Plants and Soils. Academic press. San Diego, USA. pp. 1-495.

Lessani H, Mojtahedi M (2006). Introduction to Plant Physiology. 6th Edn., Tehran University Press, Tehran, Iran, ISBN: 964-03-3568-1.

Ludlow MM, Muchow RC (1990). A critical evaluation of the traits for improving crop yield in water limited environments. Adv. Agron. 43:107–153.

Mitra J (2001). Genetics and genetic improvement of drought resistance in crop plants. Curr. Sci. 80: 758-762.

Neumann PM (1995). The role of cell wall adjustment in plant resistance to water deficits (Review and Interpretation). Crop Sci. 35:1258-1266.

# Export performance of oilseeds and its determinants in Ethiopia

**Hailegiorgis Biramo Allaro**

Haramaya University, College of Agricuture and Environmental Science, Department of Agricutural Economics
P. O. Box: 138, Dire Dawa, Ethiopia. E-mail:hailegiorgisbiramo@yahoo.com.

**This research examines export performance of oilseeds in Ethiopia using macroeconomic time series data from annual reports of NBE (National Bank of Ethiopia) over the period 1974 to 2009. Analyses of oilseeds export performance through time divulges that the country has not yet diversified the commodity composition and structure of its export, in that its export earnings depend on only a few agricultural products. The findings of the study reveal that the country needs to break away from it's heavily depends on traditional export commodities for which it is a marginal exporter, thus a price taker. The results also show that the estimated coefficients of real output (Ry) and nominal exchange rate (e) are statistically significant. This discloses that real output and nominal exchange rate have positive brunt on the export performance of oilseeds in Ethiopia.**

**Key words:** Africa, Ethiopia, export performance, oilseeds.

## INTRODUCTION

Nations' of the world differ in their resource endowments and level of technology applied in the production of goods and services. The engagement of nations in the international trade depends upon a nation's specialisation in the production of goods in which they have comparative advantages constructs room for improvement of the welfare of the society as a whole. This theory traces back to the last half of the 18th century, the time when Adam Smith realized the importance of specialization and trade in his Wealth of Nations. Subsequently, many economists advocated the contribution of international trade for welfare of nations (as the engine of growth) in the overall process of economic development (Onafowora and Owoye, 1998; Arndt, 1999). The international trade of SSA (Sub-Saharan African) countries are mainly based on exporting primary agricultural commodities in which they have comparative advantages due to cheap labor and tropical climate. In the context of Ethiopia, coffee alone constitutes more than 50% of the total agricultural export of the country. Accordingly, different studies show that diversifying the export sector towards non-traditional agricultural commodities is crucial to attain stability in export earnings of a country (Alwang and Siegel, 1994).

Agriculture is the mainstay of the Ethiopian economy. Its share accounts for more than 40% of the total GDP, 50% of foreign currency earnings and above 80% of employment creation. Both industry and services are dependent on the performance of agriculture, which provides raw materials, generates foreign currency for the import of essential inputs and food for the fast growing population. In spite of its importance in the national economy, agriculture is based on subsistence farm households, whose modes of life and operation have remained unchanged for centuries.

The objective of this study is to scrutinise oilseeds export performance in Ethiopia with the application of adopted Goldstein and Khan (1985) imperfect substitution model and annual macroeconomic time series data from the annual reports of NBE for the period 1974 to 2009. However, as results of this study reveal, despite this research attempt in examining the export performance of oilseeds in Ethiopia, there has been no study conducted on export performance of oilseeds in the country. The basis for the analyses is to validate empirically the hypothesis that exports from Ethiopia is rising since the reform was launched.

## WORLD OILSEEDS PRODUCTION AND MARKETING

The worldwide major oilseeds are: Soybean, rapeseed, cottonseed, safflower seed, groundnut, palm, copra, sesame, Castro seed, maize oil and coconut oil. The major producers of oilseeds are USA, China, Brazil, India, Malaysia, Indonesia, EU-15 Countries, Central Europe, Canada, and Argentina. According to FAO database, the world oilseeds production is 449 million tons for 2010. Ethiopia has altitudes from below sea level up to more than 4,500 m above sea level with very different climates. This enables Ethiopia to grow a wide range of oilseeds, in which it has a long tradition. Oilseeds are the second largest export earner for the country after coffee in which more than 3 million smallholders are involved in its production. Exports actually consist of sesame and Niger seed, for which there is a growing demand in the world market. Ethiopia ranks among the top 5 world producers of sesame seed and linseed. It is also an important producer of Niger seed. Castoro, linseed and safflower have good export potential. The growing demand in the world market for these specialty products and the available capacity to expand production could make oilseeds turn into one of the engines of economic growth of Ethiopia. Groundnuts, safflower, rapeseed and many other oilseeds are produced on a limited scale. The main production regions of sesame seed are Tigray, Amhara, Oromia and Benshangul Gumuzi (Bennet, 2004).

## DOMESTIC USE, IMPORT AND EXPORT OF OILSEEDS IN ETHIOPIA

Sesame seed is the main oilseed export product. Niger seed, as second, is exported mainly to the US as bird seed. In 2005 and 2006 Ethiopian exports further increased, whereby China came up as a new market. Export of sesame seed has grown in double digits each year from 1998 to 2006:50,000 tons in 1998 and more than 100,000 tons in 2006. In the second half of 2006, the main export markets for Ethiopian sesame seed were China, Israel and other countries in the Middle East and Turkey. The EU had a market share of 8% (CSA, 2006). The major importers of Ethiopian sesame seed in the EU are Greece, Germany, the Netherlands and the UK. Japan is the biggest world importer of sesame seed. Sesame oil, particularly from roasted seed, is an important component of Japanese cooking. Traditionally, this is the principal use of the seed. Japanese are very critical of quality. Because of quality problems in the past, Ethiopia is not an important exporter for Japan. However, it seems that via China, the volume of Ethiopian oilseeds exported to Japan is increasing. All other oilseed crops (soybeans, groundnuts, cottonseed etc.) grown in Ethiopia are almost entirely used domestically. Reliable figures of domestic use are not available. Most of the oil is consumed as crude oil. Ethiopia is a net importer of refined oil, mainly refined soybean and palm oil. Palm oil is mainly imported from Malaysia, Singapore and the United Arab Emirates. Italy is the major supplier (75%) of soybean oil, followed by Turkey (10%) (FAO, 2008).

## OILSEEDS EXPORT PERFORMANCE AND ITS DETERMINANTS IN ETHIOPIA

Export performance is the relative success or failure of the efforts of a firm or nation to sell domestically produced goods and services in other nations. Export performance can be described in objective terms such as sales, profits, or marketing measures or by subjective measures such as distributor or customer satisfaction. Determinants of export performance can be split into external and internal components. External components include market access/entry conditions and a country's location which include international markets. Internal components are related to supply-side conditions. Foreign demand is influenced by various elements. Firstly, it is strongly linked to geography (the structural component). Typically, countries at the center of a fast growing region are more likely to benefit than countries situated outside that region. Second, it is likely to be related to competition and trade policy (the market access/entry component), which could have, in principle, a similar impact on trade than geography. Finally, both the quantity and quality of physical infrastructures (the development component) are expected to play important roles (Lages et al., 2004).

Trade policy is always controversial in its impact on the overall economy. Advocates of free trade believe all nations benefit from eliminating import tariffs, quotas, and export subsidies. The Ethiopian government has been moving toward a freer trade policy ever since the liberalization. New trade agreements for agricultural commodities and other manufactured goods have reduced tariffs and quotas in the last 10 years. As a consequence, the most competitive and productive sectors of the economy benefit the most from a liberalized trade policy.

Access to foreign markets is a critical determinant of export performance. It relates directly to the characteristics of the trading partner countries, such as the size of their market and transport facilities, and inversely to their own internal transport costs. It also depends positively on the size of the export basket and the number of differentiated items and their prices, which in turn are affected by market entry conditions. Transborder costs, which also include tariff and non-tariff barriers, have the expected negative impact on foreign market access (Anderson, 2004, Fugazza, 2004).

Despite the worldwide fall in trade barriers that has occurred in the last two decades, export performance has varied substantially across countries. World exports increased almost 220% in twenty years. The figure jumps to 720% for East Asian and Pacific countries and falls to 80% for Sub-Saharan countries. The exports of "best performers", such as the Republic of Korea, China, Cambodia and Vietnam, have grown by more than 15% annually over the whole period. "Worst performers", mostly African and Latin American countries, have negative annual growth rate records in at least one decade. As a result of various trade negotiations and autonomous reforms, access to international markets has improved in the last twenty years. Nonetheless, it is likely that there is still much to gain from further improvements in market access conditions. Concerns have also been raised about the necessity to improve supply conditions. Supply conditions are fundamental in defining the export potential of an economy. For a given level of access to international markets, countries with better supply conditions are expected to export more (Redding and Venables, 2004).

Recently, Redding and Venables (2004) investigated the relative contribution towards export performance of international linkages relative to internal geographical factors. They found that most of the differential in export performance of various countries and over the last three decades can be due to differences in the evolution of external components. Nevertheless, they found that internal components related to supply capacity such as internal geography and institutional quality also have played a significant role in explaining the observed differential in export performance. Accounting for unobservable heterogeneity should allow the identification of any differences in the effect of and importance of export performance components, which are linked to the degree of development of the external sector itself. In other words, the techniques used here allow for the testing for non-linearities in the relationship between export performance and its components. While dynamic panel techniques would seem to be the most desirable approach, data availability is likely to restrict their implementation. In this context, cross-sectional analysis proves to be a viable alternative. Regression techniques which are able to account for unobserved heterogeneity across countries, namely quantile regressions, are used. Moreover, more emphasis is put on the determination and impact assessment of variables related to supply conditions. This is done with the aim of determining as clearly as possible what are policy implications. The study revealed important differences across countries and regions when looking at their respective determinants of export performance. External and internal components prove to have played an equal role in determining export performance for Asian countries. Their improvement in the South East and Pacific region appears to be high relative to that observed in any other region. Sub - Saharan African countries owe their export performance to the evolution of external components. The latter were strong enough to more than offset a relative deterioration of their internal production conditions. Further investigation also indicates that good internal conditions are necessary to obtain good export performance. Particular attention should be paid to the macroeconomic dimension. Good infrastructures and non-stringent institutions are also necessary to put the export sector on a durable development path.In addition; there is scope for promoting a dynamic process of diversification across and within sectors. Constant efforts to support diversification are particularly relevant for commodities exporters' when a secular downward trend is observed in volatile commodities prices (Redding and Venables, 2004).

## ETHIOPIAN OPPORTUNITIES FOR SPECIALTY OILSEEDS

Most oilseed crops -soybeans, cotton seed, rapeseed etc. - grown in Ethiopia are also grown in many other countries in large volumes. For these oilseed crops, it will be very difficult for Ethiopia to compete on the world market due to its relatively low volumes and high handling and transport costs. The Ethiopian production for these crops is mostly less than 0.1% of the world production. To achieve a beneficial market position on the world market huge efforts are required, which will be out of scope for Ethiopia in the near future. These crops, however, can be of high importance for the domestic market, as food crop. Most potential for the Ethiopian oilseeds sector concerns the following specialty Oilseeds: sesame seed, safflower seed, linseed, Niger seed and castor beans (Abate, 2006).

## METHODS

The study was persistence on secondary data sources. When a close look was made at the time-series estimates of Ethiopia's major macroeconomic variables, it shows that different sources report different values for the same variables. In order to avert this problem and ensure the consistency as well as comparatively of results annually published time- series data were extracted from the published annual reports of the NBE (National Bank of Ethiopia). Annual data were favoured to the purpose of precision for most of the variables. Long term data are not available for the variable considered in the study in the case of Ethiopia. The ideal national account data available to undertake time series based studies is from 1963 onwards which still lack accuracy.

### Data coverage

The study covers 36 years, from 1974 to 2009. Review of the different publications of the NBE (Annual Reports and Quarterly Bulletins) reveals that data are available on the export earnings,

**Table 1.** ADF unit root tests.

| Variables | ADF test statistics | Order of integration | Critical values at 5% | Critical values at 10% |
|---|---|---|---|---|
| $\log p_t^d$ | -5.725716 | I(2) | 3.55 | 3.21 |
| $\log p_t^w$ | -5.161600 | I(2) | 3.55 | 3.21 |
| $\log Ry_t$ | -6.640211 | I(2) | 3.55 | 3.21 |
| $\log e_t$ | -5.039660 | I(2) | 3.55 | 3.21 |

price (unit values), and quantities of major export commodities since 1973. The latest available data on the external trade of the country are 2009, the last year of study period (Annex Tables 1 to 3).

### Definitions of export performance

A conceptual definition of export performance addresses two parts:

Export and performance: Export is the international marketingrelated decisions and activities of internationally active firms (Cavugil and Neviv, 1981). The overtone of the word performance, in the literature sense, does not pose any problem for it is the act of carrying out or accomplishing something such as a task or action. When it comes to economics, this word has been defined in many ways and no unifying principle has underlined its quantification. However, in the context of current study, 'Export performance'is defined as:

1. The success or failure of the efforts of a nation to sell domestically produced goods and services in other nations markets (Zou and stan, 1998);
2. The composite outcome of a nation's international sales (Shoham, 1996), and
4. The three sub-dimensions which encompasses sales, profit and growth (Madsen, 1987).

### Model specification

The study signifies export performance of oilseeds in Ethiopia as a function of domestic price, world price, real output and nominal exchange rate. The analysis is expressed with the adopted Goldstein and Khan (1985) imperfect substitution model expressed as follows:

$$OilseedsE\text{export}\, performane(OlsXP) = f(P^d, p^w, Ry, e) \qquad (1)$$

where, OlsXP refers oilseeds export performance; $p^d$, domestic price; $p^w$ , world price; Ry, real out put and e, nominal exchange rate.

At estimation stage, taking logs of the variables in Equation (1) and differentiating with respect to time gives the trend of exports as:

$$\log OlsXP_t = \beta_0 + \beta_1 \log p_t^d + \beta_2 \log p_t^w + \beta_3 \log Ry_t + \beta_4 \log e_t + \varepsilon_t \qquad (2)$$

where $\beta$'s are unknown parameters to be estimated, t is time in years (1974 to 2009) and $\varepsilon$ is random terms that are independently and identically distributed with mean zero and variance[2] ($\delta^2$). To estimate Equation (2) the time-series approach was applied. The empirical results were tested using Eviews 3 and SPSS 15.

## RESULTS AND DISCUSSION

### Empirical estimation

The classical linear regression model (CLRM) and ECM (error correction model) are used to estimate the data. Prior to running the estimation, model diagnostic tests and corrections are made. These include heteroscedasticity, autocorrelation, multicollinearity and non stationary in the data. In order to detect heteroscedasticity, a plot of OLS residuals against the dependant variable ( $\log OlsXP$ ) is made (Annex Figures 1 to 4). Although the model passes ANOVA[1] test, plot of OLS residuals against time and formal test to detect autocorrelation using partial autocorrelation function (PACF) are done. The estimation begins with the testing of variables for unit roots to determine whether they can be considered as a stationary or non-stationary process. Table 1 presents the Augmented Dickey Fuller (ADF) tests of variables. The tests showed that all the variables were non-stationary at level. The variables were stationary at second difference. Critical values for tests were found to be -3.55 and -3.21 at 5 and 10%, respectively. Annex Tables (4a to 4e) gives details of unit root test outputs of variables.

To examine whether the integrated variables are cointegrated, it was modelled using variables to achieve stationarity which leads to loss of long-run information. The concept of cointegration implies that if there is a long-run relationship between two or more non-stationary variables, deviations from this long-run path are stationary. Johansen's cointegration multivariate procedure is used to establish whether the variables are cointegrated in the long run. As result, the likelihood ratio indicates one co-integrating equations at 5% significance

[1] ANOVA-Analysis of Variance

**Table 2.** Co-integration tests for logrxt, logpdt, logpwt, logryt and loget.

| Hypothesized no. of (CE) | Eigen value | Likelihood ratio | 5% critical value | 1% critical value |
|---|---|---|---|---|
| r=0 | 0.70 | 85.18 | 68.52 | 76.07 |
| r≤1 | 0.51 | 43.98 | 47.21 | 54.46 |
| r≤2 | 0.35 | 19.80 | 29.68 | 35.65 |
| r≤3 | 0.10 | 5.34 | 15.41 | 20.04 |
| r≤4 | 0.05 | 1.87 | 3.76 | 6.65 |

The test assumes linear deterministic trend in the data.

**Table 3.** Estimations results using ECM.

| Variables | Coefficient | t-statistic | Sign. |
|---|---|---|---|
| Constant | -0.0051(0.02) | -0.285472 | 0.78 |
| $\log p_t^d$ | -0.5192(0.71 | -0.731061 | 0.47 |
| $\log p_t^w$ | 0.7811(0.72) | 1.077986 | 0.29 |
| $\log Ry_t$ | 0.2678(0.12)** | 2.218024 | 0.03 |
| $\log e_t$ | 2.004(0.75)* | 2.683676 | 0.01 |
| ECT | 1.6621(3.72) | 0.446621 | 0.66 |
| $R^2$ =0.56; F= 7.41 (0.00)*(**) | | | |

*(**) shows significance at 1% (5%) significance level and the number in the brackets refers standard error.

level. In other words, it accepts alternative hypothesis of having one co-integrating vector. Since the test statistic (85.18*)* is greater than the 95% critical value (68.52) of the likelihood ratio test, it is possible to reject the null of more than one co-integrating vector (Annex Table 5). The maximum Eigen value test starts with the null hypothesis of at most r co-integrating vector against the alternative of r+1. The result for maximum Eigen value test confirms the rejection of the null hypothesis; that is, no co-integrated vectors. Therefore, both maximum Eigen value and likelihood ratio indicate that there is one co-integrating equations at 5% significance levels (Table 2). After identifying the co-integrating equation, unit root test carried for the equation indicates that the equation is integrated of order AR I (2). Furthermore, unit root test with intercept shows that the equations pass's ADF test (Annex Table 6).Then equation entered into the final ECM estimation model (Annex Table 7). The equation is:

ECT=logpdt-0.997538*LOGPWT-0.000323*LOGRYT-1.003508*LOGET +0.004729

Estimation is done for the equation using ECM. The results indicate that $\log Ry_t$ and $\log e_t$ are significant and positive while others were insignificant. The significance of real output and nominal exchange rate highlight that these variables as the imperative factors of oilseeds export performance. Thus, 0.2678 and 2.004 of $\log Ry_t$ and $\log e_t$ respectively indicate the short run impact of real output and nominal exchange rate which implies that in the short run, real output and nominal exchange rate cause export growth of oilseeds. Although, other variables are insignificant that can not explain explicitly the short run impact on export performance. The long runs estimates suggest that logpwt, logryt and loget are positively related to $\log Olsxp_t$ while logpdt negatively. The insignificance of error correction term (ECT) shows that this variable is exogenous in the given model. Since all the variables are in their logarithmic form, it is possible to attach a suppleness meaning to the coefficients. Consequently, a percentage change in $\log p_t^w$, $\log Ry_t$ and $\log e_t$ are associated with 0.7811, 0.2678 and 2.004 in the same direction with $\log olsxp_t$, while $\log p_t^d$ is associated with 0.0051 in different direction with $\log olsxp_t$. The result of $R^2$ is 0.56 (56%) which reveals that 56% of Ethiopian oilseeds export performance is caused by real output and nominal exchange rate, while 44% is by other variables which were not included in the model. Furthermore, F-statistic is significant which implies that the model; $\log OlsXP_t = \beta_0 + \beta_1 \log p_t^d + \beta_2 \log p_t^w + \beta_3 \log Ry_t + \beta_4 \log e_t + \varepsilon_t$ fit (Table 3).

**Table 4.** Coefficient diagnosis.

| Model | | Unstandardized coefficients | | Standardized coefficients | t | Sig. |
|---|---|---|---|---|---|---|
| | | B | Std. error | Beta | | |
| 1 | Constant | 4.041 | 0.208 | | 19.453 | 0.000 |
| | $\log p_t^d$ | -0.803 | 0.865 | -0.613 | -0.929 | 0.360 |
| | $\log p_t^w$ | 1.205 | 0.924 | 0.368 | 1.304 | 0.202 |
| | $\log Ry_t$ | 0.350 | 0.057 | 0.327 | 6.129 | 0.000 |
| | $\log e_t$ | 1.899 | 0.878 | 1.204 | 2.163 | 0.038 |

The constant ($\beta_0$= 4.041) states that there are others variables that were not comprised in the model to explain export performance. The estimators of $\log p_t^d$ ($\beta_1$=-0.803), $\log p_t^w$ ($\beta_2$=1.205), $\log Ry_t$ ($\beta_3$=0.350) and $\log e_t$ ($\beta_4$=1.899) signify that $\log p_t^w$, $\log Ry_t$ and $\log e_t$ positively and significantly explain oilseeds export performance of Ethiopian, while $\log p_t^d$ negatively (Table 4).

## Conclusions

This paper has evaluated the magnitude of trends of oilseeds export performance in Ethiopia over the period 1974 to 2009. This study investigates empirically the trends of export performance by analysing prices (domestic and world), real output and nominal exchange rate. The evidence from this study suggests that real output and nominal exchange rate significantly influence oilseeds export performance. It was also revealed that during the reform period, oilseeds export showed improvement. It was inferred that oilseeds export performance demands an appropriate macroeconomic incentive environment and complementary structural policies. Thus, Ethiopia's oilseeds export performance will be determined primarily by its domestic policies. It was scrutinised that despite the generally open trade regime, industrial countries tend to have restrictions on imports of agricultural products, where much of Ethiopia's export potential is concentrated. Further, it is argued that a country's oilseeds export may fail to grow as rapidly as the world average for three reasons. First, exports may be concentrated in commodity groups for which demand tends to grow relatively at a low rate. Second, export may be going mainly to relatively stagnant regions/blocs. Third, the country in question may have been unwilling or unable to compete effectively with other sources of supply in the international market. For this purpose, exports from rest of the world are treated as competitor to Ethiopia. Therefore, regional trading arrangements (within Africa) should be set to put in economic efficiency, trade, investment, and growth in the region.

## ACKNOWLEDGEMENT

The author gratefully acknowledged the constructive comments and suggestions of anonomous reviewers.

### REFERENCES

Abate T (2006). Focusing Agricultural Research to Address Development Needs; the way I see it. In Abate, T.2006a, pp. 1-20.

Alwang J, Siegel PB (1994). Portfolio Models and Planning for Export Diversification: Malawi, Tanzania and Zimbabwe, J Dev. Stud., 30(2): 405-422.

Anderson K (2004). Agriculture Trade Reform and Poverty Reduction: Implications for Sub-Saharan Africa. UNCTAD Policy Issues in International Trade and Commodities No. 22.

Arndt SW (1999). Globalization and Economic Development. J. Int. Trade Econ. Dev., 8(3): 309-318.

Bennet M (2004). Sesame. In; Salvin. S., M. Bourke and T. Byrne, 2004. The new crop industries handbook. Australian Government, RIRDC publication04/125, pp. 214-220

CSA (Central Statistical Agency of Federal Democratic Republic of Ethiopia), 2006. Annual report.

FAO (Food and Agricultural Organization) (2010). Food outlook.

Fugazza M (2004). Export Performance and its Determinants: Supply and Demand Constraints. UNCTAD Policy Issues in International Trade and Commodities No. 26.

Goldstein M, Khan M (1985). Income and Price Effects in Foreign Trade. In R. Jones and P. Kenen, eds., Handbook of International Economics 2. Amsterdam: Elsevier.

Lages L, Luis F, Cristiana RL (2004). The STEP scale: A measure of short-term export performance improvement." J. Int. Market, 12(1): 36-56.

Madsen T (1987). Empirical export performance studies: A review of conceptualizations and findings. In S. T. Cavusgil (Ed.). Advances in international marketing. Greenwich, CT: JAI Press Inc., 2: 177-198.

Onafowora OA, Owoye O (1998). Can trade liberalization Stimulate Economic growth in Africa? World Dev., 26(3): 497-506

Redding S, Venables A (2004). "Economic geography and international Inequality." J. Intern. Econ. Elsevier, 62(1): 53-82.

Shoham A (1996). Export performance: A conceptualization and empirical assessment. J. Int. Market, 6(3): 59-81.

Zou S, Stan S (1998). The Determinants of Export Performance: A Review of the Empirical Literature Between 1987 and 1997". Int. Market Rev., 15(5): 333-356.

## ANNEX

**Table 1.** Area, yield, production and export of oilseeds in Ethiopia (1974-2009).

| Year | Oilseeds | | | | | |
|---|---|---|---|---|---|---|
| | Area (000' ha) | Yield qt/ha | Production (in metric tones) | Export (in metric tones) | Value of exports (000's birr) | Share of export as percentage of production |
| 1974 | 529.20 | 4.60 | 241,300 | 103,280 | 95,862 | 43 |
| 1975 | 145.00 | 6.46 | 93,700 | 79,351 | 83,985 | 85 |
| 1976 | 308.00 | 3.30 | 108,500 | 32,120 | 31,242 | 30 |
| 1977 | 173.00 | 3.10 | 46,000 | 21,624 | 17,505 | 47 |
| 1978 | 146.50 | 4.67 | 53,900 | 13,835 | 12,245 | 26 |
| 1979 | 159.50 | 3.23 | 49,100 | 5,198 | 9,270 | 11 |
| 1980 | 187.00 | 4.77 | 891,160 | 9,305 | 13,814 | 1 |
| 1981 | 223.39 | 4.56 | 1,020,580 | 16,128 | 28,371 | 2 |
| 1982 | 230.38 | 3.58 | 823,860 | 12,173 | 19,533 | 1 |
| 1983 | 259.97 | 4.68 | 1,216,770 | 11,546 | 15,349 | 0.9 |
| 1984 | 256.19 | 3.85 | 985,290 | 33,610 | 27,859 | 3 |
| 1985 | 284.15 | 3.64 | 1,035,130 | 12,466 | 15,640 | 1 |
| 1986 | 293.20 | 3.46 | 1,014,270 | 5,630 | 7,686 | 0.5 |
| 1987 | 212.34 | 3.98 | 844,910 | 8,219 | 9,793 | 0.9 |
| 1988 | 190.38 | 4.25 | 808,670 | 17,795 | 22,014 | 2 |
| 1989 | 215.49 | 3.64 | 785,020 | 5,389 | 11,030 | 0.6 |
| 1990 | 238.00 | 4.08 | 972,930 | 6,914 | 8,388 | 0.7 |
| 1991 | 243.14 | 12.89 | 3,133,320 | 1,459 | 3,633 | 0.1 |
| 1992 | 208.15 | 4.18 | 870,030 | 176 | 383 | 0.0 |
| 1993 | * | * | * | 392 | 1,186 | * |
| 1994 | 457.67 | 4.15 | 1,901,280 | 10,188 | 44,187 | 0.5 |
| 1995 | 335.69 | 3.43 | 1,152,530 | 12,132 | 50,130 | 1 |
| 1996 | 391.58 | 4.99 | 1,952,610 | 7,832 | 41,938 | 0.4 |
| 1997 | 478.45 | 4.46 | 2,132,790 | 14,069 | 74,239 | 0.7 |
| 1998 | 410.01 | 4.48 | 1,836,960 | 66,554 | 314,660 | 4 |
| 1999 | 396.22 | 4.26 | 1,688,740 | 51,366 | 271,462 | 3 |
| 2000 | 424.26 | 4.49 | 1,902,840 | 43,131 | 255,328 | 2 |
| 2001 | 574.89 | 4.28 | 2,463,370 | 55,051 | 269,598 | 2 |
| 2002 | 438.17 | 4.88 | 2,138,160 | 76,604 | 278,738 | 4 |
| 2003 | * | * | * | 82,801 | 395,565 | * |
| 2004 | 570.78 | 5.48 | 3,128,630 | 105,945 | 712,738 | 3 |
| 2005 | 824.43 | 6.38 | 5,263,960 | 170,796 | 1,082,215 | 3 |
| 2006 | 797.34 | 6.10 | 4,866,100 | 265,649 | 1,835,270 | 5 |
| 2007 | 741.80 | 6.70 | 4,970,840 | 234,976 | 1,654,707 | 5 |
| 2008 | 707.06 | 8.73 | 6,169,280 | 152,091 | 2,037,090 | 2 |
| 2009 | 855.15 | 7.67 | 6,557,040 | 286,987 | 3.819,429 | 4 |

* (No Annual Report). Source: CSA (Central Statistics Agency of Ethiopia); various years' annual reports.

**Table 2.** Percentage share of oilseeds export value in total values of exports.

| Year | Total values (in 000 birr) | percent share of oilseeds | Year | Total value (in 000 birr) | Percent share of oilseeds |
|---|---|---|---|---|---|
| 1974 | 556,230 | 17.2 | 1992 | 318,356 | 0.1 |
| 1975 | 497,824 | 16.9 | 1993 | 948,983 | 0.1 |
| 1976 | 580,568 | 5.4 | 1994 | 1,419,468 | 3.1 |

**Table 2.** Contd.

| | | | | | |
|---|---|---|---|---|---|
| 1977 | 688,961 | 2.5 | 1995 | 2,835,179 | 1.8 |
| 1978 | 633,629 | 1.9 | 1996 | 2,607,288 | 1.6 |
| 1979 | 864,327 | 1.1 | 1997 | 3,901,671 | 2 |
| 1980 | 950,667 | 1.5 | 1998 | 4,141,582 | 7.4 |
| 1981 | 851,509 | 3.3 | 1999 | 3,637,280 | 7.5 |
| 1982 | 778,083 | 2.5 | 2000 | 3,957,802 | 6.5 |
| 1983 | 809,625 | 1.4 | 2001 | 3,866,606 | 7 |
| 1984 | 929,625 | 0.3 | 2002 | 3,864,320 | 7.2 |
| 1985 | 744,572 | 2.1 | 2003 | 4,142,356 | 9.6 |
| 1986 | 942,700 | 0.8 | 2004 | 5,176,644 | 13.8 |
| 1987 | 809,800 | 1.2 | 2005 | 7,331,258 | 14.5 |
| 1988 | 788,100 | 2.8 | 2006 | 8,685,376 | 21.1 |
| 1989 | 918,200 | 1.2 | 2007 | 10,457,615 | 15.8 |
| 1990 | 756,860 | 1.1 | 2008 | 13,643,332 | 14.9 |
| 1991 | 542,485 | 0.7 | 2009 | 15,217,730 | 25.1 |
| Total | 13,643,765 | 63.9 | Total | 96,152,850 | 159.1 |
| Average | 100.0 | 3.5 | Average | 100.0 | 8.8 |

Source: Authors' computation from data obtained from National Bank of Ethiopia.

**Table 3.** List of Ethiopian oilseeds and pulses exporters.

| S. no. | Name of exporters |
|---|---|
| 1 | Amal trading ltd co. |
| 2 | Omar and Awad Baobed |
| 3 | Bajeba private ltd co, |
| 4 | J.J kotari ltd.co. |
| 5 | Antypas and brothers ltd |
| 6 | Ambassel trading house |
| 7 | Guna trading |
| 8 | Dina trading plc |
| 9 | Yahia Sayed Omar |
| 10 | Al-eman trading |
| 11 | Ethiopian grain trading enterprise |
| 12 | Oda share co. |
| 13 | K. A.S. international trading |
| 14 | Warka trading plc |
| 15 | Nazreti trading co. |
| 16 | Taye Belay general import- export |
| 17 | Hawas agri business pvt ltd co. |
| 18 | Ali Abdu Ali import- export |
| 19 | Beyen Teka general import- export |
| 20 | Malima plc |
| 21 | Ajli international trading pvt ltd co. |
| 22 | Mandura Ethiopia |
| 23 | Asnake Addissu Negash exporter |

**Table 4a.** ADF test on $logOLSXP_t$ at 2[nd] difference with intercept and trend.

| | | | |
|---|---|---|---|
| ADF test statistic | -5.312820 | 1% Critical value* | -4.2712 |
| | | 5% Critical value | -3.5562 |
| | | 10% Critical value | -3.2109 |

*MacKinnon critical values for rejection of hypothesis of a unit root.
Augmented Dickey-Fuller test equation
Dependent Variable: D(LOGolsxpt,3)
Method: Least squares
Date: 08/16/10 time: 16:06
Sample(adjusted): 1978 2009
Included observations: 32 after adjusting endpoints

| **Variable** | **Coefficient** | **Std. error** | **t-statistic** | **Prob.** |
|---|---|---|---|---|
| D(LOGolsxpt(-1),2) | -1.737246 | 0.326991 | -5.312820 | 0.0000 |
| D(LOGolsxpt(-1),3) | 0.124239 | 0.185909 | 0.668278 | 0.5094 |
| C | -0.009994 | 0.065868 | -0.151728 | 0.8805 |
| @TREND(1974) | 0.000539 | 0.003053 | 0.176690 | 0.8610 |
| R-squared | 0.775552 | Mean dependent var | | -0.002500 |
| Adjusted R-squared | 0.751504 | S.D. dependent var | | 0.319778 |
| S. E. of regression | 0.159407 | Akaike info criterion | | -0.718237 |
| Sum squared resid | 0.711501 | Schwarz criterion | | -0.535020 |
| Log likelihood | 15.49180 | F-statistic | | 32.25012 |
| Durbin-Watson stat | 2.045163 | Prob(F-statistic) | | 0.000000 |

**Table 4b.** ADF unit root test on logpdt at 2[nd] difference with intercept and trend.

| | | | |
|---|---|---|---|
| | | 1% Critical value* | -4.2712 |
| ADF test statistic | -5.725716 | 5% Critical value | -3.5562 |
| | | 10% Critical value | -3.2109 |

*MacKinnon critical values for rejection of hypothesis of a unit root.
Augmented Dickey-Fuller test equation
Dependent Variable: D(LOGPDT,3)
Method: Least Squares
Date: 08/15/10. Time: 15:25
Sample(adjusted): 1978 2009
Included observations: 32 after adjusting endpoints

| **Variable** | **Coefficient** | **Std. error** | **t-statistic** | **Prob.** |
|---|---|---|---|---|
| D(LOGPDT(-1),2) | -2.032778 | 0.355026 | -5.725716 | 0.0000 |
| D(LOGPDT(-1),3) | 0.131845 | 0.187279 | 0.704001 | 0.4872 |
| C | -0.045336 | 0.053502 | -0.847370 | 0.4040 |
| @TREND(1974) | 0.002026 | 0.002476 | 0.818522 | 0.4200 |
| R-squared | 0.899852 | Mean dependent var | | 0.001250 |
| Adjusted R-squared | 0.889122 | S.D. dependent var | | 0.383984 |
| S. E. of regression | 0.127861 | Akaike info criterion | | -1.159285 |
| Sum squared resid | 0.457753 | Schwarz criterion | | -0.976068 |
| Log likelihood | 22.54856 | F-statistic | | 83.86183 |
| Durbin-Watson stat | 2.055184 | Prob(F-statistic) | | 0.000000 |

**Table 4c.** ADF unit root test on logpwt at $2^{nd}$ difference with intercept and trend.

| | | | |
|---|---|---|---|
| | | 1% Critical value* | -4.2712 |
| ADF test statistic | -5.161600 | 5% Critical value | -3.5562 |
| | | 10% Critical value | -3.2109 |

*MacKinnon critical values for rejection of hypothesis of a unit root.
Augmented Dickey-Fuller test equation
Dependent Variable: D(LOGPWT,3)
Method: Least squares
Date: 08/15/10 Time: 15:27
Sample(adjusted): 1978 2009
Included observations: 32 after adjusting endpoints

| Variable | Coefficient | Std. error | t-statistic | Prob. |
|---|---|---|---|---|
| D(LOGPWT(-1),2) | -1.779718 | 0.344800 | -5.161600 | 0.0000 |
| D(LOGPWT(-1),3) | 0.069771 | 0.188261 | 0.370606 | 0.7137 |
| C | -0.035330 | 0.060499 | -0.583971 | 0.5639 |
| @TREND(1974) | 0.001322 | 0.002803 | 0.471408 | 0.6410 |
| R-squared | 0.826963 | Mean dependent var | | -0.003125 |
| Adjusted R-squared | 0.808424 | S.D. dependent var | | 0.332551 |
| S.E. of regression | 0.145555 | Akaike info criterion | | -0.900050 |
| Sum squared resid | 0.593219 | Schwarz criterion | | -0.716833 |
| Log likelihood | 18.40080 | F-statistic | | 44.60516 |
| Durbin-Watson stat | 2.036182 | Prob(F-statistic) | | 0.000000 |

**Table 4d.** ADF unit root test on logryt at $2^{nd}$ difference with intercept and trend.

| | | | |
|---|---|---|---|
| | | 1% Critical value* | -4.2712 |
| ADF test statistic | -6.640211 | 5% Critical value | -3.5562 |
| | | 10% Critical value | -3.2109 |

*MacKinnon critical values for rejection of hypothesis of a unit root.
Augmented Dickey-Fuller Test Equation
Dependent Variable: D(LOGRYT,3)
Method: Least squares
Date: 08/16/10. Time: 19:12
Sample(adjusted): 1978 2009
Included observations: 32 after adjusting endpoints

| Variable | Coefficient | Std. Error | t-statistic | Prob. |
|---|---|---|---|---|
| D(LOGRYT(-1),2) | -2.020305 | 0.304253 | -6.640211 | 0.0000 |
| D(LOGRYT(-1),3) | 0.375854 | 0.177677 | 2.115378 | 0.0434 |
| C | -0.000636 | 0.064040 | -0.009929 | 0.9921 |
| @TREND(1974) | 0.000126 | 0.002970 | 0.042288 | 0.9666 |
| R-squared | 0.766518 | Mean dependent var | | -0.007187 |
| Adjusted R-squared | 0.741502 | S.D. dependent var | | 0.304909 |
| S. E. of regression | 0.155024 | Akaike info criterion | | -0.774007 |
| Sum squared resid | 0.672907 | Schwarz criterion | | -0.590790 |
| Log likelihood | 16.38411 | F-statistic | | 30.64115 |
| Durbin-Watson stat | 2.059745 | Prob(F-statistic) | | 0.000000 |

**Table 4e.** ADF test on loget at 2[nd] difference with intercept and trend.

| | | | |
|---|---|---|---|
| | | 1% Critical value* | -4.2712 |
| ADF test statistic | -5.039660 | 5% Critical value | -3.5562 |
| | | 10% Critical value | -3.2109 |

*MacKinnon critical values for rejection of hypothesis of a unit root.
Augmented Dickey-Fuller test equation
Dependent variable: D(LOGET,3)
Method: Least squares
Date: 08/24/10. Time: 22:23
Sample(adjusted): 1978 2009
Included observations: 32 after adjusting endpoints

| Variable | Coefficient | Std. error | t-statistic | Prob. |
|---|---|---|---|---|
| D(LOGET(-1),2) | -1.531692 | 0.303928 | -5.039660 | 0.0000 |
| D(LOGET(-1),3) | 0.235216 | 0.191938 | 1.225478 | 0.2306 |
| C | -0.004046 | 0.028623 | -0.141346 | 0.8886 |
| @TREND(1974) | 0.000430 | 0.001327 | 0.324168 | 0.7482 |
| R-squared | 0.623785 | Mean dependent var | | 0.003125 |
| Adjusted R-squared | 0.583476 | S.D. dependent var | | 0.107296 |
| S.E. of regression | 0.069248 | Akaike info criterion | | -2.385787 |
| Sum squared resid | 0.134267 | Schwarz criterion | | -2.202570 |
| Log likelihood | 42.17260 | F-statistic | | 15.47516 |
| Durbin-Watson stat | 2.041389 | Prob(F-statistic) | | 0.000004 |

**Table 5.** Johansen cointergration test on logolsxpt, logpdt, logpwt, logryt and logret .

Date: 08/24/10. Time: 19:15
Sample: 1974 2009
Included observations: 34
Test assumption: Linear deterministic trend in the data
Series: LOGOLSXPT LOGPDT LOGPWT LOGRYT LOGET
Lags interval: 1 to 1

| Eigenvalue | Likelihood Ratio | 5% Critical value | 1% Critical value | Hypothesized No. of CE (s) |
|---|---|---|---|---|
| 0.702286 | 85.17737 | 68.52 | 76.07 | None ** |
| 0.509022 | 43.98222 | 47.21 | 54.46 | At most 1 |
| 0.346297 | 19.79614 | 29.68 | 35.65 | At most 2 |
| 0.096985 | 5.342674 | 15.41 | 20.04 | At most 3 |
| 0.053629 | 1.874108 | 3.76 | 6.65 | At most 4 |

*(**) denotes rejection of the hypothesis at 5%(1%) significance level
L.R. test indicates 1 cointegrating equation(s) at 5% significance level

**Unnormalized cointegrating coefficients:**

| LOGOLSXPT | LOGPDT | LOGPWT | LOGRYT | LOGET |
|---|---|---|---|---|
| -1.620127 | -42.51544 | 43.03450 | 0.703449 | 44.38415 |
| -2.240353 | 11.69520 | -10.80391 | 0.949989 | -9.358385 |
| 0.070290 | -5.465979 | 7.054629 | 0.080156 | 5.116791 |
| 0.237543 | 3.504041 | -4.060736 | 0.324888 | -4.603558 |
| 0.931814 | 2.099893 | -2.551611 | 0.218778 | -3.146213 |

**Normalized cointegrating coefficients: 1 cointegrating equation(s)**

| LOGOLSXPT | LOGPDT | LOGPWT | LOGRYT | LOGET | C |
|---|---|---|---|---|---|
| 1.000000 | 26.24204(5.12476) | -26.56243 (5.12129) | -0.434194(0.05269) | -27.39548(5.13797) | -3.625986 |

**Table 5** Contd.

| | | | | | |
|---|---|---|---|---|---|
| Log likelihood | 285.7394 | | | | |
| | **Normalized cointegrating coefficients: 2 cointegrating equation(s)** | | | | |
| **LOGOLSXPT** | **LOGPDT** | **LOGPWT** | **LOGRYT** | **LOGET** | **C** |
| 1.000000 | 0.000000 | -0.384984 (0.10833) | -0.425721(0.04828) | -1.061369(0.05652) | -3.750092 |
| 0.000000 | 1.000000 | -0.997538 (0.00551) | -0.000323 (0.00246) | -1.003508 (0.00287) | 0.004729 |
| Log likelihood | 297.8324 | | | | |
| | **Normalized cointegrating coefficients: 3 cointegrating equation(s)** | | | | |
| **LOGOLSXPT** | **LOGPDT** | **LOGPWT** | **LOGRYT** | **LOGET** | **C** |
| 1.000000 | 0.000000 | 0.000000 | -0.400125(0.06344) | -1.130787 (0.07572) | -3.882873 |
| 0.000000 | 1.000000 | 0.000000 | 0.065999 (0.11225) | -1.183379 (0.13398) | -0.339322 |
| 0.000000 | 0.000000 | 1.000000 | 0.066485 (0.11228) | -0.180314(0.13401) | -0.344900 |
| Log likelihood | 305.0591 | | | | |
| | **Normalized cointegrating coefficients: 4 cointegrating equation(s)** | | | | |
| **LOGOLSXPT** | **LOGPDT** | **LOGPWT** | **LOGRYT** | **LOGET** | **C** |
| 1.000000 | 0.000000 | 0.000000 | 0.000000 | -1.933868 (0.55967) | -5.093612 |
| 0.000000 | 1.000000 | 0.000000 | 0.000000 | -1.050915 (0.18301) | -0.139617 |
| 0.000000 | 0.000000 | 1.000000 | 0.000000 | -0.046874 (0.18329) | -0.143723 |
| 0.000000 | 0.000000 | 0.000000 | 1.000000 | -2.007076 (1.36071) | -3.025900 |
| Log likelihood | 306.7934 | | | | |

ECT= logpdt-0.997538* LOGPWT-0.000323*LOGRYT-1.003508* LOGET +0.004729.

**Table 6.** ADF test on ECT at level with intercept.

| | | | |
|---|---|---|---|
| | | 1% Critical value* | -3.6353 |
| ADF test statistic | -7.115321 | 5% Critical value | -2.9499 |
| | | 10% Critical value | -2.6133 |

*MacKinnon critical values for rejection of hypothesis of a unit root.
Augmented Dickey-Fuller test equation
Dependent Variable: D(ECT3)
Method: Least squares
Date: 08/24/10 Time: 21:44
Sample(adjusted): 1976 2009
Included observations: 34 after adjusting endpoints

| **Variable** | **Coefficient** | **Std. error** | **t-statistic** | **Prob.** |
|---|---|---|---|---|
| ECT3(-1) | -1.282956 | 0.180309 | -7.115321 | 0.0000 |
| D(ECT3(-1)) | 0.035629 | 0.034403 | 1.035641 | 0.3084 |
| C | -0.000194 | 0.000763 | -0.253940 | 0.8012 |
| R-squared | 0.627061 | Mean dependent var | | -6.89E-05 |
| Adjusted R-squared | 0.603001 | S.D. dependent var | | 0.006973 |
| S. E. of regression | 0.004393 | Akaike info criterion | | -7.933344 |
| Sum squared resid | 0.000598 | Schwarz criterion | | -7.798665 |
| Log likelihood | 137.8668 | F-statistic | | 26.06180 |
| Durbin-Watson stat | 2.029297 | Prob(F-statistic) | | 0.000000 |

**Table 7.** Equation estimation with ECM.

Dependent variable: D(LOGRXT)
Method: Least squares
Date: 08/24/10. Time: 21:52
Sample(adjusted): 1975 2009
Included observations: 35 after adjusting endpoints
D(LOGOLSXPT)= C(1)+ C(2)*D(LOGPDT)+C(3)*D(LOGPWT)+C(4)
*D(LOGRYT)+C(5)*D(LOGET) +C(6)*ECT3

| | Coefficient | Std. error | t-statistic | Prob. |
|---|---|---|---|---|
| C(1) | -0.005071 | 0.017762 | -0.285472 | 0.7773 |
| C(2) | -0.519185 | 0.710181 | -0.731061 | 0.4706 |
| C(3) | 0.781067 | 0.724561 | 1.077986 | 0.2899 |
| C(4) | 0.267825 | 0.120749 | 2.218024 | 0.0345 |
| C(5) | 2.004283 | 0.746842 | 2.683676 | 0.0119 |
| C(6) | 1.662131 | 3.721573 | 0.446621 | 0.6585 |
| R-squared | 0.560853 | Mean dependent var | | 0.040857 |
| Adjusted R-squared | 0.485139 | S.D. dependent var | | 0.126011 |
| S. E. of regression | 0.090417 | Akaike info criterion | | -1.813955 |
| Sum squared resid | 0.237084 | Schwarz criterion | | -1.547324 |
| Log likelihood | 37.74421 | F-statistic | | 7.407438 |
| Durbin-Watson stat | 1.970283 | Prob(F-statistic) | | 0.000140 |

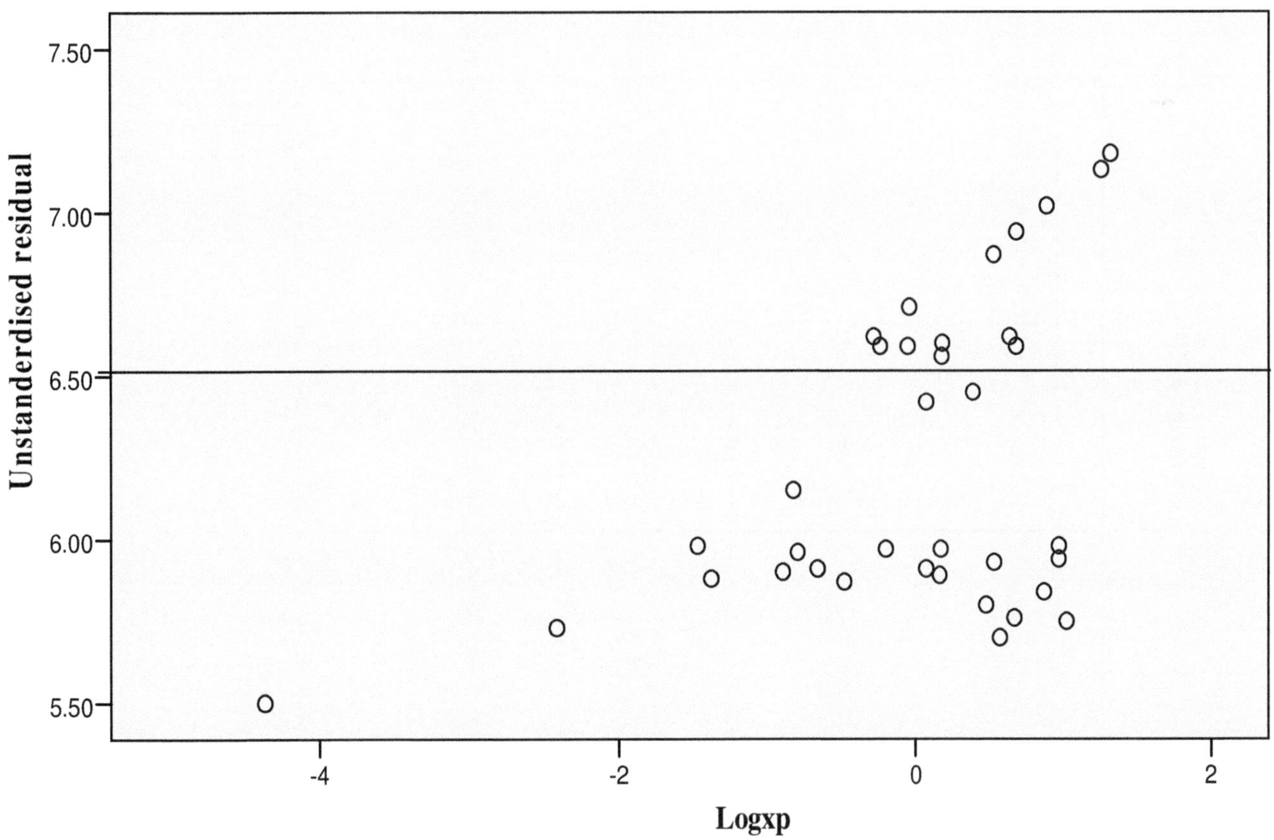


**Figure 1.** Heteroscedasticity diagnosis.

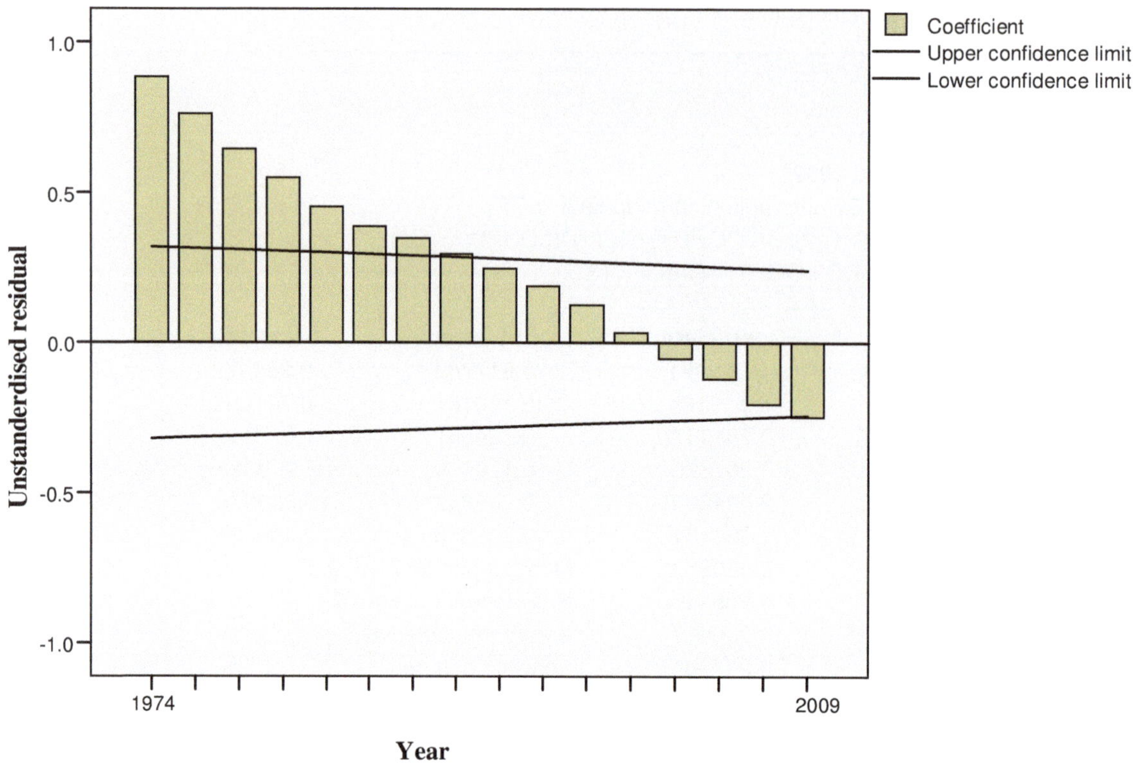


**Figure 2.** Autocorrelation diagnosis.

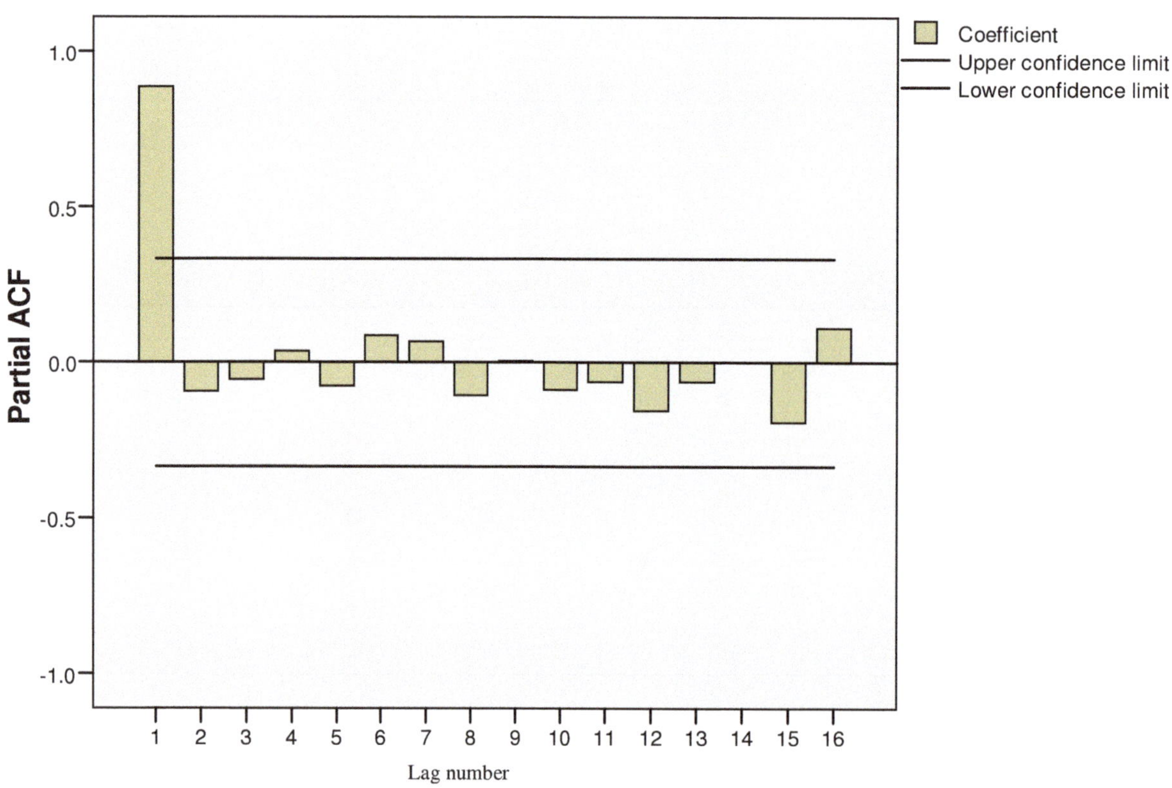


**Figure 3.** Partial Autocorrelation (before correction).

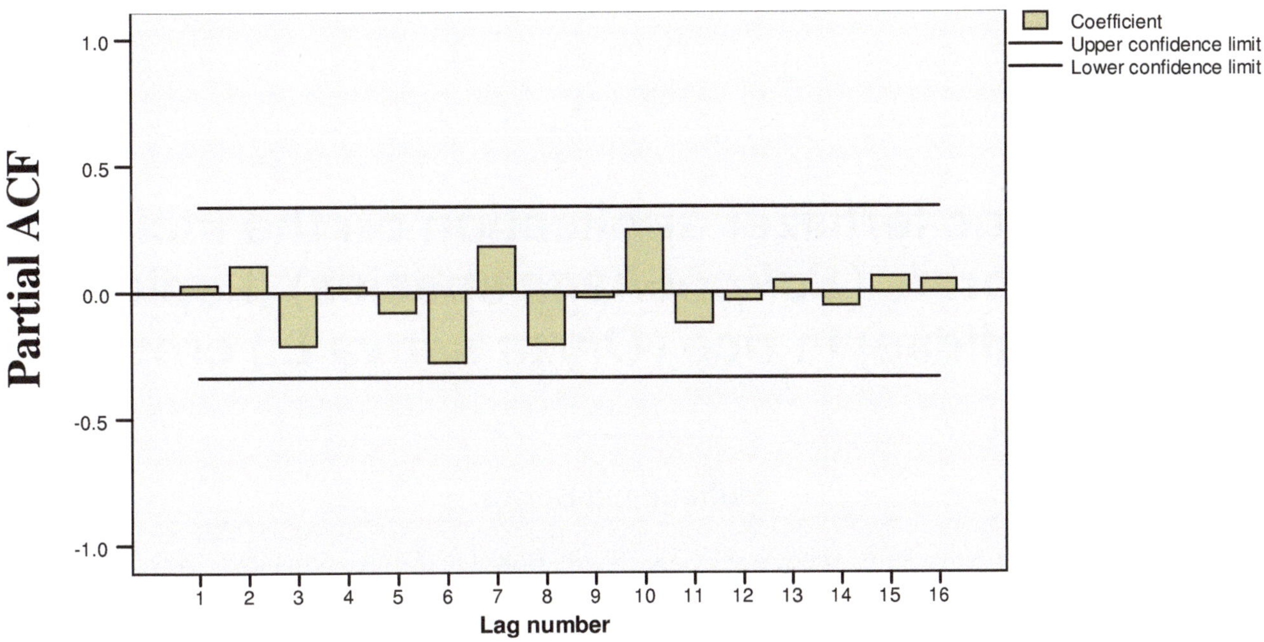


**Figure 4.** Partial autocorrelation function (after autocorrelation problem corrected).

# 3

# Effects of Zinc fertilizer application on the incidence of rice stem borers (*Scirpophaga* species) (Lepidoptera: Pyralidae) in rice (*Oryza sativa* L.) crop

**Muhammad Sarwar**

Nuclear Institute of Agriculture, Tandojam-70060, Sindh, Pakistan. E-mail: drmsarwar64@yahoo.com.

**Use of fertilizer within plant protection system can prove a key factor for pest management strategy. The effects of Zinc sulfate (33% - Zn) fertilizer on the incidence of rice stem borers (*Scirpophaga* species) (Lepidoptera: Pyralidae) in rice crop were investigated. The Zn was applied using broadcast method at three levels (20, 25 and 30 kg/ha) in the vicinity of the roots, 30 days after crop transplantation. The results of field experiment on rice cultivars Shadab and Mehak showed that Zn had significant influences on the stem borers population and paddy yield over the unfertilized control. Average yield per hectare was increased at all doses of Zn and by applying it at the rates of 20, 25 and 30 kg/ha, the differences in grain yield were not significant, but production increased significantly over the untreated crop due to varying pest prevalence. With respect to borers incidence, higher dose 30 kg Zn/ha markedly decreased infestation, while, applications at 20 and 25 kg permitted slightly more deadhearts and whiteheads incidence, but differed significantly from unfertilized control. The present study implies that due to suppressive effects of Zn fertilizer on the incidence of the rice stem borers at crop maturity stage increased grain yield, and the use of soil amendments strategy can create an unfavorable environment for pest inducing resistance through antibiosis or feeding inhibition.**

**Key words:** Rice, *Oryza sativa, Scirpophaga,* insect control, rice stem borer, Zinc.

## INTRODUCTION

Rice (*Oryza sativa* L.) is the main staple food of around half of the world's population. On a global basis, rice provides 21 and 15% per capita of dietary energy and protein, respectively (Maclean et al., 2002). One of the main problems and yields limiting factors is the attack of different kinds of insect pests in the rice fields. Rice stem borers, *Scirpophaga incertulas* (Walker) and *Scirpophaga innotata* (Walker) infest the rice crop right from nursery till harvest and cause complete loss of affected tillers (Salim and Masih, 1987). The borers *S. incertulas* or *S. innotata* could attack most of the growing stages of rice plant, beginning with seedling through tillering and up to ear setting (Ranasinghe, 1992). Their caterpillars bore into the rice stem and hollow out the stem completely. The damage symptoms vary according to the growth stage of the plant.

Yield loss is caused by a loss of bearing stems due to the production of deadhearts (the outright death of stems), stems attacked in the vegetative stage, the smaller panicles borne by compensatory nodal tillers, whiteheads (empty panicles) caused by attack in the reproductive phase, a decrease in filled grains and lowering of panicle weight from late damage (Catling, 1992). Sarwar et al. (2010) reported the infestation of 10.4% deadhearts and 19.3% whiteheads observed in aromatic rice, due to stem borers under natural field conditions. The increasing pressure on the land of the fast growing world's population has made it necessary to intensify rice production and efficient use of mineral fertilizers that can prove a quick way of boosting crop yield per unit area of land.

Progressively, the modern research is showing that the ability of a crop plant to resist or tolerate insect pests and diseases is tied to optimal physical, chemical and mainly biological properties of soils (Altieri, 2002). Soils with high organic matter and active soil biology generally exhibit good soil fertility as well as complex food webs and beneficial organisms that prevent infection. On the other hand, farming practices that cause nutrition imbalances can lower pest resistance (Magdoff and Van, 2000). Much of what we know today about the relationship between crop nutrition and pest incidence comes from studies comparing the effects of organic agricultural practices and modern conventional methods on specific pest populations. Soil fertility practices can impact the physiological susceptibility of crop plants to insect pests by either affecting the resistance of individual plant to attack or by altering plant acceptability to certain herbivores (Chau and Heong, 2005). Several studies have also been documented that show how the shift from organic soil management to chemical fertilizers has increased the potential of certain insects and diseases. Studies of plant resistance to insect pests have shown that resistance varies with the age or growth stage of the plant (Slansky, 1990). This advocates that resistance is related directly to the physiology of the plant and thus any factor that affects the physiology of the plant may lead to changes in resistance to insect pests.

Zinc is essential for the normal healthy growth and reproduction of plants, animals and humans, and when the supply of plant available Zinc is inadequate, crop yields are reduced and the quality of crop products is frequently impaired. In plants, Zinc plays a key role as a structural constituent or regulatory co-factor of a wide range of different enzymes in many important biochemical pathways and these are mainly concerned with, carbohydrate metabolism (both in photosynthesis and in the conversion of sugars to starch), protein metabolism, auxin (growth regulator) metabolism, pollen formation, the maintenance of the integrity of biological membranes, and the resistance to infection by certain pathogens (Alloway, 2004). Zinc deficiency is usually more prevalent in rice soils with a high pH and high content of organic matter or when organic manures are applied (Dobermann and Fairhurst, 2000). There is a dire need to explore the alternative means of pest management in rice, and fertilizer management seemed as the basic and unexplored point in this regard. Better growth of plants and yield depend upon balanced fertilization, which in turn may have an indirect effect on pests. Pioneer work of Padhee and Mishra (1993) has established Zinc requirement for optimum yield of rice and Zinc application has been shown to impart resistance against insect pests of rice. Shu et al. (2009) investigated the effects of Zn on reproduction of phytophagous insect *Spodoptera litura* Fabricius in artificial diets of larvae at ecological and molecular levels. A significantly shorter period of laying eggs was observed in *S. litura* exposed to 300 to 750 mg Zn/kg. The oviposition rate, fecundity and hatchability of female adults treated with 750 mg Zn/kg were significantly lower than those of the controls. Therefore, the objective of this work was to assess effects of Zinc fertilizer application against the incidence of rice stem borers in rice crop to determine its potential effectiveness to improve farmers' income and to encourage its adoption by farmers in Pakistan and other similar regions.

## MATERIALS AND METHODS

This study was carried out during 2008 crop season at the Nuclear Institute of Agriculture, Tandojam-70060, Sindh, Pakistan. The experiment was done under field conditions by planting rice varieties namely, non aromatic Shadab and aromatic Mehak. Seedlings of 30 days old were uprooted from the nursery beds carefully and transplanted at 25 × 25 cm spacing in the well prepared beds of experimental area (each replicate measuring 4.5 $m^2$). The experiment was laid out in a Randomized Complete Block Design having three replications with Zinc and varieties as factors, while, pest and yield as parameters.

Research Farm has an ample of good quality canal water available for irrigation, and irrigations were given to crop continuously after every 6 to 7 days interval. The soil is silt loam in texture having an ECe of 1.3 dS $m^{-1}$, pH of 7.9, Olsen's P -6.7 mg $kg^{-1}$, organic matter -0.9% and $CaCO_3$-10.6%. Nitrogen doses and a dose of phosphorus were uniformly applied to all treatments. A basal and starter dose of N at the rate of 30 kg $ha^{-1}$ as urea was applied to all the treatments at the time of sowing and second dose at mid tillering stage 35 days after transplanting. The phosphorus as triple super phosphate was applied at the rate of 50 kg $ha^{-1}$ to the respective treatments and uniformly spread out on the soil surface and incorporated in the soil with spades before sowing of crop. In order to explore the effects of Zinc fertilizer, it was applied alongside the crop rows by using broadcast method, three weeks after crop transplantation. The Zinc sulfate (33%) was applied at three levels (20, 25 and 30 kg/ha) in the vicinity of the crop roots. Normal cultural practices were carried out throughout the growth period, and no pesticide was used as plant protection measure.

Observations on insect pest incidence were recorded at fortnightly intervals. In each treatment, an area of 1 $m^2$ was randomly selected at three different sampling sites in each treatment, and numbers of healthy and infested tillers were observed. Total number of plants in selected area was counted and then total numbers of tillers in the unit row area were also recorded in order to calculate pest percentages infestation. Stem borers damage was recorded as whitehead percentage at maturity stage by counting number of whiteheads per 1 $m^2$ area of rice plants. The observations on percent infestation of stem borers (dead heart and whitehead) were calculated using the following formulae:

Deadheart (%) = No. of deadhearts×100/ Total no. of tillers;

Whitehead (%) = No. of white heads×100/ Total no. of tillers with panicles in sample area

The crop was sampled at maturity stage to examine grain yield, plants were separated into straw and grain, dried in sunlight to have a constant weight, and grain yield was recorded separately (per 4.5 $m^2$) by using an electronic balance. After harvesting, data on paddy yield and other yield components were also recorded. Plot yields of all treatments were converted to kg $ha^{-1}$ subsequent to data analysis. The data were checked for the need for transformation and analyzed statistically using ANOVA and treatment means were compared at 5% level of probability. Mean separation was done by LSD-test by using Statistical Product and Service Solutions (2005), and Statistix 8.1 softwares to assess the significance of the treatment effects. Standard errors of the differences were computed for mean comparisons.

## RESULTS AND DISCUSSION

### Effects of Zinc fertilizer on population of rice stem borers

Seasonal mean population of rice stem borers was different on both cultivars on most of the samplings; pest density was recorded lower on variety Shadab than that on Mehak. Population was non-significantly different between applications of different Zinc levels but differed significantly from control. The population of rice borers was significantly affected by different treatments though the highest prevalence was found in control treatment (3.09±0.16 and 4.81±0.20; 5.24±0.83 and 7.06±0.20 deadhearts and whiteheads on both varieties Shadab and Mehak, respectively). The prevalence of the borers tended to decrease with increasing doses of Zinc. The higher prevalence of deadhearts was found in the treatment of 20 kg Zinc, while, the percentage of damage was found lower with 30 kg treatment. The higher percentage of whiteheads was found with 20 kg Zinc treatment, that was statistically non-significant to that of 25 and 30 kg treatments. During the growing season, the relative prevalence of insect in the rice field environment could be ranked in the order of 20>25>30 kg Zn $ha^{-1}$ treatments (Table 1).

Zn proved to be a better mode of fertilization for the crop at all levels of application, as it allowed a balanced nutrient blend in soil and reduced the crop damage. Hence, reduced susceptibility of plant to pests may be a reflection of differences in plant health, as mediated by soil fertility management through Zinc. The soils with well Zinc fertilizer normally exhibited good soil fertility, as well as complex food webs and supported beneficial organisms that prevented pest contamination. In contrast, in control treatment, crop nutrition inequity resulted to lower pest tolerance in host. These findings are in close conformity with those reported by Magdoff and Van (2000), who observed that nutrition imbalance could lower pest resistance.

### Effects of Zinc fertilizer on grain yield and yield components

The yield of rice was increased significantly by Zinc treatments compared to control without fertilizer application. With the increase in dose level from 20 kg to 30 kg Zn $ha^{-1}$, there was corresponding increase in grain yield regardless of the two varieties. The highest grain yield of 2926.66±78.81 and 1700.00±87.36 g/ plot (4.5 $m^2$) on both varieties Shadab and Mehak, respectively was recorded at 30 kg Zn $ha^{-1}$ applied. The control treatment produced the lowest grain yield (1886.66±63.33 and 1320.00±60.62 g/ 4.5 $m^2$ plot) on both varieties (Table 1). The mean values of grain yield varied significantly in both cultivars but parameters like number of productive tillers per hill and number of grains per panicle did not vary among themselves. However, plant height, panicle length and 1000-grain weight differed significantly from one another in Zn treated and untreated plants, maximum grain yield was obtained from minimum percent infestation level, conversely, at privileged percent infestation levels, average grain yield was inferior per treatment.

The plant height response to Zn application was more pronounced, significantly higher growing efficiency were recorded (96.10 to 97.00 cm and 132.83 and 133.33 cm) with Zn and the lowest (94.00 and 130.00 cm) without Zn application in Shadab and Mehak, respectively. The low size of panicle length (25.66 and 27.16 cm) without Zn application resulted in a relatively minimum 1000 grain weight (24.00 and 22.00 g) compared to higher length (26.50 to 26.83 and 28.00 to 28.50 cm) and its effects on rice grain weight (27.33 to 28.00 and 25.33 to 26.00 g, respectively) due to Zn application on both cultivars (Table 2). The application of Zn fertilizer got an edge at all levels of application over control without fertilizer by producing significantly higher grain yield. The superiority of Zn application for grain yield may be due to improvement in soil properties to support the roots of treated plants due to Zn supply. Similar explanations have been given by Miguel and Nicholls (2003) that the ability of a crop plant to resist or tolerate insect pests and diseases is tied to optimal physical, chemical and mainly biological properties of soils. The better performance of Zn over control may also be attributed to high Zn fixing capacity of the calcareous soils and undergoing a series of reactions that gradually increased its solubility and availability to plants in many important biochemical pathways.

These findings are in close conformity with Alloway (2004), who observed that Zn is required for the synthesis of auxin (a growth regulating compound-indole acetic acid). Stunted growth and 'little leaf' are the most distinct visible symptoms of Zinc deficiency and are the result of disturbances in the metabolism of auxins. Flowering and seed production were severely depressed in Zinc deficient plants. On the other hand, application of Zinc to plants produced more grain yield than the control as it plays a key role as a structural constituent or regulatory co-factor of a wide range of different enzymes in many important biochemical pathways concerned with; carbohydrate metabolism, both in photosynthesis and in the conversion of sugars to starch, protein metabolism, auxin metabolism, pollen formation, maintenance of the integrity of biological membranes, and the resistance to infection by certain pathogens.

In case of Zinc deficiency, one or more important physiological functions are unable to operate normally and the growth of the plant is adversely affected

**Table 1.** Mean numbers of rice stem borers on rice plants receiving different Zinc doses.

| S.N | Variety | Zinc treatment (kg ha$^{-1}$) | Borers infestation (%) | | Yield/plot (4.5 m$^2$) (g) | Yield (kg ha$^{-1}$) |
|---|---|---|---|---|---|---|
| | | | Deadhearts | Whiteheads | | |
| 1. | Shadab | 20 | 1.13a ± 0.06 | 3.26a ± 0.30 | 2746.66b ± 37.11 | 6103.68 |
| 2. | | 25 | 1.03a ± 0.09 | 2.57a ± 0.25 | 2873.33b ± 61.73 | 6385.17 |
| 3. | | 30 | 0.78a ± 0.08 | 2.09a ± 0.11 | 2926.66b ± 78.81 | 6503.68 |
| 4. | | Control | 3.09b ± 0.16 | 5.24b ± 0.83 | 1886.66a ± 63.33 | 4192.57 |
| 1. | Mehak | 20 | 2.52a ± 0.27 | 3.22a ± 0.06 | 1583.33b ± 44.09 | 3518.51 |
| 2. | | 25 | 2.34a ± 0.18 | 2.88a ± 0.36 | 1656.66b ± 32.82 | 3681.46 |
| 3. | | 30 | 1.90a ± 0.18 | 2.56a ± 0.05 | 1700.00b ± 87.36 | 3777.77 |
| 4. | | Control | 4.81b ± 0.20 | 7.06b ± 0.20 | 1320.00a ± 60.62 | 2933.33 |

Means sharing similar letters in columns are non- significantly different (P= 0.05).

**Table 2.** Comparison of different yield components of rice by application of different Zinc levels.

| S.N | Variety | Zinc treatments (kg ha$^{-1}$) | Plant height (cm) | No. of productive tillers/ hill | Panicle length (cm) | No. of grains/panicle | 1000 Grain weight (g) |
|---|---|---|---|---|---|---|---|
| 1. | Shadab | 20 | 96.10a | 17.00a | 26.50a | 156.00a | 27.33a |
| 2. | | 25 | 96.50a | 17.66a | 26.63a | 156.33a | 27.66a |
| 3. | | 30 | 97.00a | 18.00a | 26.83a | 156.50a | 28.00a |
| 4. | | Control | 94.00b | 16.66a | 25.66b | 155.50a | 24.00b |
| Standard error | | | 0.804 | 0.593 | 0.162 | 0.471 | 0.304 |
| LSD value | | | 1.967 | 1.451 | 0.398 | 1.153 | 0.744 |
| 1. | Mehak | 20 | 132.83a | 15.00a | 28.00a | 125.83a | 25.33a |
| 2. | | 25 | 133.00a | 15.66a | 28.33a | 126.00a | 25.66a |
| 3. | | 30 | 133.33a | 16.00a | 28.50a | 126.33a | 26.00a |
| 4. | | Control | 130.00b | 14.66a | 27.16b | 125.50a | 22.00b |
| Standard error | | | 0.245 | 0.593 | 0.225 | 0.504 | 0.304 |
| LSD value | | | 0.600 | 1.451 | 0.552 | 1.234 | 0.744 |

Means sharing similar letters in columns are non-significantly different (P= 0.05).

(Alloway, 2004). The prevalence of the borer pests tended to decrease in Zn added treatments. It is likely to state that Zn addition can lead the plant to become less succulent and consequently with a reduction of attractiveness to insects. High levels of Zn can stimulate early vegetative growth resulting to increase the photosynthetic activity leading to a decreased insect pest infestation and elevated crop yield. Further, the indirect effects of Zn fertilization acting through changes in the nutrients composition of the rice crop have influenced plant resistance to borer insect pests. In conformity to present findings, many researchers have suggested that increasing insect pest pressure in agro-ecosystems is due to changes that have occurred in agricultural practices. For example, Meyer (2000) argues that soil nutrient availability not only affects the amount of damage that plants receive from herbivores but also the ability of plants to recover from herbivory. Miguel and Nicholls

(2003) showed that cultural methods such as crop fertilization can affect susceptibility of plants to insect pests, by altering plant tissue nutrient levels. As a consequence, current data suggest that the addition of Zinc along with other fertilizers would be the best combination for maximizing rice yield and reducing insect prevalence in rice. Hence, due to the combine action of above mentioned different factors, the resistance has developed in the rice plant against borers with the application of Zn that led to reduce pest damage and ultimately grain yield enhanced.

The present studies documented that numbers of deadhearts and whiteheads per plant were affected by different doses of Zinc application. The differences in population between these treatments were non-significant receiving different levels of Zn, but maximum population was recorded on plants receiving no Zn. Mean seasonal population of borer pests was also different on different varieties, where maximum population was observed on Mehak and minimum on Shadab. Based on the number of pest prevalence and grain yield variety, Shadab could be classified as resistant and Mehak as susceptible, under the ecological conditions of this region. Further, the lower abundance of stem herbivores in lofty Zinc grown crop, attributed an increase in the yield level, thus the use of soil amendments can create an unfavorable environment for pest, inducing resistance through anti-biosis or feeding inhibition. Zinc applications decreased the borers' populations to reduce pest damage, which is parallel to the findings of Babczynska et al. (2008) who stated that insect reproduction is influenced by various factors, including food quality and quantity, temperature, population density and female age. The metals like Zinc may disturb reproductive processes in insects. When a comparison was done, the basic developmental parameters (number of hatchlings, body mass, embryonic developmental rate) in grasshopper nymphs were additionally exposed to Zinc during diapause. Zinc application caused the decrease in hatching success and duration of embryogenesis in insects from each site. Shu et al. (2009) indicated that excess Zn made expression of vitellogenin gene down-regulated and caused poor accumulation of egg yolk, which led to a reduction in egg numbers and failure of eggs to hatch. Results of the current experiments showed that addition of Zinc along with other practices, can play an important role in the management of rice stem borers and there is a need to integrate this component into integrated pest management systems to maximize crop yield for encouraging farmers to adopt this improved technology for further improving their livelihoods. Optimizing the dose of Zn in relation with other macronutrients and its split applications are extremely essential to investigate in future because the nutrients can make plant tissue either tolerant or vulnerable to pest attack.

## REFERENCES

Alloway BJ (2004). Zinc in Soils and Crop Nutrition. International Zinc Association (IZA), 168 Avenue de Tervueren, 1150 Brussels-Belgium, p. 116.

Altieri MA (2002). Agroecology: The science of natural resource management for poor farmers in marginal environments. Agric. Ecosyst. Environ., 93(1-3): 1-24.

Babczynska A, Kozłowski M, Sawczyn T, Augustyniak M (2008). Effects of Zinc and female aging on nymphal life history in a grasshopper from polluted sites. J. Insect Physiol., 54(1): 41-50.

Catling D (1992). Rice in Deep Water. International Rice Research Institute. The Macmillan Press Ltd., pp. 419-426.

Chau LM, Heong KL (2005). Effects of organic fertilizers on insect pest and diseases of rice. Omonrice, 13: 26-33.

Dobermann A, Fairhurst T (2000). Rice: Nutrient Disorders and Nutrient Management. Potash and Phosphate Institute of Canada, and International Rice Research Institute, Los Baños, Philippines, p. 192.

Maclean JL, Dawe DC, Hardy B, Hettel GP (2002). Rice Almanac (3rd edition). CABI Publishing, Wallingford., p. 253

Magdoff F, Van EH (2000). Building Soils for Better Crops. SARE, Washington, DC., p. 230

Meyer GA (2000). Interactive effects of soil fertility and herbivory on *Brassica nigra*. Oikos, 22: 433- 441.

Miguel AA, Nicholls CI (2003). Soil fertility management and insect pests: harmonizing soil and plant health in agroecosystems. Soil Till. Res., 72: 203-211.

Padhee AK, Mishra DS (1993). Influence of Zinc application on the influence of rice leafhopper, *Cnaphalocrocis medinalis* Guenee. Indian J. Entomol., 55(3): 335-336.

Ranasinghe MASK (1992). Paddy Pests in Sri Lanka. Science education series No. 30, Natural Resources energy Science Authority of Colombo (Sri Lanka), pp. 47-59.

Salim M, Masih R (1987). Efficacy of insecticides against rice borer at NARC, Islamabad. Pak. J. Agric. Res., 8(4): 477.

Sarwar M, Ahmad N, Tofique M (2010). Tolerance of different rice genotypes (*Oryza sativa* l.) against the infestation of rice stem borers under natural field conditions. Nucleus, 47(3): 1-7.

Shu YH, Gao YY, Sun HX, Zou ZW, Zhou QA, Zhang GR (2009). Effects of Zinc exposure on the reproduction of *Spodoptera litura* Fabricius (Lepidoptera: Noctuidae). Ecotoxicol. Environ. Saf.. 72(8): 2130-2136.

Slansky F (1990). Insect nutritional ecology as a basis for studying host plant resistance. Florida Entomol., 73: 354-378.

# 4

# The effect of alternate furrow irrigation under different nutritional element supplies on some agronomic traits and seed qualitative parameters in corn (*Zea mays* L.)

**Mohammad Nasri[1]*, Mansoureh Khalatbari[2] and Hossein Aliabadi Farahani[3]**

[1]Agronomy and Plant Breeding Department, Islamic Azad University, Varamin Branch, Iran.
[2]Agricultural Researcher at Varamin, Iran.
[3]Member of Young Researchers Club, Islamic Azad University, Shahr-e-Qods Branch, Iran.

**A new irrigation method under different nutritional element supplies for maize production was designed and tested for some agronomic traits and seed qualitative parameters with a split-plot field experiment. Irrigation was applied through furrows in three ways as the main plots: alternate furrow irrigation (AFI), fixed furrow irrigation (FFI), and conventional furrow irrigation (CFI). AFI means that one of the two neighboring furrows was alternately irrigated during consecutive watering. FFI means that irrigation was fixed to one of the two neighboring furrows. CFI was the conventional way where every furrow was irrigated during each watering. Each irrigation method was further divided into five sub-treatments with different fertilizer combinations: (1) P+N (control) (2) P+N+K (3) P+N+K+Zn (4) P+N+K+Zn+B (5) P+N+K+Zn+B+Fe. The results indicate that water stress effects caused by furrow irrigation on yield may be alleviated by more frequent irrigation intervals. We concluded that AFI is a way to save water in arid areas where maize production relies heavily on repeated irrigation. Fertilized combinations influenced dry matter partitioning to seed filling. Thus, sufficient both macro and micro nutritional elements increased harvest index which was mostly due to more number of seeds per row than higher individual grain weight. Complete fertilizer combination increased total above ground biomass through more radiation use efficiency and by increasing leaf area. In order to utilize the water sources efficiently and increase corn production under limited water supply, we propose the use of circular irrigation care along with instance, K, Zn, B and Fe fertilizer.**

**Key words:** Alternate furrow irrigation, corn (*Zea mays* L.), fertilizer combination, water stress.

## INTRODUCTION

Corn (Zea mays L.) is the world's third most important cereal after wheat and rice grown primarily for grain and secondarily for fodder (Nelson, 2005). Globally, irrigated corn is 17% of total acreage producing 40% of total grain yield (Carbtree et al., 1985; Popova et al., 1998b). Corn is a high water demanding crop in all stages of its physiological development and can achieve high yields when water and nutrients are not limiting (Song and Dia, 2000; Traore et al., 2000). However, maize is very sensitive to water stress. The effects of water stress on maize include the visible symptoms of reduced growth, delayed maturity and reduced biomass and crop grain yield. For example, water stress and four days of visible wilting between the boot stage (a week prior to tasseling) and the milk stage may reduce yield by 50% or more (Earl and Davis, 2007; Fischabach and Mulliner, 1974). Drought stress decreased corn seed yield by lowering individual grain weight but nitrogen supply increased yield components (Popova and Petrova, 1993; Popova et al., 1998a). Grain yield can be reduced by yield components like: grain number, grain weight, water stress and nutrient deficiencies during the seed filling can reduce the seed yield and its efficiency (Lauer, 2006). Water stress on maize has been shown to reduce plant height, diameter of shank, leaf area index and root growth (Wilson et al., 2006).

However, the number of leaves did not decrease under water shortage in different varieties (Sepaskhah and

*Corresponding author.E-mail:farahani_aliabadi@yahoo.com

Haghighi, 1990). Deficit irrigation decreases the effective filling period of seed, but it doesn't affect the collection of dry material existing in the endosperm and the germ (Wasson et al., 2002), thus decreasing the quantity of seeds in corn. Similarly, Stikic et al. (2004) reported that corn seed yield wasn't very different between normal and deficit irrigation and that the functional components been weren't influenced notably. It has been shown that during a specific stage of plant development, controlled deficit irrigation along with adjusting the humidity of soil can result in water reservation (Monteith, 1973; Lehrsch et al., 2000). Deficit irrigation during germination and medium deficit irrigation during growth of shank can be an appropriate way to save water for corn production in the semi-drought areas of the world (Taleghani et al., 1999; Wilson, 2006). It seems vital to evaluate ways for increasing water and nutrient use efficiency and resistance to drought stress. Scientists have confirmed the effect of nutritional elements on the resistance of plant against deficit irrigation (Pritchard et al., 1999).

Nutrient elements supply, uptake and transport are impaired without sufficient water (Baker, 1997). Among the micro-nutrient elements, Bohr has the most importance in the process of seed development (Parasad and Power, 2002), and also has a metabolic role to control the biochemical reactions and protecting the plant from penetration of living and non-living factors into the cell (Popova, 1990). When the plant has access to elements such as B, K and Zn the growth of root and also the proportion of shank: root increases thus producing more carbohydrates and proteins enabling the plant to utilize the humidity of soil more efficiently specifically during the drought periods (Parasad and Power, 2002). Utilization of these elements will increase the efficiency of plant when utilizing the water (Nasri, 2008). Potassium plays a great role in drought conditions and plants which can access K much easier can sustain the deficit irrigation much better (Krauss, 1999). Marschner (1999) declared that using Zn and Fe elements increases the total amount of carbohydrate, starch and proteins in plant and this increases the resistance of plant against the deficit irrigation. At the attempts to increase food production by means of application of more fertilizers some new problems have been recognized (Popova, 1994b). Excessive amounts of nutrient elements can cause losses through leaching of ions into surface and ground waters (Tandon, 1995). Deep water percolation and chemical leaching is a recognized environmental problem with furrow irrigation (Earl and Davis, 2007).

Furrow irrigation is commonly used in arid and semi-arid zones to supply crops with water and furrow-irrigated maize has been identified as a major contributor to groundwater nitrate pollution (Kuldeep et al., 1997). As much as 40% of the available $NO_3$-N lost from the root zone from one 300 mm irrigation on a clay soil (Kang, 2003). Elements poisoning is the problem encountered with attempts to utilize drought damaged corn and unable to produce enough carbohydrate by photosynthesis to combine with nitrite ions, these plants have the highest accumulation of nitrates near the base of the stalk. If no ear is formed or if ears cannot be pollinated, these plants will accumulate the largest stores of nitrates (Popova et al., 1994a; Benjamin et al., 1998). Alternate furrow irrigation was proposed as a method to increase water use efficiency and decrease chemical leaching compared with every-furrow irrigation and with small yield losses for different crops compared with fixed furrow irrigation system (Baker et al., 1997; Mailholl et al., 2001). Fertilizer leaching is reduced when the applied fertilizer was spatially separated from the irrigation water in a furrow irrigation system (Boyer, 1999). It seems that placing fertilizer in the non-irrigated furrow of an alternate-furrow irrigation system or placing fertilizer in the row with either alternate- or every-furrow irrigation has the potential to decrease fertilizer leaching and nutrient elements poisoning without reducing crop productivity (Sepaskhah and haghighi, 1991; Kajdi and Pocsia, 2003). The objective of the present study was to evaluate the response of maize growth; yield and quality parameters to different irrigation management and nutrient supply with a view to reducing irrigation applied and extend the farming lands with a minimum of yield loss.

## MATERIALS AND METHODS

A new irrigation method under different nutritional element supplies for maize production was designed and tested for some agronomic traits and seed qualitative parameters with complete randomized block experimental design with split-plot arrangement with three replications. This study was conducted under various irrigation strategies and fertilizer combinations with the corn hybrid "KSC704". The field experiment was carried out in Varamin, Iran, in 2007. The site is located at 19':35"N latitude, 39':51"E longitude, with an altitude of 1000 m above sea level. This region has a semi-arid climate (<200 mm annual rainfall). The soil was a montmorillonite clay loam, low in nitrogen (0.4 $mg.kh^{-1}$), low in organic matter (7 $mg.kg^{-1}$) and alkaline with a pH of 7.4 and electrical conductivity of $EC = 0.87\ dS\ m^{-1}$. Irrigation was applied through furrows in three ways as the main plots: alternate furrow irrigation (AFI), fixed furrow irrigation (FFI), and conventional furrow irrigation (CFI). AFI means that one of the two neighboring furrows was alternately irrigated during consecutive watering. FFI means that irrigation was fixed to one of the two neighboring furrows. CFI was the conventional way where every furrow was irrigated during each watering. Each irrigation method was further divided into five sub-treatments with different fertilizer combinations: (1) P+N (control) (2) P+N+K (3) P+N+K+Zn (4) P+N+K+Zn+B (5) P+N+K+Zn+B+Fe. So, each block was consisted of 15 irrigation × fertilizer combination patterns. A subplot size of 4.5 × 5 m, having six rows five meter long each was used. Uniformity of sowing depth was achieved by using a hand dibbler to make holes of 5-7 cm deep. The space between rows was 75 cm wide and sowing was done in the top side of hills at the rate of 20 seeds per square meter. Before anthesis, all the experimental units were irrigated uniformly when the water soil content reached 75% of the amount of available soil water content (SWC) corresponding to the difference between the SWC at field capacity (FC) and at wilting point (WP).

After anthesis, control and under-water stress plots were irrigated when the soil water content decreased to 75 and 25% of field capacity – wilting point (FC-WP), respectively. Soil water content was controlled with two rectangular electrodes embedded in a block of gypsum with wires attached to each electrode, extruded from the gypsum block, and connected to the tensiometer. Gypsum blocks located at two depths (20 and 40 cm) in the soil profile near the fibrous root zone were used to determine the timing of the irrigation depending on the water regime. For installation of the gypsum blocks, the hole was augured at angles of 45 degrees to the horizontal plane to prevent preferential water penetration down the backfilled augured hole. After soil analysis, the required fertilizer was consisted of 300 kg $ha^{-1}$ urea, 60 kg $ha^{-1}$ triple phosphate and 120 kg $ha^{-1}$ potassium sulphate. All of these fertilizers except urea were added when plowing. One third of nitrogenous fertilizer was used at sowing and the remaining was used at the beginning of germination. Required Fe was extracted from iron sequestering fertilization source (13% Fe) whereas Zn and B were extracted from their commercial resource and according to advices of laboratory, The Zn, Fe and B micro-fertilizers with 4000, 8000 and 6000 ppm condensations concentrations respectively were used through three stages and the beginning of implantation.

The fertilizer solution was sprayed by back-hanging cylinders and steadily at the days without wind blowing. During growing season, weed removal was performed manually three times within experimental blocks whereas between replications, a rotavator machine was used. Before harvesting, yield components such as the number of plants per $m^{-2}$, the number of spikes per plant, number of rows and the number of grains in each row were recorded. Grain yield was calculated in each split-plot after grain moisture reached 14% and the weight of each grain was determined after counting and finally the harvest index was calculated by ratio of grain yield to total above ground biomass. Within each plot, an area of 4 $m^2$ was hand harvested to determine grain yield, leaf area index and total above ground biomass. Dry weights were recorded after the plant material had been oven-dried at 70°C for 48 h. Oil and protein percentage of seed were determined with infra red grain analyzer (IRGA) and oil and protein yields were calculated by multiplying grain yield to oil and protein percentage of seed, respectively. Data were statistically analyzed using analysis of variance technique appropriate for randomized complete block-design with fertilizer combinations factor split on irrigation strategy. Duncan's multiple range tests ($p < 0.05$) was applied for mean separation when F values were significant.

## RESULTS AND DISCUSSION

The results declared that the differences between various levels of grain yield, biological yield, harvesting index, 1000 grains weight, the quantity of grains per row and the quantity of spikes per plant among the various irrigation levels were notable (Table 1). Although the quantity of rows in each spike wasn't influenced irrigation levels, but fertilization cares caused a notable difference for this quantity (Tables 1 and 3). According to the Table 1, the average of various irrigation levels and also fertilization cares couldn't influence the quantity of plants per square meter. Because the quantity of plants per SM depends on the implantation density and the implantation density in this research was equal in all splits, the mentioned characteristic wasn't influenced by the searched cares. Thus, we conclude that AFI is a way to save water in arid areas where maize production relies heavily on repeated irrigation. In other hand, if alternate-furrow irrigation is used to increase water use efficiency in furrow-irrigated fields, placing the fertilizer in the non-irrigated furrow of the alternate-furrow irrigation system could decrease N availability because of drier soil conditions in the no-irrigated furrow. Row placement of N fertilizer seems to be beneficial in both alternate-furrow and every-furrow irrigation applications. Fertilized combinations influenced dry matter partitioning to seed filling.

Water stress decreased the grain yield mainly through decreasing the number of grains per cluster and in a lesser degree by decrease in 1000-seed weight (Tables 2 and 3). Sufficient of both macro and micro nutritional elements increased harvest index which was mostly due to more number of seeds per row than higher individual grain weight (Tables 2 and 3). Complete fertilizer combination increased total above ground biomass through more radiation use efficiency and not by increasing leaf area (Tables 2 and 3). So, regarding to mentioned consequences, in order to utilize the water sources efficiently and increase the farming volume and the resistance of plant in this area, we can perform circular irrigation care along with instance, K, Zn, B and Fe fertilizer. It has been showed that there is a significant linear relation between dry matter production and the absorbed radiation. Radiation use efficiency (RUE) significantly was affected by irrigation strategy. Thus, conventional irrigation and alternate furrow irrigation showed higher light use efficiency than fix furrow irrigation method (0.136 vs. 0.127 $Kcal^{-1}$ $m^{-2}$). Highest (0.168 $Kcal^{-1}$ $m^{-2}$) and lowest (0.104 $Kcal^{-1}$ $m^{-2}$) amounts of radiation use efficiency were resulted from complete (P+N+K+Zn+B+Fe) and control (N+P) fertilizer combinations, respectively (Table 2).

Lack of Fe (P+N+K+Zn+B vs. complete fertilizer combination) or presence of K (N+P+K vs. N+P) did not change RUE (Table 2). In the present study, the highest number of spikes per plant was resulted from alternate furrow irrigation and conventional irrigation with 11.54 and 11.18 spikes average respectively and the B5 fertilization care acquired the highest quantity with an average about 11.37 spikes with in comparison with the instance care with an average about 9.90 spikes per plant 13% increase. The interactive effects of irrigation levels and fertilization cares on the quantity of spikes per each plant were notable and the normal and circular irrigational cares and also B5 fertilization care with an average about 12.16 and 11.75 spikes per plant respectively have been positioned in the A statistical grade and the fixed irrigation and instance fertilization care were positioned in the last grade with an average of 8.10 spikes per plant. The effects of irrigations and the performance of nutrients on the yield of corn have been reported by many scientists (Tandon, 1995) and the quantity of spikes per each plant is included in these reports. If the plant can access to required water and nutrients, then can grow much faster and the corn plant can have more extensive photosynthesis level and finally can nurture more spikes.

The results show that the quantity of grains per row is simply affected by the nutritional and irrigational cares.

**Table 1.** Mean squares of some agronomic traits and seed qualitative parameters

| S.O.V | df | Mean Squares | | | | | | |
|---|---|---|---|---|---|---|---|---|
| | | Grain yield (kg ha$^{-1}$) | Seed oil percentage (%) | Oil yield (kg ha$^{-1}$) | Seed protein percentage (%) | Protein yield (kg ha$^{-1}$) | Leaf area index | Radiation use efficiency (Kg kcal$^{-1}$m$^{-2}$) |
| Block | 2 | 589882.3ns | 0.688 ns | 3.25ns | 1.557ns | 678.21ns | 0.375ns | 0.039ns |
| Irrigation (A) | 2 | 21582314.21** | 4.251 * | 29.45* | 17.251** | 7774.34* | 5.893* | 0.354* |
| Error(a) | 4 | 374249.44 | 1.182 | 18.22 | 0.559 | 5250.44 | 1.525 | 0.051 |
| Fertilizer (B) | 4 | 6579938.11** | 14.285** | 151.803** | 3.997* | 22635.15** | 29.354** | 0.128** |
| A×B | 8 | 33414111.39** | 10.425** | 119.42** | 34.829** | 2747.71** | 17.421** | 0.268** |
| Error(b) | 24 | 253928.21 | 0.921 | 12.29 | 0.529 | 1589.51 | 0.980 | 0.020 |
| C.V | | 15.89 | 4.88 | 7.19 | 3.49 | 12.92 | 6.12 | 7.80 |

| S.O.V | df | Plants per square meter | Number of spikes per plant | Number of rows per spike | Number of grains per row | 1000 grains weight (g) | Total above ground biomass (kg ha$^{-1}$) | Harvest Index (%) |
|---|---|---|---|---|---|---|---|---|
| Block | 2 | 0.023ns | 0.072ns | 0.691* | 69.21ns | 1201.49ns | 368541.55ns | 1.238ns |
| Irrigation (I) | 2 | 0.45ns | 3.98* | 1.27ns | 29.88** | 2485.47** | 1897921.35** | 34.11** |
| Error(a) | 4 | 374249.44 | 1.182 | 18.22 | 0.559 | 5250.44 | 1.525 | 0.051 |
| Fertilizer (F) | 4 | 1.03ns | 2.48* | 6.99** | 135.42** | **71.2142 | **26.1241392 | 19.25** |
| I×F | 8 | 0.78ns | 15.81** | 5.88** | 75.42** | **47.2701 | **35.998841 | 11.37** |
| Error(b) | 24 | 253928.21 | 0.921 | 12.29 | 0.529 | 1589.51 | 0.980 | 0.020 |
| C.V (%) | | 3.48 | 3.21 | 11.11 | 14.25 | 45.8 | 45.19 | 6.14 |

*, ** means significant in 0.05 and 0.01 level of probability respectively and NS: non-significant.

The normal and circular cares with an average between 30.56 - 31.36 grains excluded the most quantity and the fixed irrigation was at the minimum level with an average about 26.24 grains per row. Because the quantity of grains per row is one of the most important bases of yielding components and is related to the general function directly, more the quantity of grains per row, better the function grain. The quantity of grains per row, in addition to congenital characteristics, relates to various factors such as suitable irrigation, existence of enough nutrients and suitable insemination time and climatic situation. The lack of each of these factors leads to emptiness of spike grains and decreases the quantity of grains per row (Marschner, 1999). Deficit irrigation will lead to a delay in insemination and the distribution of granules will encountered with problem (Wasson et al., 2002; Koleva, 1973; Denmead and Shaw, 2002) and finally the number of grains per row and total number of grains will decrease. Generally, the drought tension will ban the action of fundamental cells specifically in the burst stage and eventually influence the quantity of grains in spike (Lauer, 2006).

Regarding to acquire consequences in this inspection, it is clear that the products which received the most water and fertilizers had the most number of grains per row which is conformed to other's findings. The weight of 1000 grains is influenced completely by the interactive irrigation and fertilization cares and the difference was so clear. The most weight of 1000 grains was

**Table 2.** Means of some agronomic traits and seed qualitative parameters.

| Treatments | Grain yield ($kg\ ha^{-1}$) | Seed oil percentage (%) | Oil yield ($kg\ ha^{-1}$) | Seed protein percentage (%) | Protein yield ($kg\ ha^{-1}$) | Leaf area index | Radiation use efficiency ($Kg\ kcal^{-1}m^{-2}$) |
|---|---|---|---|---|---|---|---|
| CFI (A1) | 11156a | 6.4a | 714a | 12.55c | 1400b | 3.55a | 0.136a |
| AFI (A2) | 10944a | 6a | 656.65a | 14.65b | 1603.3a | 3. 25ab | 0.135a |
| FFI (A3) | 9056.6b | 4.7b | 425.7b | 15.86a | 1431b | 2.8b | 0.127b |
| P+N (B1) | 8930.6e | 5.29b | 472.5c | 13.11b | 1171b | 2.79c | 0.104c |
| P+N+K (B2) | 9639d | 5.576ab | 537.5bc | 13.94b | 1343.7c | 2.97bc | 0.111bc |
| P+N+K+Zn (B3) | 10104.6c | 5.71ab | 577b | 14.9a | 1505.5b | 3.125b | 0.125b |
| P+N+K+Zn+B (B4) | 11356.6b | 6.11a | 694a | 14.8a | 1681a | 3.28ab | 0.150a |
| P+N+K+Zn+B+Fe (B5) | 11896.6a | 6.156a | 732.35a | 14.65a | 1743a | 3.46a | 0.168a |

CFI, AFI and FFI: conventional furrow irrigation, alternate furrow irrigation and fixed furrow irrigation, respectively. Means with the same letter in each column have not statistically significant difference.

**Table 3.** Means of some agronomic traits and seed qualitative parameters

| Treatments | Plants per square meter | Number of spikes per plant | Number of rows per spike | Number of grains per row | 1000 grains weight (g) | Total above ground biomass ($kg\ ha^{-1}$) | Harvest Index (%) |
|---|---|---|---|---|---|---|---|
| CFI (A1) | 6.6a | 11.54a | 17.6a | 31.36a | 346.8a | 22514a | 49.3a |
| AFI (A2) | 8.6a | 11.18a | 17.6a | 30.56a | 342.2a | 22234a | 49.08a |
| FFI (A3) | 6.4a | 9.18a | 16.8a | 26.24b | 319.2b | 19570b | 46.16b |
| P+N (B1) | 6.3a | 9.9b | 16b | 24.1e | 317c | 19617.3e | 45.4 b |
| P+N+K(B2) | 6.3a | 10.23b | 17.3ab | 26.56d | 323c | 20452.3d | 47ab |
| P+N+K+Zn(B3) | 7a | 10.7ab | 17.3ab | 28.96c | 333.6bc | 21370c | 47.1ab |
| P+N+K+Zn+B(B4) | 6.3a | 10.97ab | 18a | 32.5b | 343.3b | 22287b | 51.1 a |
| P+N+K+Zn+B+Fe(B5) | 7a | 11.37a | 18a | 34.76a | 363.3a | 23470a | 50.5 a |

CFI, AFI and FFI: conventional furrow irrigation, alternate furrow irrigation and fixed furrow irrigation, respectively. Means with the same letter in each column have not statistically significant difference.

acquired from normal and circular irrigation with an average of between 342.2 and 346.8 gr in comparison with the fixed irrigation of an average of 319.2 gr was 8% better. The water tension in corn, because of leaf welter, decreases the photosynthesis and transposition of the materials and excludes the growth of seed (Nelson, 2003) and so the weight of 1000 grains declines. The simple effects of fertilization cares also change the weight of 1000 grains. B5 fertilization care and also B1 instance care with averages of 363.3 and 317 gr, respectively acquired the most and least weight of 1000 grains. Marschner (1999), after his inspections, announced that using Fe and Zn increases the total carbohydrate, starch and protein of grain and so increases the weight of 1000 grain. The inspections declared that the total carbohydrate, starch and protein of grain have increased using Fe and Zn and thus the resistance of plant against the deficit irrigation increase (Marschner, 1999).

Deficit irrigation in the regeneration stage of corn leads to the death of granule tube in spike and using some elements such as enough Zn, B and K in the soil leads to lessening this damage and their processes in the soil increase the resistance of plant against the possible damages resulted from deficit irrigation (Lauer, 2006; Boyer, 1999). Practicing the fertilization care in this inspection resulted in almost 25% increase in the grain yield. Regarding to the vital role of B, Zn, Fe and K in the plant, this event is predictable, specifically from the results of interactive effects which describe that the highest grain yield acquires from circular and normal irrigation and B5 fertilization care which itself describes the importance of mentioned elements in this study. Using the mentioned elements in the normal condition increases the growth of root and also in this condition the plant utilizes the humidity of soil more efficiently (Pandey and Agarwal, 2004; Simuneh et al., 1999).

The highest biologic yield was acquired from normal irrigation with an average about 22514 kg/ha which wasn't so different from the circular irrigation and both were in the same statistic grade. But the fixed irrigation with an average about 19750 kg/ha was about 13% less efficient than perfect irrigation and positioned in the second place. Also the fertilization levels categorized the biologic yield into 5 statistic groups and B5 fertilization care, with an average about 23470 kg/ha was positioned at a grade and the instance fertilization care with an average about 19617.3 kg/ha was positioned at E grade. Also their interactive effects on the biologic yield was notable and B5 fertilization care and normal irrigation with averages about 24920 and 24760 kg/hectare respectively was in the statistic grade and the fixed irrigation with an average about 18431 kg/hectare occupied the lat position. The most surprising result was that AFI maintained high grain yield with up to 50% reduction in irrigation amount and as a result, WUE for irrigated water was substantially increased (Table 2).

Thus, Alternate furrow irrigation and conventional furrow irrigation resulted in same grain yield, leaf area index, radiation use efficiency, kernel oil percentage and oil yield but alternate furrow irrigation significantly increased kernel protein percentage and protein yield by 14.33 (12.55 to 14.65%) and 12.66% (1400 to 1603 kg $ha^{-1}$), respectively (Tables 2 and 3). Grain yield was significantly increased by 17.2% (1094 to 9056 kg $ha^{-1}$) under alternate furrow irrigation vs. fixed furrow irrigation which was due to 6% more radiation use efficiency, but leaf area index did not change (Table 2). These results indicate that water stress effects caused by furrow irrigation on yield may be alleviated by more frequent irrigation intervals.

## REFERENCES


Artiola JF (1991). Nonuniform leaching of nitrate and other solutes in a furrow-irrigated, sludge amended field. Commun. Soil Sci. Plant Anal., 22:1013-1030.

Barker DM, Raine SR, Robertson MJ (1997). A Preliminary. Investigation of Alternate Furrow Irrigation for Sugar Cane Production. Proc. Aus. Soc. Sugar Cane Tech., 20: 302-309.

Benjamin JG, Porter LK, Duke HR, Ahuja LR, Butters G (1998). Nitrogen movement with furrow irrigation method and fertiliser band placement. Soil Sci. Soc. Am. J., 62: 1103-1108.

Boyer JS (1999). Relationship of water potential to growth of leaves plant. Physiol., 43: 1056-1062.

Carbtree RJ, Yassin AA, Kargiugou I, Mcnew RW (1985). Effect of Alternate Furrow Irrigation: Water Conservation on the Yield of two Soybean Cultivars. Agric. Water Manage., 10(3): 253-264.

Denmead OT, Shaw RH (2002). The effects of soil moisture stress at different stages of growth on the development and yield of corn. Agron. J., 52: 272-274.

Earl HG, Davis RF (2007). Effect of drought stress on leaf and whole canopy radiation use efficiency and yield of maize. Agron. J., 95: 688-698.

Fischbach PE, Mulliner HR (1974). Every-other furrow irrigation of corn. Trans. ASAE, 17: 426-442.

Kajdi F, Pocsia K (2003). Effect of irrigation on the yield potential, protein yield of oilseed rape cultivars". Acta Ovarinsis, 35: 65-72.

Kang SZ (2003). After reflect of regulated deficit irrigation on maize Trans. Chinese Soc. Agric. Eng., 16(4): 58-60.

Krauss A (1999). Role of potassium in nutrient deficiency. 4th national congress of soil science, Islambad, Pakistan.

Kuldeep SR, Bhobria D, Singh K (1997). Effect of copper and zinc on the dry matter yield and nutrient composition of wheat. Haryana Agric. Univ. J. Res., 17(4): 312-318.

Koleva S (1973). Water retention curves at different soil density level. Soil Science and Agrochemistry.

Lauer J (2006). What happen with in the corn plant when drought occurs? Wisconsin Crop Manager., 10(22): 225-228.

Lehrsch GA, Sojka RE, Westermann DT (2000). Nitrogen placement, row, spacing, and furrow irrigation water positionong effects on corn yield. Agron. J., 92: 1266-1275.

Mailhol JC, Ruelle P, Nemeth I (2001). Impact of fertilisation practices on nitrogen leaching under irrigation. Irrig. Sci., 20: 139-147.

Marschner H (1999). Mineral nutrition of higher plants. Academic Press.

Monteith J (1973). Solar radition and productivity in tropical ecosyster J. Apple. Ecol., 9: 747-766.

Nasri M, Khalatbari M (2008). The effect of water stress on changing macro elements (N, P, K) in rapeseed genotypes leaf (*Brassica napus*). Agro environment Antalai.

Nelson RL (2005). Tassel emergence and pollen shed. Corny news network.

Pandey R, Agarwal MR (2004). Water stress – induced changes in proline contents and Nitrate Reductase Activity in Rice under Light and our k conditions. Phisiol. Mol. Biol. Plants, 4: 53-57.

Parasad R, Power JF (2002). Soil fertility for sustain able agriculture .CRC Press LTC, Australia.

Popova Z (1990). Water application efficiency in surface irrigation. PhD thesis. Univercity of Architecture Civil engineering and Geodesy Sofia.

Popova Z, Petrova I (1993). Impact of Modeling in Furrow Irrigation Erosion, International symposium "Farm lands erosion in temperate plains, environment and hills", Saint-Cloud, Paris, Elsevier Science Publishers, pp. 347-355.

Popova Z, Varlev I, Kutev V (1994a). Soil erosion in continuous and surge furrow irrigation, Transactions of the 15th International Congress of Soil Science, Volume 2b, Commission I. Acapulco, Mexico, pp. 25-26.

Popova Z, Varlev I, Gospodinov I (1994b). Surge irrigation as an environment friendly technology. Regional European Conference of ICID, Varna, Bulgaria, May, pp. 241-350.

Popova Z, Ikonomova E, Varlev I, Kutev V (1998a). Alternative Furrow Irrigation under the conditions of cinnamonic forest red soils (part two): deep percolation and environmental impact. Transactions of the International Conference "Hydro-melioration" 98, Sofia.

Popova Z, Varlev I, Kutev V, Ikonomova E (1998b). Irrigation and cropping techniques to prevent natural water pollution. Papers of the 1st Inter-Regional Conference "Environment - Water: Innovative

Issues in Irrigation and Drainage, Lisbon, pp. 6-13.
Pritchard FM, Norton RM, Eagles HA, Nicolas M (1999). The effect of Environment on Victorian Canola Quality. 10th International oil seed crops.
Sepaskhah H, Haghighi K (1991). Comparison of Usual and Alternate Furrow Irrigation in Sugar Beet Crop. University of Shiraz, Iran.
Simunek J, Sejna M, Van Genuchten M (1999). The HYDRUS-1D and HYDRUS-2D codes for estimating unsaturated soil hydraulic and solutes transport parameters. Agron. Abstr,, p. 357.
Song FB, Dia JY (2000). Effect of drought stress on growth and development of female inflorescence and yield of maize .Journal of Jillian Agric. Univ., 22(1): 17-22.
Stikic R, Davies WJ (2004). Stomata reactions of two different maize lines to osmotic ally induced drought stress. Biological Plant Arum, 43(3): 399-405.
Taleghani DF, Hashemi Dezfuli A, Majidi E (1999). Study of Water and Nitrogen Use Efficiency in two Plant Spacing of Sugar Beet. Research Report of Sugar Beet Seed Institute.
Tandon K (1995). Micronutrients in soil, crops, and fertilizers. Fertilizer Development and consultation Organization, New Delhi, India.
Traore SB, Carlson RE, Pilcher CD, Rice ME (2000). B t and NON BT maize growth and development as affected by temperature and drought stress. Agron. J., 92(5): 1027-1035.
Wasson JJ, Schumacher R, Wicks TE (2002). Maize water content and solute potential at three stages of development. University of Illinois, Dept.
Wilson DR, Ston PG Gillespie RN (2006). Drought effects on water use, growth and yield sweet corn .Proceedings of the 9th Australian Agronomy Conference.

# Studies on taxonomy of a new mite species in the genus *Caloglyphus* berlese (Acarina: Acaridae) recognized in Pakistan

**Muhammad Sarwar[1]* and Muhammad Ashfaq[2]**

[1]Pakistan Atomic Energy Commission, Nuclear Institute of Agriculture, Tando Jam-70060, Pakistan.
[2]University of Agriculture, Faisalabad-38040, Pakistan.

**This paper provides information relevant to categorization of one mite taxon *Caloglyphus captus* in the genus *Caloglyphus berlese* (Acarina: Acaridae) recognized in Pakistan which is new to science. This** species **exhibits unique characters where coxal fields II, III and IV are closed and all dotted. Data provided cover an outline of literature, taxonomic description, morphological features, differentiating and resemblance remarks with closely related species as well as identification key to recognize different taxa of the genus. The locality of collection, collector name, date of collection and the name of host are mentioned.**

**Key words:** Acarina, Acaridae, *Caloglyphus,* new mite hypopus, taxonomy, Pakistan.

## INTRODUCTION

The acaroid mites' species in the Acaridae family (Tyroglyphidae) are cosmopolitan infesting many types of foods and stored products. Mites that belong to the cheese mite group and may be encountered in certain processed foods are *Caloglyphus* species. These are also encountered in places where food residues may accumulate and remain wet, such as in out of the way places in food processing plants (Rodriguez, 1972). Acaroid mites' infestation is a well known problem for stored grain, often influencing quality and hygienic condition of the grain. However, acaroid mite can survive in many environments including the storehouse, human and animal bodies (Li et al., 2003). The process of building taxonomic system may serve as a powerful analytical tool, which can help researchers to identify gaps in information and ways to resolve pest problems. Originally, Berlese in 1923, described *Caloglyphus* as a subgenus of *Tyroglyphus*; he designated *Tyroglyphus mycophagus* Mégnin, (1874) as a type species of genus *Caloglyphus* for a single species (hypopus), afterward, several species of genus *Caloglyphus* have been referred from numerous expanses of world. Zakhvatkin (1941) arranged a comprehensive review of this genus and portrayed 4 new species and redescribed 6 species with improved descriptions. Nesbitt (1944 and 1949) and Samsinak (1966) supplemented 1, 3 and one new species to this genus, respectively. Mahunka (1973, 1974, 1978 and 1979) described 2, 1, 2 and one new species, respectively from his area of study. Hughes (1976) prepared an excellent accumulation of knowledge to this genus. Tseng and Hsieh (1976) redescribed one species with improved depiction. Samsinak (1980) amended the tribe *Caloglyphini*, re-established the genus *Caloglyphus* and illustrated one new species. Channabasavanna et al. (1981), Rao et al. (1982) and Ashfaq and Chaudhri (1983) incorporated 1, 1 and four new species, respectively in this genus. Samsinak (1988) pointed out 1 new species of the tribe *Caloglyphini*. Zou and Wang (1989), Sevastyanov and Radi (1991), Sher et al. (1991), Klimov (1996) and Eraky (1999) supplemented 1, 3, 2, 1 and one new species, respectively to this genus. Klimov (2000) reviewed acaroid mites of the tribe *Caloglyphini* with description of a new species. Klimov and Oconnor (2003) published phylogeny, historical ecology and systematic of some mites including full descriptions of each taxon, keys and biological

*Corresponding author. E-mail: drmsarwar64@yahoo.com.

informations. Sarwar and Ashfaq (2004, 2006), and Sarwar et al. (2005, 2009) in their study recognized and expressed six new species from this global stretch.

Current investigations on the occurrence of mites fauna in the urban environment, has encountered one species that is classified as new to science. A sequential analysis of the genus, description and illustration of main body characters for the new taxon and resemblance remarks with closely related species are given. The terms of body parts and idiosomal chaetotaxy follow Griffiths et al. (1990) and terms of leg chaetotaxy and solenidiotaxy follow Griffiths (1970). All measurements taken from whole specimen are given in micrometers (µm).

## CALOGLYPHUS CAPTUS, SP. NOV.

### Hypopus

This new species has been described from hypopial stage. Gnathosoma consists of gnathosoma fused pedipalpi, 30 µm long, (basal part 20 µm long, distal part 10 µm), 2 segmented, rounded at base, parallel laterally, bifurcated anteriorly, one pair aristae, 34 µm long, two pairs small setae. (Gnathosoma Figures 2 and 3). Dorsal idiosoma consists of the body 283 mµ long, 215 mµ wide, divided into propodosomal and hysterosomal shields. Propodosomal shield 75 µm long, 185 µm wide, with well defined rostral projection antero-medially, dotted, latero-medially with broken striations; setae *vi, ve, sci, sce* and *scs*, each one pair, simple, measuring 17, 9, 4, 1 and 26 mµ in length, respectively; *sci-sci* 35 µm, *sce-sce* 83 µm and *sci-sce* 26 µm apart; setae *sci* and *sce* forming a straight line, middle in position. Hysterosomal shield 233 mµ long, 215 µm wide, dotted, anterior, lateral and posterior margins with dots and broken striations, lateral margins turn towards venter. Hysterosomal shield with eleven pairs setae, two pairs visible pores. Setae d1 = d2 = d3 = d4 = 4 µm; hi 11 µm, he 7 µm; la 5 µm, lp1 = lp2 = 11 µm; sae 40 µm, sai 11 µm long; d1 - d1 89 µm, d2 - d2 59 µm, d3 - d3 69 µm, d4 - d4 67 µm; d1 - d2 60 µm, d2 - d3 60 µm, d3 - d4 57 µm and la - la 120 µm apart. Hysterosomal shield anterior margin overlapping propodosomal shield posterior margin up to 25 µm, overlapping area with transverse and broken striations (Dorsal idiosoma Figure 1).

Ventral idiosoma consists of apodeme 1 (ap1) Y-shaped continuing with sternum 1 (st1), sternum 1 (st1) free, slightly bifid posteriorly, 45 mµ long. Apodeme 2 (ap2) not free, meeting Apodeme 3 (ap3) and Apodeme 4 (ap4) forming broad, rounded tip. Apodeme 3 (ap3) meeting Apodeme 4 (ap4). Apodeme 4 (ap4) not meeting medially. Apodeme 4 (ap4) and Apodeme 5 (ap5) meeting anteriorly, making broad, rounded tip, not meeting with same structure from other side. Sternum 2 (st2) continuing with Apodeme 4 (ap4) and apodeme 5 (ap5), double-lined, 40 mµ long. Apo deme 5 (ap5), sternum 2 (st2) and posterior line between sternum 2 (st2) and apodeme 5 (ap5) making a closed, dotted and triangular area. Metasternal seta (mts) 1 pair, small, each seta in encircled area of apodeme 4 (ap4) and apodeme 5 (ap5). Seta hv 1 pair, simple, 9 µm long. Coxal fields I open, II, III and IV closed, all dotted. Ventral shield separated from genital shield. Genital shield, dotted, genital slit elongated, 2 pairs genital suckers and 1 pair paragenital seta (pr) messed to genital disc (gdi3). Coxal discs di1 and di2 present, conoids. Genital disc (gdi3) kidney-shaped without radial striations (Figures 2 and 4). Suctorial shield 55 µm long, 65 mµ wide, dotted, concave medio-anteriorly, rounded posteriorly having 1 pair suckers below, sclerotized piece, pointed, bifurcatedly arising from latero-posterior margin on either sides, anterior suckers 1 pair, oval, anal suckers 1 pair, rounded, anal suckers not equal to anterior suckers, 1 pair lateral and 1 pair posterior conoids, 2 pairs clear areas, representing vestigial suckers towards periphery (Figure 5). Suctorial shield separated from posterior end of body by 32 mµ, a distance smaller than suctorial shield length (Figure 2) (Ventral idiosoma Figures 2, 4 and 5).

Legs are strong and stout, I-IV measuring 115, 100, 85 and 75 µm in length, respectively (trochanter base to tarsus tip). Setae and solenidia on legs I-IV segments: coxae 0-0-0-0, trochanters 1-1-1-0, femora 1-1-0-0, genua 3-3-1-1, tibiae 3-3-2-2, tarsi 13-11-9-8. Tarsi I and II 37 and 34 µm long, respectively. Seta *vF* on femora I and II 36 mµ and 38 mµ long, respectively, absent on femora III and IV .Seta *e* on tarsi I-IV measuring 28, 25, 15 and 15 µm long, respectively. Seta *mG* on genua I, lancet-like, on II, a spine; *hT* on tibiae I and II lancet-like, 15, 12, 20 and 15 µm long, respectively. Seta ó on genu I, simple, on genu II, a solenidion, and 30 and 12 mµ long, respectively. Tarsi I and II each with a solenidion *w1* 24 and 25 µm long, respectively. Tarsi III and IV are short and stout. Seta ö on tibiae I and II 65 and 36 µm long, respectively. Seta *ba* on tarsus I 25 µm long. Tarsi I-IV provided with 3 leaf-like + 1 spoon-shaped; 4 leaf-like + 1 spoon-shaped; 4 leaf-like + 1 spoon-shaped; 4 leaf-like + 1 spoon-shaped setae, respectively. Seta *d* is on leg IV tarsus 30 mµ long (Legs Figure 2).

### Types

Holotype, hypopus, collected from Dera Ismail Khan from maize (*Zea mays* L.) on 16.10.1994 (Sarwar) and deposited in Acarology Research Laboratory, Department of Agricultural Entomology University of Agriculture Faisalabad, Pakistan.

### Remarks

This new species is comparable to *Caloglyphus faisalabadiensis* Sher, Ashfaq and Parvez due to following points:

1. Setae *sci* and *sce* posterior in position forming a semi-circular line in *C. faisalabadiensis* but middle in position

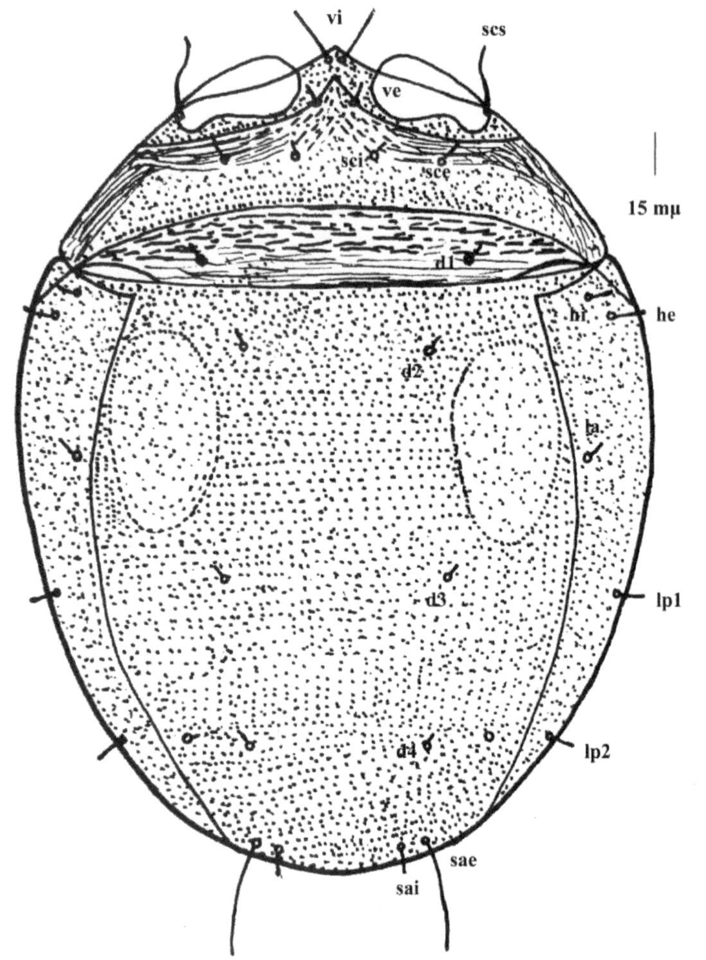

**Figure 1.** Dorsal side.

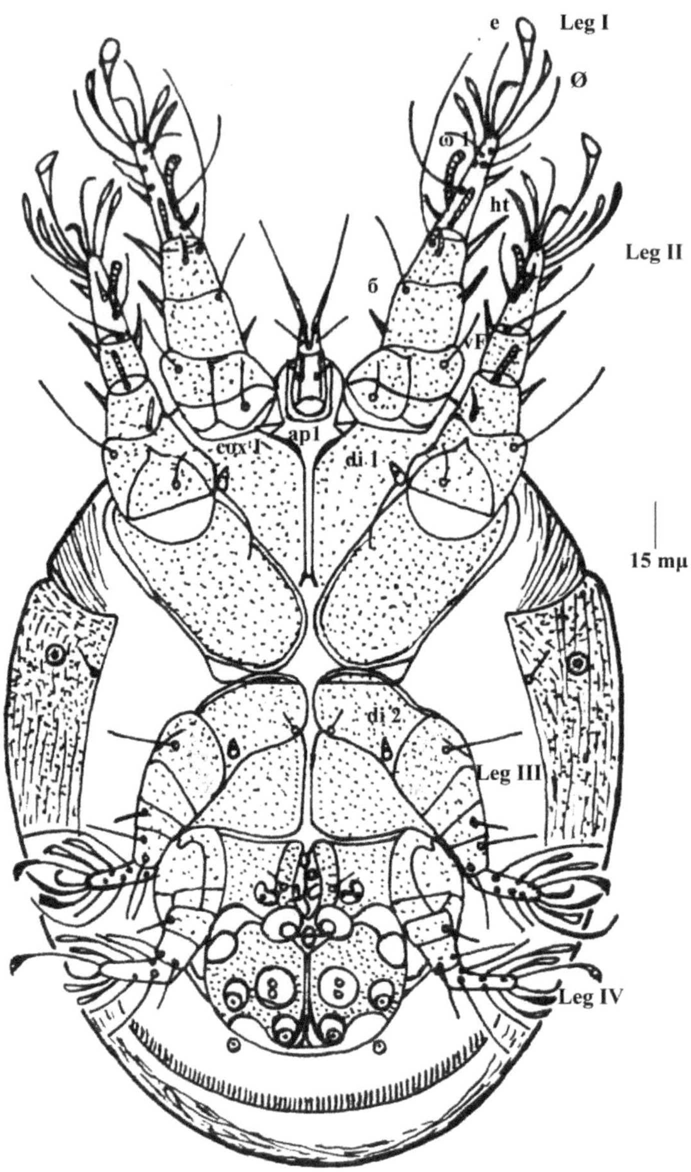

**Figure 2.** Ventral side.

forming a straight line in this new species.
2. Coxal field II open in *C. faisalabadiensis* but closed in this new species.
3. Coxal discs (*di1, di2*) simple in *C. faisalabadiensis* but conoids in this new species.
4. Paragenital seta (*pr*) antero-medial to genital disc (*gdi3*) in *C. faisalabadiensis* but messed in this new species.

This new species is also nearer to *Caloglyphus hadros* Sarwar, Ashfaq and Akbar, but following characters separate them:

1. Dorsal body smooth in *C. hadros* but dotted in this new species.
2. Gnathosoma notched posteriorly in *C. hadros* but not notched in this new species.
3. Coxal field II open in *C. hadros* but closed in this species.
4. Tarsi I-III with 4-5-3 leaf-like setae, respectively in *C. hadros* but with 3-4-4 leaf-like setae, respectively in this new species.

This new species differs from *Caloglyphus mycophagus* Megnin, due to having the following features:

1. Sternum 2 (st2) not meeting genital shield posteriorly in *C. mycophagus* but meeting in this new species.
2. Gnathosoma with 1 pair small setae in *C. mycophagus* but with 2 pairs small setae in this new species.
3. Coxal disc (di1) replaced by a seta in *C. mycophagus* but not so in this new species.

This new species and *C.berlesei* Michael (1903) have the following deviating characters in them:

1. Sternum 2 (*st2*) absent in *C. berlesei* but present in this new species.

**Figure 3.** Gnathosoma.

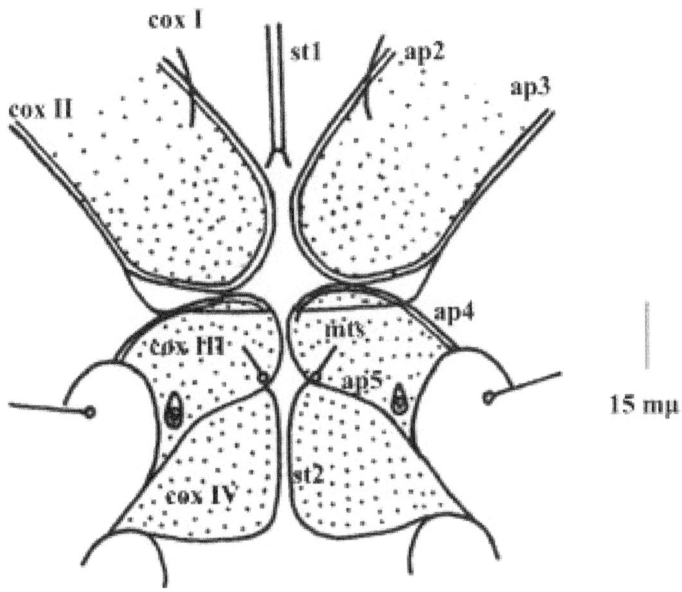


**Figure 4.** Coxal fields.

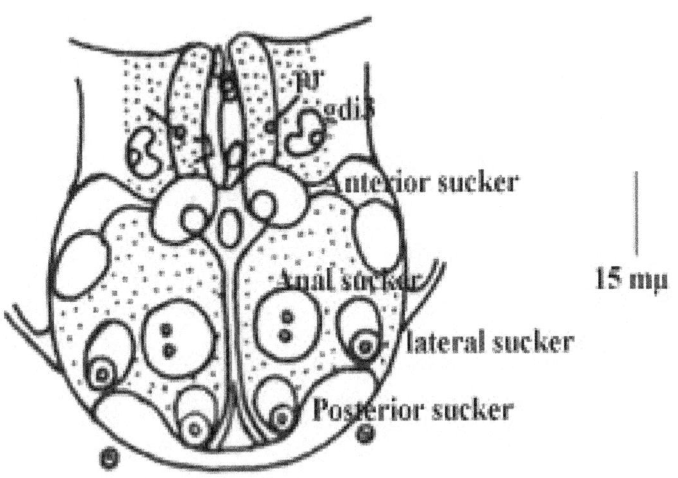


**Figure 5.** Suctorial shield.

2. Sternum 1 (st1) bifid posteriorly in this new species but not so in *C. berlesei*.
3. Metasternal seta (mts), paragenital seta (*pr*) and seta *hv* absent in *C. berlesei* but present in this new species.

**Key to the species of genus *Caloglyphus* (hypopus)**

1. Sternum 2 (*st2*) present ..........................................................................2
- Sternum 2 (*st2*) absent ..........................................26

2. Sternum 2 (*st2*) meeting genital shield posteriorly ..........................................................................3
- Sternum 2 (*st2*) not meeting genital shield posteriorly............................................................19

3. Apodeme 2 (*ap2*) meeting apodeme 3 (*ap3*)..................................................................4
- Apodeme 2 (*ap2*) not meeting apodeme 3 (*ap3*)..............................................................7

4. Dorsum dotted; sternum 1 (*st1*) bifid posteriorly ..........................*C. captus* Sarwar and Ashfaq, sp. nov.
- Dorsum not dotted; sternum 1 (*st1*) not bifid posteriorly..............................................................5

5. Propodosomal setae (*sci,sce*) forming a straight line shaped...................................*C. austerus* Sarwar et al.
- Propodosomal setae (*sci,sce*) not forming a straight line.....................................................................6

6. Apodeme 2 (*ap2*) meeting apodeme 4 (*ap4*); metasternal seta (*mts*) present.........................*C. spinipes* Mahunka
- Apodeme 2 (*ap2*) not meeting apodeme 4 (*ap4*); metasternal seta (*mts*) absent..............*C. rodionovi* A. Z.

7. Apodeme 3 (*ap3*) meeting apodeme 4 (*ap4*)..................................................................12
- Apodeme 3 (*ap3*) not meeting apodeme 4 (*ap4*)..................................................................8

8. Gnathosomal lateral margins parallel.................................................................9
-Gnathosomal lateral margins not parallel............................................................ 11

9. Sternum 1 (*st1*) bifid posteriorly; paragenital seta (*pr*) bifid.....................*C. multaniensis* Ashfaq and Chaudhri
- Sternum 1 (*st1*) not bifid posteriorly; paragenital seta (*pr*) not bifid...........................................................10

10. Tarsus I with 4 leaf-like setae; externalventral seta of genu II simple..................*C. tshernyshevi* Zakhvatkin
- Tarsus I with 5 leaf-like setae; external ventral seta of genu II spine-like............*C. agrios* Sarwar et al. (2000)

11. Setae *sci* and *sce* forming straight line; apodemes 4

(*ap4*) not meeting medially.*C. opacatus* Ashfaq and Chaudhri
- Setae *sci* and *sce* not forming straight line; apodemes 4 (*ap4*) meeting medially..............*C. trigonellum* Sher et al.

12. Gnathosoma notched posteriorly..........................................................................13
- Gnathosoma not notched posteriorly..........................................................................14

13. Setae *sci* and *sce* of equal size; apodemes 4 (*ap4*) meeting medially.........*C. merisma* Ashfaq and Chaudhri
- Setae *sci* and *sce* not of equal size; apodemes 4 (*ap4*) not meeting medially...*C. hadros* Sarwar et al.

14. Dorsal body setae simple..........................................................................15
- Dorsal body setae serrate..........*C. oviformis* Mahunka

15. Gnathosoma much below anterior margin of propodosoma; coxal field III opened...........*C. moniezi* A.Z.
- Gnathosoma not much below anterior margin of propodosoma; coxal field III not open.......................16

16. Gnathosoma parallel laterally, distal fork separated from basal joint; genital disc (*gdi3*) kidney-shaped..........................................................................18
- Gnathosoma not parallel laterally, distal fork not separated from basal joint; genital disc (*gdi3*) not kidney-shaped..........................................................................17

17. Hysterosomal shield smooth; trochanter I without seta; ó on genu II spine-like...*C. muscarius* Sevastyanov and Radi
- Hysterosomal shield dotted; trochanter I with 1 seta; ó on genu II a solenidion............*C. kenos* Sarwar and Ashfaq.

18. Hysterosomal shield smooth; sternum 1 (*st1*) not bifid posteriorly; coxal discs (*di1, di2*) conoids...........................................*C. bradys* Sarwar et al.
- Hysterosomal shield dotted; sternum 1 (*st1*) bifid posteriorly; coxal discs (*di1, di2*) not conoids....*C. faisalabadiensis* Sher, Asfaq and Parvez

19. Ventral shield separated from genital shield..........................................................................22
- Ventral shield not separated from genital hield..........................................................................20

20. Apodemes 4 (*ap4*) meeting medially...........................................*C. polyphyllae* Zakhvatkin
- Apodemes 4 (*ap4*) not meeting medially..........................................................................21

21. Suctorial shield rounded......*C. mandzhur* Zakhvatkin
- Suctorial shield longer than wide..........................................*C. rhizoglyphoides* Zakhvatkin

22. Gnathosoma with 1 pair small setae..........................................................................23
- Gnathosoma not with 1 pair small setae..........................................................................25

23. Coxal disc (*di1*) replaced by a seta..........................................*C. mycophagus* Megnin
- Coxal disc (*di1*) not replaced by a seta..........................................................................24

24. Sternum 2 (*st2*) free from either sides; coxal field III open..........................................*C. baloghi* Mahunka
- Sternum 2 (*st2*) free posteriorly; coxal field III closed..........................................*C. conus* Mahunka

25. Apodeme 2 (*ap2*) meeting apodeme 3(*ap3*); suctorial shield with radial striations posteriorly.................................*C. bifurcatus* Mahunka
- Apodeme 2 (*ap2*) not meeting apodeme 3(*ap3*); suctorial shield without radial striations posteriorly.........*C. forficularis* Sevastyanov and Radi

26. Gnathosoma extended beyond the body.......................................................................... 27
- Gnathosoma not extended beyond the body..........................................................................28

27. Apodemes 4 (*ap4*) meeting medially, metasternal seta (*mts*), paragenital seta (*pr*) and seta *hv* present........................... *C. morosus* Ashfaq and Chaudhri
- Apodemes 4 (*ap4*) not meeting medially; metasternal seta (*mts*), paragenital seta (*pr*) and seta *hv* not present.......................................... *C. berlesei* Michael

28. Apodeme 3 (*ap3*) meeting apodeme (*ap4*)..........................................................................31
- Apodeme 3 (*ap3*) not meeting apodeme (*ap4*)......... 29

29. Gnathosoma reduced; coxal field III opened............................ *C. caroli* Channabasavanna et al.
-Gnathosoma well developed; coxal field III closed..........................................................................30

30. Coxal field II open; legs strong and stout.................................... *C. geotruporum* Zakhvatkin
- Coxal field II closed; legs fat and small..........................................*C. spinitarsus* Hermann

31. Coxal field III opened; genital disc (*gdi3*) and suctorial shield with radial striation...*C. clemens* Sarwar and Ashfaq
- Coxal field III closed; genital disc (*gdi3*) and suctorial shield without radial striation.................*C. cingentis* Sarwar and Ashfaq

**REFERENCES**

Ashfaq M, Chaudhri WM (1983). Four new (Hypopi) species of the genus

*Caloglyphus* Berlese from Pakistan (Acarina: Acaridae). Pakistan Entomol., 5(1-2): 61-78.
Berlese A (1923). Centuria sesta di Acari Nuovi. Redia, 15: 237-262.
Channabasavanna GP, Krishna NS, Rao, Ranganath HR (1981). A new *Caloglyphus* (Astigmata: Acaridae) from poultry litter in India with taxonomic comments on the genus. Indian J. Acarol., 6(1/2): 57-63.
Eraky SA (1999). Five new hypopial nymphs (Acari: Acaridae and Histiostomatidae) described from different habitats. Folia Entomologica Hungrica, 60: 45-56.
Griffiths DA (1970). A further systematic study of the genus *Acarus* L. 1758 (Acaridae: Acarina) with a key to species. Bulletin of British Museum (Natural History), Zoology Series, 19: 85-118.
Griffiths DA, Atyeo WT, Norton RA , Lynch CA (1990). The idiosomal chaetotaxy of astigmatid mites. J. Zool., 220: 1-32.
Hughes AM (1976). The Mites of Stored Food and Houses. Technical Bulletin, No. 9, Ministry of Agriculture Food and Fisheries, London, p. 400.
Klimov PB, Oconnor BM (2003). Phylogeny historical ecology and systematics of some mushroom associated mites of the genus *Sancassania* (Acari: Acaridae) with new generic synonymies. Invertebrate Systematics, 17: 469-514.
Klimov PB (2000). A review of acarid mites of the tribe *Caloglyphini* (Acaridae: Acariformes) with description of a new genus and species from Siberia and Russian Far East. Vestink Zoologii Ukraine, 34(4-5): 27-35.
Klimov PV (1996). A new species of acarid mite from the genus *Caloglyphus* (Acari: Acaridae) from the Russian Far East. Zoologischeskii Zhurnal, 75(4): 613-619.
Li CP, Cui YB, Wang J, Yang QG, Tian Y (2003). Acaroid mite, intestinal and urinary acariasis. World J. Gastroenterol., 9(4): 874-877.
Mahunka S (1973). Auf insekten lebende Milben (Acari: Acarida, Tarsonemida) aus Afrika II. Acta Zoologica Hungrica, 19(3-4): 289-337.
Mahunka S (1974). Auf insekten lebende Milben (Acari: Acarida, Tarsonemida) aus Afrika III. Acta Zoologica Hungrica, 20(1-2): 137-154.
Mahunka S (1978). Schizoglyphidae fam. n. and new taxa of Acaridae and Anoetidae (Acari: Acarida). Acta Zoologica Hungrica, 24(1-2): 107-131.
Mahunka S (1979). The examination of myrmecophilous Acaroidea mites based on the investigations of Dr. C. W. Rettenmeyer (Acari: Acaroidea) II. Acta Zoologica Hungrica, 25: 311-356.
Megnin P (1874). Mémoire sur les *Hypopus* (Dugès), Acariens parasites encore nommés *Homopus*, Koch, et *Trichodactylus*, L. Dufour; détermination de leur position zoologique et de leur rôle physiologique. *J. Anat. Physiol. Normales Pathol. Homme Anim.* 10: 225-254.
Michael AD (1903). British Tyroglyphidae. *Ray Society* London, 2: 183.
Nesbitt HHJ (1944). Three new mites of the subfamily Rhizoglyphinae. Canadian Entomolog., 76(2): 21-27.
Nesbitt HHJ (1949). Six new Mexican mites of the sub family Rhizoglyphinae, Acarina. Pan Pacific Entomolog., 25(2): 57-70.
OConnor BM (2003). Systematics and ecology of North American bee-associated mites: potential threats to native and introduced pollinators. *Ann Arbor* Michigan, USA, pp. 1079-48109.
Rao NSK., Ranganath HR, Channabasavanna GP, Krishna NS, Rao and Rao NSK (1982). *Caloglyphus karnatakaensis* sp. Nov. (Acari: Acaridae) from India with taxonomic comments on the genus C*aloglyphus. Indian* J. Acarol., 7(1): 37-43.
Rodriguez JG (1972). Inhibition of Acarid Mite Development by Fatty Acids. pp., 637-650. In: insect and Mite Nutrition- Significance and Implication in Ecology and pest management. Edited by J.G. Rodriguez. North- Holland Publ. Co., Amsterdam, p. 702.
Samsinak K (1966). Die Neuerrichtung der Gattung *Cosmoglyphus* Oudmans 1932 gleichzeitig ein Beitrag zum Problem der "Copra itch". *Zoologischer Anzeiger*, 176(1): 27-42.
Samsinak K (1980). *Caloglyphus rodriguezi* new species with taxonomic remarks on the tribe C*aloglyphini* (Acari: Acaridae). *Mitteilungen aus dem Zoologischen Museum* Berlin, 56(2): 201-206.
Samsinak K (1988). *Sancassania ultima* a new mite of the tribe *Caloglyphini* (Acari: Acaridae). *Entomologie Mitteilungen aus dem Zoologischen Museum Hamburg,* 9(133): 159-164.
Sarwar M, Ashfaq M (2004). Two new *Caloglyphus* Berlese mites (Astigmata: Acaridae) recorded in Pakistan. Pakistan J. Sci. Industrial Res., 47(6): 455-461.
Sarwar M, Ashfaq M (2006). A new mite pest (Acarina: Acaridae) detected from stored commodity in the Punjab province of Pakistan. *Acarologia*, XLVI, 1-2: 115-120.
Sarwar M, Ashfaq M Akbar S (2005). Numerical taxonomy of two new mites species of the genus *Caloglyphus* Berlese (Acaridae) from Pakistan. Pakistan J. Sci. Industrial Res., 48(5): 345-353.
Sarwar M, Ashfaq M, Nadeem S (2009). On the identity of new acarid mites in genus *Caloglyphus* Berlese occurring in Asian expanse (Pakistan) (Acarina: Acaridae). J. Agric. Biolo. Sci., 1(1): 38-47.
Sevastyanov VD, Radi GKKK (1991). New species of the mite family Acaridae (Sarcoptiformes) from Lower Egypt. Entomol. Rev., 8: 139-146.
Sher F, Ashfaq M, Parvez A (1991). Two new (hypopi) species of genus *Caloglyphus* Berlese (Acarina: Acaridae) from Pakistan. Pakistan Entomol., 13(1-2): 27-34.
Tseng Y, Hsieh SA (1976). A new record of acarid mite *Caloglyphus mycophagus* (Megnin) from Taiwan (Acarina: Astigmata). Taiwan Sugar Research Institute, 74: 47-52.
Zakhvatkin AA (1941). Fauna of USSR *Arachnoidea* VI (1) *Tyroglyphoidea (Acari).* Zoological Institute of the USSR Academy of Sciences, New Series, No. 28. English Translation 1959, Rateliffe A, Hughes AM, American Institute of Biological Sciences, p. 573.
Zou P, Wang XZ (1989). A new species and two new records of Acaridae associated with edible fungi from China (Acarina: Acaroidea). Acta Agriculturae, Shanghai 5(3): 21-24.

# Response of nitrogen and foliar spray of nutrient mixture on yield attributes and yield of wheat (*Triticum aestivum* L.)

**Latief Ahmad*, M. Kaleem and R. A. Bhat**

Allahabad Agricultural Institute-Deemed University Allahabad,-U.P

**A field experiment was conducted to study the effect of nitrogen and foliar spray of nutrient mixture on growth and yield of wheat (*Triticum aestivum L*). The obtained data showed that both grain and straw yield, and different yield attributing characters *viz.,* number of spikes plant$^{-1}$, spike weight and grain spike$^{-1}$ were significantly maximum with application of 120 kg Nha$^{-1}$ and the least with application of 30 kg Nha$^{-1}$. Foliar spray of 2% DAP + 1% KCl + nutrient mixture ($F_2$) significantly increased the periodic plant height , dry weight plant$^{-1}$ and number of tillers plant$^{-1}$ over 2% DAP + 1% KCl + 100 ppm maleic hydrazide ($F_4$). Foliar spray of 2% DAP + 1% KCl + 100 ppm salicylic acid 100 ppm ($F_3$) and 2% DAP + 1% KCl ($F_1$) also recorded significantly higher values of these growth characters over $F_4$ treatments. Both grain and straw yield and different yield contributing characters *viz.,* spikes plant$^{-1}$, spike weight, grains spike$^{-1}$ and test weight were significantly higher with foliar spray of 2% DAP + 1% KCl + 100 ppm nutrient mixture ($F_2$) and significantly least with foliar spray of 2% DAP + 1% KCl + 100 ppm maleic hydrazide ($F_4$).**

**Key words:** Nitrogen, nutrient mixture, foliar spray, yield, yield attributes

## INTRODUCTION

Wheat is the second most important food grain of India next only to rice and it is a staple diet of people. It contributes 35% of the total food grain production of the country. In India, wheat is cultivated over an area of 28.17 million hectares with a production and productivity of 73.70 million tones and 26.17 qha$^{-1}$, respectively (Anonymous, 2007). India alone produces 13% of world's wheat. Green revolution has enabled India to make about four fold increase in food production during last 50 years, whereas before green revolution annual wheat imported touched 10 million tones and India was a beggar bowl. Uttar Pradesh, an important wheat growing state of India, has an area of 9 million hectares under wheat cultivation with a production of22.51 million tones and productivity of 25.02 qha$^{-1}$ (Anonymous, 2005). The lower productivity could be attributed to the fact that under intense cereal-cereal cropping system and immense use of inorganic fertilizers, especially nitrogen, there has been great depletion of soil fertility.

Nitrogen driven agriculture has caused mining of phosphorous, potassium, sulphur and micronutrients. This is evident from the fact that U.P soils have shown to be deficient in micronutrients *viz;* zinc, copper, iron, manganese and boron by 64.0, 9.0, 6.0 and 24.0%, respectively (Takkar et al., 1997). The rice-wheat-jawar rotation having an economic produce of 6.0 tones /ha/ year removes from the soil zinc, copper, iron and manganese to the tune of 606, 282, 4603 and 1238 grams/ha/year, respectively (Rattan et al., 1999). Urea is generally being applied as basal and as top dressing at critical growth stages of the crop and under ideal

*Corresponding author. E-mail: electricdream2003@yahoo.co.in.

conditions, the percent recovery of nitrogen is not more than 40%, meaning a loss of 60% of applied nitrogen in the form of urea. However, the foliar application of urea on the crop can substantially improve its recovery due to elimination of losses of nitrogen in the soil.

A favorable balance of macro and micronutrients is required for optimum crop production. However, the nutrient imbalances can occur due to non judicious and liberal use of major nutrient and presence of low levels of micronutrients. Zinc is known to be involved in the synthesis of Indole-3-acetic acid thereby indirectly involved in elongation of stems, whereas manganese plays an active role in the photolysis of water in the light reaction of photosynthesis. Boron functions in cell wall formation; transport of sugars, flower retention and pollen formation thereby improving grain production (Henry and Cathey, 2009). Maleic hydrazide, a known growth inhibitor, has been found to be involved in the improvement of growth at very low concentrations (Henry and Cathey, 2009). Similarly salicylic acid improves the transport and uptake of ions, induces changes in chloroplast structure and is involved in growth and development, photosynthesis and respiration (Hayat et al., 2005). Since the deficiency of micronutrients *viz.,* zinc boron and manganese is widely noticed in wheat so their foliar spray in a mixture can improve the wheat yields.

## MATERIAL AND METHODS

The experiment was conduct at Crop Research Farm, Department of Agronomy, Allahabad Agricultural Institute-Deemed University Allahabad- India during *rabi* 2007 to 2008 and 2008 to 2009. The soil of the experimental field was sandy loam in texture, low in available nitrogen 221 kg $ha^{-1}$, medium in available phosphorous 14.4 kg $ha^{-1}$ and potassium 253.0 kg $ha^{-1}$ with pH 7.9 (Alkaline) and 0.27% organic carbon. The experiment was conducted on Variety PBW-443, comprising of two factors (four nitrogen levels *viz.* $N_1$:30kg/ha $N_2$:60kg $ha^{-1}$ $N_3$:90kg $ha^{-1}$ $N_4$:120kg $ha^{-1}$ and four foliar spray of nutrient mixture *viz* $F_1$: 2% DAP + 1% KCl. $F_2$: 2% DAP + 1% KCl + nutrient mixture $F_3$: 2% DAP + 1% KCl + 100 ppm salicylic acid $F_4$: 2% DAP + 1% KCl + 100 ppm maliec hydrazide) was laid out in $4^2$ factorial randomized block design replicated thrice. (NB: Nutrient mixture = 0.5% $MgSO_4$, 0.2% $ZnSO_4$, 0.2% $MnSO_4$ and 0.1% Borax, DAP = Di-ammonium phosphate and MOP= Murate of potash). A uniform dose of phosphorous and potassium at the rate of 60 and 40 kg $P_2O_5$ and $K_20$/ha, respectively and half dose of nitrogen as per treatments was applied as basal at the time of sowing. Remaining half dose of nitrogen was applied in two equal splits, one each at 30 DAS and tillering stage as per treatment. Nitrogen, phosphorous and potassium was applied through urea, DAP and MOP. Foliar application of different nutrient mixtures was applied at full vegetative growth to each plot as per treatment.

## RESULTS AND DISCUSSION

### Yield attributes

Results over two consecutive years shows that spikes $plant^{-1}$ were significantly more over application of 30 kg $Nha^{-1}$ due to application of 120 kg $Nha^{-1}$ but the spikes $plant^{-1}$ did not vary significantly among 120, 90 and 60 kg $Nha^{-1}$ application. Among different nitrogen levels, maximum number of spikes $plant^{-1}$ (3.42) were recorded with application of 120 kg $Nha^{-1}$, followed by application of 90 kg $Nha^{-1}$ (3.24), 60 kg $Nha^{-1}$ (3.21) and the least number of spikes $plant^{-1}$ was observed with application of 30 kg $Nha^{-1}$ (Table 1). Singh et al. (1995) reported significant improvement in the effective tillers $plant^{-1}$ with nitrogen application up to 120 $kgha^{-1}$. Pooled data of two years shows that spikes $plant^{-1}$ were significantly higher with application of foliar spray of 2% DAP + 1% KCl + nutrient mixture ($F_2$) over 2% DAP + 1% KCl + 100 ppm maleic hydrazide ($F_4$) but the spikes per plant did not vary significantly amongst $F_2$, $F_3$ (2% DAP + 1% KCl + 100 ppm salicylic acid) and $F_1$ (2% DAP + 1% KCl) treatments. Maximum number of spikes (3.41) were recorded with $F_2$ treatment followed by $F_3$ (3.29) and $F_1$ (3.25) and the least number of spikes $plant^{-1}$ (3.15) was found in $F_4$ treatment. The significant increase in the spikes $plant^{-1}$ recorded by $F_2$ treatment could be attributed to better growth and development attained by the crop due to additional application of micronutrients, known to be involved in photosynthesis, protein metabolism and energy transfer reactions. Mohammad (1994) reported improvement in the number of spikes in wheat with foliar spray of zinc, iron and other micronutrients.

Grain weight $spike^{-1}$ was significantly influenced by different nitrogen levels and foliar spray of nutrient mixture during both years and in pooled data over two years. Pooled data of two years shows that the grain weight per spike was significantly higher over 30 kg $Nha^{-1}$ due to application of 120 , 90 and 60 kg $Nha^{-1}$ but the grain weight $spike^{-1}$ did not differ significantly amongst 120, 90 and 60 kg $Nha^{-1}$ application. The higher nitrogen nutrition to the crop increased the vegetative growth by better photosynthesis through higher leaf area thereby resulting in better translocation of photosynthesis from source to sink (Table 2). Among different nitrogen levels, maximum grain weight $spike^{-1}$ (1.96 g) was observed with application of 120 kg $Nha^{-1}$ followed by application of 90 kg $Nha^{-1}$ (1.95 g), 60 kg $Nha^{-1}$ (1.93 g) and the least grain weight $spike^{-1}$ (1.87 g) was observed with application of 30 kg $Nha^{-1}$. Singh et al. (1995) reported significant improvement in the grain weight $spike^{-1}$ with increasing nitrogen doses.

Pooled data of two years shows that grain weight $spike^{-1}$ improved significantly with foliar spray of 2% DAP + 1% KCl + nutrient mixture ($F_2$) over 2% DAP + 1% KCl + 100 ppm ($F_4$). This could be attributed to the better growth and development attained by the crop due to additional application of micronutrients known to be involved in photosynthesis and other energy transfer reactions. Besides, the lowest grain weight $spike^{-1}$ recorded with $F_4$ treatment could be due to growth inhibitory effect of maleic hydrazide application even at lower concentration.

**Table 1.** Effect of different levels of nitrogen and foliar spray of nutrient mixture on number of spikes $plant^{-1}$.

| Nitrogen levels (kg $ha^{-1}$) | Foliar spray of nutrient mixture | | | | | | | | | | | | | | |
|---|---|---|---|---|---|---|---|---|---|---|---|---|---|---|---|
| | 2007-08 | | | | | 2008-09 | | | | | Pooled | | | | |
| | $F_1$ | $F_2$ | $F_3$ | $F_4$ | Mean | $F_1$ | $F_2$ | $F_3$ | $F_4$ | Mean | $F_1$ | $F_2$ | $F_3$ | $F_4$ | Mean |
| $N_{30}$ | 3.14 | 3.32 | 3.16 | 3.02 | 3.16 | 3.07 | 3.32 | 3.19 | 3.05 | 3.16 | 3.10 | 3.32 | 3.17 | 3.02 | 3.15 |
| $N_{60}$ | 3.18 | 3.36 | 3.20 | 3.07 | 3.20 | 3.22 | 3.37 | 3.24 | 3.10 | 3.23 | 3.20 | 3.36 | 3.22 | 3.08 | 3.21 |
| $N_{90}$ | 3.32 | 3.50 | 3.34 | 3.21 | 3.34 | 3.33 | 3.49 | 3.36 | 3.22 | 3.35 | 3.32 | 3.49 | 3.35 | 3.21 | 3.34 |
| $N_{120}$ | 3.39 | 3.55 | 3.41 | 3.28 | 3.41 | 3.42 | 3.55 | 3.44 | 3.30 | 3.43 | 3.40 | 3.55 | 3.42 | 3.29 | 3.42 |
| Mean | 3.26 | 3.44 | 3.28 | 3.14 | | 3.26 | 3.44 | 3.31 | 3.17 | | 3.25 | 3.41 | 3.29 | 3.15 | |
| | | F-test | SEd ± | CD (P=0.05) | | | F-test | SEd ± | CD (P=0.05) | | | F-test | SEd ± | CD (P=0.05) | |
| Nitrogen levels | | S | 0.132 | 0.27 | | | S | 0.127 | 0.26 | | | S | 0.127 | 0.26 | |
| Foliar spray of nutrient mixture | | S | 0.132 | 0.27 | | | S | 0.127 | 0.26 | | | S | 0.127 | 0.26 | |
| Interaction effect | | NS | 0.264 | - | | | NS | 0.254 | - | | | NS | 0.254 | - | |

F1 = 2% DAP + 1% KCl ; N1 = 30 Kg $ha^{-1}$, F2 = 2% DAP + 1% KCl + Nutrient mixture *; N2 = 60 Kg $ha^{-1}$ , F3 = 2% DAP + 1% KCl + 100 ppm Salicylic acid , N3 = 90 Kg $ha^{-1}$, F4 = 2% DAP + 1% KCl + 100 ppm Maleic hydrazide; N4 = 120 Kg ha-1, * Nutrient mixture prepared by 0.5% MgSO4, 0.25% MnSO4, 0.25% ZnSO4 and 0.1 % Boric acid.

**Table 2.** Effect of different levels of nitrogen and foliar spray of nutrient mixture on spike length (cm).

| Nitrogen levels (kg $ha^{-1}$) | Foliar spray of nutrient mixture | | | | | | | | | | | | | | |
|---|---|---|---|---|---|---|---|---|---|---|---|---|---|---|---|
| | 2007-08 | | | | | 2008-09 | | | | | Pooled | | | | |
| | $F_1$ | $F_2$ | $F_3$ | $F_4$ | Mean | $F_1$ | $F_2$ | $F_3$ | $F_4$ | Mean | $F_1$ | $F_2$ | $F_3$ | $F_4$ | Mean |
| $N_{30}$ | 15.16 | 15.23 | 15.17 | 15.09 | 15.16 | 15.18 | 15.23 | 15.18 | 15.1 | 15.17 | 15.17 | 15.23 | 15.17 | 15.09 | 15.16 |
| $N_{60}$ | 15.22 | 15.29 | 15.23 | 15.10 | 15.21 | 15.22 | 15.28 | 15.23 | 15.15 | 15.22 | 15.22 | 15.28 | 15.23 | 15.12 | 15.21 |
| $N_{90}$ | 15.27 | 15.34 | 15.28 | 15.20 | 15.27 | 15.28 | 15.34 | 15.28 | 15.21 | 15.28 | 15.27 | 15.34 | 15.28 | 15.20 | 15.27 |
| $N_{120}$ | 15.27 | 15.34 | 15.28 | 15.20 | 15.27 | 15.29 | 15.35 | 15.30 | 15.22 | 15.29 | 15.28 | 15.34 | 15.29 | 15.21 | 15.28 |
| Mean | 15.23 | 15.3 | 15.26 | 15.15 | | 15.24 | 15.3 | 15.25 | 15.17 | | 15.23 | 15.3 | 15.24 | 15.15 | |
| | | F-test | SEd ± | CD (P=0.05) | | | F-test | SEd ± | CD (P=0.05) | | | F-test | SEd ± | CD (P=0.05) | |
| Nitrogen levels | | NS | 0.156 | | | | NS | 0.196 | | | | NS | 0.176 | | |
| Foliar spray of nutrient mixture | | NS | 0.156 | | | | NS | 0.196 | | | | NS | 0.176 | - | |
| Interaction effect | | NS | 0.312 | | | | NS | 0.392 | | | | NS | 0.352 | - | |

F1 = 2% DAP + 1% KCl; N1 = 30 Kg $ha^{-1}$, F2 = 2% DAP + 1% KCl + Nutrient mixture *; N2 = 60 Kg $ha^{-1}$, F3 = 2% DAP + 1% KCl + 100 ppm Salicylic acid; N3 = 90 kg $ha^{-1}$ F4 = 2% DAP + 1% KCl + 100 ppm Maleic hydrazide; N4 = 120 Kg $ha^{-1}$, *Nutrient mixture prepared by 0.5% MgSO4, 0.25% MnSO4, 0.25% ZnSO4 and 0.1 % Boric acid.

Mohamed (1994) reported improvement in the grain weight $spike^{-1}$ of wheat with micronutrient application. Maximum grain weight $spike^{-1}$ (1.96 g) was recorded with $F_2$ treatment followed by $F_3$ (1.92 g), $F_1$ (1.92 g) and the lowest grain weight $spike^{-1}$ (1.90 g) was observed with $F_4$ treatment (Table 3). However, the grain weight $spike^{-1}$ recorded with $F_2$ did not differ statistically with $F_3$ and $F_1$ as well as the grain weight $spike^{-1}$ recorded with $F_3$, $F_1$ and $F_4$ was also statistically similar.

Grain number $spike^{-1}$ increased significantly and consistently with increase in nitrogen application up to 90 kg $Nha^{-1}$. However, further increase in nitrogen dose upto 120 kg $Nha^{-1}$ didn't differ statistically with 90 kg $Nha^{-1}$ dose. This could be

**Table 3.** Effect of different levels of nitrogen and foliar spray of nutrient mixture on grain weight spike$^{-1}$ (g).

| Nitrogen level (kg ha$^{-1}$) | Foliar spray of nutrient mixture | | | | | | | | | | | | | | |
|---|---|---|---|---|---|---|---|---|---|---|---|---|---|---|---|
| | 2007-08 | | | | | 2008-09 | | | | | Pooled | | | | |
| | $F_1$ | $F_2$ | $F_3$ | $F_4$ | Mean | $F_1$ | $F_2$ | $F_3$ | $F_4$ | Mean | $F_1$ | $F_2$ | $F_3$ | $F_4$ | Mean |
| $N_{30}$ | 1.86 | 1.9 | 1.87 | 1.85 | 1.87 | 1.86 | 1.91 | 1.86 | 1.86 | 1.87 | 1.86 | 1.9 | 1.86 | 1.85 | 1.87 |
| $N_{60}$ | 1.92 | 1.96 | 1.93 | 1.91 | 1.93 | 1.92 | 1.97 | 1.92 | 1.91 | 1.93 | 1.92 | 1.96 | 1.92 | 1.91 | 1.93 |
| $N_{90}$ | 1.95 | 1.99 | 1.96 | 1.94 | 1.96 | 1.94 | 1.99 | 1.94 | 1.94 | 1.95 | 1.94 | 1.99 | 1.95 | 1.94 | 1.95 |
| $N_{120}$ | 1.96 | 2.0 | 1.97 | 1.95 | 1.97 | 1.95 | 2.0 | 1.95 | 1.95 | 1.96 | 1.95 | 2.0 | 1.96 | 1.95 | 1.96 |
| Mean | 1.92 | 1.96 | 1.93 | 1.91 | | 1.92 | 1.97 | 1.92 | 1.91 | | 1.92 | 1.96 | 1.92 | 1.91 | |
| | | F-test | SEd ± | CD (P=0.05) | | | F-test | SEd ± | CD (P=0.05) | | | F-test | SEd ± | CD (P=0.05) | |
| Nitrogen levels | | S | 0.024 | 0.05 | | | S | 0.024 | 0.05 | | | S | 0.029 | 0.06 | |
| Foliar spray of nutrient mixture | | S | 0.024 | 0.05 | | | S | 0.024 | 0.05 | | | S | 0.029 | 0.06 | |
| Interaction effect | | NS | 0.048 | - | | | NS | 0.048 | - | | | NS | 0.058 | - | |

$F_1$ = 2% DAP + 1% KCl ;$N_1$ = 30 Kg ha$^{-1}$, $F_2$ = 2% DAP + 1% KCl + Nutrient mixture * ;$N_2$ = 60 Kg ha$^{-1}$, $F_3$ = 2% DAP + 1% KCl + 100ppm Salicylic acid; $N_3$ = 90 Kg ha$^{-1}$ $F_4$ = 2% DAP + 1% KCl + 100ppm Maleic hydrazide; $N_4$ = 120 Kg ha$^{-1}$, * Nutrient mixture prepared by 0.5% $MgSO_4$, 0.25% MnSO4, 0.25% $ZnSO_4$ and 0.1 % Boric acid.

attributed to the fact that nitrogen helped to produce better vegetative growth as a result of which photosynthetic area increased thereby more food material synthesized contributed to the improvement in grains spike$^{-1}$. Chander and Pandey (1996) also observed that spikes per square meter and filled grains spike$^{-1}$ increased significantly with nitrogen application up to120kg ha$^{-1}$. Amongst different nitrogen levels, maximum number of grains spike$^{-1}$(50.2) were recorded with application of 120 kg Nha$^{-1}$, followed by 90 kg Nha-$^{1}$ (49.6), 60 kg Nha$^{-1}$ (48.6) and the least number of grains spike$^{-1}$ (47.1) was observed with application of 30 kg Nha$^{-1}$. Number of grains spike$^{-1}$ was significantly higher with foliar spray of2% DAP + 1% KCl + nutrient mixture ($F_2$) over 2% DAP + 1% KCl + 100 ppm salicylic acid ($F_3$), 2% DAP + 1% KCl ($F_1$) and 2% DAP + 1% KCl + 100 ppm maleic hydrazide ($F_4$). This could be attributed to the fact that application of micronutrients in combination with macronutrients may have improved the photosynthetic ability of crop thereby more food material synthesized contributed to the improvement in number of grains spike$^{-1}$. The lowest grains spike$^{-1}$ recorded with $F_4$ treatment may be the result of growth inhibitory effect of maleic hydrazide.

Amongst different foliar spray of nutrient mixtures maximum grains spike$^{-1}$ (50.0) were observed with $F_2$ treatment followed by 49.1 with $F_3$ treatment, 49.0 with $F_1$ treatment and the least number of grains spike$^{-1}$ (47.4) with $F_4$ treatment (Table 4). Both $F_3$ and $F_2$ treatment, at par with one another, recorded significantly more grains spike$^{-1}$ than $F_4$ treatment. Significantly higher grains spike$^{-1}$ recorded with $F_3$ over $F_4$ treatment may be due better growth and development, photosynthesis, ion uptake and transport attained by the crop due to application of salicylic acid. Data over two years indicated that 1000-grain weight increased significantly with nitrogen application at 120 kg ha$^{-1}$ over 60 and 30 kg Nha$^{-1}$. The nitrogen levels of 120 and 90 kg Nha$^{-1}$ did not differ significantly with one another and application of nitrogen beyond 90 kg Nha$^{-1}$ didn't affect the 1000-grain weight. This increase in the 1000-grain weight could be attributed to the fact that nitrogen helped to produce better vegetative growth as a result of which photosynthetic area increased and thus more food material synthesized contributed to the improvement in the grain weight. Amongst different nitrogen levels applied, significantly maximum 1000-grain weight (40.8 g) was recorded with 120 Kg Nha$^{-1}$, followed by 40.7 g with 90 kg Nha$^{-1}$ and 40.5 g with 60 kg Nha$^{-1}$. Least 1000-grain weight (40.08) was recorded with nitrogen application at 30 kg Nha$^{-1}$ (Table 5).

Significantly, highest 1000-grain weight was recorded with foliar spray of 2% DAP + 1% KCl + nutrient mixture ($F_2$). The significant increase in 1000-grain weight as recorded with $F_2$ treatment could be attributed to the fact that application of micronutrients in combination with macronutrients may have improved the photosynthesis thus, more food material synthesized contributed to the improvement in 1000-grain weight compared to

**Table 4.** Effect of different levels of nitrogen and foliar spray of nutrient mixture on grains spike$^{-1}$.

| Nitrogen levels (kg ha$^{-1}$) | Foliar spray of nutrient mixture | | | | | | | | | | | | | | |
|---|---|---|---|---|---|---|---|---|---|---|---|---|---|---|---|
| | 2007-08 | | | | | 2008-09 | | | | | Pooled | | | | |
| | $F_1$ | $F_2$ | $F_3$ | $F_4$ | Mean | $F_1$ | $F_2$ | $F_3$ | $F_4$ | Mean | $F_1$ | $F_2$ | $F_3$ | $F_4$ | Mean |
| $N_{30}$ | 47.2 | 48.1 | 47.8 | 45.6 | 47.2 | 47.3 | 48.3 | 47.4 | 45.7 | 47.2 | 47.2 | 48.2 | 47.6 | 45.6 | 47.1 |
| $N_{60}$ | 48.6 | 49.6 | 48.7 | 47.0 | 48.5 | 48.8 | 49.8 | 48.9 | 47.3 | 48.7 | 48.7 | 49.7 | 48.8 | 47.1 | 48.6 |
| $N_{90}$ | 49.6 | 50.6 | 49.7 | 48.0 | 49.5 | 49.8 | 50.8 | 49.9 | 48.3 | 49.7 | 49.7 | 50.7 | 49.8 | 48.1 | 49.6 |
| $N_{120}$ | 50.3 | 51.3 | 50.4 | 48.7 | 50.2 | 50.3 | 51.3 | 50.4 | 48.8 | 50.2 | 50.3 | 51.3 | 50.4 | 48.7 | 50.2 |
| Mean | 48.9 | 49.9 | 49.1 | 47.3 | | 49 | 50 | 49.1 | 47.8 | | 49 | 50 | 49.1 | 47.4 | |
| | | F-test | SEd ± | CD (P=0.05) | | | F-test | SEd ± | CD (P=0.05) | | | F-test | SEd ± | CD (P=0.05) | |
| Nitrogen levels | | S | 0.411 | 0.84 | | | S | 0.441 | 0.9 | | | S | 0.426 | 0.87 | |
| Foliar spray of nutrient mixture | | S | 0.411 | 0.84 | | | S | 0.441 | 0.9 | | | S | 0.426 | 0.87 | |
| Interaction effect | | NS | 0.822 | - | | | NS | 0.882 | - | | | NS | 0.852 | - | |

$F_1$ = 2% DAP + 1% KCl; $N_1$ = 30 kg ha$^{-1}$, $F_2$ = 2% DAP + 1% KCl + Nutrient mixture * ; $N_2$ = 60 kg ha$^{-1}$, $F_3$ = 2% DAP + 1% KCl + 100 ppm Salicylic acid; $N_3$ = 90 kg ha$^{-1}$, $F_4$ = 2% DAP + 1% KCl + 100ppm Maleic hydrazide ;$N_4$ = 120 kg ha$^{-1}$, *Nutrient mixture prepared by 0.5% MgSO4, 0.25% MnSO4, 0.25% ZnSO4 and 0.1 % Boric acid.

**Table 5.** Effect of different levels of nitrogen and foliar spray of nutrient mixture on 1000 grain weight (g).

| Nitrogen levels# (kg ha$^{-1}$) | Foliar spray of nutrient mixture | | | | | | | | | | | | | | |
|---|---|---|---|---|---|---|---|---|---|---|---|---|---|---|---|
| | 2007-08 | | | | | 2008-09 | | | | | Pooled | | | | |
| | $F_1$ | $F_2$ | $F_3$ | $F_4$ | Mean | $F_1$ | $F_2$ | $F_3$ | $F_4$ | Mean | $F_1$ | $F_2$ | $F_3$ | $F_4$ | Mean |
| $N_{30}$ | 39.9 | 40.3 | 39.9 | 39.6 | 39.9 | 39.9 | 40.5 | 40 | 39.8 | 40.0 | 39.9 | 40.4 | 39.9 | 39.7 | 40.0 |
| $N_{60}$ | 40.4 | 40.9 | 40.9 | 40.2 | 40.6 | 40.4 | 41.0 | 40.6 | 40.3 | 40.6 | 40.4 | 40.9 | 40.6 | 40.2 | 40.5 |
| $N_{90}$ | 40.6 | 41.1 | 40.6 | 40.4 | 40.7 | 40.7 | 40.8 | 40.8 | 40.6 | 40.7 | 40.6 | 40.9 | 40.7 | 40.5 | 40.7 |
| $N_{120}$ | 40.8 | 41.2 | 40.8 | 40.5 | 40.8 | 40.8 | 40.9 | 40.9 | 40.7 | 40.8 | 40.8 | 41.0 | 40.8 | 40.6 | 40.8 |
| Mean | 40.4 | 40.9 | 40.5 | 40.2 | | 40.4 | 40.8 | 40.6 | 40.3 | | 40.4 | 40.8 | 40.5 | 40.2 | |
| | | F-test | SEd ± | CD (P=0.05) | | | F-test | SEd ± | CD (P=0.05) | | | F-test | SEd ± | CD (P=0.05) | |
| Nitrogen levels | | S | 0.318 | 0.65 | | | S | 0.245 | 0.5 | | | S | 0.294 | 0.6 | |
| Foliar spray of nutrient mixture | | S | 0.318 | 0.65 | | | S | 0.245 | 0.5 | | | S | 0.294 | 0.6 | |
| Interaction effect | | S | 0.636 | 1.3 | | | S | 0.49 | 1 | | | S | 0.588 | 1.2 | |

$F_1$ = 2% DAP + 1% KCl; $N_1$ = 30 kg ha$^{-1}$, $F_2$ = 2% DAP + 1% KCl + Nutrient mixture*;$N_2$ = 60 kg ha$^{-1}$, $F_3$ = 2% DAP + 1% KCl + 100 ppm Salicylic acid; $N_3$ = 90 Kg ha$^{-1}$, $F_4$ = 2% DAP + 1% KCl + 100 ppm Maleic hydrazide; $N_4$ = 120 Kg ha$^{-1}$,* Nutrient mixture prepared by 0.5% $MgSO_4$, 0.25% MnSO4, 0.25% $ZnSO_4$ and 0.1 % Boric acid.

F3, F1 and F4 treatments. Significantly,1000-grain weight recorded with $F_4$ (2% DAP + 1% KCl + 100 ppm maleic hydrazide) treatment may be as a result of the growth inhibitory effect of maleic hydrazide.Amongst different foliar spray of nutrient mixture, significantly maximum 1000-grain weight (40.8 g) was observed by $F_2$ treatment followed by 40.5 g with $F_3$ (2% DAP + 1% KCl + 100 ppm salicylic acid) treatment, 40.4 g with

**Table 6.** Effect of different levels of nitrogen and foliar spray of nutrient mixture on grain yield ($qha^{-1}$).

| Nitrogen levels ($Kg\ ha^{-1}$) | Foliar spray of nutrient mixture | | | | | | | | | | | | | | |
|---|---|---|---|---|---|---|---|---|---|---|---|---|---|---|---|
| | 2007-08 | | | | | 2008-09 | | | | | Pooled | | | | |
| | $F_1$ | $F_2$ | $F_3$ | $F_4$ | Mean | $F_1$ | $F_2$ | $F_3$ | $F_4$ | Mean | $F_1$ | $F_2$ | $F_3$ | $F_4$ | Mean |
| $N_{30}$ | 41.6 | 44.2 | 41.8 | 40.8 | 42.1 | 41.4 | 44.2 | 41.7 | 40.6 | 42.0 | 41.5 | 44.2 | 41.7 | 40.6 | 42.0 |
| $N_{60}$ | 43.7 | 46.3 | 43.9 | 42.9 | 44.2 | 43.6 | 46.4 | 43.9 | 42.8 | 44.2 | 43.6 | 46.3 | 43.9 | 42.8 | 44.1 |
| $N_{90}$ | 45.3 | 47.9 | 45.5 | 44.5 | 45.8 | 45.3 | 48.0 | 45.6 | 44.4 | 45.8 | 45.3 | 47.9 | 45.5 | 44.4 | 45.8 |
| $N_{120}$ | 46.5 | 49.1 | 46.7 | 45.7 | 47.0 | 46.6 | 49.3 | 46.9 | 45.7 | 47.1 | 46.5 | 49.2 | 46.8 | 45.7 | 46.8 |
| Mean | 44.3 | 46.9 | 44.5 | 43.5 | | 44.2 | 47 | 44.5 | 43.4 | | 44.2 | 46.9 | 44.5 | 43.4 | |
| | | F-test | SEd ± | CD (P=0.05) | | | F-test | SEd ± | CD (P=0.05) | | | F-test | SEd ± | CD (P=0.05) | |
| Nitrogen levels | | S | 0.534 | 1.09 | | | S | 0.603 | 1.23 | | | S | 0.441 | 0.9 | |
| Foliar spray of nutrient mixture | | S | 0.534 | 1.09 | | | S | 0.603 | 1.23 | | | S | 0.441 | 0.9 | |
| Interaction effect | | S | 1.068 | 2.18 | | | S | 1.206 | 2.46 | | | S | 0.882 | 1.8 | |

$F_1$ = 2% DAP + 1% KCl; $N_1$ = 30 kg $ha^{-1}$, $F_2$ = 2% DAP + 1% KCl + Nutrient mixture *; $N_2$ = 60 kg $ha^{-1}$, $F_3$ = 2% DAP + 1% KCl + 100 ppm Salicylic acid; $N_3$ = 90 Kg $ha^{-1}$, $F_4$ = 2% DAP + 1% KCl + 100ppm Maleic hydrazide; $N_4$ = 120 Kg $ha^{-1}$, * Nutrient mixture prepared by 0.5% $MgSO_4$, 0.25% MnSO4, 0.25% $ZnSO_4$ and 0.1 % Boric acid.

$F_1$(2%DAP + 1% KCl) treatment and least (40.2 g) with $F_4$ treatment.

## Yield

Grain and straw yield increased significantly with nitrogen application at 120 $kgha^{-1}$ over 90, 60 and 30 kg $Nha^{-1}$. N120 level marked grain yield superiority of 2.18, 6.12 and 14.9% and straw yield superiority of 2.70, 6.13 and 10.93% over $N_{90}$, $N_{60}$ and $N_{30}$ kg levels, respectively (Tables 6 and 7). Different nitrogen levels increased the nutrient content in the plants that lead to increase in vegetative growth. Besides, nitrogen is an involved in cell division and cell elongation. The increase in vegetative growth is evident from the plant height, tiller production and dry matter accumulation and the increase in different yield contributing characters *viz.,* spikes/plant, spike length, grains/spike thereby consequently improving the straw and grain yield of crop. The increase in grain and straw yield with application of nitrogen has also been reported by Akthar (2001); Naeem (2001) and Jatoi (2003). Comparatively lower grain and straw yield obtained with 30 kg $Nha^{-1}$ could be attributed to poor nutrition to the crop because of insufficient nitrogen uptake.

Grain and straw yield of 46.9 and 74.7 $qha^{-1}$, respectively was realized with application of $F_2$ treatment followed by $F_3$ treatment recording 44.5 and 72.7 $qha^{-1}$ of grain and straw yield, respectively, whereas the lowest grain and straw yield of 43.4 and 70.1 $qha^{-1}$, respectively was recorded with $F_4$ treatment. $F_2$ treatment marked grain and straw yield superiority of 5.39, 6.10 and 8.06% and 2.75, 2.60 and 6.56% over $F_3$, $F_1$ and $F_4$ treatments, respectively (Tables 6 and 7). Comparatively higher grain yield recorded with $F_3$ treatment over $F_1$ and $F_4$ treatments could be due to the fact that salicylic acid plays a role in growth and development, photosynthesis, ion uptake and transport. Seed treatment / foliar spray of salicylic acid induces reduction in sodium absorption and toxicity which is further reflected in low membrane injury, high water content and high dry matter production (El-Tayeeb, 2005).

Both $F_3$ and $F_1$ also recorded significantly higher straw yield over $F_4$ treatment. Significantly, lowest yield obtained with $F_4$ treatment even at lower concentration could be due to the fact that maleicessential constituent of plant tissue and thus is hydrazide may have limited the growth and development of the crop and due to its inhibitory effect (Henry and Cathey, 2009).

The interaction effect for grain and straw yield between nitrogen levels and foliar spray of nutrient mixture was found significant. The highest grain and straw yield of 49.2 and 78.1$qha^{-1}$, respectively was recorded with the treatment combination $N_4F_2$.

## Conclusion

Both grain and straw yield and different yield attributing characters *viz.,* number of spikes $plant^{-1}$, spike weight and grain $spike^{-1}$ were significantly

**Table 7.** Effect of different levels of nitrogen and foliar spray of nutrient mixture on straw yield ($qha^{-1}$).

| Nitrogen levels (Kg $ha^{-1}$) | Foliar spray of nutrient mixture | | | | | | | | | | | | | | |
|---|---|---|---|---|---|---|---|---|---|---|---|---|---|---|---|
| | 2007-08 | | | | | 2008-09 | | | | | Pooled | | | | |
| | $F_1$ | $F_2$ | $F_3$ | $F_4$ | Mean | $F_1$ | $F_2$ | $F_3$ | $F_4$ | Mean | $F_1$ | $F_2$ | $F_3$ | $F_4$ | Mean |
| $N_{30}$ | 68.3 | 70.3 | 68.5 | 65.8 | 68.2 | 69.3 | 70.9 | 69.3 | 66.5 | 69.0 | 68.8 | 70.6 | 68.9 | 66.1 | 68.6 |
| $N_{60}$ | 71.4 | 73.4 | 71.6 | 68.9 | 71.3 | 72.6 | 74.2 | 72.6 | 69.3 | 72.2 | 72.0 | 73.8 | 72.1 | 69.1 | 71.7 |
| $N_{90}$ | 73.3 | 75.8 | 73.9 | 71.2 | 73.6 | 75.0 | 76.6 | 75 | 72.2 | 74.7 | 74.3 | 76.2 | 74.4 | 71.7 | 74.1 |
| $N_{120}$ | 75.9 | 77.9 | 76.1 | 73.6 | 75.9 | 76.7 | 78.3 | 76.8 | 73.9 | 76.4 | 76.3 | 78.1 | 76.4 | 73.7 | 76.1 |
| Mean | 72.5 | 74.3 | 72.5 | 70 | | 73.4 | 75 | 73.4 | 70.5 | | 72.8 | 74.7 | 72.7 | 70.1 | |
| | | F-test | SEd ± | CD (P=0.05) | | | F-test | SEd ± | CD (P=0.05) | | | F-test | SEd ± | CD (P=0.05) | |
| Nitrogen levels | | S | 1.029 | 2.1 | | | S | 0.809 | 1.65 | | | S | 0.906 | 1.85 | |
| Foliar spray of nutrient mixture | | S | 1.029 | 2.1 | | | S | 0.809 | 1.65 | | | S | 0.906 | 1.85 | |
| Interaction effect | | NS | 2.058 | - | | | NS | 1.618 | - | | | NS | 1.812 | - | |

$F_1$ = 2% DAP + 1% KCl; $N_1$ = 30 Kg $ha^{-1}$, $F_2$ = 2% DAP + 1% KCl + Nutrient mixture*; $N_2$ = 60 Kg $ha^{-1}$, $F_3$ = 2% DAP + 1% KCl + 100 ppm Salicylic acid; $N_3$ = 90 Kg $ha^{-1}$, $F_4$ = 2% DAP + 1% KCl + 100ppm Maleic hydrazide; $N_4$ = 120 Kg $ha^{-1}$, * Nutrient mixture prepared by 0.5% $MgSO_4$, 0.25% MnSO4, 0.25% $ZnSO_4$ and 0.1 % Boric acid.

maximum with application of 120 kg $Nha^{-1}$ application and the least with application of 30 kg $Nha^{-1}$. Foliar spray of 2% DAP + 1% KCl + nutrient mixture ($F_2$) significantly increased the periodic plant height , dry weight $plant^{-1}$ and number of tillers $plant^{-1}$ over 2% DAP + 1% KCl + 100 ppm maleic hydrazide ($F_4$). Foliar spray of 2% DAP + 1% KCl + 100 ppm salicylic acid 100 ppm ($F_3$) and 2% DAP + 1% KCl ($F_1$) also recorded significantly higher values of these growth characters over $F_4$ treatments. Both grain and straw yield and different yield contributing characters *viz.,* spikes $plant^{-1}$, spike weight, grains $spike^{-1}$ and test weight were significantly higher with foliar spray of 2% DAP + 1% KCl + 100 ppm nutrient mixture ($F_2$) and significantly least with foliar spray of 2% DAP + 1% KCl + 100 ppm maleic hydrazide ($F_4$).

The interaction effects for various characters studied during the course of investigation were found mostly significant. The treatment combination $N_{120}$ X 2% DAP + 1% KCl + 100 ppm nutrient mixture ($F_2$) produced the highest values for different characters.

**REFERENCES**


Akthar MM (2001). Effect of varying levels of nitrogen on growth and yield performance of two new wheat cultivars. M.Sc Thesis SKUAST-K 84-86

Anonymous (2005) Fertilizer Statistics of India, FAI, New Delhi.

Anonymous (2007). Agriculture Research Data Book, Ministry of Agric, Govt. of India.

Chander S, Pandey J (1996) Effect of herbicide and nitrogen on yield of scented rice under different rice cultivars. Indian J. Agron. 41(2):209-214

El-Tayeeb MA (2005) Response of barley grains to the interactive effect of salinity and salicylic acid. Plant Growth Regul. 45:215-224.

Henry M, Cathay (2009) Physiology of growth retarding chemicals. Ann. Rev. Plant Physiol. 15(1):2-3.

Jatoi SA (2003) Effect of different nitrogen levels and placement on yield and yield attributes of wheat. Thesis M.Sc. (Hons.) SKUAST-K. pp. 89-91.

Mohamed, KA (1994). The effect of foliage spray of wheat with Zn, Cu, Fe and urea on yield, WUE and nutrients uptake at different levels of soil salinity. Assiut. J. Agric. Sci 25(3):179-189.

Naeem, M (2001). Growth, radiation use efficiency and yield of new cultivars of wheat under variable nitrogen rates. Summary, Thesis or, Dissertation, Non – Conventional, Bibliography. p. 128.

Rattan RK, Saharen N, Datta, SP (1999) Micronutrient depletion of Indian soils, extent and remedies. Fertilizer News 44(2):35-40.

Takkar PN, Singh MV, Ganeshmurthy AN (1997). A critical review of plant nutrient supply needs efficiency and policy issues of Indian Agriculture for the year 2000. Micronutrients and trace Elements. pp. 238-264,

Singh T, Singh KN, Bali, Amarjit, Bhatt GN (1995) Response of wheat (Triticum aestivum) genotypes to nitrogen and phosphorous. Skuast J. Res. 21(2):45-56.

Hayat S, Fariduddin Q, Ali B, Ahmad A (2005). Effect of Salicylic acid on growth and enzyme activities of wheat seedlings. Acta Agronomica Hungarica. 53(4):433-437

# A study of morphological basis of corn (*Zea mays* L.) yield under drought stress condition using correlation and path coefficient analysis

**Maryam Ashofteh Beiragi[1], Mohsen Ebrahimi[2], Khodadad Mostafavi [3]*, Mohammad Golbashy[4], Saied Khavari Khorasani[5]**

[1]Department of Agronomy and Plant Breeding, College of Agriculture, Zabol University, Zabol, Iran.
[2]Department of Agronomy & Plant Breeding- College of Abouraihan - University of Tehran, Iran.
[3]Islamic Azad University- Karaj Branch Iran.
[4]NanoBioTechnology- University of Tehran Iran.
[5]Scientific membrane of khorasan Razavi Agricultural Research and Natural Resources Center Mashhad, Iran.

**In order to study the morphological basis of corn yield under normal irrigation and drought stress condition, an experiment was carried out on basis of complete randomized block design with three replications and 28 new hybrids of corn; in addition 6 commercial hybrid (as control) under normal irrigation and drought stress condition in Khorasan Razavi Agricultural Research and Natural Resources Center, Mashhad, Iran on 2009. Genotypic and phenotypic correlation coefficients and path coefficients were estimated to determine the association and direct and indirect effects of different characters on total yield. Results of analysis of variance (ANOVA) showed that in both conditions there were significant differences between all hybrids for all traits. In normal irrigation condition KSC500 hybrid and in stress condition N-11 hybrid was better than others in grain yield trait. Under stress condition, number of kernels per row had the highest correlation with grain yield, while under normal irrigation ear diameter had the highest correlation with grain yield. According to this study, kernel No./ear is more closely and significantly related with total yield. This emphasized that selection based on the characters which enhance kernel number per ear will be more effective in improving yield under drought stress condition.**

**Key words:** Corn hybrids, phenotypic correlation, genotypic correlation, drought stress, irrigation.

## INTRODUCTION

Iran is very much less in yield due to drought stress at pollination stage, which leads to pollen desiccation and seed setting. Among various abiotic and biotic stress factors, drought stress is an important cause for genotype by environment interactions in maize across years, locations (Löffler et al., 2005) and most likely within individual fields (Bruce et al., 2002).

Drought is one of the most important abiotic stress factors (Bruce et al., 2002), which affects almost every aspects of plant growth (Aslam et al., 2006). Drought is a permanent constraint to agricultural production in many developing countries, and an occasional cause of losses of agricultural production in developed ones (Ceccarelli and Grando, 1996). The best option for crop production, yield improvement and yield stability under drought stress conditions is to develop drought tolerant crop varieties. One of the main goals in breeding programs is selection of the best genotypes under drought stress conditions (Richards et al., 2002). However, low heritability of drought tolerance and lack of effective selection approaches limit development of resistant crop cultivars to environmental stress (Kirigwi et al., 2004). Several reports of physiological, morphological and molecular traits have been suggested for improving the drought and salinity tolerance of crops that many of them potentially

*Corresponding author. E-mail: mostafavi@kiau.ac.ir.

**Table 1.** Open pollinate population of maize used in this study and 6 hybrids from Iran.

| Iranian hybrids | Open pollinate population | Open pollinate population | Open pollinate population |
|---|---|---|---|
| KSC704 | Th 91A 1354-G 42 O NTR-2 | Th 91A 1353-G 41 Q NTR-1 | Th 92B 6270-10-POb-44G2 |
| KDC370 | Th 97B 6088-POb-91 CD | Th 88A 1344-S87 P 69Q | Th 89B 6324-Rio-Hater(1)-8561 |
| KSC250 | Th 94A 1126-Side-9245 | Th 91A 1305 Comp-1-112 | Th 93B 6020--Pob-47-cC5 |
| KSC302 | Th 94A 1128-Acress 9245 | PR 91A 1306 Comp-1-54 | PR 91B 5301 EDS 90620 Flint |
| KSC400 | Th 87B 6089-Pob-92 C0 | Th 94A 1122--E | PR 93B 5212-c peel.16 C21 |
| KSC500 | Th 94A 1128 Acress 9245 | Th 93A 1121- Sakha-9134 | Th 83A 1321 R-4-Acress -8569 |

Per ear will be more effective in improving yield.

applicable to maize. Several recent reviews are available (Barker et al., 2005; Flowers, 2004, Munns, 2002). No exact figures on yield and economic losses in maize due to drought are available. Heisey and Edmeades (1999) estimated that 20 to 25% of the global maize planting area is affected by drought in any given year. In maize, grain yield reduction caused by drought ranges from 10 to 76% depending on the severity and stage of occurrence (Bolaòos et al., 1993). Drought stress coinciding with flowering delays silking and results in an increase of anthesis-silking interval (Bolaòos et al., 1993); this usually associates with reduction in grain number and yield (Edmeades et al., 1993).

Correlation of particular character with other characters contributing to seed yield is of great importance for indirect selection of genotypes for higher seed yield. Path coefficient analysis helps partitioning the correlation coefficient into its direct and indirect effects. This experiment was conducted to estimate the genotypic and phenotypic correlations and direct and indirect contributions of different traits to seed yield under normal and water stress.

## MATERIALS AND METHODS

The present study was conducted to evaluate the effects of drought stress on maize hybrids (34 hybrids) in yield and yield component and estimate the genotypic and phenotypic correlations and direct and indirect contributions of different traits to grain yield, in Khorasan Razavi Agricultural Research and Natural Resources Center, Mashhad, I.R. Iran during 2009. The hybrids consisted of 28 maize hybrids which were obtained from 18 famous open pollinate population and 6 Iranian hybrids of single cross groups (Table 1). Two independent experiments were laid out in randomized complete block design (RCBD) with three replications at both normal and drought stress conditions. The hybrids were grown in two row plots, with 3.15 m length and 0.75 cm between rows. The plant density was 75000 plant/ha. Fertilizer was used based on soil test. Irrigation was done based on 50 and 80% allowed water depletion for non-stress and stress conditions, respectively.

Data were recorded on 10 competitive plants of each plot and grain yield (kg $ha^{-1}$) was calculated for the entire plot. The data recorded were subjected to correlation analysis to estimate genotypic and phenotypic correlation coefficients between different traits following the method described by Kwon and Torrie (1964). Data were statistically analyzed using ANOVA appropriate for RCBD with SAS ver. 9.1 and Path coefficients were determined following Dewey and Lu (1957) using genotypic correlation with Path 2 software's. Means were compared using Duncan's multiple range test at 0.05 level of probability when the F values were significant (Steel and Torrie, 1984).

## RESULTS

Results of ANOVA showed significant differences among hybrids in all of traits at both, normal irrigation and drought stress condition (P≤0.01) (Table 2), which demonstrates high diversity among combinations that enabled us to screen and select drought tolerant hybrids. Corn hybrids showed different sensitivity to drought. Among all hybrids, KSC500 (13.79 ton/ha) and KSC302 (12.89 ton/ha) had the highest and H11 (5.69 ton/ha) and KSC250 (4.51 ton/ha) produced the highest yields in optimal and stress conditions, respectively. The other researcher showed that drought stress declined grain yield and its component (Reca et al., 2001; Seghatoleslami et al., 2008).

In general, means comparisons of maize hybrids indicated that under water deficit condition the maximum ear weight, kernel no. per row, total kernel no. per ear and kernel percentage in ear was obtained in H11 hybrid (data not shown). Westgate and Boyer (1985) found that water stress during the critical period of silking to early grain filling inhibits photosynthesis and consequently lowers the carbohydrate reserves to levels that are insufficient to support optimum reproductive development. Such effects explain the observations made in this study concerning the reduction of kernel number in ear in the non-stress condition versus drought stress condition (Table 2). Other research shown that under water stress condition, a maize plant will be able to make a better use of available water if vegetative growth is restricted early in the season (Shekoofa and Emam, 2006).

Results of this experiment indicated that yield component such as ear no. per plant, 300- kernel weight, row no. per ear, kernel no. per row and ear length were affected by water deficit condition. The percent of total yield reduction in stress condition was 71.54% (Table 2). Water stress reduced both Plant height (-32.8%) and ear

**Table 2.** Analysis of variance of investigated traits in corn hybrids under both normal and drought stress condition.

| | Normal condition | | | | | Stress condition | | | | | |
|---|---|---|---|---|---|---|---|---|---|---|---|
| | Replication | Genotype | Error | CV (%) | Mean | Replication | Genotype | Error | CV (%) | Mean | Variation percentage |
| Plant height (cm) | 1140.37** | 749.19** | 140.02 | 5.08 | 232.73 | 4242.03** | 479.05** | 89.81 | 6.05 | 156.4 | -32.80 |
| Ear height (cm) | 49.23ns | 494.06** | 58.4 | 7.21 | 105.97 | 1096.7** | 231.9** | 30.7 | 7.53 | 73.57 | -30.57 |
| Stem diameter (mm) | 22.47** | 9.98** | 3.65 | 9.71 | 19.68 | 13.96* | 11.31** | 2.87 | 10.37 | 16.33 | -17.02 |
| leaves No. | 7.04** | 2.17** | 0.18 | 3.35 | 12.94 | 5.45** | 2.63** | 0.17 | 3.22 | 13.11 | 1.31 |
| upper leaves No. | 0.05ns | 0.52** | 0.03 | 3.34 | 5.82 | 0.0003ns | 0.35** | 0.02 | 2.74 | 5.73 | -1.55 |
| Ear No. in plant | 0.0004ns | 0.04** | 0.01 | 12.53 | 1.06 | 0.0006 ns | 0.19** | 0.02 | 12.43 | 1.34 | 26.42 |
| 10 ear weight (Kg) | 1.55** | 0.22** | 0.03 | 7.54 | 2.61 | 0.03 ns | 0.1** | 0.01 | 13.86 | 0.84 | -67.82 |
| 10 cob weight (Kg) | 0.05** | 0.03** | 0.003 | 11.36 | 0.5 | 0.002 ns | 0.01** | 0.001 | 12.55 | 0.28 | -44 |
| 300 kernel weight (gr) | 364.28** | 235.59** | 69.07 | 9.9 | 83.93 | 175.47 * | 155.45** | 48.35 | 9.62 | 72.26 | -13.9 |
| Row No./ear | 0.11ns | 5.78** | 0.55 | 4.54 | 16.38 | 5.46 ** | 4.76** | 0.83 | 7.65 | 11.96 | -26.98 |
| kernel No./row | 90.87** | 20.09** | 3.3 | 4.57 | 39.73 | 4.95 ns | 33.88** | 5.49 | 13.05 | 17.95 | -54.82 |
| total kernel No./ear | 30842.82** | 9396.63** | 2072.36 | 7 | 649.77 | 5452.08 * | 9965.13** | 1694.37 | 18.77 | 219.24 | -66.26 |
| Ear length (cm) | 14.87** | 6.42** | 0.6 | 4.66 | 16.68 | 3.81 * | 6.92** | 0.95 | 8.12 | 12.04 | -27.82 |
| ear diameter (mm) | 32.46** | 14.37** | 1.68 | 2.64 | 49.1 | 9.66 ns | 25.03** | 9.28 | 8.1 | 37.59 | -23.44 |
| cob diameter (mm) | 4.88* | 9.33** | 1.38 | 4.2 | 27.99 | 0.08 ns | 16.43** | 3.88 | 8.48 | 23.22 | -17.04 |
| kernel depth (mm) | 3.04** | 2.13** | 0.51 | 6.78 | 10.55 | 0.99 ns | 2.25** | 0.39 | 8.64 | 7.25 | -31.28 |
| Total yield (ton/ha) | 28.46** | 7.12** | 2.64 | 15.62 | 10.4 | 4.18 ** | 2.58** | 0.38 | 20.98 | 2.96 | -71.54 |

*, ** Significant at 0.05 and 0.01 probability levels, respectively.

height (-30.57%). Results of this experiment showed that drought stress affected total leaves number positively (Table 1), and this was associated with the reduction in mean plant height (-32.8%) and stem diameter (-17.02%).

Our results concur partly with observations made by Choukan *et al* (2007) who reported that the total yield decreased with increasing water deficit. The measurement of total yield components showed that in drought stress condition total yield decline was mainly due to reduction of kernel no. per row and total kernel no. per ear (Shoa Hoseini *et al* 2007). Grain weight reduction under drought stress condition might be a result of kernel depth reduction.

Combined statistical analysis of data revealed that irrigation condition and their interaction (condition by hybrids) had significant differences for all of measured traits (data not shown). Study of phenotypic correlation between studied variables and total yield in both conditions showed that all variables were positively correlated with total yield, except stem diameter (-0.02ns) and total leaves number (-0.04ns) under drought condition. In normal irrigation condition, the highest correlations were belonged to ear diameter and total yield (0.68**). The correlation between plant height and ear height in drought stress was 0.98; the best of all variables studied. This finding was in agreement with the results of Hoseini et al. (2007). Result of this study showed that, kernel no. per row could be used as an important trait for prediction of total yield under drought stress. This finding was in agreement with the results of Iafari et al. (2009) and Choucan et al. (2007).

The strongest correlation of total yield in drought stress condition was observed with kernel number per row at both genotypic (rg = 0.837) and phenotypic (rp = 0.798) levels (Table 3). Also in drought stress condition, total yield was positively and significantly correlated at genotypic level with kernel number per row (0.837), total kernel number per ear (0.779), ear length (0.615) and kernel percentage (0.624). The results are in accordance with the findings of Ali (1994) and Golbashy et al. (2009 b).

Negative but non-significant genotypic correlations of total yield in stress condition were observed with plant height, ear height and stem diameter and other measured traits were positive correlated with total yield. At phenotypic level, ear no./plant (0.399), 300 kernel weight (0.286) and row no./ear (0.511) exhibited positive but non-significant correlation with total yield. Total yield was positively and non-significantly correlated, at phenotypic level, with other measured traits than stem diameter and leaves number.

The highest correlation at genotypic level at drought stress condition was observed between kernel no. per row and total kernel no. per ear (rg = 0.95). The results are at par with the finding of Golbashy et al. (2009 a). Stem diameter exhibited negative correlation with total yield at both genotypic and phenotypic levels.

Path coefficients in drought stress condition (Table 4) revealed that kernel percentage per ear had maximum direct effect (0.4712) on total yield. Other researcher

**Table 3.** Phenotypic ($r_P$) and genotypic ($r_G$) correlation coefficients among different characters of maize under normal (on diameter) and drought stress (diameter under) condition.

| | Traits | | 1 | 2 | 3 | 4 | 5 | 6 | 7 | 8 | 9 | 10 | 11 | 12 | 13 |
|---|---|---|---|---|---|---|---|---|---|---|---|---|---|---|---|
| 1 | Plant height | rP | 1 | 0.623 | 0.221 | -0.124 | -0.132 | 0.421 | -0.036 | 0.069 | 0.032 | 0.077 | 0.350 | 0.057 | 0.375 |
| | | rG | | 0.509* | 0.141 | 0.040 | -0.254 | 0.400 | 0.041 | -0.310 | -0.191 | -0.130 | 0.290 | -0.061 | 0.171 |
| 2 | Ear height | rP | 0.801** | 1 | 0.254 | 0.251 | -0.044 | 0.085 | 0.043 | 0.139 | 0.128 | 0.111 | 0.234 | -0.095 | 0.299 |
| | | rG | 0.745** | | 0.250 | 0.451 | 0.042 | -0.091 | 0.093 | -0.040 | 0.038 | 0.040 | 0.185 | -0.234 | 0.278 |
| 3 | Stem diameter | rP | 0.224* | 0.347** | 1 | 0.191 | 0.176 | 0.106 | 0.201 | 0.060 | 0.237 | 0.300 | 0.086 | -0.096 | 0.226 |
| | | rG | 0.251 | 0.428 | | 0.547* | 0.237 | 0.138 | 0.200 | -0.317 | -0.021 | 0.367 | -0.230 | -0.350 | 0.043 |
| 4 | Leaves No. | rP | 0.033 | 0.359** | 0.324** | 1 | 0.028 | -0.297 | 0.257 | -0.125 | 0.115 | 0.062 | -0.083 | -0.165 | 0.050 |
| | | rG | 0.466 | 0.742** | 0.518* | | -0.033 | -0.222 | 0.330 | 0.049 | 0.378 | 0.214 | 0.113 | -0.083 | 0.343 |
| 5 | Ear No./plant | rP | 0.083 | -0.050 | -0.053 | 0.063 | 1 | -0.051 | 0.185 | -0.018 | 0.148 | -0.163 | 0.135 | 0.051 | 0.431 |
| | | rG | 0.190 | 0.017 | 0.066 | 0.126 | | -0.512* | 0.360 | -0.108 | 0.286 | -0.512* | 0.207 | 0.144 | 0.236 |
| 6 | 300 kernel weight | rP | 0.249* | 0.135 | 0.135 | -0.084 | -0.105 | 1 | -0.485 | 0.249 | -0.217 | 0.318 | 0.181 | -0.062 | 0.175 |
| | | rG | 0.074 | 0.003 | 0.222 | 0.115 | -0.082 | | -0.634** | 0.193 | -0.503 | 0.345 | 0.012 | -0.036 | -0.033 |
| 7 | Row No./ear | rP | -0.096 | -0.096 | -0.140 | -0.107 | -0.094 | -0.065 | 1 | -0.219 | 0.690 | -0.344 | 0.222 | 0.055 | 0.199 |
| | | rG | -0.267 | -0.165 | -0.210 | 0.000 | -0.217 | -0.178 | | -0.425 | 0.741** | -0.510 | 0.340 | 0.047 | 0.445 |
| 8 | Kernel No./row | rP | 0.081 | 0.113 | -0.028 | 0.040 | 0.124 | 0.111 | 0.690 | 1 | 0.552 | 0.638 | 0.271 | 0.243 | 0.396 |
| | | rG | -0.006 | 0.117 | -0.025 | 0.196 | 0.198 | 0.057 | 0.631** | | 0.291 | 0.596** | 0.094 | 0.224 | 0.354 |
| 9 | Total kernel No./ear | rP | 0.024 | 0.043 | -0.052 | -0.022 | 0.063 | 0.035 | 0.861** | 0.956 | 1 | 0.181 | 0.381 | 0.220 | 0.462 |
| | | rG | -0.102 | 0.010 | -0.072 | 0.132 | 0.065 | -0.063 | 0.836** | 0.950** | | -0.095 | 0.441 | 0.214 | 0.732** |
| 10 | Ear length | rP | 0.071 | 0.140 | -0.019 | 0.270** | 0.355** | 0.099 | 0.190 | 0.634** | 0.491 | 1 | -0.135 | -0.213 | 0.163 |
| | | rG | 0.302 | 0.402 | 0.155 | 0.390 | 0.435 | 0.174 | -0.003 | 0.642** | 0.452 | | -0.449 | -0.347 | -0.102 |
| 11 | Kernel depth | rP | -0.159 | -0.216* | -0.086 | -0.172 | -0.188 | 0.195* | 0.390** | 0.254 | 0.323 | 0.055 | 1 | 0.609 | 0.613 |
| | | rG | -0.410 | -0.434 | -0.273 | -0.201 | -0.391 | 0.349 | 0.497 | 0.261 | 0.344 | -0.013 | | 0.608** | 0.683** |
| 12 | Kernel percentage | rP | -0.137 | -0.199* | -0.104 | -0.219* | -0.056 | 0.192 | 0.646** | 0.595 | 0.647 | 0.137 | 0.573 | 1 | 0.297 |
| | | rG | -0.467 | -0.387 | -0.311 | -0.188 | -0.030 | 0.172 | 0.645 | 0.595** | 0.667** | 0.024 | 0.738** | | 0.401 |
| 13 | Total yield(ton/ha) | rP | 0.192 | 0.111 | -0.021 | -0.042 | 0.399** | 0.286** | 0.511** | 0.798 | 0.755 | 0.540 | 0.303 | 0.612 | 1 |
| | | rG | -0.125 | -0.086 | -0.062 | 0.188 | 0.481 | 0.224 | 0.473 | 0.837** | 0.779** | 0.615** | 0.256 | 0.624** | |

*, ** Significant at 0.05 and 0.01 probability levels, respectively.

**Table 4.** Direct (bold diagonal values) and indirect effects of different character on total yield in maize under normal irrigation condition. (Path analysis using genotypic correlation).

| | | Total value | 1 | 2 | 3 | 4 | 5 | 6 |
|---|---|---|---|---|---|---|---|---|
| 1 | Ear diameter | 0.8369 | 0.8689 | -0.0584 | -0.0747 | -0.0012 | 0.0108 | 0.0914 |
| 2 | Ear No./plant | 0.2363 | 0.3511 | -0.1446 | -0.0431 | 0.001 | 0.0534 | 0.0184 |
| 3 | kernel No./row | 0.3540 | -0.1623 | 0.0156 | 0.3997 | -0.0004 | 0.0832 | 0.0182 |
| 4 | flag leaf height | 0.4975 | 0.3122 | 0.0441 | 0.0496 | -0.0033 | 0.0011 | 0.0938 |
| 5 | cob percentage | -0.4011 | -0.0253 | 0.0208 | -0.0894 | 0.000009 | -0.3719 | 0.0647 |
| 6 | 10 ear weight | 0.5013 | 0.4079 | -0.0137 | 0.0374 | -0.0016 | -0.1236 | 0.1947 |

Residual effects: 0.315.

reported the similar results in their findings (Golbashy et al., 2009; Hoseini et al., 2007). Maximum positive indirect effect on total yield was exhibited by total kernel No./ear through 10 ear weight (0.3968). Also maximum negative indirect effect was observed by 300 kernel weight through ear no. per plant (-0.0321). The indirect effect of 300 kernel weight and kernel percentage on total yield through all other characters under study is positive except ear no. per plant which is negative (Table 5). This emphasizes that selection on the base of kernel percentage and 10 ear weight will be more effective in improving total yield of maize under drought conditions.

## DISCUSSION

Ear length showed positive and significant correlation with most of the traits under drought stress condition.

**Table 5.** Direct (bold diagonal values) and indirect effects of different character on total yield in maize under drought stress condition. (Path analysis using genotypic correlation)

| | | Total value | 1 | 2 | 3 | 4 | 5 | 6 |
|---|---|---|---|---|---|---|---|---|
| 1 | Total kernel No./ear | 0.836 | 0.0674 | 0.0779 | 0.0007 | 0.2812 | 0.3968 | 0.0130 |
| 2 | Ear No./plant | 0.480 | 0.0134 | 0.3922 | -0.0009 | -0.0144 | 0.0816 | 0.0088 |
| 3 | 300 kernel weight | 0.223 | 0.0038 | -0.0321 | 0.0114 | 0.0811 | 0.1560 | 0.0035 |
| 4 | Kernel percentage | 0.624 | 0.0401 | -0.0119 | 0.002 | 0.4712 | 0.1212 | 0.0005 |
| 5 | 10 ear weight | 0.736 | 0.0569 | 0.0681 | 0.0038 | 0.1217 | 0.4704 | 0.0158 |
| 6 | Ear length | 0.614 | 0.0433 | 0.1705 | 0.002 | 0.0113 | 0.3676 | 0.0202 |

Residual effects: 0.315.

When the ear length is more, ultimately there will be more kernel no. per row, therefore the correlation between ear length and kernel no. per row is significant and positive. Ultimately as a result of more kernel no. per row, total kernel no. will be more, increasing the yield. This is exhibited by significant and positive correlation of total kernel no. with total yield at genotypic level. The significant and positive correlation of total yield with total kernel No./ear at genotypic level and positive correlation with ear length and kernel percentage at both genotypic and phenotypic level and their positive direct effects on total yield showed that kernel No./ear is more closely and significantly related with total yield. Ear length and kernel percentage also had positive direct effects on total yield.

## Conclusion

In this experiment, drought stress had significant effects on maize hybrids yield and its components. KSC500 (13.8 ton/ha) and KSC250 (12.9 ton/ha) hybrids were the best genotypes under normal condition and H11 (5.7 ton/ha) and KSC250 (4.51 ton/ha) showed the best behaviour under drought stress condition. Finally it is concluded that total yield mainly depends upon the kernel no. per ear. This emphasized that selection based on the characters which enhance kernel number

## ACKNOWLEDGEMENTS

The authors wish to thank Mehdi Zarabi, Maryam Ashofteh beiragi and Mehdi Ghorchiani for reading of the manuscript and invaluable discussion.

### REFERENCE

Ali A (1994). Evaluation for stress related traits in aestivum spp. M.Sc.Thesis, Deptt. Pl. Br. Genet., Univ. Agric., Faisalabad–Pakistan

Aslam M, Khan IA, Saleem M, Ali Z (2006). Assessment of water stress tolerance in different maize accessions at germination and early growth stage. Pak. J. Bot., 38(5): 1571-1579.

Barker T, Campos H, Cooper M, Dolan D, Edmeades GO, Habben J, Schussler J, Wright D, Zinselmeier C (2005). Improving drought tolerance in maize. Plant Breed. Rev., 25: 173–253.

Bolaòos J, Edmeades GO, Martinetz L (1993). Eight cycles of selection for drought tolerance in lowland tropical, maize. III. Responses in drought adaptive physiological and morphological traits. Field Crops Res., 31: 269–286.

Bruce WB, Edmeades GO Barker TC (2002). Molecular and physiological approaches to maize improvement for drought tolerance. J. Exp. Bot., 53: 13–25.

Ceccarelli S, Grando S (1996). Drought as a challenge for the plant breeder. Plant Growth Reg., 20: 149-155

Choucan R, Heidari E, Mohamadi M, Hadadi H (2007). Evaluation of drought tolerance in grain maize hybrids using drought tolerance indices. (In Persian). Seed Plant Improvement J. 24: 543-562.

Dewey DR, Lu KH (1957). A correlation and path coefficient analysis of components of crested wheat grass seed production. Agron. J., 51: 515–8

Edmeades GO, Bolaòos J, Hernandez M, Bello S (1993). Causes for silk delay in a lowland tropical maize population. Crop Sci., 33: 1029–1035.

Flowers TJ (2004). Improving crop salt tolerance. J. Exp. Bot., 55: 307–319.

Golbashy M, Ebrahimi M, Khavari KS, Choucan R (2010). Evaluation of drought tolerance of some corn (Zea mays L.) hybrids in Iran. Afr. J. Agric. Res.5(19): 2714-2719..

Golbashy M, Khavari KS, Ebrahimi M, Choucan R (2009b). Study of response of corn hybrids to limited irrigation. 11th Iranian Crop Science Congress Tehran, 24-26 July 2010. University of Shahid Beheshti. (In Persian), p. 218.

Heisey PW, Edmeades GO (1999). Maize production in drought-stressed environments: Technical options and research resource allocation. Part 1 of CIMMYT 1997/1998 World Facts and Trends; Maize Production in Drought-Stressed Environments: Technical Options and Research Resource Allocation. Mexico, D.F.: CIMMYT.

Jafari A, Paknejad F, Jami AL-Ahmadi M (2009). Evaluation of selection indices for drought tolerance of corn (*Zea mays* L.) hybrids. Intern. J. Plant Prod.

Kirigwi FM, van Ginkel M, Trethowan R, Seaes RG, Rajaram S, Paulsen GM (2004). Evaluation of selection strategies for wheat adaptation across water regimes. Euphytica, 135: 361-371.

Kwon SH, Torrie JH (1964). Heritability of interrelationship among traits of two soybean population. Crop Sci., 4: 196–8.

Löffler CM, Wei J, Fast T, Gogerty J, Langton S, Bergman M, Merrill B,

Cooper M (2005). Classification of Maize Environments Using Crop Simulation and Geographic Information Systems. Crop Sci. 45:1708–1716.

Munns R (2002). Comparative physiology of salt and water stress. Plant Cell Environ., 25: 239–250.

Reca J, Roldan J, Alcaide M, Lopez R, Camacho E (2001). Optimisation model for water allocation in deficit irrigation systems. I. Description of the model. Agric. Water Manag., 48:103-116.

Richards RA, Rebetzke GJ, Condon AG, van Herwaarden AF (2002). Breeding opportunities for increasing the efficiency of water use and crop yield in temperate cereals. Crop Sci., 42: 111- 121.

Seghatoleslami MJ, Kafi M, Majidi E (2008). Effect of drought stress at different growth stage on yield and water use efficiency of five proso

millet (*Panicum Miliaceum L.*) genotypes. Pak. J. Bot., 40(4): 1427-1432.
Shekoofa A, Emam Y (2006). Maize (*Zea mays* L.) Growth and yield response to Ethephon application under water stress conditions. Iran Agric. Res.
Shoa hoseini M, Farsi M, Khavari KS (2007). Study effect of water deficit stress on yield and yield components if some corn hybrids using path analysis. majale danesh keshavarzi (In Persian). 18(1): 71-85.
Steel RGD, Torrie JH (1984). Principles and procedures of statistics: a biometrical approach. McGraw Hill Co. New York, NY, USA.
Westgate ME, Boyer JS (1985). Carbohydrate reserves and reproductive development at low leaf water potential in maize. Crop Sci., 25: 762-769.

# Effect of forced-air artificial intermittent drying on cocoa beans in South-Western Nigeria

**D. O. Oke* and K. F. Omotayo**

Department of Agricultural and Bio-Environmental Engineering, School of Engineering, Federal Polytechnic, P. M. B 5351, Ado – Ekiti, Ekiti State, Nigeria.

**Drying experiments were carried out to investigate the effect of forced-air, artificial intermittent drying system on quality of fermented cocoa beans harvested in south-western Nigeria. The beans were dried naturally by using traditional sun drying method and artificially, by hot air inside an oven at air temperatures of 35, 40, 45, 50 and 55°C respectively. Fermented beans were exposed to 9 h of drying and 15 h of resting daily. The following quality parameters were assessed; free fatty acid (FFA), pH, moisture content and acetic acid for each drying temperatures and sun drying. From the test results, the free fatty acid and acetic acid levels increases with increase in drying temperature, also, the pH level decreases with increase in drying temperature. Optimum bean-quality was obtained for cocoa beans dried at 45°C oven temperature. A predictive equation was established that shows the relationship between oven temperature (°C) and free fatty acid level (mg/g) of the beans. Generally, it was also discovered that there is no known adverse effect on the quality of the beans dried under forced-air artificial intermittent drying system.**

**Key words:** Effect, temperature, cocoa beans, forced-air, artificial, intermittent, free fatty acid, acetic acid quality, convection.

## INTRODUCTION

Cocoa (*Theobroma cacao* L.) originated from Amazonian region of Brazil and grown in tropical countries like Nigeria, Ghana, Ivory Coast, Brazil, Malaysia, Venezuela and Indonesia (Beckett, 1994). Cocoa tress grows in a limited geographical zone, of approximately 20 degrees to the north and south of equator (Buijsse et al., 2006). It was introduced into Nigeria in 1874 through Equatorial Guinea by the Portuguese traders and has been grown extensively in the south-western part of the country as the main cash crop (Opeke, 1987). Cocoa trees is mainly consumed as chocolates and widely used in beverages, cosmetics, pharmaceuticals and toiletry products. It is also associated with health benefits such as anti-carcinogenic, anti-athergenic, anti-ulcer, anti-thrombotic, anti-inflammatory, immune modulating, anti-microbial, vasodilatory, and analgesic (Porter, 2006;Taubert et al., 2007). Cocoa processing consists of two major steps namely fermentation and drying. Fresh cocoa beans are usually fermented using the heap, tray or box methods for 5 to 7 days depending on the condition of the beans (Opeke, 1987). During fermentation, the temperature of the beans will rise from ambient to about 50 to 55°C due to the exothermic oxidation reaction (Wood and Lass, 1985). Fermentation is the initial step needed in the development of various flavour precursors in the beans (Hii et al., 2009). Drying techniques vary among the farmers and it ranges from the natural sun drying technique to the artificial hot-air technique. Currently, sun drying is mostly used in Nigeria by cocoa farmers due to its simplicity, low cost set up and requires only direct sunlight which is abundant (Opeke, 1987). Drying is usually terminated when the dried beans' moisture content is as low as 7.0% (wet basis).

Artificial, hot-air drying of fermented cocoa beans has

*Corresponding author. E-mail: dddoke1970@gmail.com or dddoke1970@yahoo.com.

been reported in literatures, they mostly concentrated on the removal of moisture from the beans at the shortest possible time (Nganhou et al., 2003; Wan Daud et al., 1996; Hii et al., 2002). This resulted in high drying rate but does not allow for the completion of needed oxidative reaction and acid diffusion process (McDonald et al., 1981). The large size of cocoa beans compared with most granular materials also result in high moisture gradient in the beans during and after drying when artificial and continuous systems are used. This has resulted in broken and cracked beans which are susceptible to fatty acid development due to high fat content of cocoa beans and the consequent reduction in bean quality. The focus of this work was to exploit the combined benefits in the traditional sun-drying practices and the forced-air artificial intermittent drying at low temperature, at about 35 to 55°C, in the drying of fermented cocoa beans. The effects of temperature on the quality of the dried beans were also investigated.

## MATERIALS AND METHODS

### Research philosophy

The philosophy of this research work is to explore the possibility of using artificial intermittent drying system to dry fermented cocoa beans, without the limiting effect posed by weather conditions in traditional sun drying system and also, without compromising the quality of the final product.

### Sample preparation

Fresh cocoa beans were obtained from Obafemi Awolowo University, Teaching and Research Farm and fermented using heap-basket method for 5 days. The harvested cocoa pods were processed for fermentation the same day they were harvested. After 5 days of heap-basket fermentation, the fermented beans were divided into six streams dried at varied temperature (35, 40, 45, 50 and 55°C and sun drying temperature). These were replicated thrice and their average values were recorded.

### Drying methods

#### *Sun drying*

The fermented cocoa beans where spread thinly on a concrete slab (50 x 50 x 15 cm) inclined at angle of 10° to ease water drainage in case of accidental rain wetting. The beans were turned every 2 h to ensure uniformity and improve moisture diffusion from inner core of the beans to the outer surface. Ambient temperature and relative humidity was measured using Casella-hand sling hygrometer. Drying was conducted between 8.00 to 17.00 h daily.

#### *Artificial intermittent forced-air drying*

The beans were artificially dried using an air-ventilated oven (Galle Kamp Oven Model 300 series) at temperatures 35, 40, 45, 50 and 55°C respectively. The cocoa beans were spread thinly in single layer on a meshed sample tray (0.4 × 0.4 cm). The average relative humidity of the oven were 38, 29, 21, 14 and 8% respectively. Heat was provided by the heater embedded at the base of the oven and hot air flow through the samples. The exhaust air escaped through the ventilation hole at the top of the oven. The beans were mixed every 2 h to ensure uniformity. Drying was conducted for 9 h daily and the beans were left to rest at room temperature overnight. This was done to help moisture redistribution in the beans before next day drying commences. This is to reduce bean breakage associated with drying without tempering (Steffe and Singh, 1982).

### Test procedure

#### *Moisture content*

A known weight of beans was dried in the oven at 103°C for 16 as recommended by International Standard Organization (ISO) table of standard 2291 – 1972 (E) and AOAC (1984). The moisture content (MC) of the beans was determined with reference to wet weight of the beans using equation (1).

$$\text{MC (wet basis)} = \frac{\text{wo-wd}}{\text{Wo}} \times 100\% \qquad \text{(Hii et al., 2009)} \ldots\ldots\ldots (1)$$

Where the terms $w_o$ and $w_d$ refers to initial and dry weight of the beans respectively. The equilibrium moisture content (EMCs) were determined by drying until no further change in weight was observed by the beans in each heat treatment. The EMC values determined were 6.2, 6.3, 6.5, 6.9, 7.1 and 7.6% (wet basis) for sun drying and oven drying temperatures 35, 40, 45, 50 and 55°C respectively.

#### *Quality assessment of dried beans*

The dried beans were examined for colour, taste, and occurrence of slate or mould. Quality assessment was based on the pH, acetic acid value and free fatty acid (FFA) level of the dried beans. The pH and acidity were determined according to Duncan et al. (1989). Ten grammes of the nibs were homogenized in 200 ml distilled water; the homogenate was filtered and the pH of the supernatant was measured using a digital pH meter. A 25 ml aliquot was titrated to pH of 8.0 with sodium hydroxide (0.01M) to determine the acidity of the beans. The free fatty acid content was determined according to AOAC (1984). One gramme of grounded cocoa nib was poured into a conical flask (250 ml); 50 ml of 0.5 M ethanonic potassium hydroxide were added to each of the samples. Heat was applied to dissolve the solution and allowed to cool. One to two drops of phenolphthalein indicator was added and the solution was titrated with 0.5 M HCl until the colour changed. The percentage FFA was then determined as the percentage molar equivalent of the KOH titre to the mass of cocoa nib sample.

The acetic acid content was determined according to AOAC (1984). A forty grammes grounded cocoa nib was mixed with 80 ml of de-ionized water in a conical flask. The mixture was distilled and allowed to cool. One to two drops of phenolphthalein indicator was added and the solution was titrated with 0.5 M potassium hydroxide (KOH) until the colour changed. The percentage FFA was then determined as the percentage molar equivalent of the KOH titre to the mass of cocoa nib sample.

### Statistical analyses

All experimental treatments were conducted in three replicates. The experimental data were subjected to manual regression analysis and correlation between the oven temperature and free fatty acid levels of the beans were calculated.

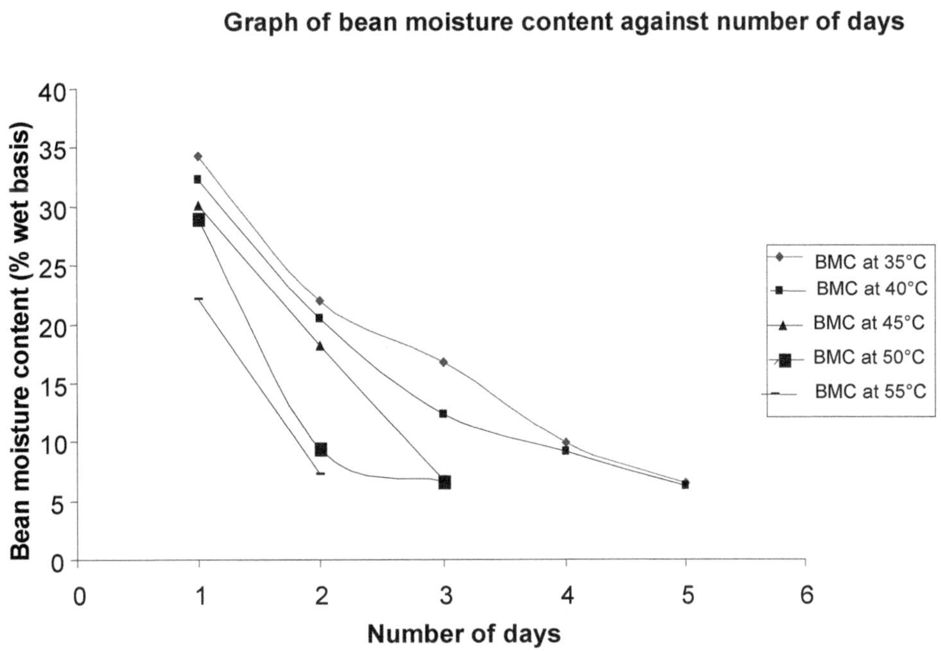


**Figure 1.** Bean moisture content against drying days in artificial drying.

## RESULTS AND DISCUSSION

### Moisture content fluxes

The moisture content of the cocoa beans at the time of the experiment was found to be 53.4% (wet basis). Although, the beans were apparently wet there was no free surface water on the bean. Moisture variation was observed in the oven for a 24 period, starting from 8 to 17 h of drying period and 15 h of resting (Figure 1). The relative humidity in the oven ranged from an average of 74% at 8 am to a maximum of 35% at 5 pm. The corresponding range of ambient temperature was 25 to 30° C. The prevailing ambient relative humidity was a minimum of 58% during the daytime whilst, it was relatively stable at 98% during the night (rest period). The moisture loss was high for the first two days in all the samples (Figure 1), the loss was marginal in the third day (between 2 and 8%) for beans dried at 35 to 50°C. After the third day, most beans have reached the storage moisture content (7% wet basis) except the sample dried at 35°C that reaches the storage moisture content on the fifth day of drying. This is consistent with what was reported by Fagunwa et al. (2009).

### Physical and chemical analyses of dried cocoa beans

The physico-chemical characteristics of the beans, using the forced – air artificial intermittent drying system are shown in Table 1. In all the tested attributes, the artificial intermittently dried samples compared favourably with the sun-dried samples. The qualities of both the forced-air artificial intermittently dried and sun-dried samples are within the acceptable level in international standards with respect to free fatty acid, acetic acid, pH, and taste flavour. All the samples under forced convection in the oven were properly dried showing dark reddish beans.

Test results show that the beans under forced-air artificial intermittent drying is more acidic (pH value of 4.7 to 5.2, acetic acid value of 3.9 to 8.7 mg/g and mildly bitter taste) against the corresponding attributes of 5.3 for pH and 3.5 mg/g and mildly bitter taste for sun dried beans. This may be due to a sealing of the beans testa under the vortex of flowing air, thus, preventing the diffusion of the acetic acid from the beans (Fagunwa et al., 2009). Consequently, a balance of the diffusion of moisture and its removal by the flowing air will present an optimum condition. Future research should investigate the quality of by-products (chocolate, confectioneries, etc.) from forced-air artificial intermittently dried beans and compare same with by-products from sun dried beans.

There was no difference in the flavour of the sun-dried and the forced-air artificial intermittently dried samples at the five different oven temperatures. Such observations had been reported by Thien et al. (1994) in the comparison of the qualities of sun-dried and air blown cocoa beans.

### Conclusions

It can be concluded from this experiment that forced-air

**Table 1.** Quality test results.

| Condition | Free fatty acid value (mg/g) | Acetic acid value (mg/g) | pH value | Taste | Colour |
|---|---|---|---|---|---|
| Sun dried | 0.40 | 3.50 | 5.30 | Mildly bitter | Light brown |
| Oven dried (35°C) | 0.41 | 3.90 | 5.20 | Mildly bitter | Very light brown |
| Oven dried (40°C) | 0.46 | 4.70 | 5.10 | Mildly bitter | Light brown |
| Oven dried (45°C) | 0.48 | 6.50 | 5.00 | Mildly bitter | Dark brown |
| Oven dried (50°C) | 0.55 | 7.30 | 4.80 | Bitter | Dark brown |
| Oven dried (55°C) | 0.61 | 8.70 | 4.70 | Bitter | Dark brown |

**Table 2.** Statistical data analysis.

| Drying temperature (°C) (x) | FFA mg/g (y) | xy | $x^2$ | $y^2$ |
|---|---|---|---|---|
| 35.00 | 0.41 | 14.35 | 1225 | 0.1681 |
| 40.00 | 0.46 | 18.40 | 1600 | 0.2116 |
| 45.00 | 0.48 | 21.60 | 2025 | 0.2304 |
| 50.00 | 0.55 | 27.50 | 2500 | 0.3025 |
| 55.00 | 0.61 | 33.55 | 3025 | 0.3721 |
| 225.00 | 2.50 | 115.40 | 10375 | 1.2847 |

artificial intermittent drying is appropriate for drying of cocoa beans to safe moisture level (7% wet basis) between 2 to 5 days depending on the oven temperature (35 to 55°C). The essential quality attributes of the beans are comparable with the product from the traditional sun drying system. From data analyses, an equation was established that predict the impact of oven temperature on the free fatty acid level of the beans (Table 2). This help in establishing a safe drying temperature for the beans.

$$y = m\,x + c \text{ (AOAC, 1984)} \quad (1)$$

$$\text{Slope } (m) = \frac{n\sum xy - \sum x \sum y}{n\sum x^2 - (\sum x)^2} \quad (2)$$

From Table 2,

$m = 0.0098$

$$c = \frac{\sum x - m\sum y}{n} \quad (3)$$

Therefore, $c = 0.061$

Thus, the regression equation,

$$y = 0.0098x + 0.061 \quad (4)$$

Correlation equation,

$$r = \frac{n\sum xy - \sum x \sum y}{\sqrt{\{n\sum x^2 - (\sum x)^2\}(n\sum y^2 - (\sum y)^2\}}} \quad (5)$$

$= 0.98633$

This is highly and truly correlated. Equation 6 can be used to predict the level of free fatty acid (y) in mg/g in the dried beans for any oven temperature (x) in degree Celsius.

## REFERENCES

AOAC (1984) Official Methods of Analysis, Association of Official Analytical Chemists, 4th Edition, Washington, D.C. USA.

Beckett ST (1994). Industrial Chocolate Manufacture and Use (2nd Ed.). Blackley Academic and Professional, Glasgow, UK.

Buijsse B, Feskens EJ, Kok FJ, Kromhout D (2006). Cocoa intake, blood pressure and cardiovascular mortality: the Zutphen elderly Study, Arch. Med., 166(4): 411-417.

Duncan RJ, Godfrey EG, Yap TN, Pettipher GL, Tharu-marajah T (1989) Improvement of Malaysian cocoa bean flavour by modification of harvesting, fermentation and drying method. The Planter (Malaysia), pp. 57-65.

Fagunwa AO, Koya OA, Faborode MO (2009). "Development of an Intermittent Solar Dryer for Cocoa Beans". Agricultural Engineering International: the CIGR E. J., 2: 1-14.

Hii CL, Tukimon MB (2002). Evaluation of fermentation techniques practical by the cocoa smallholders. The Planter, 78(910): 13-22.

Hii CL, Law CL, Cloke M, Suzannah M (2009). Thin layer drying kinetics of cocoa and dried product quality. J. Bio-sys. Eng. Elsevier Ltd, pp. 153-161.

International Standard Organisation (1972). Table of Standard, 2291-1972(E).

McDonald CR, Lass RA, Lopez ASF (1981). Cocoa drying-a review. Cocoa Grower's Bulletin, 31: 5-41.

Nganhou J, Njomo D, Benet JC, Augier F, Berthomieu G (2003).

Perfecting a method of micro-analysis of water and acetic acid in a cocoa bean in the course of drying: applying to determine transportation coefficients. Heat and mass transfer, 39: 797-803.
Opeke LK (1987). Tropical Tree Crops. John Willey and Sons, Chichester, pp. 67-213.
Porter LL (2006). Benefits of cocoa polyphenols. Manufacturing Confectioner, 86(6): 49-53.
Steffe JF, Singh RP (1982). Diffusion Coefficients for predicting rice drying behaviour. J. Agric. Eng. Res., 27: 489-493.
Taubert D, Roesen R, Schomig E (2007). Effect of cocoa and tea intake on blood pressure: a meta-analysis. Arch Intern. Med., 167(7): 626-634.
Wan Daud WR, Meor Tabib MZ, Ibrahim MH (1996). Characteristic drying curves of cocoa beans. Drying Technol., 14: 2387-2396.
Wood GAR, Lass RA (1985). Cocoa ($4^{th}$ Edition). Longman Inc., New York, USA.

9

# Line × tester analysis for seed yield and yield components in spring and winter type varieties of oil seed rape

**V. Rameeh**

Agriculture and Natural Resources Research Center of Mazandran, Sari, Iran. E-mail: valvandi225@yahoo.com.

**Line × tester analysis of two testers of spring type and six lines of winter type oil seed rape varieties were used to estimate combining ability and heterosis of yield components and seed yield. Significant mean squares of treatments for yield components and seed yield, indicating significant genetic variations among the genotypes including parents and their crosses. Parents vs crosses mean square which indicate average heterosis was significant for all the traits except pods per plant. Line × tester mean square was significant for all the traits except pods per plant. High narrow sense heritability estimate for 1000-seed weight indicating the prime importance of additive genetic effects for this trait. Significant positive general combining ability (GCA) effects of seed yield were observed for parents which had positive GCA effects of pods per plant. Most of the crosses with significant positive specific combining ability (SCA) effects for pods per plant had significant positive SCA effects of seed yield. Most of the crosses had significant positive over spring parent heterosis of seed yield, indicating winter type of oil seed rape varieties are suitable candidates for improving this trait using combination method.**

**Key words:** Heterosis, narrow sense heritability, seed yield, oil seed rape.

## INTRODUCTION

The *Brassica* species specially *Brassica napus* L. is one of the most important world sources of vegetable oil and due to spring and winter types of these species it is possible to be tolerant in the different climatic conditions (Downey and Rimer, 1993; Huang et al., 2010). Due to restricted genetic basis of spring type of rapeseed varieties, winter type of rapeseed varieties can be suitable candidate for increasing yield potential and genetic variation of spring type in winter x spring types or semi-winter type combinations (Qian et al., 2007). Butruille et al. (1999) investigated the possibility to increase yield of spring material by using winter germplasm. In other reports only spring material with spring material (Brandle and McVetty, 1990; Enqvist and Becker, 1991; Rameeh, 2010) or winter material with winter material was crossed (Amiri-Oghana et al., 2009). In oil seed rape breeding for hybrid and open pollinated varieties, general and specific combining ability effects (GCA and SCA) are important indicators of the potential of inbred lines in hybrid combinations. The line × tester analysis is one of the efficient methods of evaluating large number of inbreds as well as providing information on the relative importance of GCA effects of lines and testers and also SCA effects of pairs of parental genotypes for interpreting the genetic basis of important plant traits (Mather and Jinks, 1982). Estimation of genetic parameters for yield components can be important for indirect selection for seed yield. Although combining ability studies in oilseed *Brassica* spp. are scanty, most of these studies emphasized the preponderance effect of GCA on yield and most of the yield components indicating the importance of additive gene action (McGee and Brown, 1995; Wos et al., 1999). On the other hand, Pandey et al. (1999) reviewed evidences for the presence of significant SCA effects for yield and yield components. Ramsay et al. (1994) reported that variation for both GCA and SCA were responsible for dry matter yield and other quantitative traits in *B. napus*. Significant GCA and SCA effects were reported for pods per main axis, pods per plant, length of pod, number of seeds per pods, 1000-seed weight and seed yield in *B. napus* (Leon, 1991; Thakur and Sagwal, 1997; Rameeh, 2010), but in other study (Singh et al., 1995) the importance of additive genetic effects for pods per plant and 1000-seed weight was emphasized. Thakur and Sagwal (1997) while examining the genetic control of seed yield in oilseed rape found both additive and non additive gene effects to be involved.

Oil seed rape (*B. napu*) is usually classified as a largely self-pollinated species, significant levels of heterosis related yield and yield components have been obtained in F1 hybrids of both the spring and winter forms (Downeyand Rimer, 1993; Nassimi et al., 2006; Teklewold and Becker, 2005). Based on applied genetic material (parents and their crosses), genetic parameters which estimated in different studies will be different. The objective of the present study was to determine the importance of genetic parameters and heterosis for yield components, and seed yield in order to select suitable breeding program for rapeseed breeding winter and spring types lines and testers, respectively.

## MATERIALS AND METHODS

The material under study consisted of six winter oil seed rape (*B. napus* L.) cultivars including Zarfam, Licord, Talayeh, Okapi, Modena and Opera as Lines which were crossed with two spring testers including H1 and H2 based on line × tester crossing system during 2006 to 2007. Twelve $F_1$s along with their parents were grown in a randomized complete block design with three replications at Biekol Agriculture Research Station, located in Neka, Iran (13°, 53' E longitude and 43° 36' N latitude, 15 m above sea level) during winter 2007 to 2008. Each plot was consisted of four rows 5 m long and 40 cm apart. The distance between plants on each row was 5 cm resulting in approximately 300 plants per plot. The soil was classified as a deep loam soil (Typic Xerofluents, USDA classification) contained an average of 280 g clay $kg^{-1}$, 560 g silt $kg^{-1}$, 160 g sand $kg^{-1}$, and 22.4 g organic matter $kg^{-1}$ with a pH of 7.3. Soil samples were found to have 45 kg $ha^{-1}$ (mineral N in the upper 30 cm profile. Fertilizers were applied at the rates of 100: 50: 90 kg/ha of N: P: K, respectively. All the plant protection measures were adopted to make the crop free from insects. The pods per plant was recorded based on 20 randomly plants of each plot. The number of seeds per pod was recorded based on 10 plants/plot taking all pods including the ramifications in each plot. Seed yield (adjusted to kg/ha) was recorded based on two middle rows of each plot. Data for the genotypes were subjected to line × tester analysis (Mather and Jinks, 1982) to estimate general combining ability (GCA) and specific combining ability (SCA). A *t*-test was used to test whether the GCA and SCA effects were different from 0. For each hybrid and each studied trait, the difference between hybrid and the mean of each spring and winter type parent was computed separately. A least significant difference (LSD) was used to test whether these differences were different from 0.

## RESULTS

### Line×tester analysis

Significant difference was detected among the genotypes including parents and their crosses for pods per plant, seeds per pod, 1000-seed weight and seed yield (Table 1). There were significant differences among crosses for all the traits except pods per plant indicating the existence of genetic variability among the crosses for the traits except pods per plant. Variances among the spring genotypes, used as testers, were greater than winter types genotypes and also it was significant for all the traits. Parents vs crosses which indicating average heterosis was significant for all the traits except pods per plant. High narrow-sense heritability estimates for 1000-seed weight indicating the prime importance of additive genetic effects for these trait (Table 1).

### General and specific combining abilities

None of the lines and testers had significant positive GCA effects for pods per plant. The tester H1 and the lines including Modena and Opera had high positive GCA effects for pods per plant (Table 2). Modena had significant positive GCA effect for seeds per pod. Significant positive GCA effect of 1000-seed weight was related to H1 and Licord. Significant positive GCA effect of seed yield were displayed by Modena and Opera.

The crosses including Zarfam x H1, Licord x H2 with significant positive SCA effects for pods per plant were considered as suitable combinations for this trait. For seeds per pod, the crosses including Zarfam x H1 and Okapi x H2 with significant positive SCA effects were merit combinations. Licord x H2 with significant positive SCA effect for 1000-seed weight was the best combination for this trait. The crosses including Zarfam x H1 and Licord x H2 with significant positive SCA effects for seed yield were good combinations for this trait. Both of the crosses with significant positive SCA effects for seed yield had significant positive SCA effect of pods per plant (Table 3).

### Heterosis

None of the crosses had significant positive over winter parent heterosis for pods per plant. The crosses including Talayeh x H1, Talayeh x H2, Okapi x H1 and Okapi x H2 with positive over winter parent heterosis for pods per plant were suitable combinations for this trait. The crosses including Modena x H1 and Modena x H2 with significant positive over winter parent heterosis for seeds per pod were merit combinations for improving this trait. None of the crosses had significant positive over winter parent heterosis for 1000-seed weight and the crosses including Modena x H1 and Licord x H2 with positive over winter parent heterosis for this trait were good combinations. The crosses including Okapi x H1, Opera x H1, Licord x H2, Okapi x H2 and Opera x H2 were displayed significant positive over winter parent heterosis

**Table 1.** Analysis of variance for yield components and seed yield of rapeseed (*B. napus* L.) based on line x tester method.

| S.O.V | df | M.S | | | |
|---|---|---|---|---|---|
| | | Pods per plant | Seeds per pod | 1000 seed weight | Seed yield |
| Replication | 2 | 68.17 | 1.49 | 0.67** | 1947658.84 |
| Treatment | 19 | 314.67* | 14.08** | 0.47** | 808909.56** |
| Parents | 7 | 461.30* | 6.85 | 0.53** | 704977.64 |
| Parents vs crosses | 1 | 227.69 | 132.13** | 0.39* | 4422301.14** |
| Crosses | 11 | 229.28 | 7.94* | 0.44** | 546557.91** |
| Lines | 5 | 16.68 | 2.64 | 0.48** | 1748.27 |
| Testers | 1 | 975.91** | 30.77** | 1.15** | 3144397.13** |
| Line x tester | 5 | 292.54 | 8.68* | 0.26* | 571799.71** |
| Error | 38 | 151.24 | 3.51 | 0.10 | 109944.88 |
| Narrow-sense heritability | | - | - | 0.90 | - |

*, ** Significant at p<0.05 and 0.01, respectively.

**Table 2.** Estimates of GCA effects for yield components and seed yield in two testers and six lines of *B. napus*.

| Genotypes | Parents | Pods per plant | Seeds per pod | 1000-seed weight | Seed yield |
|---|---|---|---|---|---|
| Testers | 1-H1 | 1.52 | 0.61 | 0.26** | 15.59 |
| | 2-H2 | -1.52 | -0.61 | -0.26** | -15.58 |
| Lines | 3-Zarfam | 1.15 | -1.37 | 0.17 | -314.36* |
| | 4-Licord | -11.13** | -0.32 | 0.20* | -262.87 |
| | 5-Talayeh | 1.63 | 0.55 | -0.08 | -142.49 |
| | 6-Okapi | 0.13 | -0.28 | 0.04 | -84.20 |
| | 7-Modena | 3.55 | 1.65* | -0.34** | 454.79** |
| | 8-Opera | 4.70 | -0.23 | 0.01 | 349.16* |

*, ** Significant at p<0.05 and 0.01, respectively.

**Table 3.** Estimates of SCA effects for yield components and seed yield in crosses of two testers and six lines of rapeseed.

| Crosses | Pods per plant | Seeds per pod | 1000-Seed weight | Seed yield |
|---|---|---|---|---|
| 1-Zarfam x H1 | 10.33* | 1.85** | 0.17 | 534.66** |
| 2-Licord x H1 | -10.39* | -0.91 | -0.36** | -397.42* |
| 3-Talayeh xH1 | 2.61 | 0.39 | 0.04 | 62.20 |
| 4-Okapi xH1 | -1.02 | -1.67** | -0.07 | -168.31 |
| 5-Modena x H1 | -3.94 | 0.19 | 0.23 | -18.31 |
| 6-Opera x H1 | 2.41 | 0.14 | -0.02 | -12.82 |
| 7-Zarfam x H2 | -10.33* | -1.85** | -0.17 | -534.66** |
| 8-Licord xH2 | 10.39* | 0.91 | 0.36** | 397.42* |
| 9-Talayeh x H2 | -2.61 | -0.39 | -0.04 | -62.20 |
| 10-Okapi xH2 | 1.02 | 1.67** | 0.07 | 168.31 |
| 11-Modena x H2 | 3.94 | -0.19 | -0.23 | 18.31 |
| 12-Opera x H2 | -2.41 | -0.14 | 0.02 | 12.82 |

*, ** Significant at p<0.05 and 0.01, respectively.

for seed yield as good combinations (Table 4).

None of the crosses had significant positive over spring parent heterosis for pods per plant and the crosses including Zarfam x H1, Modena x H2, Opera x H1 and Opera x H2 with positive over spring parent heterosis for this trait were merit combinations. Out of 12 crosses, 6

**Table 4.** Heterosis over winter parent for yield components and seed yield in crosses of two testers and six lines of rapeseed.

| Crosses | Pods per plant | Seeds per pod | 1000-seed weight | Seed yield |
|---|---|---|---|---|
| 1-Zarfam x H1 | 3.20 | 3.33 | -0.33 | 142.36 |
| 2-Licord x H1 | -9.80 | 2.70 | 0.12 | 13.10 |
| 3-Talayeh xH1 | 22.53 | 3.00 | -0.04 | 352.77 |
| 4-Okapi xH1 | 12.60 | 0.20 | -0.04 | 528.70* |
| 5-Modena x H1 | -18.13 | 5.57** | 0.31 | 232.67 |
| 6-Opera x H1 | 11.87 | 2.83 | -0.23 | 1360.98** |
| 7-Zarfam x H2 | -20.50 | -1.57 | -1.19 | -958.13** |
| 8-Licord xH2 | 7.93 | 3.30 | 0.32 | 776.77** |
| 9-Talayeh x H2 | 14.27 | 1.00 | -0.63 | 197.21 |
| 10-Okapi xH2 | 11.60 | 2.33 | -0.43 | 834.15** |
| 11-Modena x H2 | -13.30 | 3.97* | -0.67* | 238.12 |
| 12-Opera x H2 | 4.00 | 1.33 | -0.71** | 1355.45** |

*, ** Significant at $p<0.05$ and 0.01, respectively.

**Table 5.** Heterosis over spring parent for yield components and seed yield in crosses of two testers and six lines of rapeseed.

| Crosses | Pods per plant | Seeds per pod | 1000-seed weight | Seed yield |
|---|---|---|---|---|
| 1-Zarfam x H1 | 19.00 | 5.37** | 0.62* | 1050.69** |
| 2-Licord x H1 | -14.00 | 3.67 | 0.12 | 170.10 |
| 3-Talayeh xH1 | 11.77 | 5.83** | 0.24 | 750.10** |
| 4-Okapi xH1 | 6.63 | 2.93 | 0.25 | 577.88* |
| 5-Modena x H1 | 7.13 | 6.73** | 0.17 | 1266.87** |
| 6-Opera x H1 | 14.63 | 4.80 | 0.27 | 1166.73** |
| 7-Zarfam x H2 | 1.97 | 2.13 | 0.16 | 216.87 |
| 8-Licord xH2 | 10.40 | 5.93 | 0.72** | 1200.43** |
| 9-Talayeh x H2 | 10.17 | 5.50** | 0.05 | 861.21** |
| 10-Okapi xH2 | 12.30 | 6.73** | 0.27 | 1150.00** |
| 11-Modena x H2 | 18.63 | 6.80** | -0.41 | 1538.99** |
| 12-Opera x H2 | 13.43 | 4.97 | 0.19 | 1427.88** |

*, ** Significant at $p<0.05$ and 0.01, respectively.

crosses had significant positive over spring parent heterosis for seeds per pod. The crosses including Zarfam x H1 and Licord x H2 with significant positive over spring parent heterosis for 1000-seed weight were good combinations for improving this trait. Out of 12 crosses, 10 crosses had significant positive over spring parent heterosis for seed yield (Table 5).

## DISCUSSION

Significant differences among the genotypes for seed yield and yield components indicate significant genetic variation in parents and their crosses for these traits. The crosses and lines had no significant genetic variations for pods per plant but the testers had enough genetic variation for this trait. Average heterosis was significant for all the traits except pods per plant and due to low narrow- sense heritability estimates for this trait, it seemed most of non additive genetic effects of this trait was related to epistasis. High narrow-sense heritability estimates for 1000-seed weight indicating the prime importance of additive genetic effects for this trait and therefore the efficiency of breeding based on selection method for this trait will be high. In earlier studies (McGee and Brown, 1995; Wos et al., 1999) were reported high narrow-sense heritability estimates for 1000-seed weight.

Significant GCA effect of seed yield was observed for Modena and Opera and also both of these parents had significant positive GCA effects for pods per plant, therefore GCA effect of pods per plant can be suitable indicator for prediction of GCA effect of seed yield. Similarly, significant GCA effects were reported for pods per main axis, pods per plant, length of pod, number of seeds per pods, 1000-seed weight and seed yield in *B. napus* (Leon, 1991; Thakur and Sagwal, 1997; Rameeh, 2010).

All of crosses with significant positive SCA effect for seed yield had significant positive SCA effect of pods per plant, therefore SCA effects of pods per plant is good indicator for seed yield improving. Due to yield components compensatory, different crosses were suitable for different yield components improving. For pods per plant, Zarfam x H1, Licord x H2 and for seeds per pod the crosses including Zarfam x H1 and Okapi x H2 and for 1000-seed weight Licord x H2 were suitable combinations. Similarly, Pandey et al. (1999) reviewed evidences for the presence of significant SCA effects for yield and yield components.

Over winter parent heterosis variation for yield components were less than SCA effects for these traits, so none of the crosses had significant over winter parent heterosis for pods per plant. The crosses including Modena x H1 and Modena x H2 with significant over winter parent heterosis for seeds per pod had at least one parent with significant positive GCA effect of this trait. Most of crosses had significant positive over winter parent heterosis for seed yield because of less yield potential of winter type parents in humidity and warm region of this experiment. Among yield components, over spring parent heterosis was significant for seeds per pod in the most of the crosses. For seed yield, most of the crosses had significant positive over spring parent heterosis, indicating winter type of oil seed rape varieties are merit candidates for improving this trait using combination method. Similarly, in earlier studies (Becker, 2005; Downey and Rimer, 1993; Nassimi et al., 2006; Teklewold and Becker, 2005) significant levels of heterosis related yield and yield components have been obtained in F1 hybrids of both the spring and winter.

In conclusion among all yield components, 1000-seed weight was more heritable, so the efficiency of selection for this trait will be high. Due to more variation of SCA effects than heterosis of crosses for all yield components, selection of merit combinations based on SCA effect is easier than heterosis. Most of the crosses had significant positive over spring parent heterosis of seed yield, indicating winter type of oil seed rape varieties are suitable candidates for improving this trait using combination method.

## ACKNOWLEDGEMENTS

Author wish to thank Agricultural and Natural Resources Research Center of Mazandaran and Seed and Plant Institute Improvement (SPII) for providing genetic materials and facility for conducting the experiment.

## REFERENCES

Amiri-Oghana H, Fotokianb MH, Javidfar F, Alizadeh B (2009). Genetic analysis of seed yield, days to flowering and maturity in oilseed rape (*Brassica napus* L.) using diallel crosses. Int. J. Plant Prod., 2: 19-26.

Brandle JE, McVetty PBE (1990). Geographical diversity parental selection and heterosis in oilseed rape. Can. J. Plant Sci., 70: 935-940. 10.4141/cjps90-115.

Butruille DV, Guries RP, Osborn TC (1999). Increasing yield of spring oilseed rape hybrids through introgression of winter germplasm. Theor. Appl. Genet., 39: 1491-1496.

Downey RK, Rimer SR (1993). Agronomic improvement in oilseed brassicas. Adv. Agron. 50: 1-150.

Enqvist GM, Becker HC (1991). Heterosis and epistasis in rapeseed estimated from generation means. Euphytica, 58: 31-35.

Huang Z, Laosuwan P, Machikowa T, Chen Z (2010). Combining ability for seed yield and other characters in rapeseed. Suranaree J. Sci. Technol., 17: 39-47.

Leon J (1991). Heterosis and mixing effects in winter oilseed rape. Crop Sci., 31: 281-284.

Mather K, Jinks JL (1982). Biometrical Genetics, 3rd edn. Chapman & Hall, London.

McGee KP, Brown J (1995). Investigation of F1 hybrids performance in fall- and spring-planted canola. In Proc. 9th Int. Rapeseed Confr., 4–10 July, 1995. Cambridge, U.K., pp. 116-118.

Nassimi AW, Ali RS, Hassan G, Ali N (2006). Combining ability analysis for maturity and other traits in rapeseed (*Brassica napus* L.). J. Agron., 5(3): 523-526.

Pandey LD, Singh B, Sachan JN (1999). Brassica hybrid research in India: Status and prospect. Paper 263. In Proc. 10th Int. Rapeseed Confr. [CD-ROM]. 26–29 Sept. 1999, Canberra, Australia.

Qian W, Sass O, Meng J, Li M, Frauen M, Jung C (2007). Heterotic patterns in rapeseed (*Brassica napus* L.): I. Crosses between spring and Chinese semi-winter lines. Theor. Appl. Genet., 115: 27-34. DOI: 10.1007/s00122-007-0537-x.

Rameeh V (2010). Combining ability and factor analysis in F2 diallel crosses of rapeseed varieties. Plant Breed. Seed Sci., 62: 73-83. DOI: 10.2478/v10129-011-0006-1.

Ramsay LD, Bradshaw JE, Kearsey MJ (1994). The inheritance of quantitative traits in swedes (*Brassica napus* L. spp. rapifera) diallel analysis of dry matter yield. J. Genet. Breed., 48: 253–257.

Singh D, Malik V, Singh H (1995). Gene action of seed yield and other desirable characters in oilseed rape (*Brassica napus* L.). Ann. Biol. Luddhiana, 11(1-2): 94-97.

Teklewold A, Becker HC (2005). Heterosis and combining ability in a diallel cross of Ethiopian Mustard inbred lines. Crop Sci., 45: 2629-2635.

Thakur HL, Sagwal JC (1997). Heterosis and combining ability in oilseed rape (*Brassica napus* L.). Indian J. Genet., 57: 163-167.

Wos H, Bartkowiak-Broda I, Budzianowski G, Krzymanski J (1999). Breeding of winter and spring oilseed rape hybrids at Malyszyn. Paper 544. In Proc. 10th Int. Rapeseed Confr. [CD-ROM]. 26–29 Sept. 1999, Canberra, Australia.

# 10

# Effects of potassium fertilization on population build up of rice stem borers (lepidopteron pests) and rice (*Oryza sativa* L.) yield

**Muhammad Sarwar**

Nuclear Institute of Agriculture, Tandojam-70060, Sindh, Pakistan. E-mail: drmsarwar64@yahoo.com.

**The effects of potassium (K) application rates on population build up of rice stem borers and rice yield were evaluated using cultivar "Shua-92". The treatments were consisted of K applications at 40, 50 and 60 kg/ha at the top dressing, and no K fertilizer application along with normal doses of nitrogen and phosphorous. The results showed that compared with the untreated control, K fertilizer significantly reduced the rate of rice borers' infestation and increased paddy yield. The treatments 50 and 60 kg K significantly reduced the incidence of pest and increased the grain yield of rice, but none significantly followed by application at 40 kg compared to control treatment. This assures that at low level of K application, biomass of rice plants were significantly less than the plants grown in the higher levels, but, further increase in the amount of K from 50 to 60 kg/ha did not significantly increase the pest growth and biomass of rice plants. Based on observations, it is possible to make sure that by considering cost benefit ratio application of K by adopting the rate of 50 kg/ha might be the most effective strategy in inhibiting rice pest's incidence. The results suggest that application of K fertilizer could be useful in the recovery of plants damage when attacked by larvae of borers which can be safely and successfully managed to contribute larger volume of yield and reduce the environmental pollution.**

**Key words:** Plant nutrition, K fertilizer, herbivore, insect pests, rice, *Oryza sativa.*

## INTRODUCTION

The stem borers of the families Pyralidae and Crambidae (Lepidoptera) such as the yellow stem borer *Scirpophaga incertulas*, white stem borer *Scirpophaga innotata*, striped stem bore *Chilo suppressalis*, gold-fringed stem borer *Chilo auricilius*, dark-headed stem borer *Chilo polychrysus*, and pink stem borer *Sesamia inferens* are among the major insects of rice. Among them, *S. incertulas* is the most dominant species in the tropical regions of Asia, while, *C. suppressalis* is the major species in temperate countries. The young stem borer larvae feed within the leaf sheath, and older larvae feed inside the stem and vascular tissues (Alinia et al., 2000). In Pakistan, Sarwar et al. (2010) reported an infestation of 10.4% dead hearts and 19.3% white heads in aromatic rice, due to stem borers under natural field conditions.
Therefore, there is the need to develop optimized stem borers management practices for increasing rice production.

Three mechanisms of host plant resistance are generally recognized (Smith, 1989): antixenosis, or non-preference and antibiosis, which reduce insect survival, reproduction and invasion, where the plant compensates for insect feeding. If plant tolerance was a mechanism of resistance to pest then it is expected that adding K could increase plant vigor and further increase resistance to pest invasion. The plant nutrient status is an indicator of host plant quality, which plays an important role in the population dynamics of many herbivores (Sarwar, 2011). Sustainable agriculture requires both economic and environmental sustainability. One way to reduce agricultural production cost is balanced applications of fertilizer. For increasing food and fiber demands of large and growing populations, development of high yield and high quality crops will result in larger areas of K deficiency unless immediate increases in potash supplies are made available. Until this occurs, the efficiency of N and P fertilizer use will remain low because of imbalances in NPK. As a result, soil K will be further depleted, economic returns to farmers reduced and the cost of restoring soil K to productive levels will become extremely expensive for future generations (Jin, 1997). Potassium in rice producing soils is one of the limiting factors for increasing rice yield (Yang et al., 2003). There is a considerable decrease in available K due to increased cropping intensity and lower K application rates (Zhang et al., 2004).

As agricultural fertilizations entail significant economic and environmental costs, among the fertilizers, potassium plays a vital role in crop growth and metabolism. Potassium is noted for increasing fiber strength and quality in plant (Read et al., 2006). As an enzyme activator, potassium has been implicated in over 60 enzymatic reactions, which are involved in many processes in the plant such as photosynthesis, respiration, carbohydrate metabolism, translocation and protein synthesis (Pettigrew, 2008). Potassium also plays an important role in the maintenance of osmotic potential and water uptake during plant development, and its shortage will result in poorer quality and lowered yields (Oosterhuis, 2001).

Recently, most attention has been paid on the relationship between K content in host plants and changes in performances of herbivores. In an earlier study, K has been proved to improve the rice plant's tolerance to adverse climatic conditions, lodging, insect pests and diseases (Tiwari, 2002). Although potassium is one of the major plant nutrients underpinning crop yield and quality, use of K fertilizer for pest suppression has not been so popular among farmers. The objectives of present research were to investigate the effects of different K doses in combination with N and K fertilization on rice yield and to determine the effects of fertilization on injury of rice stem borers. It is hypothesized that (i) addition of K may increase biological yield in the field; and (ii) applied K can decrease pest density. Consequently, new strategy considering the possible synergy between specific mineral nutrients could permit us to control plant insect pests.

## MATERIALS AND METHODS

An experiment was conducted at Nuclear Institute of Agriculture, Tandojam, Sindh, Pakistan, to determine the effects of potassium application rates on population build up of rice stem borers and rice yield. The soil of experimental site is loamy textured with organic matter 0.9%, an ECe of 1.3 dS $m^{-1}$, pH of 7.9, Olsen's P 6.7 mg $kg^{-1}$, and $CaCO_3$ 10.6%. Soil was plowed and harrowed when its mellowness was considered physically acceptable. A high-yielding commercial rice cultivar "Shua-92"was used for carrying out the current experimentation. One month old rice seedlings were transplanted in standing water having one seedling per hill keeping 25 × 25 cm plant to plant and row to row distance. Seedlings were thinned by hand by leaving one vigorous plant per hill at the two-leaf stage. Throughout the growing season, crop was irrigated regularly. Other management practices, including insect and weed control were conducted according to local agronomic practices unless otherwise indicated.

### Experiment design and procedure

A randomized complete block experimental design was used with three replications and each experimental unit measuring an area of 3 $m^2$. Four treatments were used: (1) Potassium applications of 40, 50 and 60 kg $ha^{-1}$ separately at the top dressing and (2) no K fertilizer application during the growing season. The customary sources of nitrogen (N), phosphorous (P) and potassium (K) were urea, single super phosphate and potassium sulfate, respectively, applied by broad cast method. The standard amount of N fertilizer recommended on rice in the region was 60 kg $ha^{-1}$. The N fertilizer was applied in 2 splits; 30 kg $ha^{-1}$ during transplantation and 30 kg $ha^{-1}$ at 30 days after transplantation. All treatments received a basal rate of 50 kg $ha^{-1}$ P according to local practice.

During growing season, data on stem borers were started after one month of transplanting at fortnightly intervals basis. An area of 1 $m^2$ from each of treatments was randomly selected to quantify the incidence of stem borers as dead hearts $hill^{-1}$. At the later stage of borers infestation data were recoded on white heads basis. Then percent dead hearts and white heads were thus calculated. Pest samples were taken toward the middle of every treatment to reduce edge effects. Specimens of common rice borer pests were identified in the field and counted. If identification of specimens could not be determined in the field, they were brought to the laboratory for further identification. At the end of the crop production season, plants from each treatment were manually harvested, threshed and weighed for yield of the respective treatments.

### Statistical analysis

Data collected for grain yield and pest density were tabulated, and difference between treatment means were separated using the least significant difference (LSD) test at the 5% probability level (P= 0.05). An analysis of variance was performed using Statistix 8.1 data processing system.

## RESULTS AND DISCUSSION

Application of K fertilization significantly affected rice plant growth and yield, and their effects were documented on pest density and grain among treatments. The arthropod pest damage and yield observed during the growing season are shown in (Table 1).

### Pest invasion due to different fertility levels in rice field

Among rice stem borers, *S. incertulas* was the most dominant species at the experimental site. Potassium fertilizer for rice reduced the rate of borer's infestation and increased rice yield. Potassium fertilization treatments of 40, 50 and 60 kg $ha^{-1}$ significantly reduced the incidence of pest with 3.05, 2.40 and 2.64% dead hearts, respectively, compared with no K fertilizer application (4.33% dead hearts) during the growing season. Potassium fertilization significantly affected white heads formation, pest prevalence was very low and irregular in 50 and 60 kg $ha^{-1}$ treatments (3.58 and 3.37% white heads, respectively, but not significantly different

**Table 1.** Impacts of soil potassium on population build up of rice stem borers and grain yield within rice field.

| S. No. | Potassium treatments (kg ha$^{-1}$) | Stem borers infestation (%) | | Yield/ plot (gm/ 3 m$^2$) | Yield (kg ha$^{-1}$) |
|---|---|---|---|---|---|
| | | Dead hearts | White heads | | |
| 1 | 40 kg | 3.05$^b$ | 5.37$^b$ | 1913.00$^b$ | 6376.66 |
| 2 | 50 kg | 2.64$^{bc}$ | 3.58$^c$ | 2287.00$^a$ | 7623.33 |
| 3 | 60 kg | 2.40$^c$ | 3.37$^c$ | 2317.00$^a$ | 7723.33 |
| 4 | Control | 4.33$^a$ | 7.12$^a$ | 1690.00$^c$ | 5633.33 |
| LSD value | | 0.619 | 0.561 | 219.4 | |

Means, in each column followed by at least one letter in common are not significantly different at the 5% level of probability ($p \geq 0.05$).

from one another) compared with 40 kg treatment (5.37% white heads). White heads were abundant (7.12%) in no K fertilizer application than in the other treatments throughout the season. Pest incidence was low throughout the period in the higher K treatment, while, significantly lower in the minimum K during growing season. Not surprisingly, K fertilization had a significant effect on the insect pest populations in all cases.

### Paddy yield due to different fertility levels in rice field

Yield tended to be quite responsive to plant K levels, when nitrogen and phosphorus were combined with K fertilization; it was the best treatment combination for higher yield formation among the treatments. But the yield variability was not significantly observed among 50 and 60 kg ha$^{-1}$ treatments, although the differences were significant from 40 kg treatment and no K used treatment. Numerically higher yields were found in the K treatments with 1913.0, 2287.0 and 2317.0 gm per 3 m$^2$ (6376.66, 7623.33 and 7723.33 kg ha$^{-1}$) than in the no K treatment (1690.0 gm per 3 m$^2$) (5633.33 kg ha$^{-1}$) (P= 0.23). Thus, at low level of K application, biomass of rice plants were significantly less than the plants grown in the higher levels, but, further increase in the amount of K from 50 to 60 kg/ hectare did not significantly increase yield. Thus, resistance to insect pest in rice plants was enhanced due to K application and rice plants receiving high amounts of potassium accumulated high grain yield.

This study had added new information on the common perception that K fertilizer application is a most important practice to suppress pest population and for improving yield components. Lower pest damage on high dose K applied plants can arise from (1) pest's non preference due to high quality nutrients in plants; (2) higher pest mortality by high K level in plants; and (3) plant's chemical composition such as higher contents of defensive compounds that caused higher pest mortality. This demonstrates the capability of mineral nutrient to induce the activation of specific defense mechanisms developed by plants against the attack of pest. Similar results had also been observed previously where a number of studies have shown the potential of certain mineral nutrients to provoke the creation of definite defense system by plant which is not in favor of the pest due to presence of mono-potassium phosphate (Reuveni and Reuveni, 1998) and K (Sarwar et al., 2011). Role of potassium in imparting resistance is well documented, culm strength of rice plants was noted increased with improved K nutrition, and increased stalk strength and decreased lodging are associated with proper K nutrition (DeDatta and Mikkelson, 1985). On the other hand, K deficiency in rice can make tissues vulnerable to pest attack by reducing grain yield. Vaithilingam and Baskaran (1985) examined the mechanism of induced resistance to insect pests in rice plants with enhanced potassium application. Increase in the application of K in the culture solution increased K, but decreased N, P, Mg, Si, Zn and soluble proteins in the rice plants. Deficiency of K in rice plants increased intake and assimilation of food, honey dew excretion, growth index, adult longevity and population build up of pest. While, application of high dose of K to rice plants decreased adult life and population build up of the insect (Salim, 2002).

Potassium fertilization in this study generally seemed to increase plant K level and promoted rice plant growth, as had been shown in other studies (Dudt and Shure, 1994; Stiling and Moon, 2005). These changes have the potential to affect pest populations (Bi et al., 2001; Chau and Heinz, 2006), suggesting possible compensatory reproduction by the plant, so, the yield was positively affected. Singh et al. (2003) reported that K has been known to impart resistance against diseases and a high concentration of K ions in the cell sap restricts attack by insects.

The study by Dong et al. (2011) found a strong positive relationship between K fertilization and grain abundance and suggested that K fertilization was extremely important for maintaining high yield. The yield advantage was attributed to increases in the tillering capacity of the rice plant and spikelet production. Potassium played a critical role in plant physiology, it provided regulatory control over such processes as transpiration, starch synthesis, sucrose translocation, respiration, and lipid synthesis (Tisdale et al., 1985). Previous studies had indicated that the amount, rate and quality of food consumed by immature herbivores influenced their fitness, growth rates, developmental duration, final body weights, dispersal abilities and probabilities of survival. Similarly, the amount, rate and quality of food consumed by adults influenced their performance in mating success as well as timing and extent of reproduction and disposal ability (Slansky and Scriber, 1985). However, a rise or decrease in a single nutrient sometimes may not be reflected in the yield. As such, K fertilization effects on plant growth and pest populations can be offset by other factors in the field. These factors may include natural enemies, environmental factors and the health and defensive responses of the plant to insect pest herbivore (Hacker and Bertness, 1995). Some more recent studies have also shown significant yield changes due to uses of K as stated by Quampah et al. (2011) that K fertilizer increased the yield of rice. The present .finding suggests that K application should be emphasized for high-yielding rice production. This could reduce the input of fertilizers, and decrease the application of pesticides due to lower occurrence of pests, reduce losses of yield and finally increase farmers' income to sustainable development of rice protection and production.

## REFERENCES


Alinia F, Ghareyazie B, Rubia L, Bennett J, Cohen MB (2000). Effect of plant age, larval age, and fertilizer treatment on resistance of a *cry1Ab*-transformed aromatic rice to Lepidopterous stem borers and foliage feeders. J. Econ. Entomol., 93: 484-493.

Bi JL, Ballmer GR, Hendrix DL, Henneberry TJ, Toscano NC (2001). Effects of cotton nitrogen fertilisation on *Bemisia argentifolii* populations and honeydew production. Entomol. Exp. Appl., 99: 25-36.

Chau A, Heinz K (2006). Manipulating fertilization: a management tactic against *Frankliniella occidentalis* on potted Chrysanthemum. Entomol. Exp. Appl., 120: 201-209.

DeDatta JG, Mikkelson DS (1985). Potassium nutrition in rice. In: Potassium in Agriculture. R. D. Munson, Ed. Am. Soc. Agron. Madison, WI, pp. 665-699.

Dong H, Xiangqiang K, Li W, Wei T, Dongmei Z (2010). Effects of plant density and nitrogen and potassium fertilization on cotton yield and uptake of major nutrients in two fields with varying fertility. Field Crops Res., 119: 106-113.

Dudt JF, Shure DJ (1994). The influence of light and nutrients on foliar phenolics and insect herbivory. Ecol., 75: 86-98.

Hacker SD, Bertness MD (1995). A herbivore paradox: Why salt marsh aphids live on poor-quality plants. Am. Nat., 145: 192-210.

Jin JY (1997). Soil Potassium and Potassium Fertilizer Use in Northern China. Better Crops International, 11(1): 16-17.

Oosterhuis D (2001). Physiology and nutrition of high yielding cotton in the USA. Informac¸ Oes Agronômicas Piracicaba, 95: 18-24.

Pettigrew WT (2008). Potassium influences on yield and quality production for maize, wheat, soybean and cotton. Physiol. Plantarum, 133: 670-681.

Quampah A, Wang RM, Shamsi IH, Jilani G, Zhang Q, Hua S, Xu H (2011). Improving water productivity by potassium application in various rice genotypes. Int. J. Agric. Biol., 13: 9-17.

Read JJ, Reddy KR, Jenkins JN (2006). Yield and .fiber quality of Upland cotton as induced by nitrogen and potassium nutrition. Eur. J. Agron., 24: 282-290.

Reuveni R, Reuveni M (1998). Foliar fertilizer therapy- a concept in integrated pest management. Crop Protection, 17: 111-118.

Salim M (2002). Effects of potassium nutrition on growth, biomass and chemical composition of rice plants and on host-insect interaction. Pakistan J. Agric. Res., 17(1): 14-21.

Sarwar M (2011). Effects of Zinc fertilizer application on the incidence of rice stem borers (*Scirpophaga* species) (Lepidoptera: Pyralidae) in rice (*Oryza sativa* L.) crop. J. Cereals and Oilseeds, 2(5): 61-65.

Sarwar M, Ahmad N, Nasrullah, Tofique M (2010). Tolerance of different rice genotypes (*Oryza sativa* l.) against the infestation of rice stem borers under natural field conditions. The Nucleus, 47(3): 253-259.

Sarwar M, Ahmad N, Tofique M (2011). Impact of Soil Potassium on Population Buildup of Aphid (Homoptera: Aphididae) and Crop Yield in Canola (*Brassica napus* L.) Field. Pakistan J. Zool., 43(1): 15-19.

Singh B, Singh Y, Imas P, Chang XJ (2003). Potassium nutrition of the rice- wheat cropping system. Advances in agronomy, Academic press, 81: 203-259.

Slansky FJ, Scriber JM (1985). Food consumption and utilization. In: Kerkut G. A, Gilbert L. I. Comprehensive Insect Physiology, Biochemistry and Pharmacology. Pergamon: Oxford, 4: 87-163.

Smith CM (1989). Plant resistance to insects. A fundamental approach. John Wiley, New York, pp. 201.

Stiling P, Moon DC (2005). Quality or quantity: the direct and indirect effects of host plants on herbivore and their natural enemies. Oecologia, 142: 413-420.

Tisdale SL, Nelson WL, Beaton JD (1985). Soil fertility and fertilizers. Macmillan, Publishing Co., New York, pp. 754.

Tiwari KN (2002). Nutrient Deficiency Symptoms in Rice. Rice Production (Special Supplement Publication, Ed: D.L. Armstrong, International Program Saskatoon, Saskatchewan, Canada. Better Crops Int., 16: 23-25.

Vaithilingam C, Baskaran P (1985). Induced resistance to insect pests in rice with enhanced potassium application. *In*: Role of potassium in crop resistance to insect pests. Potash Res. Inst. India, pp. 43-51.

Yang XE, Liu JX, Wang WM, Li H, Luo AC, Ye ZQ, Yang Y (2003). Genotypic differences and some associated plant traits in potassium internal use efficiency of lowland rice (*Oryza sativa* L.). Nutr. Cyc. Agroecosys, 67: 273-282.

Zhang FR, Hao XY, Wang R, Xü Y, Kong XB (2004). Changes in soil properties in southern Beijing Municipality following land reform. Soil Till. Res.,75: 143-150.

# Development of model equations for selecting optimum parameters for dry process of shea butter extraction

**A. M. Olaniyan[1]* and K. Oje[2]**

Department of Agricultural and Biosystems Engineering, Faculty of Engineering and Technology, University of Ilorin, P. M. B.1515, Ilorin 240003, Kwara State, Nigeria.

**Shea butter is the fat content of the kernel of shea nut *(Vitellaria paradoxa)* which grows naturally from the wild and uncultivated state in most parts of Africa. The fat is used as edible oil and for raw material in the production of soaps, pomade, drugs and medicinal ointments. Wet extraction process is the method used in shea butter processing industry among women in African rural and urban communities. Apart from the low yield (below 20%), this wet extraction process is associated with environmental polluting effluents as by products. Experiments on dry extraction of shea butter from shea kernel were carried out using an instrumented piston-cylinder rig in conjunction with the TESTOMETRIC Universal Testing Machine (Model M500–50 KN). Shea butter was mechanically expressed when pressures of 1.5, 2.9, 5.8 and 8.8 MPa were applied at the rate of 2.50, 5.00, 7.50 and 10.00 mm/min on crushed shea kernel heated at 50, 70, 90 and 110°C. Measurements were made of oil yield, oil recovery efficiency and process loss during the mechanical expression process. The measured effects of heating temperature, applied pressure and loading speed on oil yield, oil recovery efficiency and process loss were examined using a $4^3$ factorial experiment in randomized complete block design. Model equations were developed by employing multiple regression analysis using SPSS 11.0 package. Further analysis by optimization process revealed optimum heating temperature, applied pressure and loading rate of 82.24°C, 9.69 MPa and 2.50 mm $min^{-1}$ respectively. These combinations gave 35.39% oil yield, 58.62% oil recovery efficiency and 2.83% process loss. This information provides useful data for developing a process line for dry extraction of shea butter from shea kernel.**

**Key words:** Shea butter, dry extraction, model equations, optimization, oil yield.

## INTRODUCTION

Shea nut (SN) is contained in the fruit of shea tree (ST) plant (*Vitellaria paradoxa)* which grows in the wild and uncultivated state in most parts of African savannah zones. Shea fruit (SF) contains a green epicarp, a fleshy mesocarp and a relatively hard shell (endocarp) which encloses the shea kernel (embryo). Shea kernel (SK) contains about 60% edible fat (shea butter) and the residual product, from which the butter is extracted (shea cake, SC), is an excellent ingredient for livestock feed production. Apart from this, shea butter (SB) is gaining popularity in food, soap, cosmetic, pharmaceutical, medical and engineering industries for the production of cooking oil, toilet soaps, pomade, drugs, ointments and metal cutting fluids respectively. These products, in Africa, have a lot of potential for export and as a foreign exchange earner.

According to Addaquay (2004), in the traditional wet extraction method of SB processing, SK is pounded with pestle and mortar to break it into grits. The grits are then roasted and grounded into a paste to facilitate easy extraction of the fat. The process is continued by kneading the paste in or with, water to capture the fat into an emulsion, boiling the mixture to separate the fat and skimming off the fat. The final cooling process leads to SB. This process is practiced by rural women in Mali and Burkina Faso where, according to Addaquay (2004), 80% of SB is produced traditionally. Wet extraction process is

*Corresponding author. E-mail: amolan397@hotmail.com

tedious, laborious and time consuming which is associated with low yield and environmental polluting effluents (such as waste water, slurry and shea paste) as by products. Some modified and improved methods invented in some parts of West Africa, though slightly improved the yield, are also based on wet extraction process with its associated problems.

Dry extraction process makes use of oil extraction equipment like hydraulic presses or screw expellers. This process has gained a worldwide popularity and acceptance in modern vegetable oil industry but, unfortunately, has not been widely extended to SB extraction. Relevant data like extraction temperature and pressure are needed for the operation of a dry extraction process and these are not readily available. Therefore, the objective of the work reported in this paper was to determine the optimum extraction parameters for dry process of shea butter extraction from shea kernel.

## MODEL DEVELOPMENT

Ott (1977) and Babatunde (1997) described methods of multiple regression of a dependent variable Y ($X_1$, $X_2$) where $X_1$ and $X_2$ are independent variables in a 2-factor factorial design as having any of the following response equations for use in optimizing Y ($X_1$, $X_2$). The possible response equations have the following combination of $X_1$ and $X_2$:

$$Y_1 = b_o + b_1X_1 + b_2X_2 + b_3X_1^2 + b_4X_2^2 + b_5X_1X_2 \quad ...1$$

$$Y_2 = b_o + b_1X_1 + b_2X_2 + b_3X_1^2 + b_4X_1^2X_2 + b_5X_1X_2 \quad ...2$$

$$Y_3 = b_o + b_1X_1 + b_2X_2 + b_3X_2^2 + b_4X_2^2X_1 + b_5X_1X_2 \quad ......3$$

and $$Y_4 = b_o + b_1X_1 + b_2X_2 + b_3X_1^2 + b_4X_2^2 + b_5X_1X_2 + b_6X_2^2X_1 + b_7X_1^2X_2 + b_8X_1^2X_2^2 \quad ......4$$

where, $b_o$, $b_1$, ........, $b_8$ are constants; $Y_1$, ...., $Y_4$ are dependent variables; and Equation 4 is a combination of Equations 1, 2 and 3.

Model equations were determined from the above response equations based on regression coefficient, coefficient of variation, standard error of estimate and F test. These parameters were determined by the statistical analysis of variance (ANOVA) and multiple regression analysis using the Statistical Package for Science and Social Sciences (SPSS 11.0 computer software package).

Extreme values of Y ($X_1$, $X_2$) were obtained by methods described by Stephenson (1975) as follows. The coordinates of the extreme values of Y ($X_1$, $X_2$), the maximum and minimum values were obtained by solving the partial differentials of Y ($X_1$, $X_2$) simultaneously as follows:

$$\frac{\partial Y(X_1, X_2)}{\partial X_1} = 0 \quad ......5$$

$$\frac{\partial Y(X_1, X_2)}{\partial X_2} = 0 \quad ......6$$

The nature of the function Y ($X_1$, $X_2$) is determined by the criteria given as:

$$\frac{\partial^2 Y(X_1, X_2)}{\partial X_1 \partial X_2} - \frac{\partial^2 Y(X_1, X_2)}{\partial X_1^2}\frac{\partial^2 Y(X_1, X_2)}{\partial X_2^2} \angle 0 \quad .... 7$$

$$\frac{\partial^2 Y(X_1, X^2)}{\partial X_1^2} \angle 0 \quad ......8$$

Y ($X_1$, $X_2$) is a maximum when Equation 7 is true and the right hand side of Equation 8 is less than zero; Y ($X_1$, $X_2$) is a minimum when Equation 7 is true and the right hand side of Equation 8 is greater than zero; Y ($X_1$, $X_2$) is a saddle point or give no information when the right hand side of Equation 8 is equal to, or greater than, zero at the coordinates defined by the solution to the simultaneous Equations 5 and 6. For Y ($X_1$) or Y ($X_2$), the coordinates of the extreme values of Y ($X_1$) or Y ($X_2$) are defined by the solution of the Equations 9 and 10:

$$\frac{dY(X_1)}{dX_1} = 0, for Y(X_1) \quad ......9$$

$$\frac{dY(X_2)}{dX_2} = 0 \; for Y(X_2) \quad .....10$$

where the nature of the extreme values are given by the positive or negative signs of the second differential of Y ($X_1$) or Y ($X_2$) given by $\frac{\partial^2 Y}{\partial X_1^2}$ and $\frac{\partial^2 Y}{\partial X_2^2}$ respectively. The function Y ($X_1$, $X_2$) is a maximum when the second differential is negative and minimum when it is positive at the values of $X_1$ or $X_2$ given by the solutions of Equations 9 or 10 for Y ($X_1$) or Y ($X_2$) respectively. Babatunde (1997) has used this same method for the development of model equation for selecting disc and tilt angles of disc plough for optimal operation.

### LABORATORY EXPERIMENTS

Laboratory experiments on dry extraction of SB from crushed SK were carried out in the Engineering Materials Testing Laboratory, Technical and Scientific Services Complex, National Centre for Agricultural Mechanization (NCAM), Idofian, Ilorin, Nigeria. The average room temperature of the laboratory was about 30°C throughout the period of experimentation.

### Experimental equipment

The experimental equipment consists of a piston-cylinder rig in conjunction with a Universal Testing Machine (Model M500-50 KN, TESTOMETRIC Company Ltd., England, United Kingdom). The Universal Testing Machine (UTM) is of 50 KN capacity and its vital parts includes the control console, load frame, crosshead, load cell, personal computer and the printer. The piston-cylinder rig, which was developed by Olaniyan and Oje (2007), is made up of a compression piston, a press cage cylinder, a supporting platform and an oil collecting pan. The piston serves as the pressing ram and it distributes pressure from the UTM evenly on the oilseed sample in the press cage cylinder.

A 605 W electric band heater was installed to enfold the press cage cylinder and hence serve as a heating device for extraction process. The rig was adequately instrumented with a temperature transducer to control the extraction temperature while the pressure for extraction was obtained from the UTM. The temperature transducer is a system of thermocouple connected to an Electronic Temperature Controller (Model JTC–902, Japan). The temperature range of the Electronic Temperature Controller is 0 to 400°C with the voltage of 110/220 V, frequency of 50/60 Hz and output of 840 W. In operation, the heat sensor (thermocouple probe) is inserted into the crushed SK sample through a hole drilled on the side of the press cage cylinder 70 mm height from the base. The arrangement of the equipment during extraction is as shown in Figure 1.

### Experimental procedure

A sample of 200 g of crushed SK sample was weighed and transferred into the press cage cylinder. The sample was heated inside the press cage cylinder with the aid of the temperature-controlled band heater at 50°C for 30 min. Using the jug mode of the UTM, the compression piston was moved down to touch the sample in the press cage cylinder. The sample was then pre-compacted to a height of 70 mm inside the press cage cylinder by UTM loading at a constant speed of 10 mm min$^{-1}$. After pre-compaction, the crushed SK sample was compressed by the UTM through the compression piston at a constant loading speed of 2.50 mm min$^{-1}$ to a pressure 1.5 MPa for 10 min. The oil expressed was collected in the oil collecting pan placed below the drainage area and weighed. All weight measurements were carried out using a Triple Beam Balance (Model M2610 g, OHAUS, New Jersey, USA). After expression, the compression piston was lifted well above the press cage cylinder by the jug mode of the UTM. The press cage cylinder (with the residual cake inside) was unscrewed and the residual cake was extruded into the cake extruding die. The experiment was repeated for the three other heating temperature levels of 70, 90 and 110°C; three other applied pressure levels of 2.9, 5.8 and 8.8 MPa; and three other loading speed levels of 5.00, 7.50 and 10.00 mm min$^{-1}$. Each experimental trial was replicated three times making a total of 192 treatment combinations that were carried out.

### Measured parameters

Oil yield was calculated as the ratio of the weight of oil expressed to the weight of the sample before expression. It was mathematically expressed by Adeeko and Ajibola (1989) as stated in Equation 11:

$$O_Y = \frac{W_{OE}}{W_{US}} X\ 100\% \qquad \ldots\ldots\ldots\ 11$$

Oil recovery efficiency was calculated as the ratio of the weight of oil expressed to the total weight of oil in the crushed SK sample before expression. It was mathematically expressed by Adeeko and Ajibola (1989) as shown in Equation 12:

$$O_E = \frac{W_{OE}}{xW_{US}} X\ 100\% \qquad \ldots\ldots\ldots 12$$

Process loss was calculated as the difference between the weight of the sample before expression and the sum total of the weights of oil expressed and residual cake after expression divided by the weight of the sample before expression. It was mathematically denoted by Olaniyan and Oje (2007) as stated in Equation 13:

$$O_L = \frac{W_{US} - (W_{RC} + W_{OE})}{W_{US}} X\ 100\% \qquad \ldots\ldots 13$$

where, $O_Y$ = oil yield (%); $O_E$ = oil recovery efficiency (%); $O_L$ = process loss (%); $W_{US}$ = weight of crushed SK sample before expression (g); $W_{RC}$ = weight of residual cake after expression (g); $W_{OE}$ = weight of oil expressed (g); and x = oil content of SK = 0.60 or 60%.

### Data analysis

Based on the four response equations for an experiment with two independent variables given by Ott (1977) and shown in Equations 1 to 4, the data obtained for oil yield, oil recovery efficiency and process loss from the laboratory experiments were subjected to multiple regression analysis using SPSS 11.0 computer software package.

## RESULTS AND DISCUSSION

### Model equations

The data analysis was based on a two-factor (heating temperature and applied pressure) multiple regression technique. The third experimental factor (loading rate) was not included in the analysis since its best value (2.50 mm min$^{-1}$) has been established from the experimental results (Olaniyan and Oje, 2007). From the regression analysis, the following regression models were obtained for oil yield, oil recovery efficiency and process loss as shown in Equations 14 to 25:

$$Y_1 = -11.782 + 0.7225X_1 + 3.175X_2 - 4.319E\text{-}03X_1^2 - 0.163X_2^2 + 2.563X_1X_2 \qquad \ldots\ldots 14$$

$$Y_2 = -15.077 + 0.894X_1 + 2.822X_2 - 5.397E\text{-}03X_1^2 + 2.269E\text{-}04X_1^2X_2 - 3.374E\text{-}02X_1X_2 \qquad \ldots\ldots 15$$

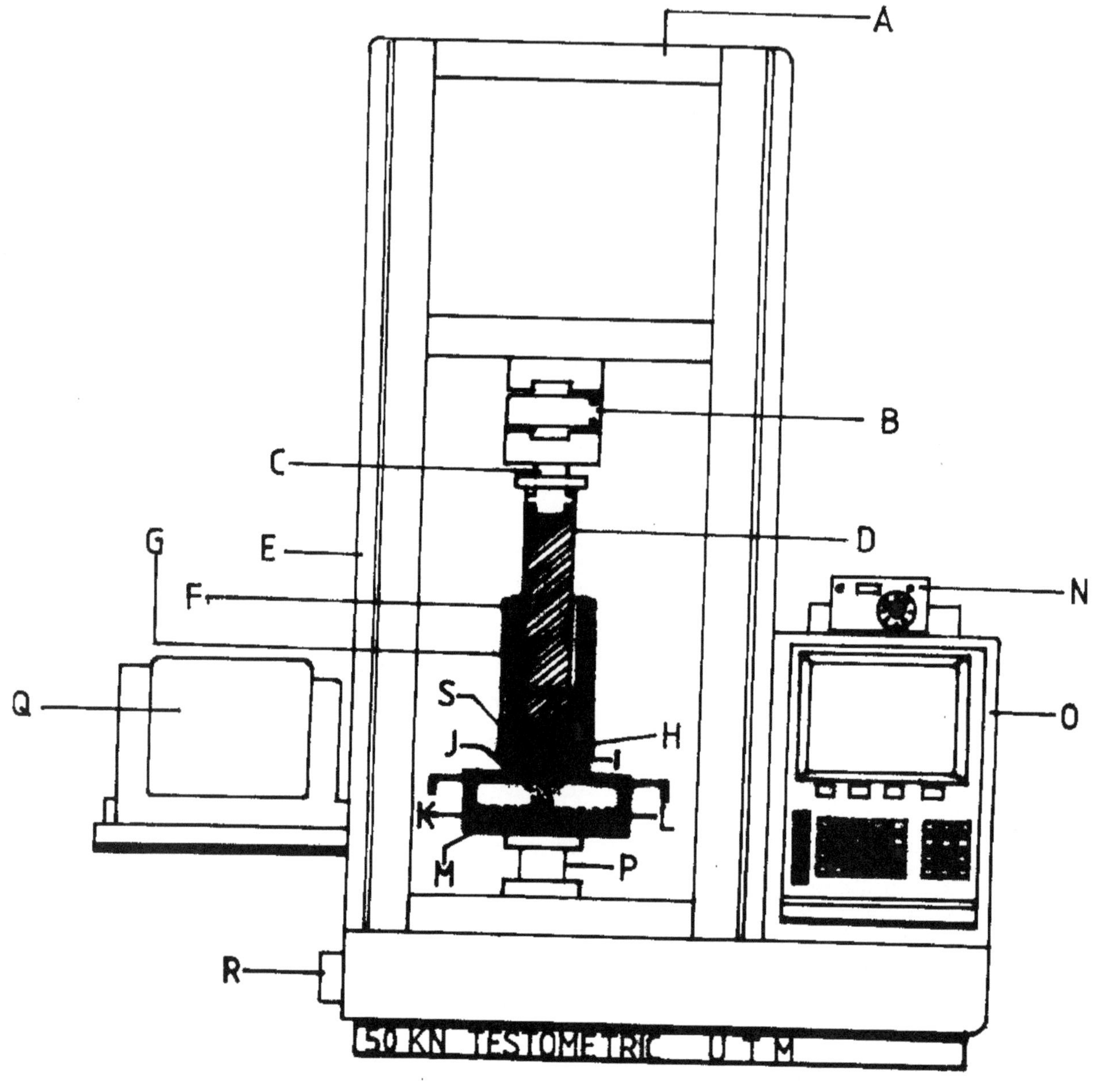


**Figure 1.** The mechanical expression rig: A - crosshead; B - load cell of the UTM; C - upper attachment of the UTM; D - compression piston; E - frame of the UTM; F - press cage cylinder; G - heating device; H - thermocouple probe from the electronic temperature controller; I - oilseed cake; J – drainage channel; K - supporting platform; L - oil collecting pan; M - oil expressed; N - electronic temperature controller; O - control console; P - lower attachment of the UTM; Q - computer system with printer; R - switch of the UTM; S - wire mesh.

$$Y_3 = 7.834 + 0.104X_1 + 6.415X_2 - 0.476X_2^2 + 3.911E\text{-}03X_2^2X_1 - 3.794E\text{-}02X_1X_2 \quad ......16$$

$$Y_4 = -19.114 + 0.834X_1 + 5.023X_2 - 4.563E\text{-}03X_1^2 - 0.211X_2^2 - 3.303E\text{-}03X_2^2X_1 - 2.381E\text{-}04X_1^2X_2 + 4.518E\text{-}05X_1^2X_2^2 \quad ......17$$

$$Y_5 = -25.999 + 1.351X_1 + 6.011X_2 - 8.012E\text{-}03X_1^2 - 0.296X_2^2 - 3.344E\text{-}03X_1X_2 \quad ......18$$

$$Y_6 = -21.188 + 1.371X_1 + 3.101X_2 - 8.138E\text{-}03X_1^2 + 2.650E\text{-}05X_1^2X_2 - 7.584E02X_1X_2 \quad ......19$$

$$Y_7 = 13.299 + 0169X_1 + 10.414X_2 - 0.721X_2^2 + 5.315E\text{-}03X_2^2X_1 - 5.838X_1X_2 \quad ......20$$

$$Y_8 = -30.462 + 1.356X_1 + 8.253X_2 - 7.422E\text{-}03X_1^2 - 0.499X_2^2\ X_2^2X_1 - 6.824E\text{-}04 - 3.637E\text{-}04X_1^2X_2 + 3.783E\text{-}05X_1^2X_2^2 \quad ......21$$

**Table 1.** Analysis of variance of multiple regression of oil yield as a function of heating temperature and applied pressure.

| Model | Source | Sum of squares | DF | Mean square | F | Sig. | $R^2$ |
|---|---|---|---|---|---|---|---|
| 1 | Regression | 1304.447 | 5 | 260.889 | 88.972 | 0.000 | 0.914* |
| | Residual | 123.155 | 42 | 2.932 | | | |
| | Total | 1427.602 | 47 | | | | |
| 2 | Regression | 1268.266 | 5 | 253.653 | 66.861 | 0.000 | 0.888* |
| | Residual | 159.366 | 42 | 3.794 | | | |
| | Total | 1427.602 | 47 | | | | |
| 3 | Regression | 1172.422 | 5 | 234.484 | 38.584 | 0.000 | 0.821* |
| | Residual | 255.181 | 42 | 6.076 | | | |
| | Total | 1427.602 | 47 | | | | |
| 4 | Regression | 1319.454 | 7 | 188.493 | 69.716 | 0.000 | 0.924* |
| | Residual | 108.149 | 40 | 2.704 | | | |
| | Total | 1427.602 | 47 | | | | |

*Significant at $P \leq 0.05$.

$$Y_9 = -3.763 + 6.873E\text{-}02X_1 + 1.807X_2 + 4.219E\text{-}04 - 6.365E\text{-}02X_2^2 - 1.616E\text{-}02X_1X_2 \quad ......22$$

$$Y_{10} = -8.010 + 0.216X_1 + 2.294X_2 - 5.003E\text{-}04X_1^2 + 1.941E\text{-}04X_1^2X_2 - 4.723E\text{-}02X_1X_2 \quad ......23$$

$$Y_{11} = -4.431 + 0.133X_1 + 0.802X_2 + 3.345E\text{-}02X_2^2 - 1.214E\text{-}03X_2^2X_1 - 3.395E\text{-}03X_1X_2 \quad ......24$$

$$Y_{12} = -3.244 + 9.176E\text{-}02X_1 + 0.468X_2 + 7.980E\text{-}05X_1^2 + 0.128X_2^2 - 3.279E\text{-}03X_2^2X + 6.760E\text{-}06X_1^2X_2 + 1.018E\text{-}05X_1^2X_2^2 \quad ...........25$$

where, $Y_1$ - $Y_4$ = oil yield (%); $Y_5$ - $Y_8$ = oil recovery efficiency (%); $Y_9 - Y_{12}$ = process loss (%); $X_1$ = heating temperature (°C); $X_2$ = applied pressure (MPa).

The existence and sufficiency of the regression models given in Equations 14 to 25 were examined and shown in the analysis of variance (ANOVA) of the multiple regression models shown in Tables 1 to 3. The analysis was carried out using SPSS 11.0 computer software package. The ANOVA tables indicate that the highest F-values, the highest correlation coefficient ($R^2$) and the lowest standard error of estimate were consistently obtained from model 4, which formed the basis of Equations 17, 21 and 25 respectively for oil yield, oil recovery efficiency and process loss. This implies that the model variables fit the data well. Also the high values of the regression sum of squares (RSS) as against the low values of error (residual) sum of squares (ESS) infer that model 4 accounted for most of the variation in the dependent variable, Y ($X_1$, $X_2$). The model is significant at 95% confidence level in each case and this is also evidence that it fits the data well and, therefore readily applicable.

## Optimization process

Based on the selected model equations for oil yield Equation 17, oil recovery efficiency Equation 21 and process loss Equation 25, critical values of the process parameters were determined through partial differential equations and the results are as presented in Table 4. The table shows that, in order to maximize oil yield, a temperature of 86.62°C and a pressure of 10.42 MPa should be selected for extraction process. To maximize oil recovery efficiency, a temperature of 82.24°C and a pressure of 9.69 MPa should be used. However, in order to minimize process loss, a temperature of 55.51°C and a pressure of 7.15 MPa should be selected. Therefore, for optimal operation of a dry process of SB extraction, a temperature of 82.24°C and a pressure of 9.69 MPa should be selected for extraction.

## Model validation

By substituting different values of heating temperature and applied pressure into the model equations, the expected values of oil yield, oil recovery efficiency and process loss were predicted. The predicted and measured values were subjected to a paired sample t-test using SPSS 11.0 computer software package and

**Table 2.** Analysis of variance of multiple regression of oil recovery efficiency as a function of heating temperature and applied pressure.

| Model | Source | Sum of squares | DF | Mean square | F | Sig. | $R^2$ |
|---|---|---|---|---|---|---|---|
| 1 | Regression | 3399.634 | 5 | 679.927 | 145.794 | 0.000 | 0.946* |
| | Residual | 195.872 | 42 | 4.664 | | | |
| | Total | 3595.506 | 47 | | | | |
| 2 | Regression | 3270.643 | 5 | 654.129 | 84.569 | 0.000 | 0.910* |
| | Residual | 324.863 | 42 | 7.735 | | | |
| | Total | 3595.506 | 47 | | | | |
| 3 | Regression | 2927.421 | 5 | 585.484 | 36.807 | 0.000 | 0.814* |
| | Residual | 668.085 | 42 | 15.907 | | | |
| | Total | 3595.506 | 47 | | | | |
| 4 | Regression | 3420.635 | 7 | 488.662 | 111.776 | 0.000 | 0.951* |
| | Residual | 174.871 | 40 | 4.372 | | | |
| | Total | 3595.506 | 47 | | | | |

*Significant at P ≤ 0.05.

**Table 3.** Analysis of variance of multiple regression of process loss as a function of heating temperature and applied pressure.

| Model | Source | Sum of squares | DF | Mean square | F | Sig. | $R^2$ |
|---|---|---|---|---|---|---|---|
| 1 | Regression | 149.466 | 5 | 29.893 | 12.479 | 0.000 | 0.598* |
| | Residual | 100.609 | 42 | 2.395 | | | |
| | Total | 250.075 | 47 | | | | |
| 2 | Regression | 145.783 | 5 | 29.157 | 11.742 | 0.000 | 0.533* |
| | Residual | 104.291 | 45 | 2.483 | | | |
| | Total | 250.075 | 47 | | | | |
| 3 | Regression | 149.182 | 5 | 29.836 | 12.420 | 0.000 | 0.597* |
| | Residual | 100.892 | 42 | 2.402 | | | |
| | Total | 250.05 | 47 | | | | |
| 4 | Regression | 151.257 | 7 | 21.608 | 8.747 | 0.000 | 0.605* |
| | Residual | 98.817 | 40 | 2.470 | | | |
| | Total | 250.075 | 47 | | | | |

* Significant at P ≤ 0.05.

**Table 4.** Critical points of heating temperature and applied pressure for optimal measured parameters.

| Measured parameter (%) | Heating temperature (°C) | Applied pressure (MPa) | Nature of critical points |
|---|---|---|---|
| Oil yield | 86.62 | 10.42 | Maximum |
| Oil recovery efficiency | 82.24 | 9.69 | Maximum |
| Process loss | 55.51 | 7.15 | Minimum |

the result is as shown in Table 5. From the table, the correlation coefficients were 0.956, 0.972 and 0.773 for oil yield, oil recovery efficiency and process loss respectively. These high levels of correlation show that

**Table 5.** Paired sample t-test of the predicted and measured values of oil yield, oil recovery efficiency and process loss.

| Parameter | Paired sample | Standard deviation | Correlation coefficient | t - value |
|---|---|---|---|---|
| Oil yield | Predicted – measured | 1.62 | 0.956 | 0.186 |
| Oil recovery efficiency | Predicted – measured | 2.04 | 0.972 | 0.039 |
| Process loss | Predicted – measured | 1.46 | 0.773 | -0.001 |

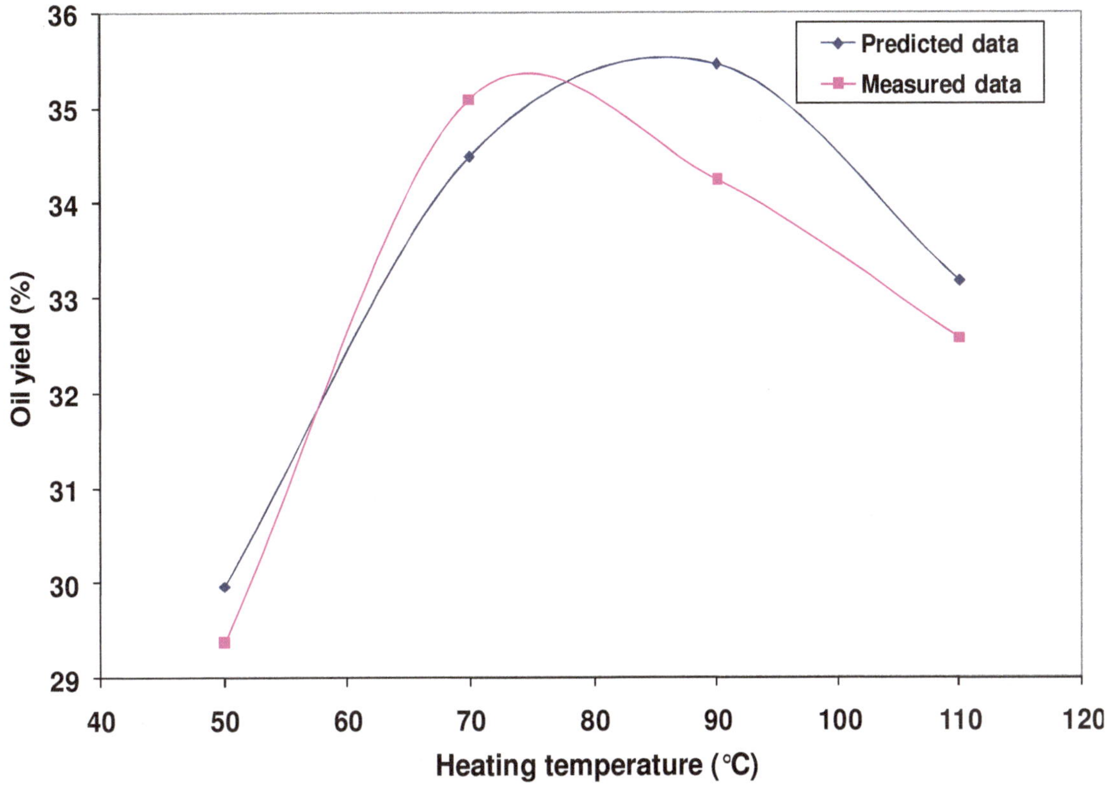


**Figure 2.** Effect of heating temperature on oil yield.

the predicted values compared favourably with the measured values. Therefore, the model can be used to select the best process parameters for optimal operation of a dry process of SB extraction. A careful observation of predicted and measured responses of oil yield, oil recovery efficiency and process loss to extraction temperature and pressure (Figures 2 to 7) also show high level of correlation which confirms that the model is applicable.

## Practical applications

This study establishes process parameters for industrial extraction of shea butter from shea kernel. The dry process of shea butter extraction can be used for the production of good quality shea butter products for the purpose of domestic and industrial use. The quality of preosmosed carrots is much superior to the product dehydrated with the convectional method of convective dehydration. The osmotically dehydrated carrots can be used for cooking as vegetables after rehydration or can be added directly into soups, stews or casseroles before cooking. If the product is blanched before osmotic dehydration, the process can be used successfully for the preparation of carrot candy.

## Conclusion

The result of the experiment and analysis showed the effect of heating temperature and applied pressure on oil yield, oil recovery efficiency and process loss. It was indicated by the result that an optimum heating

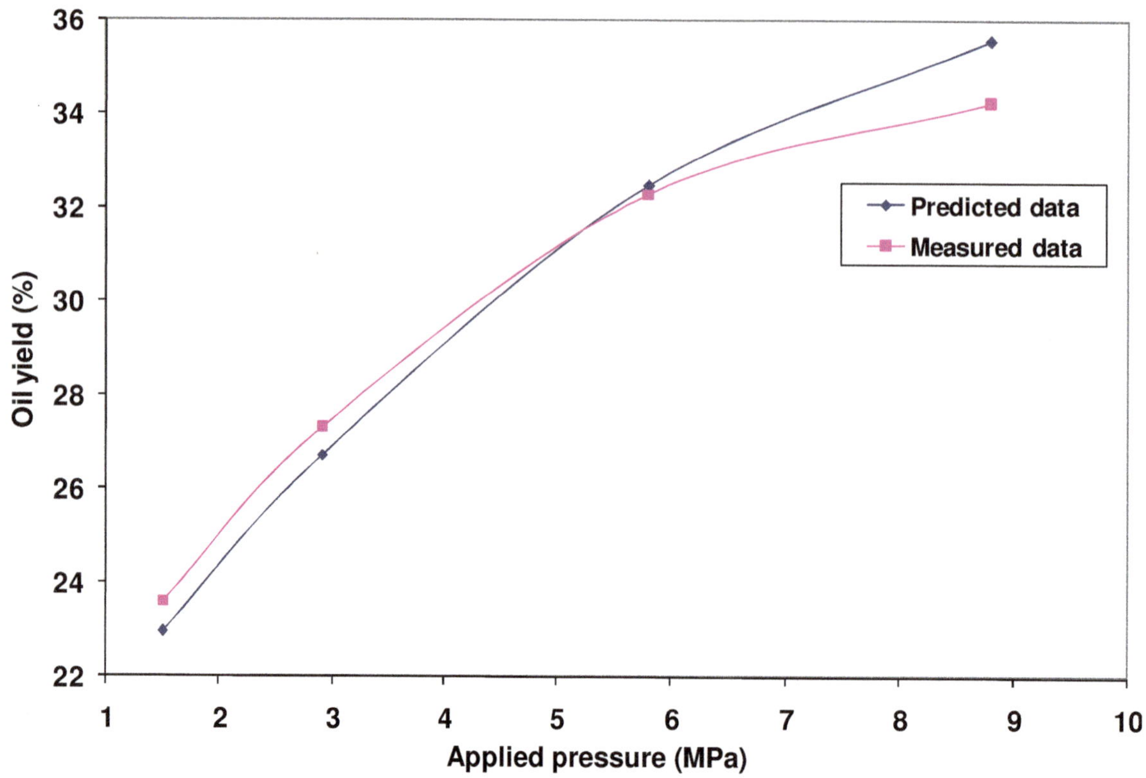


**Figure 3.** Effect of applied pressure on oil yield.

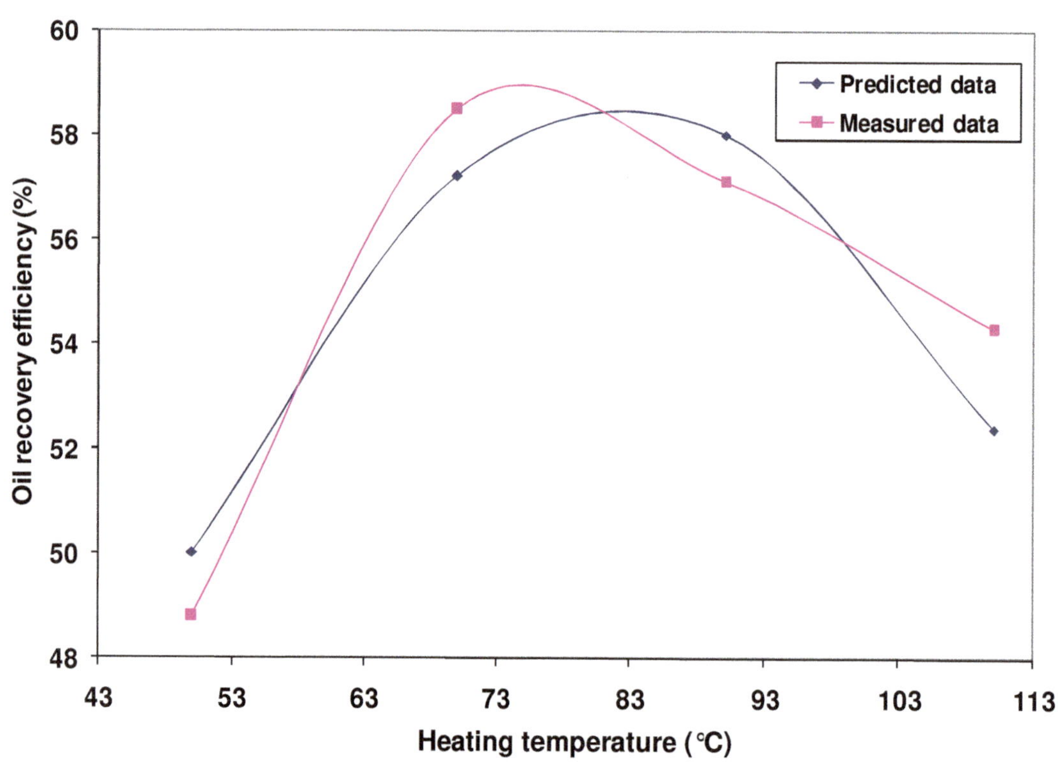


**Figure 4.** Effect of heating temperature on oil recovery efficiency.

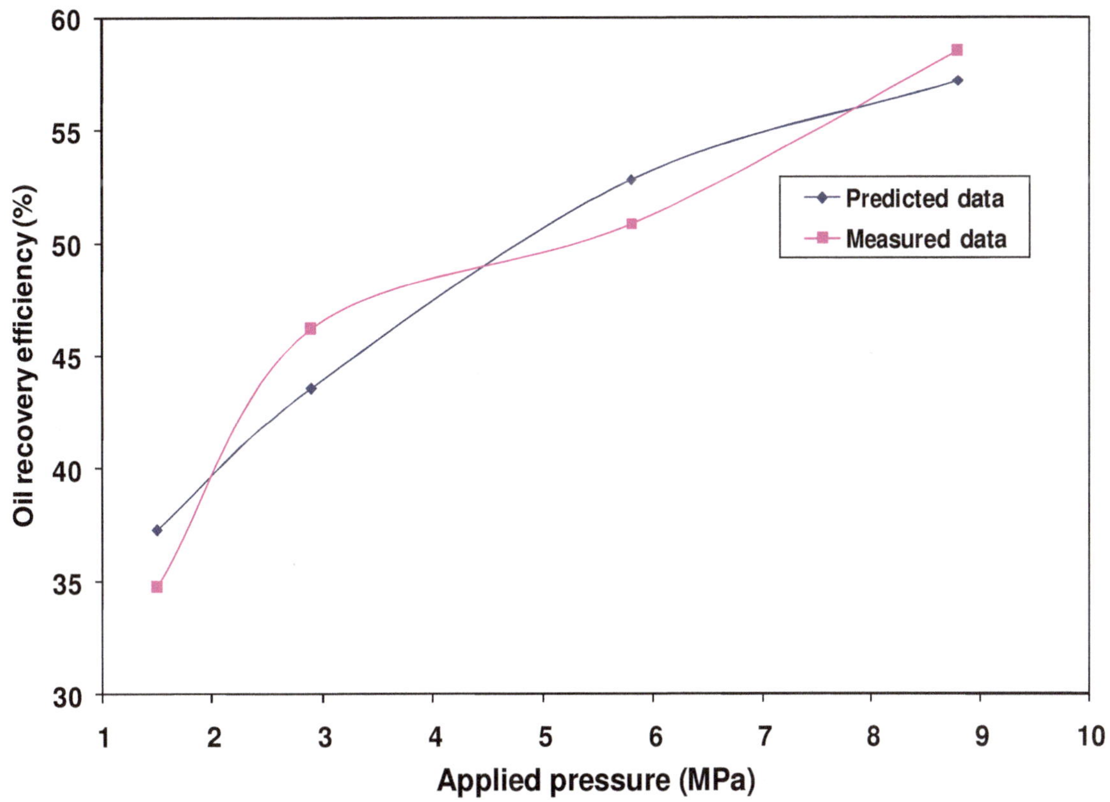


**Figure 5.** Effect of applied pressure on oil recovery efficiency.

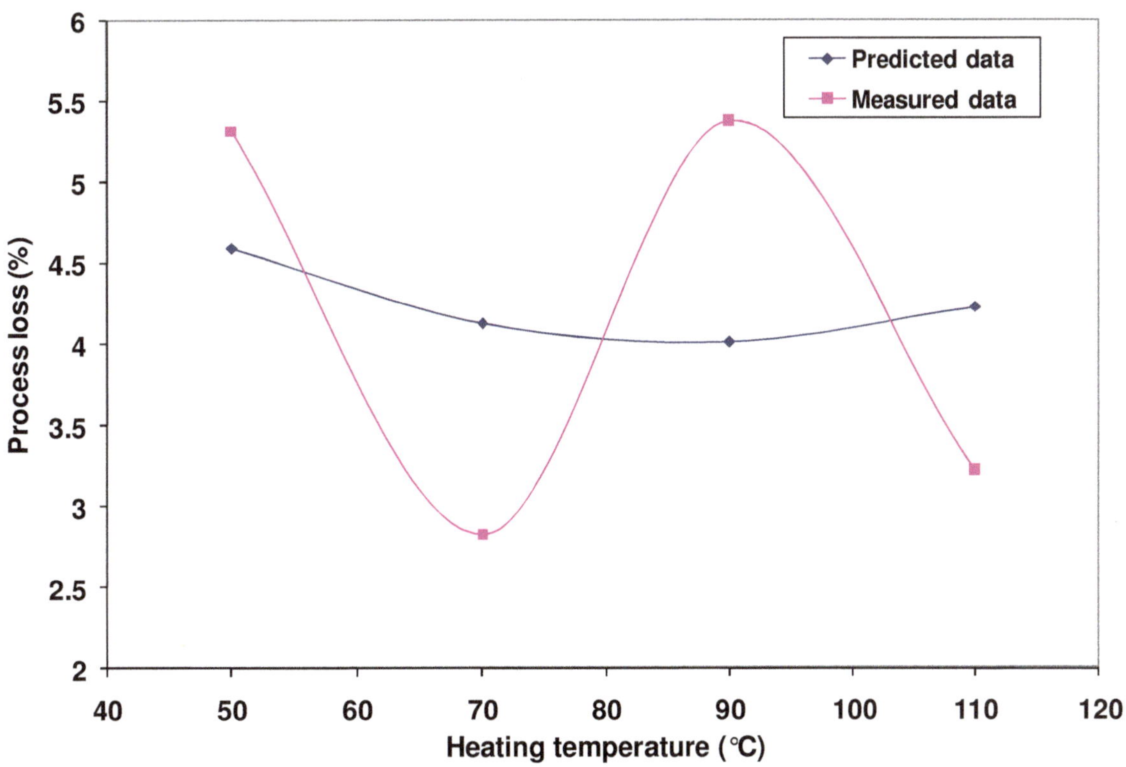


**Figure 6.** Effect of heating temperature on process loss.

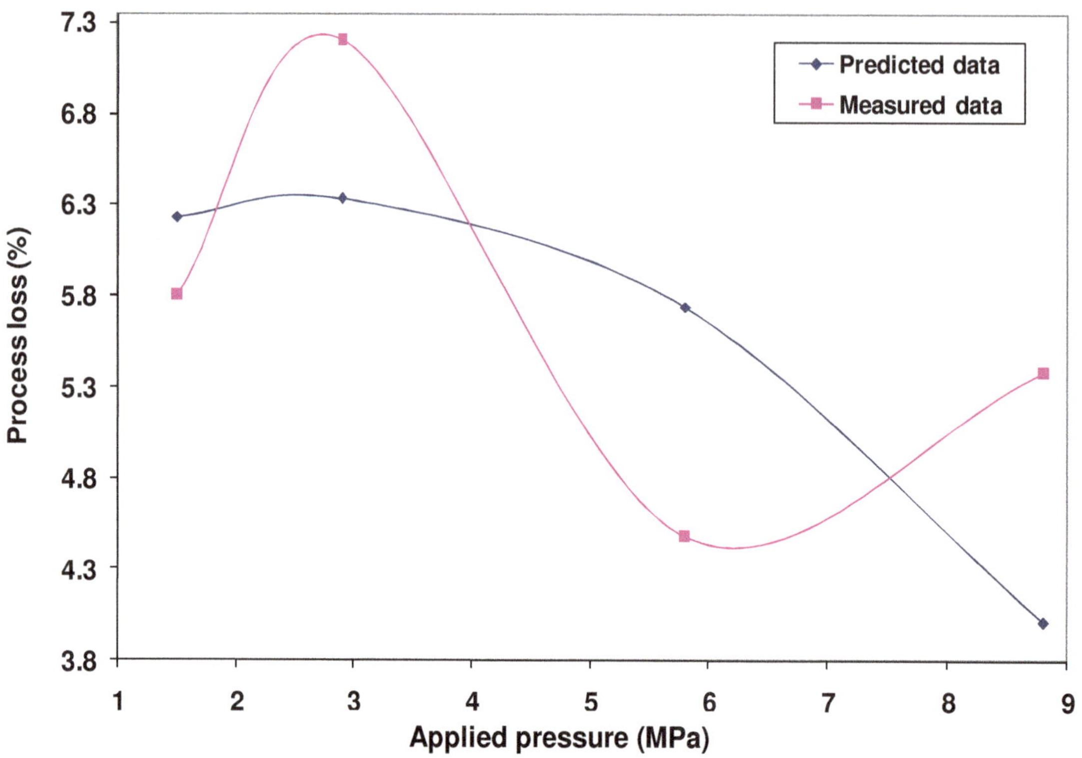


**Figure 7.** Effect of applied pressure on process loss.

temperature, applied pressure and loading rate of 82.24°C, 9.69 MPa and 2.50 mm $min^{-1}$ respectively should be used for dry process of SB extraction. These combinations gave 35.39% oil yield, 58.62% oil recovery efficiency and 2.83% process loss.


## ACKNOWLEDGEMENTS

The authors sincerely acknowledge the support of the University of Ilorin through the Teaching and Research Material Grant at the discretion of the Head of Agricultural and Biosystems Engineering Department and National Centre for Agricultural Mechanization (NCAM), Idofian, Ilorin, for laboratory experiments.


## REFERENCES


Addaquay J (2004). The shea butter value chain: refining in West Africa. WATH Technical Report No. 3.The United States Agency for International Development, USA.

Adeeko KA, Ajibola OO (1989). Processing factors affecting yield and quality of mechanically expressed groundnut oil. J. Agric. Eng. Res., 5: 31-43

Babatunde OO (1997). Development of model equation for selecting disc and tilt angles of disc plough for optimal operation. Proceedings of the Nigerian Branch of Intern. Soil Res. Organ., pp. 169-178

Olaniyan AM, Oje K (2007). Development of mechanical expression rig for dry extraction of shea butter from shea kernel. J. Food Sci. Technol., 44(5): 465-470

Ott L (1977). An introduction to statistical methods and data analysis. Wadsworth Publishing Company Inc, California, USA, pp. 459-518.

Stephenson G (1975). Mathematical models for science students (2nd edition) Longman Group Ltd., London, pp. 168-174.

12

# Assessment of genetic diversity of non-basmati rice of Jammu and Kashmir using microsatellite markers

**Ritu Mahajan, Syed Tabia, Gourav Raina and Neha Mangotra**

School of Biotechnology, University of Jammu, Jammu, India.

**Rice (*Oryza sativa* L.) is one of the oldest domesticated crop species in the world and is widely recognized as an ideal model plant for the study of grass genetics and genome organization. Rice improvement depends on the conserved use of genetic variability and diversity in plant breeding programmes. Phenotypic variability and genetic diversity was studied in 13 non basmati rice samples collected from Jammu and Kashmir. The length of the seeds varied from 0.5 to 0.8 cm and the seed width varied from 0.1 to 0.3 cm. The weight of the seed varied from 0.36 to 0.77 g. A total of 94 alleles were amplified in 13 rice samples using six simple sequence repeat (SSR) markers. The analysis and dendrogram construction was performed using the DARWin 5.0 software. The genetic dissimilarity index calculated between samples ranged from 0.00 to 0.63. The generated dendrogram based on the dissimilarity matrix using the neighbor-joining approach of the unweighted pair group with arithmetic mean (UPGMA) method showed three distinct clusters.**

**Key words:** *Oryza sativa*, diversity, characterization, molecular markers.

## INTRODUCTION

Rice (*Oryza sativa*), one of the agronomically and nutritionally important cereal crop in the grass family (Poaceae), is the principal staple food in developing countries. It is no longer a luxury food, but has become the cereal that constitutes a major source of calories for the urban and the rural dwellers (Sasaki and Burr, 2000). In the mid-1960s, the Green revolution resulted in a dramatic increase in food production as a result of a combination of advances in plant breeding and production-oriented intensification of agriculture, stimulated by the implementtation of enabling policies. This technological revolution integrated the development and use of modern high-yielding varieties and led to an increase in both level and efficiency of agricultural production.

Rice plays a pivotal role in Indian economy, being the staple food for two third of the population. However, its productivity per unit area compared to the world average is low. This may possibly be due to narrow range of genetic variability and lack of adequate genetic information regarding the inheritance of quantitative traits controlling important economic characters. Molecular markers are powerful tools in the assessment of genetic variation and in the elucidation of genetic relationships within and among species (Thomson et al., 2007). Among different polymerase chain reaction (PCR) based markers, microsatellites are highly polymorphic, more reproducible, co-dominant and distributed throughout the genome. More than 2200 microsatellite markers have been mapped to specific locations in rice genome (McCouch et al., 2002). These markers have been utilized for many purposes, including genome mapping, gene tagging, estimation of genetic diversity, varietal differentiation and purity testing (Nagaraju et al., 2002).

In order to increase rice productivity, high yielding and disease resistant varieties should be developed. Unlike high-yielding varieties (whose variability is limited due to homozygosity), the landraces maintained by farmers are endowed with tremendous genetic variability as they are not subjected to subtle selection over a long period of time. Further crop improvement depends on the conserved

*Corresponding author. E-mail: ritufeb@gmail.com.

**Table 1.** List of SSR markers of rice.

| Oligomers | Molecular weight | Size range | Tm | Sequence 5′ - 3′ |
|---|---|---|---|---|
| RM 223F | 6222 | 139-163 | 63.9 | GAGTGAGCTTGGGCTGAAAC |
| RM 223R | 6182 | | 64.9 | GAAGGCAAGTCTTGGCACTG |
| RM 210F | 5505 | 141-165 | 64.4 | TCACATTCGGTGGCATTG |
| RM 210R | 6179 | | 63.1 | CGAGGATGGTTGTTCACTTG |
| RM 284F | 6565 | 141-149 | 60.7 | ATCTCTGATACTCCATCCATCC |
| RM 284R | 6093 | | 64.5 | CCTGTACGTTGATCCGAAGC |
| RM 85F | 6463 | 85-107 | 63.1 | CCAAAGATGAAACCTGGATTG |
| RM 85R | 5534 | | 62.2 | GCACAAGGTGAGCAGTCC |
| RM 42F | 6062 | 161-171 | 64.0 | ATCCTACCGCTGACCATGAG |
| RM 42R | 6139 | | 64.0 | TTTGGTCTACGTGGCGTACA |
| RM515F | 6240 | 205-231 | 63.9 | TAGGACGACCAAAGGGTGAG |
| RM515R | 5962 | | 64.0 | TGGCCTCCTCTCTCTCTCTC |

use of genetic variability and diversity in plant breeding programmes and the use of new biotechnological tools. There is a wide genetic variability in varieties leaving a wide scope for future crop improvement.

## MATERIALS AND METHODS

### Plant material and growth experiment

The experimental material consisted of 13 cultivated non-basmati varieties of rice (*O. sativa*). Seeds of these varieties were collected from different areas of Jammu and Kashmir. The varieties collected were from Jammu (KJ-1, KJ-2 and JJ); Baderwah (Chi and Jap); Rajouri (Ro and Gi), Samba (SS); Udhampur (S-1 and S-2), Kathua (Kat), Mera Sahib (MS) and Kashmir (Kas). Since the temperature conditions were unfavorable in January (9 - 10°C) for the growth of rice plants in field conditions, the plants were therefore grown in laboratory conditions. For this, 50 seeds from each collected rice samples were taken and soaked in water overnight for 24 h. Afterward, the seeds were placed in sterilized Petri plates (50 seeds/per plate) on a wet sterilized filter paper soaked in distilled water. The Petri plates were then kept in an incubator at 25°C for 3 days till the seeds germinated. Subsequently, the germinated seedlings were transferred from the Petri plates to the small plastic pots for their further growth to young plantlets.

### Phenotypic evaluation of rice grain

Twenty rice seeds taken from thirteen non-basmati samples were dehulled and their length and width, weight, ratio of length: width and variation in colour were calculated.

### Molecular marker analysis

DNA isolation was carried out using cetyltrimethylammonium bromide (CTAB) method as modified by Saghai-Maroof et al. (1984). Six simple sequence repeat (SSR) markers RM223, RM210, RM284, RM42, RM85 and RM515 were selected from the linkage map of Temnykh et al. (2000) to study the diversity (Table 1). The PCR reaction mixture contained 75 ng template DNA, 12.3 μL ddH2O, 2.5 μl 10X PCR buffer, 2.5 μl of 100 mM dNTPs, 1 μl of 5 μM forward and reverse primer each and 0.3 μL Taq polymerase (5 U/μl). Template DNA was initially denatured at 94°C for 4 min followed by 35 cycles of PCR amplification following: 1 min of denaturation at 94°C, 1 min of primer annealing at 55°C and 2 min of primer extension at 72°C. Final 7 min incubation at 72°C was allowed to complete primer extension. The amplified products were electrophoretically resolved on a 2.5% agarose gel in 0.5X Tris-acetate-EDTA (TAE) and visualized under UV light after staining with 0.1% ethidium bromide. The bands representing particular alleles at the microsatellite loci were scored manually.

### Data collection and diversity analysis

To measure the allelic diversity of the SSR markers, the polymorphism information content (PIC) for each marker was calculated according to the formula: $PIC = 1 - \Sigma(Pi\ j)^2$, where Pi j is the frequency of the ith allele in the jth population for each SSR locus (Botstein et al., 1980). The dissimilarity matrix was used for clustering of genotypes based on the unweighted neighbor-joining method and the analysis was performed using DARWin 5.0 (http://darwin.cirad.fr), (Perrier et al., 2003). Confidence limits of different clades were tested by bootstrapping 500 times to assess the repetitiveness of genotype clustering (Felsenstein, 1985).

## RESULTS AND DISCUSSION

Fifty seeds from each non basmati rice sample were kept in plates with moist sterilized filter paper at different growth conditions. The seeds showed no germination when grown at room temperature (9 to 10°C in January) at 24 h, however, when grown at 25°C it showed growth of varying degrees after 48 h. The samples showed best

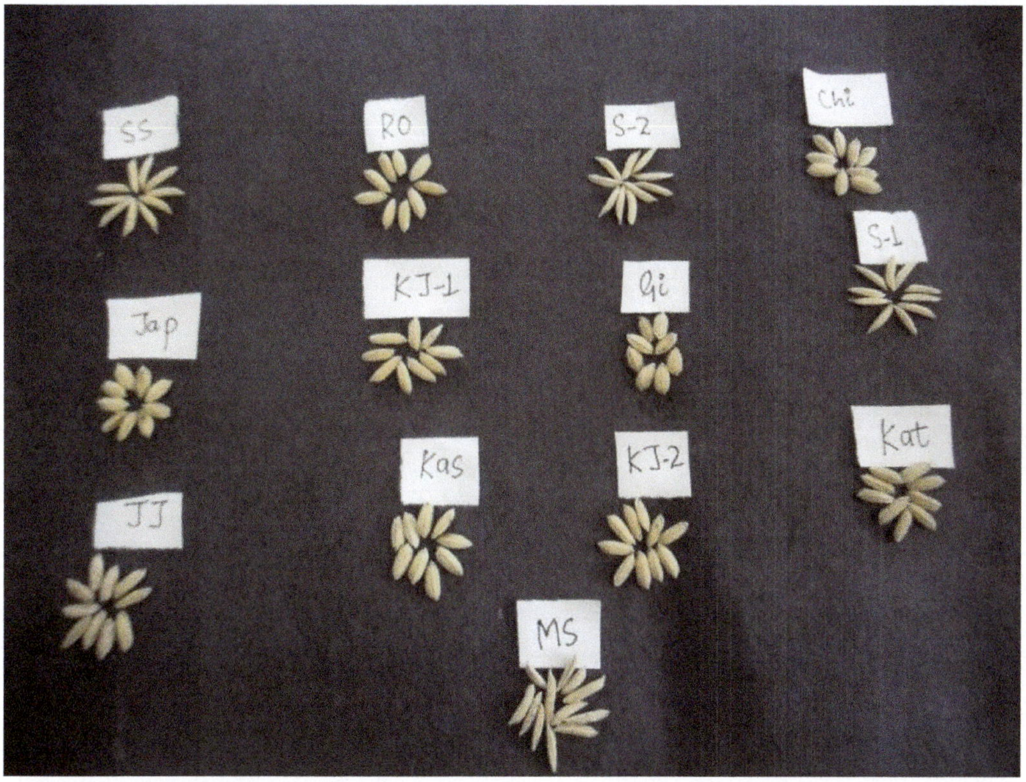


**Figure 1.** Shapes and sizes of different rice samples collected from Jammu and Kashmir.

germination at 30°C after 3 days of incubation.

### Phenotypic evaluation of rice seeds

The rice genetic resources showed a great diversity for all the measured grain morphological characters (Figure 1). The variation for grain length ranged from 0.5 to 0.8 cm (Table 2 and Figure 2). The longest grain size was for sample SS and smallest was for Jap and Gi. The variation for grain width ranged from 0.1 to 0.3 cm, with the largest grain width for sample Jap and smallest for SS and S-1. The length to width ratio varied from 8.0 (for SS and S-1) to 1.83 (for Jap and Gi). The grain weight varied from 0.36 to 0.77 g for Kat and SS, respectively (Table 2). The long grain types were present in few locations (on the basis of seed length), whereas maximum medium length grain types were recorded while, the short grain types were not found.

On the basis of length to width ratio, long and medium grain types were found distributed in samples of all locations. However, the short grain types were only recorded in samples from the region of Bhaderwah, Rajouri and Kashmir. Samples from Udhampur and Samba showed the highest grain length and length to width ratio. Samples collected from Jammu showed comparatively little shorter grain length than those collected from Udhampur, but the grain size was larger than the rice samples collected from Kathua (Figure 1). It was observed that the longer grains had less width. The J&K rice cultivars with more width (0.3 cm or more) were shorter in length and samples with less width (0.1 cm) were longer in length. Similar results in seed shape variation were also found by Siddiqui et al. (2007).

### Molecular analysis

Six microsatellite markers were chosen from Temynk et al. (2000) for diversity analysis of 13 non-basmati rice samples. The PIC value of each marker ranged from 0.6 to 0.9. The amplified product was scored on the basis of presence and absence of bands (Figure 3). A total of 94 alleles were amplified in 13 rice samples using 6 SSRs. Out of 94 alleles, 26 were monomorphic and the rest 68 were polymorphic. SSR markers RM 85 and RM 210 amplified all monomorphic alleles and marker RM 515, RM 223, RM42, RM 284 resulted in polymorphic alleles. The number of alleles generated by each marker varied from two (RM 284, RM 42 and RM 515) to three (RM 223). The number of alleles observed in the present study corresponded well with the earlier report of Siwach et al. (2004) among basmati and non-basmati rice varieties from India.

PIC is the reflection of allele diversity and frequency among the varieties and varied greatly for all SSR loci

**Table 2.** Phenotypic variation in average length, width, ratio of length: width, weight and seed colour in 13 non-basmati samples.

| S/N | Variety code | Place of collection | Average weight (g) | Average length (cm) | Average width (cm) | Ratio length/ width (cm) | Colour |
|---|---|---|---|---|---|---|---|
| 1 | Kat | Kathua | 0.77 | 0.60 | 0.20 | 3.0 | Dark brown |
| 2 | (RO) | Rajouri | 0.54 | 0.60 | 0.20 | 3.0 | Light greenish |
| 3 | (Gi) | Rajouri | 0.66 | 0.55 | 0.30 | 1.8 | Golden brown |
| 4 | (KJ-1) | Jammu | 0.62 | 0.70 | 0.20 | 3.5 | Golden brown |
| 5 | Jap | Bhaderwah | 0.71 | 0.55 | 0.30 | 1.8 | Straw golden |
| 6 | Kas | Kashmir | 0.68 | 0.70 | 0.25 | 2.8 | Golden brown |
| 7 | S-1 | Udhampur | 0.58 | 0.80 | 0.10 | 8.0 | Bright golden |
| 8 | Chi | Bhaderwah | 0.76 | 0.60 | 0.25 | 2.4 | Light brown |
| 9 | S-2 | Udhampur | 0.62 | 0.75 | 0.15 | 5.0 | Light Golden |
| 10 | KJ-2 | Jammu | 0.50 | 0.65 | 0.20 | 3.2 | Golden brown |
| 11 | SS | Samba | 0.36 | 0.80 | 0.10 | 8.0 | Straw golden |
| 12 | JJ | Jammu | 0.59 | 0.60 | 0.20 | 3.0 | Reddish brown |
| 13 | MS | Mera Sahib | 0.40 | 0.65 | 0.15 | 4.3 | Light golden |

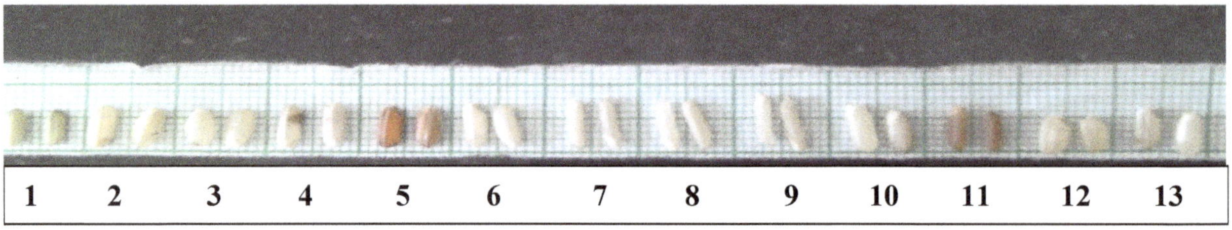


**Figure 2.** Length and width of different rice samples (1= Ro, 2= Kas, 3=Gi, 4=JJ, 5=KJ-2, 6=KJ-1, 7=SS, 8=S-2, 9=S-1, 10=Kat, 11=MS, 12=Jap, 13=Chi).

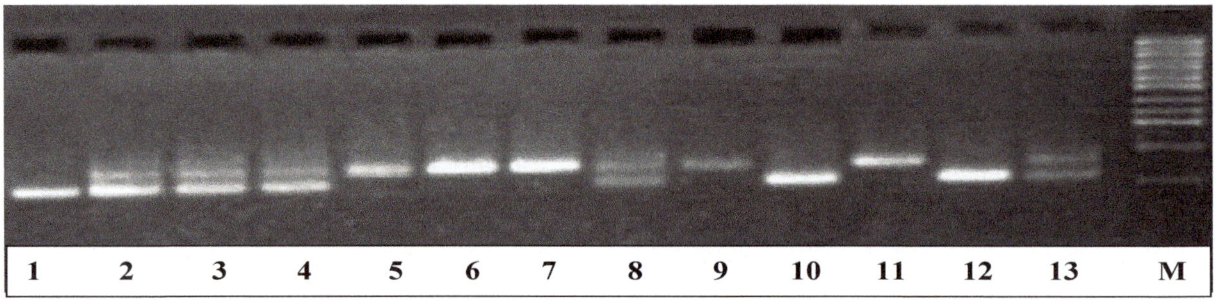


**Figure 3.** PCR amplification of thirteen non Basmati Rice samples with SSR marker RM 284. 1=Kat, 2= Ro, 3= Gi, 4= KJ-1, 5= Jap, 6= Kas, 7= S-1, 8= Chi, 9= S-2, 10= KJ-2, 11= SS, 12= JJ, 13=MS, M= Marker (250 bp).

tested. The PIC values varied from 0.34 (RM 42) to 0.72 (RM 223), whereas PIC values varied from 0.259 to 0.782 with an average of 0.571 in the studies conducted in basmati and non basmati rice of Pakistan by Rabbani et al. (2010). Similarly, the PIC values ranged from 0.09 to 0.60 and the overall frequency of biallelic markers was high in the rice landraces collected from Tamil Nadu (Vanniarajan et al., 2012). The lower level of polymorphism may be attributed to a narrow genetic diversity as some of the local varieties are product of the same location Ablett et al. (2006) stated that though SSRs have a potentially high polymorphic frequency, the level of polymorphism was found to be 26% when tested in 10 wheat varieties using 126 markers.

## Analysis of genetic dissimilarity

Genetic dissimilarity was calculated from the matrix of binary data using software DARWin 5.0, where "0" and "1" were standardized as the least and maximum of dissimilarity respectively. To ascertain the statistical

**Table 3.** Diversity matrix between 13 different non-Basmati rice samples as revealed by SSR markers.

| | 1 | 2 | 3 | 4 | 5 | 6 | 7 | 8 | 9 | 10 | 11 | 12 |
|---|---|---|---|---|---|---|---|---|---|---|---|---|
| 2 | 0.45 | | | | | | | | | | | |
| 3 | 0.27 | 0.16 | | | | | | | | | | |
| 4 | 0.63 | 0.16 | 0.33 | | | | | | | | | |
| 5 | 0.38 | 0.14 | 0.14 | 0.28 | | | | | | | | |
| 6 | 0.45 | 0.00 | 0.16 | 0.16 | 0.14 | | | | | | | |
| 7 | 0.50 | 0.07 | 0.23 | 0.23 | 0.06 | 0.07 | | | | | | |
| 8 | 0.00 | 0.45 | 0.27 | 0.63 | 0.38 | 0.45 | 0.50 | | | | | |
| 9 | 0.33 | 0.23 | 0.38 | 0.38 | 0.20 | 0.23 | 0.14 | 0.33 | | | | |
| 10 | 0.40 | 0.09 | 0.27 | 0.27 | 0.23 | 0.09 | 0.16 | 0.40 | 0.33 | | | |
| 11 | 0.63 | 0.16 | 0.33 | 0.00 | 0.28 | 0.16 | 0.23 | 0.63 | 0.38 | 0.27 | | |
| 12 | 0.63 | 0.16 | 0.33 | 0.00 | 0.28 | 0.16 | 0.23 | 0.63 | 0.38 | 0.27 | 0.00 | |
| 13 | 0.45 | 0.00 | 0.16 | 0.16 | 1.14 | 0.00 | 0.07 | 0.45 | 0.23 | 0.09 | 0.16 | 0.1 |

1=Kat, 2= Ro, 3= Gi, 4= KJ-1, 5= Jap, 6= Kas, 7= S-1, 8= Chi, 9= S-2, 10= KJ-2, 11= SS, 12= JJ, 13=MS.

strength of genetic relationships identified through this analysis, bootstrapping of the data (500 permutations) was performed. The dissimilarity coefficients were used for cluster analysis based on the unweighted neighbor-joining method and a dendrogram was generated with the aim of analyzing the relationship between rice samples. The genetic dissimilarity index calculated between samples ranged from 0.00 to 0.63 (Table 3). The lowest value 0.1 was obtained between Jaya (JJ) collected from Jammu and MS collected from Mera Sahib, while the highest dissimilarity value calculated was 0.63 between the samples collected from Jammu (K -448) and China (Chi) from Bhaderwah.

## Cluster analysis

The tree generated showed no divergence between samples collected from Samba and Jammu (Figure 4). A dendrogram was generated by unweighted pair group with arithmetic mean (UPGMA) to show the genetic relationships of the samples studied and is presented in Figure 5. The dendrogram indicated that the samples collected from the same region clustered together and were clearly separated into three distinct clusters:

1. Cluster I contained samples collected from Kashmir, Jammu and Mera Sahib
2. Cluster 2 contained samples from Rajouri, Udhampur (Sharbati-1) and Bhaderwah
3. Cluster 3 contained a single rice sample collected from Udhampur (Sharbati-2)

Though both the samples of Sharbati were collected from Udhampur, the difference could be due to mixing of seeds or because it is a new or different variety (Figure 5). Adéoti et al. (2011) also reported the genetic diversity based on UPGMA cluster analysis of dissimilarity data in different accessions of leafy vegetables and observed the coefficient of genetic dissimilarity between 0.01 and 0.63. Similarly, Sharma et al. (2010) reported cluster analysis using Rogers' genetic distance in different maize accessions. Their results revealed distinct genetic identity of the 'Sikkim Primitives' from the rest of the accessions in India, including Sikkim.

## Conclusion

The characterization and quantification of genetic diversity within closely related crop germplasm has long been a major goal, as it is essential for a rational use of genetic resources.

In all, the analysis of genetic variation among breeding materials is of fundamental interest to plant breeders as it contributes immensely to selection, monitoring of germplasm, and also prediction of potential genetic gains (Chakravarti and Rambabu, 2006).

Although a large number of aromatic and coarse cultivars of rice are available in Jammu and Kashmir, no systematic analysis has been carried out so far for genetic diversity, identification and discrimination. In the present study, the evaluation of the pattern and the extent of genetic variability and relatedness among traditional varieties to improved cultivars of rice were conducted using SSR markers.

The results obtained would help in the identification and differentiation of various cultivars being cultivated and/or for exports and will also contribute to maximize the selection of diverse parent cultivar and broaden the germplasm base in the future of rice breeding programs. In addition, it will help in identifying efficient strategies for the sustainable management of the genetic resources of rice crops.

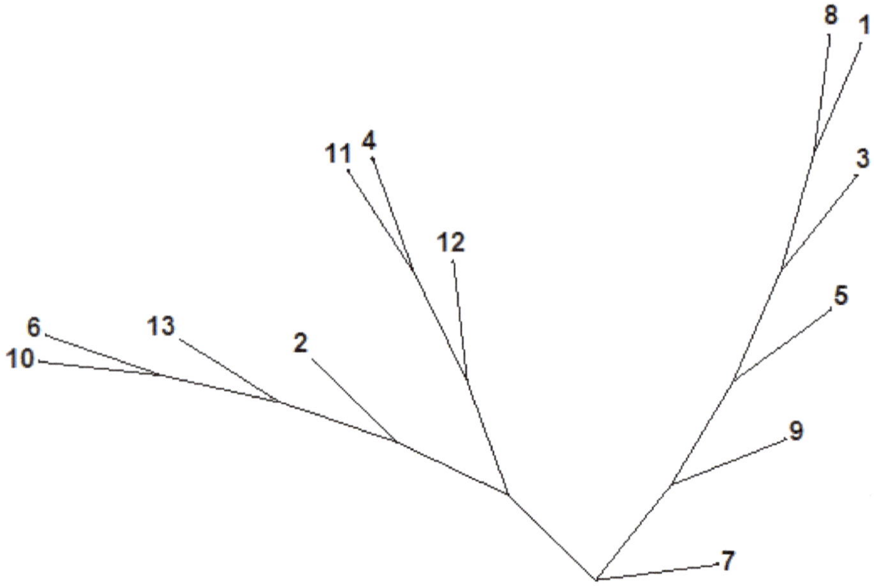


**Figure 4.** Tree based on neighborhood joining method showing genetic dissimilarity between (1=Kat, 2= Ro, 3= Gi, 4= KJ-1, 5= Jap, 6= Kas, 7= S-1, 8= Chi, 9= S-2, 10= KJ-2, 11= SS, 12= JJ, 13=MS).

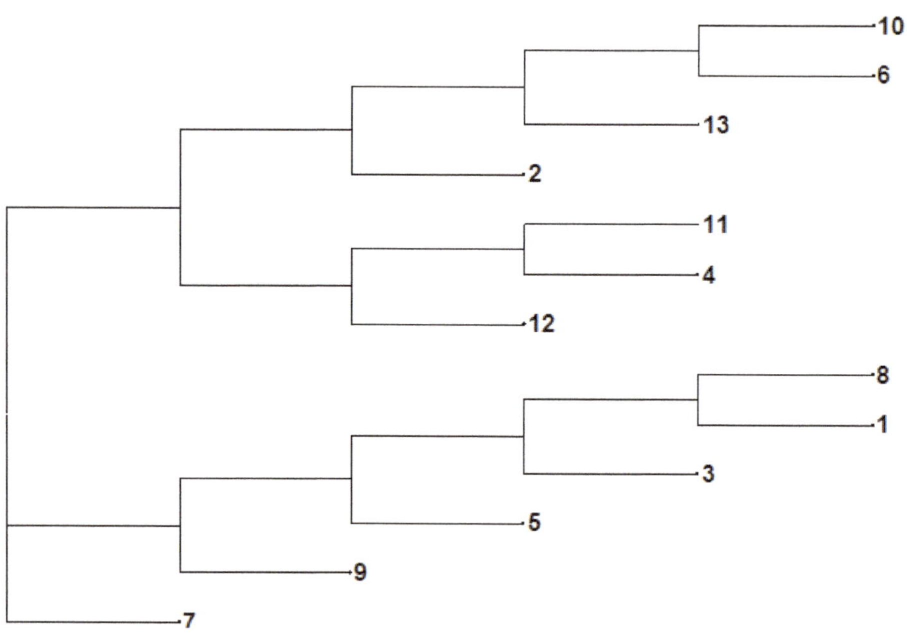


**Figure 5.** UPGMA based dendrogram for all the thirteen non basmati rice samples (1=Kat, 2= Ro, 3= Gi, 4= KJ-1, 5= Jap, 6= Kas, 7= S-1, 8= Chi, 9= S-2, 10= KJ-2, 11= SS, 12= JJ, 13=MS).

## REFERENCES

Ablett G, Hilland H, Henry R (2006). Sequence polymorphism discovery in wheat microsatellite flanking regions using pyrophosphate sequencing. Mol. Breed.17:281-289.

Adéoti K, Rival A, Dansi A, Ahohuendo BC, Santoni S, Beule T, Nato A, Henry Y, Ahanchédé A, Vodouhè R, Hounhouigan DJ, Sanni A (2011). Assessment of genetic diversity among accessions of two traditional leafy vegetables (*Acmella uliginosa* (L.) and *Justicia tenella* (Nees) T.) consumed in Benin using amplified fragment length polymorphism (AFLP) markers. Afr. J. Biotechnol. 10(15):2811-2819.

Botstein D, White RL, Skolnick ME, Davis RW (1980). Construction of a genetic linkage map in man using restriction fragment length polymorphism. Am. J. Hum. Genet. 32:314-331.

Chakravarti KB, Rambabun N (2006). SSR marker based DNA fingerprinting and diversity study in rice (*Oryza sativa*. L). Afr. J.

Biotechnol. 5:684-688.
Felsenstein J (1985). Confidence limits in phylogenies: an approach using bootstrap. Evolution 39:783-791.
McCouch S, Teytelman L, Xu Y, Lobos K, Clare K, Walton M, Fu B, Maghirang R, Li Z, Xing Y, Zhang Q, Kono I, Yano M, Fjellstrom R, DeClerck G, Schneider D, Cartinhour S, Ware D, Stein L (2002). Development of 2,240 new SSR markers for rice (*Oryza sativa* L.). DNA Res. 9:199-207.
Nagaraju J, Kathirvel M, Kumar RR, Siddiq EA, Hasnain SE (2002). Genetic analysis of traditional and evolved Basmati and non-Basmati rice varieties by using fluorescence-based ISSR-PCR and SSR markers. Proc. Natl. Acad. Sci. USA. 99:5836-5841.
Perrier X, Flori A, Bonnot F (2003). Data analysis methods. In: Hamon, P., Seguin, M., Perrier, X. and Glaszmann, J. C. (eds.), Genetic diversity of cultivated tropical plants. Enfield, Science Publishers. Montpellier pp. 43-76.
Rabbani MA, Masood MS, Shinwari ZK, Shinozaki KY (2010). Genetic analysis of basmati and non-basmati Pakistani rice (*Oryza sativa* L.) cultivars using microsatellite markers. Pak. J. Bot. 4:2551-2564.
Saghai-Maroof HA, Soliman KM, Jorgensen AR, Allard RW (1984). Ribosomal DNA spacer length polymorphism in barley: Mendelian inheritance, chromosomal location and population dynamics. Proc. Natl. Acad. Sci. USA. 81:8014-8018.
Sasaki T, Burr B (2000). International rice genome sequence project: the effort to complete the sequence of rice genome. Curr. Opin. Plant. Biol. 3:138-141.
Sharma L, Prasanna BM, Ramesh B (2010). Analysis of phenotypic and microsatellite-based diversity of maize landraces in India, especially from the North East Himalayan region. Genetica 138:619–631.
Siwach PS, Jain NS Chowdhury VK, Jain RK (2004). Allelic diversity among Basmati and non-Basmati long-grain indica rice varieties using microsatellite markers. J. Plant Biochem. Biotechnol. 13:25-32.
Temnykh S, Park WD, Ayres N, Cartinhour S, Hauck N, Lipovtiesich L (2000). Mapping and genome organization of microsatellite sequences in rice (*Oryza sativa* L.). Theor. Appl. Genet. 100:697–712.
Thomson, MJ, Septiningsihn EM, Suwardjo F, Santoso TJ, Silitonga TS, McCouch SR (2007). Genetic diversity analysis of traditional and improved Indonesian rice (*Oryza sativa* L.) germplasm using microsatellite markers. Theor. Appl. Genet. 114:559-568.
Vanniarajan C, Vinod KK, Pereira A (2012). Molecular evaluation of genetic diversity and association studies in rice (*Oryza sativa* L.). J. Genet. 91:9–19.

# 13

# Management of rice stem borers (Lepidoptera: Pyralidae) through host plant resistance in early, medium and late plantings of rice (*Oryza sativa* L.)

**Muhammad Sarwar**

Pakistan Atomic Energy Commission, Nuclear Institute of Agriculture, Tando Jam-70060, Pakistan. E-mail: drmsarwar64@yahoo.com.

**This study was conducted to determine the effects of host plant resistance in early, medium and late sown varieties of rice [2 non aromatic IR8 (P) and Sharshar, and 2 aromatic Basmati-370 (P) and Mehak] due to the incidence of rice stem borers. Treatments comprised the crop sown on 3 different dates at the fortnightly intervals starting from the last week of June till the end of July to note the rate of stem borer's infestation. The data on the incidences of insect pests and yield performances of rice varieties were studied. Rice yellow stem borer *Scirpophaga incertulas* (Walker) (Lepidoptera: Pyralidae), was the most important insect pest of rice, attacking all stages of the crop causing substantial losses in early, medium and late-sown crops, and degree of stem borer infestation depended upon the planting time. The pest incidence was the least on early sown crop as compared to medium and late sown crops. Similarly, the highest yield was obtained in early sown crop, and the crop sown after this date showed drastic reductions in yield. Among different varieties tested, Sharshar showed best results in holding least pest infestation and increased grain yield approached by, IR8 (P), Mehak and Basmati-370 (P). Hence, use of tolerant rice varieties and their early sowing under agro climatic conditions of this region is recommended to protect the crop from borer's invasion.**

**Key words:** Rice, yellow stem borer, resistance, planting time, *Scirpophaga, Oryza sativa.*

## INTRODUCTION

Rice (*Oryza sativa* L.) is the dominant staple food in the developing world. More than 90% of the world's rice is produced and consumed in Asia (Schoenly et al., 1998). Several insects feed on rice, but stem borers are considered the most important rice pests, in particular *Scirpophaga incertulas* (Walker) and *S. innotata* (Walker) (Lepidoptera: Pyralidae) (Sigsgaard, 2000). Stem borer *S. incertulas* usually comprised more than 90% of the borer population in rice. The onset of flooding and stem elongation provided a more favorable environment for *S. incertulas*. The rice borers' activity increased steadily during the first 3 to 4 months of flooding, to average 23% damaged stems by the flowering stage. Borer's activity continued at about the same level as the water receded; to reach maximum annual levels of 38 t o 44% damaged stems at the late-ripening stage. At harvest, 60% of the fields were at outbreak level (> 40% damaged stems) (Catling et al., 1984). It is the serious pest species of rice throughout the Orient, and abundant both on lowland rice and upland rice attacking young plant even in the nursery stage (Litsinger et al., 1987). In Pakistan, rice crop is attacked by about 70 species of insect pests and out of these, rice stem borers are by far the most pervasive and injurious insects to rice. These borers vary in severity of damage and population intensity (Hashmi, 1994). The rice stem borer *S. incertulas*, infesting the plant from seedling to maturity, is one of the main problems and yields limiting factors in the rice fields (Sarwar et al., 2007; Sarwar, 2011).

Farmers depend upon a great deal of insecticide applications, even though a lot of insecticide applications are not effectual (Sarwar et al., 2005). Therefore, the challenge before the agricultural scientists today is to produce insect resistant plants. Insect resistant plants have the ability to withstand the effects of an insect by becoming resistant to its ill effects by means of genetic manipulation (Sarwar et al., 2010). Plant breeders have produced some new rice cultivars that partially resist the attacks from the borers. Rice breeding programs often emphasized selection for insect resistant rice varieties and made much progress. Research demonstrated that certain rice cultivars had less borers' damage than more susceptible varieties (Khan et al., 2005). We certainly need new technology to accomplish this task as the prevailing technologies alone do not seem to be adequate. To achieve this aim, pest control will have to rely on integrated management practices which include crop planting techniques and insect resistant plants (Baloch and Abdullah, 2011) to improve productivity and sustainability. This study is intended to assess the stem borer infestation and grain yield of four rice cultivars grown in diverse planting times because of problems that past studies could not solve appropriate date of sowing which is being addressed here. The study is projected to endow with appropriate information for implementing Integrated Pest Management (IPM) practices for controlling the stem borers in rice crop.

## MATERIALS AND METHODS

A field study to determine the incidence of rice stem borers on different varieties of rice in the command area of Nuclear Institute of Agriculture, Tandojam, Sindh, Pakistan, was carried out during the year 2006. Study was conducted to determine the effects of host plant resistance in early (last week of June), medium (second week of July) and late sowings (fourth week of July) of rice varieties [2 non aromatic IR8 (P) and Sharshar, and 2 aromatic Basmati-370 (P) and Mehak] due to the incidence of rice stem borers. Seeding technique comprised of direct seeding on well prepared flat beds. After 30 days interval, rice transplanting were made starting from the last week of June till the end of July, so as to have a sequence of crop growth for recording the natural incidence of borers. The rice varieties were sown according to randomized complete block design with three replicates each having 4.5 $m^2$ plot size. Plant to plant and row to row distance of 30 cm was maintained during planting and thinning of crop. All agronomic practices followed were uniform in whole rice field under trial. Recommended dose of synthetic fertilizers were:

1) Nitrogen 60 kg $ha^{-1}$, half at the time of sowing and half at the time of 1st irrigation.
2) Phosphorous 50 kg $ha^{-1}$ at the time of transplanting.

In each treatment, a random area of 1 $m^2$ was selected every time and observed for recording the pest incidence at 15 days interval from 30 days after transplanting till crop harvest. Borer incidence was recorded by counting the healthy and infested plants in each variety to calculate the percentage infestation of the pest as follows:

$$\text{Percentage of stem borer infestation} = \frac{\text{Number of infested tillers}}{\text{Total tillers}} \times 100$$

For this trial no chemical or bio-control agent was used, but, planting times and varietals resistance of the test varieties to rice stems borers were observed. For this purpose, data regarding deadhearts and white ear heads were an indicator of rice stem borers' incidence. Ranks to different varieties were assigned arbitrarily on the basis of percent infestation (deadhearts and white ear heads). The ranking for percent deadhearts and white ear heads were put together to determine the comparative resistance level of different varieties. The yield of paddy was recorded after harvesting and threshing the crop. The observations thus recorded were tabulated in tables and data were subjected to analysis of variance using Statistix 8.1 version. Significant differences in means were separated using Duncan's multiple range test ($P= 0.05$).

## RESULTS AND DISCUSSION

Analyses of variance showed that no variety had been observed with consistent number of particular insect pest among the tested varieties, which may suggest varietals potential of test plant to support population of pest. The planting times influenced stem borer infestation and effects of sowing dates and cultivars on pest infestation were obvious. From these studies, it was found that early sown rice crop (last week of June) was the most resistant having the lowest borer infestation among other plantings. Non-aromatic IR8 and Sharshar beard 3.21%, 2.47% deadhearts, 6.05 and 4.24% whiteheads and higher production of 2313 and 2807 g per 4.5 $m^2$, respectively. While, aromatic Basmati-370 and Mehak with 5.44%, 4.75% deadhearts, 8.94 and 7.45% whiteheads and production of 1500 and 1823 g per 4.5 $m^2$, respectively, had higher borer infestation. In the trail, medium sown, Basmati-370 was the highest infested cultivar (11.35% deadhearts, 9.94% whiteheads and production 1367 g per 4.5 $m^2$) and significantly higher than Mehak (8.97% deadhearts, 9.02% whiteheads and production 1433 g per 4.5 $m^2$), IR8 (6.79% deadhearts, 6.97% whiteheads and production 1810 g per 4.5 $m^2$) and Sharshar (5.44% deadhearts, 5.27% whiteheads and production 2440 g per 4.5 $m^2$) which were the lowest stem borer infested cultivars. The mean percent infestation of stem borers under natural conditions on late sown crop also varied significantly $P< 0.05$ on different varieties. On the basis of deadhearts and whiteheads, variety Sharshar was found most resistant with the minimum 8.73 and 9.42% damage and 1383 g per 4.5 $m^2$ yield followed by IR8 with 12.8 and 11.02% damage and 1207 g yield. The variety Basmati-370 was found to have the maximum 20.0% deadhearts and 16.46% whiteheads with 633.3 g yield, followed by Mehak with 16.19 and 12.94% damage and 773.3 g yield, therefore, considered as the most susceptible varieties under the present studies (Table 1).

The mean overall seasonal incidence of stem bores on all rice varieties tested in the present studies is given in Table 2. There were significant differences in the effects of planting time on pest invasion of the above-mentioned insect species on all rice varieties. However, on early sowing date, the infestation (3.96% deadhearts, 6.67% whiteheads) was significantly low compared with medium (8.13% deadhearts, 7.82% whiteheads) and late sowing(14.45% deadhearts, 12.46% whiteheads) treatments. In general, early planting had the highest

**Table 1.** Effects of host plant resistance in early, medium and late sown varieties of rice.

| Varieties | Deadhearts (%) | Whiteheads (%) | Yield (g) (4.5 m$^2$) |
|---|---|---|---|
| **Early sown crop** | | | |
| IR8 | 3.21$^c$ | 6.05$^c$ | 2313$^b$ |
| Sharshar | 2.47$^c$ | 4.24$^d$ | 2807$^a$ |
| Basmati-370 | 5.44$^a$ | 8.94$^a$ | 1500$^d$ |
| Mehak | 4.75$^b$ | 7.45$^b$ | 1823$^c$ |
| LSD value | 0.63 | 0.93 | 256.2 |
| **Medium sown crop** | | | |
| IR8 | 6.79$^c$ | 6.97$^c$ | 1810$^b$ |
| Sharshar | 5.44$^d$ | 5.27$^d$ | 2440$^a$ |
| Basmati-370 | 11.35$^a$ | 9.94$^a$ | 1367$^c$ |
| Mehak | 8.97$^b$ | 9.02$^b$ | 1433$^c$ |
| LSD value | 0.87 | 0.83 | 167.5 |
| **Late sown crop** | | | |
| IR8 | 12.84$^c$ | 11.02$^c$ | 1207$^b$ |
| Sharshar | 8.73$^d$ | 9.42$^d$ | 1383$^a$ |
| Basmati-370 | 20.04$^a$ | 16.46$^a$ | 633.3$^c$ |
| Mehak | 16.19$^b$ | 12.94$^b$ | 773.3$^c$ |
| LSD value | 2.97 | 0.71 | 175.9 |

Means within columns followed by different letters are significantly different (α= 0.05).

**Table 2.** Performance of all rice varieties in different dates of sowing (pooled data).

| Time of sowing | Deadhearts (%) | Whiteheads (%) | Yield (g) (4.5 m$^2$) |
|---|---|---|---|
| Early sowing | 3.96 c | 6.67 c | 2111 a |
| Medium sowing | 8.13 b | 7.82 b | 1763 b |
| Late sowing | 14.45 a | 12.46 a | 999.2 c |
| LSD value | 2.23 | 1.10 | 267.8 |

Means within columns followed by different letters are significantly different (α= 0.05).

grain yields followed by mid and late plantings with 2111, 1763 and 999.2 g grain yield per 4.5 m$^2$ under organic management. Severity of stem borer infestation varied with all varieties, pest incidence observed was the lowest on Sharshar (non aromatic) Mehak (aromatic), and highest on IR8 (non aromatic) and Basmati-370 (aromatic).

The present studies undertaken elaborated that the dominant species of borer was yellow stem borer (*S. incertulas*) and degree of its infestation depended upon the planting time and host plant resistance. Observations from stem borer infestation, displayed that the yellow stem borer infestation on each rice cultivar was not similar, Basmati-370 was highly infested compared to Sharshar that had low infestation. This implies that all of the rice cultivars tested have variable resistant gene to yellow stem borer. Our findings further demonstrated that in comparison between germplasm types the aromatic varieties were more sensitive than non aromatic varieties. These considerable variations were found due to more attractiveness of borers by aromatic varieties for oviposition and this contributed to higher borers' damage. High stem borer damage to aromatic varieties was apparently due to elevated feeding activities of the larvae especially the fifth larvae which are most voracious. Stem borer resistance to non aromatic varieties can be due to antibiosis (reduced longevity and reproduction, increased mortality of insect), antixenosis (reduced plant attractiveness), and tolerance (ability to withstand and recover from damage).

Rice plant resistance to the stem borers may be attributed through two physical characteristics that caused direct mortality and other sublethal effects to the borer young larvae: - 1) Tight oppression of the leaf sheath around the stalk to prevent larval movement (susceptible cultivars might have leaf sheaths that loosen rapidly as the plant grows) and 2) premature hardness of the internodes to reduce penetration and feeding of larvae. Our findings are supported by Zhu et al. (2002) who investigated that resistant varieties caused mortality or inhibited the growth of stem borers, and resistance was highly correlated with smaller interval vascular bundles and larger width of the leaf sheath ridge. These findings are also in accordance with that of Padhi and Sen (2002) who found that in wild rice, non-preference is attributed to very narrow pith, and slender, hard and tough stems. Padhi (2004) recorded lower borer incidence, larval survival and sugar content, but higher amount of total phenols, orthodehydroxy phenol and silica on tolerant varieties indicating their resistance to yellow stem borer as compared to the susceptible check. Irrespective of varieties, the sugar content of the stem tissue was the highest at maximum tillering stage, thus, also had the higher borer incidence. However, the amount of phenolic compounds and silica increased significantly with the age of the plant, but the survival of the borer larvae decreased as crop age advanced. So, the degree of resistance in rice varieties to *S. incertulas* was influenced by the presence of high silica content (Chand and Murlirangan, 2000) and high phenolic compounds (Carbonari and Martins, 1998). Consequently, it is recommended that rice genotypes having high phenolic compounds and silica with lower sugar content could be utilized in the breeding programme for developing borers' resistant varieties.

Three important peaks of yellow stem borer were found from June to August, but the earlier planting had lower pest incidence than the later two peaks. Definitely, the rice crop at the early sowing was the off season planting, which was 15 days earlier from the farmers' crop growing season. At that time, most of rice fields were still fallow, so, in June, number of yellow stem borer damage was low indicating that pest was in larval and pupal stages, which then emerged to become adult moths in the medium and late plantings, therefore, stem borers concentrated more in the rice plants grown at the later two sowings. The higher damage at both later sowings was due to moth emergence from previous rice planting. Higher yellow stem borer invasions in July and August could be presumed due to wider spread of pest moths because larger rice crop area was available in these months which caused heavy infestation in the vegetative stage of rice crop than grown in June. This entails that later stages of rice plantings were more pronounced to exceedingly stem borer infestation than earlier or off season sowing. This correspond with the finding of Suharto and Usyati (2005) where the study implied that adjustment of planting time is the most feasible effort to reduce stem borer infestation because none of the rice cultivars tested were able to minimize damage under heavily infestation of yellow stem borer. Similar findings of Abraham et al. (1992) back up the present work stating that the activity of stem borer of first generation appeared after transplanting. Being a monophagous pest, Srivastava et al. (2003) reported that the population of stem borer increased in subsequent generation caused serious damage and expectedly is one of the main reasons for successive increases in pest population.

The grain yield of all rice cultivars grown at the early period was higher than that of the medium and late plantings. The rice crop of the later two plantings was heavily infested with stem borer during vegetative stage. Even though affected rice plants had produced new tillers from healthy hills, they grew slower. Thus, yield of a cultivar depends upon its yield potential, but biotic (insect pest) and abiotic threats (climate during growing season) could influence its yield potential. Severity of stem borer infestation varied with all varieties, incidence was observed the lowest on non aromatic (Sharshar and IR8) and highest on aromatic (Mehak and Basmati-370) varieties. Similar reports have been made by Khan et al. (2010) while determining the tolerance of rice varieties to manage borers that fine varieties (Basmat-aromatic) were highly susceptible to the yellow stem borer compared to the coarse varieties tested (non aromatic).

## Conclusion

Stem borers incidence at study site peaked during July, necessitating the initiation of control strategies since June every year. There is hope that plant varietals resistance can be used to control stem borers. The varietals resistance for stem borers may originates from two sources: - 1) natural variations present in cultivated rice and their non-cultivated relatives, and 2) resistance from other organisms or plants, e.g. bacteria or non-related plants that can be inserted into rice using genetic transformation, and current genetically engineered plant breeding technology can facilitate this trait transfer to make plants toxic to insect pests.

### REFERENCES


Abraham CC, Thomas B, Karunakaran K, Gopalkrishnan R (1992). Effect of planting season and the associated weather conditions on the incidence of the rice stem borer. Agric. Res. J., 19: 141-151.

Baloch SM, Abdullah K (2011). Effect of Planting Techniques on Incidence of Stem Borers (*Scirpophaga* spp.) in Transplanted and Direct Wet-Seeded Rice. Pakistan J. Zool., 43(1): 9-4.

Carbonari JJ, Martins JFDH (1998). Chemical and morphological factors of rice plants associated with insect resistance. Agropecuria clima temperado, 1: 107-115.

Catling HD, Islam Z, Pattrasudhi R (1984). Seasonal occurrence of the yellow stem borer *Scirpophaga incertulas* (Walker) on deepwater rice in Bangladesh and Thailand. Agriculture, Ecosystems and Environ., 12(1): 47-71.

Chand DS, Muralirangan MC (2000). Silica of rice cultivars, *Oryza sativa* (Linn.) versus feeding by *Oxya nitidula* (Wlk.). Uttar Pradesh J. Zool., 20: 29-35.

Hashmi AA (1994). Insect pests of paddy crop. *In* A. A. Hashmi (Ed.) Insect Pest Management: Cereal and Cash Crops. PARC, Islamabad, pp. 1-61.

Khan RA, Junaid AK, Jamil FF, Hamed M (2005). Resistance of different basmati rice varieties to stem borers under different control tactics of IPM and evaluation of yield. Pak. J. Bot., 37(2): 319-324.

Khan SM, Ghulam M, Hina M (2010). Screening of six rice varieties against Yellow stem borer, *Scirpophaga incertulus* Walker. Sarhad J. Agric., 26(4): 591-594.

Litsinger JA, Barrion AT, Soekarna D (1987). Upland rice insect pests:

their ecology, importance and control. IRRI Research Paper Series Number 123. The International Rice Research Institute, Philippines, p. 41.

Padhi G (2004). Biochemical basis of resistance in rice to yellow stem borer, *Scirpophaga incertulas* Wlk. Madras Agric. J., 91(4-6): 253-256.

Padhi G, Sen P (2002). Evaluation of wild rice species against yellow stem borer (*Scirpophaga incertulas* Walk.). J. App. Zool. Res., 13(2/3): 147-148.

Sarwar M (2011). Effects of Zinc fertilizer application on the incidence of rice stem borers (*Scirpophaga* species) (Lepidoptera: Pyralidae) in rice (*Oryza sativa* L.) crop. J.Cereals and Oilseeds, 2(5): 61-65.

Sarwar M, Ahmad N, Nasrullah, Tofique M (2010). Tolerance of different rice genotypes (*Oryza sativa* L.) against the infestation of rice stem borers under natural field conditions. The Nucleus, 47(3): 253-259.

Sarwar M, Akbar A, Ahmad N, Khan GZ, Bux M, Tofique M (2007). Field Performance of Systemic Foliar and Granular Insecticides against Rice Stem Borers (*Scirpophaga* spp) in Rice Crop. Proce. 26th Pakistan Conger. of Zoology, Multan, March 1-3, 27: 89-94.

Sarwar M, Ali A, Ahmad N, Tofique M (2005). Expediency of Different Botanical Products Intended for Managing the Population of Rice Stem Borers. Proce. 25th Pakistan Conger. of Zoology, March 1-3, Sindh Agric. Uni., Tandojam, 25: 15-23.

Schoenly KG, Justo HD, Barrion AT, Harris M, Bottrell DG (1998). Analysis of invertebrate biodiversity in a Philippine farmer's irrigated rice field. Environ. Entomol., 27(5): 1125-1136.

Sigsgaard L (2000). Early season natural biological control of insect pests in rice by spiders- and some factors in the management of the cropping system that may affect this control. Proc. 19th European Colloquium of Arachnology, Århus, 17-22 July 2000, pp. 57-64.

Srivastava SK, Salim M, Rehman A, Singh A, Garg DK, Prasad CS, Gyawali BK, Jaipal S, Kamal NQ (2003). Stem Borer of Rice-Wheat Cropping System: Status, Diagnosis, Biology and Management. Rice-Wheat Consortium Bulletin Series. Rice-Wheat Consortium for the Indo-Gangetic Plains, New Delhi, India, p. 273.

Suharto H, Usyati N (2005). The stem borer infestation on rice cultivars at three planting times. Indonesian J. Agric. Sci., 6(2): 39-45.

Zhu ZP, Borromeo AM, Cohen MB (2002). Comparison of stem borer damage and resistance in semi-dwarf indica rice varieties and prototype lines of a new plant type. Field Crop Res., 75(1): 37-45.

# GC-MS of volatile components of fermented wheat germ extract

**Oluwatosin K. Yusuf[1]* and Clement O. Bewaji[2]**

[1]Trypanosomosis Research Unit, Department of Biochemistry, Federal University of Technology, P. M. B. 65, Minna, Nigeria.
[2]Department of Biochemistry, University of Ilorin, Ilorin, Nigeria.

**Wheat has been used as food and folk medicine. We have earlier reported that the administration of ethyl acetate extract of ferrmented wheat at 300 mg/kg body weight have anti-trypanosomal activity. To better understand the active constituent for anti-trypanosomal properties, the crude extract was partially purified using column chromatography to give fractions A, B and C, which were further characterized by gas chromatographic – mass spectral (GC/MS) analysis. The fractions identified 12, 14 and 23 compounds respectively. The major volatile compounds were classified into oxygenated hydrocarbon (manool, hexadecyl acetate and butyl dodecanoate) and n-hydrocarbon (1- eicosene).**

**Key words:** Wheat seeds, volatile compound, GC-MS, Retention index (KI).

## INTRODUCTION

As source of remedies for the treatment of many diseases dated back to prehistory and people of all continents have this old tradition of using medicinal plants. About 25% of prescribed medicines in industrialized countries derived directly or indirectly from plants (Raskin et al., 2002). In developing countries where medicines are quite expensive, investigation on antimicrobial activities from ethanomedicinal plants may still be needed. It is obvious that these phytochemicals will find their way in the antimicrobial drugs prescribed by physicians (Cowan, 1999). Notably in West Africa, new drugs are not often affordable. Wheat (*Triticum* spp) is a world -wide cultivated grass from the area of Middle East. Globally after maize, wheat grain is a staple food used to make flour for leavened, flat and steamed breads, cookies, cakes, pasta and for fermentation to make beer (Palmer and John, 2001). Wheat kernel contains 2 to 4% germ and most nutrients with the exception of starch are concentrated in the germ, the germ is the richest known natural source of tocopherol, abundant in B-group vitamins and protein of high biological value (Tsen, 1985) and its oil of favourable fatty acid pattern (Paul et al., 1987), because of their beneficial nutritional values, wheat germ and wheat bran are frequently used in human food supplements.

Therefore, they are part of the regular western diet. Fermented wheat germ extract called avemer has been reported to control cell growth and proliferation mainly by inhibiting ribonucleotide reductase needed to make new DNA to support replication (Sukkar and Edoardo, 2004). It had also been reported that avemer limit the access to glucose, needed to make the ribose sugar for DNA and RNA for new cancer cells (Boros et al., 2002; Boros et al., 1997). We have earlier reported that the administration of ethylacetate extract of wheat seed at 300 mg/kg body weight to *Trypanosoma brucei*-infected rats was able to reduce parasitaemia and extended the life span of infected treated rats when compared with infected untreated rats (Yusuf and Ekanem, 2010), but the actual volatile constitute has not been fully studied. In this study, the volatile constituent of fermented wheat seeds extract was characterized using GC-MS analysis. GC-MS analysis has been proved to be of great utility in the analysis of compound.

## EXPERIMENTAL

### Collection of plant material

Wheat seeds were collected from Minna Central Market in the month of March/April 2008 and authentication was carried out at the

*Corresponding author. E-mail: toscue@yahoo.com.

**Table 1.** Compounds present in fraction A of fermented wheat extract.

| KI | Compound | % |
|---|---|---|
| 642 | Benzene | 5.64 |
| 1329 | 1,3- Benzodioxole-5- carbaldehyde | 1.27 |
| 1224 | 5- hydroxymethyl-2-furancarboxaldehyde | 1.09 |
| 1494 | Ethyl undecanoate | 3.64 |
| 1676 | Methyl-z(1R,2S)-3-oxo-2-(z)-pent-2-cyclopentyl-acetate | 8.55 |
| 1786 | Butyl dodecanoate | 1.27 |
| 1994 | 1-Eicosene | 13.77 |
| 2009 | Hexadecyl acetate | 12.19 |
| 2500 | Pentacosane | 4.91 |
| 2900 | Nonacosane | 9.10 |
| 2370 | 9-Tetracosene | 3.46 |
| 2082 | 1-Octadecan-1-ol | 3.64 |
| | **Classification of compounds** | |
| | n-hydrocarbon | 36.90 |
| | Oxygenated hydrocarbon | 19.50 |
| | Aromatic hydrocarbon | 8.50 |
| | Total | 64.90 |

The compounds were identified by the combination of both the mass spectra and retention indices on DB – 5 capillary coated column. Values (%) represent percentage composition, KI represent retention index.

Federal College of Forestry, Ibadan, Oyo state.

#### Plant preparation

Wheat germ powder of 70 g was fermented using 30 g of *Saccharomyces cerevisiae* (baker's yeast) for 48 h and the paste was extracted using 250 ml ethyl acetate. The filtrate was concentrated using rotary evaporator and stored at room temperature. The resulting extract was subjected to column chromatography using silical gel (60 to 120 mesh), eluting with a step gradient of n-hexane, n-hexane – ethyl acetate, ethyl acetate, ethyl acetate – methanol and methanol. Thin layer chromatography was performed with precoated silica gel GF- 25- UV 254 plates and detection was done by spraying with iodine to give three (3) fractions (A, B and C). The essential oils of the fractions were studied.

#### Gas chromatography-mass spectrometry analyses

Agilent 6890N GC was interfaced with a VG Analytical 70 - 250s double-focusing mass spectrometer. Helium was used as the carrier gas. The MS operating conditions were: Ionization voltage 70 eV, ion source 250°C. The GC was fitted with a 30 m x 0.32 mm fused capillary silica column coated with DB-5. The GC operating parameters were identical with those of the GC analysis. The percentage compositions of the oil were computed in each case from GC peak areas and are shown in Table 1. Retention indices for all the compounds were determined according to the Kovats method relative to the n-alkanes series. The identification of the compounds was done by comparison of retention indices and by matching their fragmentation patterns in mass spectra with those of published mass spectra data (Jennings and Shibamoto, 1980; Adams, 1995; Joulain et al., 1998; Koenig et al., 2004). In a few cases, identification of components was carried out by means of commercial libraries (Wiley, NIST05 and Hochmuth) (Itakura et al., 2001).

## RESULTS AND DISCUSSION

The results of Gas chromatography/ Mass spectroscopy (GC/MS) analysis of the various fractions of wheat extract are presented in Tables 1 to 3 representing fraction A, B and C respectively.

### Fraction A

The result of GC/MS analysis of the fractions A of wheat extract showed the presence of twelve (12) compounds corresponding to 64.90% of the total fraction (Table 1). The compound comprises of n- hydrocarbon (36.90%), oxygenated hydrocarbon (19.50%), aromatic hydrocarbon (8.50%). The major compound: Novacosane (9.10%), 1-eicosene (13.77%) and benzene (5.64%). The prominent among the oxygenated hydrocarbons are hexadecyl acetate (12.19%), ethyl undecanoate (3.64%) and octadecan-1-ol (3.64%). While that of the aromatic hydrocarbon is methyl-z-(1R,2S)-3-oxo-2-(z)-pent-2-cyclopentyl-acetate (8.50%) (Table 1).

### Fraction B

The GC/MS analysis of the fractions B of wheat extract

**Table 2.** Compounds present in fraction B of fermented wheat extract.

| KI | Compound | % |
|---|---|---|
| 1494 | Ethyl undecanoate | 4.24 |
| 1676 | Methyl-z(1R,2S)-3-oxo-2-(z)-pent-2-cyclopentyl-acetate | 5.03 |
| 1922 | Dibutyl phthalate | 5.32 |
| 1786 | Butyl dodecanoate | 10.15 |
| 1994 | 1-Eicosene | 5.62 |
| 2056 | Manool | 15.05 |
| 2200 | Octadec-9-enoic acid | 7.98 |
| 2456 | Tetracosan-1-ol | 5.41 |
| 2370 | 9-Tetracosene | 5.03 |
| 2195 | 1-Docosene | 3.15 |
| 2900 | Nonacosane | 7.19 |
| 2830 | 10- Demethyl squalene | 1.28 |
| 2852 | Hexacosan-1-ol | 1.28 |
| | 5- Octadene | 5.75 |
| | **Classification of compounds** | |
| | n-hydrocarbon | 28.10 |
| | Oxygenated hydrocarbon | 49.50 |
| | Aromatic hydrocarbon | 5.00 |
| | Total | 82.60 |

The compounds were identified by the combination of both the mass spectra and retention indices on DB – 5 capillary coated column. Values (%) represent percentage composition, KI represent retention index.

revealed the presence of fourteen (14) compounds corresponding to 82.60% of the total fraction (Table 2). The compound comprises of n-hydrocarbon (28.10%), oxygenated hydrocarbon (49.50%), aromatic hydrocarbon (5.00%). The major compound 9-novacosane (7.19%), 1-eicosene (5.62%) and 5-octadene (5.75%). The prominent among the oxygenated hydrocarbon are: Manool (15.05%), butyl dodecanoate (10.15%) and octadec -9-enoic acid (7.98%). While that of aromatic hydrocarbon is methyl-z-(1R, 2S)-3-oxo-2-(z)-pent-2-cyclopentyl-acetate (5.03%) (Table 2).

## Fraction C

The GC/MS analysis of the fractions C of wheat extract showed the presence of twenty three (23) compounds corresponding to 72.10% of the total fraction (Table 3). The compound comprises of n-hydrocarbon (35.70%), oxygenated hydrocarbon (29.60%), aromatic hydrocarbon (5.20%) and nitrogen containing hydrocarbon (1.60%). The major compound 1-eicosene (6.03%), pentacosane(5.535) and tetracosene (4.79%). The prominent among the oxygenated hydrocarbon are butyl dodecanoate (5.78%) octadecanoic acid (4.98%) and 5-tetradecencyl acetate (3.49%). While that of aromatic hydrocarbon is methyl-z(1R,2S)-3-oxo-2-(z)-pent-2-cyclopentyl-acetate (5.19%) and nitrogen containing hydrocarbon is 2- phenyl acetonitrile (1.60%) (Table 3).

It worth mentioning that, a major compound such as manool, which was detected in ethyl acetate wheat extract has been previously reported as part of the oil constituent of Pinus contorta bark and Tsuga heterophylla wood (Conner and Rowe, 1976); Salvia stenophylla (Jequier et al., 1980) and Picea ajanensis (Caeov et al., 1981). Manool is a labdane alcohol group of diterpenoids.

Diterpenes from many species are well known for their biological activity. Labdane derivatives have been reported to show significant antileishmanial activity (Kayser and Kiderlen, 1998) Also, compounds such as hexadecyl acetate, and butyl dodecanoate were previously detected in Frieseomelitta varia and Frieseomelitta silvestrii (Cruz et al., 2002) and old plum (Vele et al., 2005). These are ester and retroester derivatives.

Endogenous fatty acid amides and esters are the subject of growing interest in pharmacology, since the discovery of anandamide, the ethanolamide of arachidonic acid (Se´verine et al., 2003). Also, 1- eicosene can be classified as simple aliphatic. Simple aliphatic have been previously reported to show significant suppression of Typanosoma parasitaemia in vivo with daily (Perez et al., 1994). Therefore, it could be concluded that the volatile compound of ethyl acetate extract of fermented wheat can be useful in the management of African trypanosomiasis.

**Table 3.** Compounds present in fraction C of fermented wheat extract.

| KI | Compound | % |
|---|---|---|
| 642 | Benzene | 3.29 |
| 1065 | Phenyl ethanone | 5.00 |
| 1197 | 1,2-Benzenediol | 1.20 |
| 1140 | 2- Phenyl acetonitrile | 1.60 |
| 1224 | 5- hydroxymethyl-2-furancarboxaldehyde | 0.80 |
| 1274 | 1-(4-ethylphenyl)-ethanone | 1.40 |
| 1392 | 1-Tetradecene | 2.80 |
| 1494 | Ethyl undecanoate | 3.29 |
| 1676 | Methyl-z(1R,2S)-3-oxo-2-(z)-pent-2-cyclopentyl-acetate - | 5.19 |
| 1790 | (z)- 5- Tetradecenyl acetate | 3.49 |
| 1786 | Butyl dodecanoate | 5.78 |
| 1793 | 1- Octadene | 4.59 |
| 1994 | 1- Eicosene | 6.03 |
| 2200 | Octadecanoic acid | 4.89 |
| 2370 | 9-Tetracosene | 4.09 |
| 2400 | Tetracosene | 4.77 |
| 2900 | Nonacosane | 3.59 |
| 2862 | Hexacosan-1-ol | 0.47 |
| 2456 | Tetracosan-1-ol | 1.70 |
| 2600 | Hexacosane | 1.00 |
| 2500 | Pentacosane | 5.53 |
| | 4-hdroxy-2,5-dimethylfuran-3(H) one | 0.47 |
| | 3,4,5-Trimethoxybenzaldehyde | 0.98 |
| | **Classification of compounds** | |
| | n-hydrocarbon | 35.70 |
| | Oxygenated hydrocarbon | 29.60 |
| | Aromatic hydrocarbon | 5.20 |
| | Nitrogen containing hydrocarbon | 1.60 |
| | Total | 72.10 |

The compounds were identified by the combination of both the mass spectra and retention indices on DB – 5 capillary coated column. Values (%) represent percentage composition, KI represent retention index.


## ACKNOWLEDGEMENTS

Authors are grateful to Dr Oladosu of Chemistry department, University of Ibadan for his assistance in GC-MS analysis.


## REFERENCES


Adams RP (1995). Identification of essential oil components by gas chromatography-mass spectrometry. Allured Publishing Corporation, Carol Stream, Illinois, USA.

Boros LG, Cascante M, Paul Lee WN (2002). Metabolic profiling of cell growth and death in cancer: applications in drug discovery. Drug Discov Today, 7: 364-372.

Boros LG, Puigjaner J, Cascante M, Lee WN, Brandes JL, Bassilian S (1997). Oxythiamine and dehydroepiandrosterone inhibit the nonoxidative synthesis of ribose and tumor cell proliferation. Cancer Res., 57: 4242-8, 1997

Caeov NS, Chirkova MA, Titova TF, Raidugin VA, Pentegova VA (1981). Labdane and cembrane diterpenoids from the oleoresin of picea ajanesis Chemistry of natural compounds. 17( 2): 137-139.

Conner AH, Rowe JW (1976).Differentiating manool and 13-epimanool with NMR chiral shift reagents. Phytochem. 15(12) 1949-1951.

Cowan MM (1999). Plants products as anti-microbial agents. Clin. Microbiol. Rev., (12): 564-582.

Cruz Lo´pez, Patricio EFLRA, Maile R, Morgan ED (2002). Secretions of stingless bees: cephalic secretions of two Frieseomelitta species. J. Insect Physiol., 48: 453–458.

Itakura Y, Ichikawa M, Mori Y, Okino R, Udayama M, Morita T (2001). How to Distinguish Garlic from the Other Allium Vegetables J. Nutr., 131: 963S-967S.

Jennings W, Shibamoto T (1980). Qualitative analysis of flavour volatiles by gascapillary chromatography. Academic Press, New York.

Jequier C, Nicollier G, Tabacchi R, Garnero J (1980). Constituents of the essential oil of Salvia stenophylla. Phytochemistry, 19(3): 461-462

Joulain D, Koenig WA (1998). The atlas of spectral data of sesquiterpenehydrocarbons. E-B Verlag, Hamburg, Germany.

Koenig WA, Joulain D, Hochmuth D (2004). Terpenoids and Related Constituents of Essential Oils. Library of Massfinder 3. Dr. Detler

Hchmuth: Hamburg,Germany

Kayser O, Kiderlen AF (1998). In proceedings of the international congress of parasitology. Monduzzi Editore Bologna, Italy. pp. 925 - 929

Palmer J, John J (2001). How to brew. Defenestrative Publishing company, pp. 223.

Paul AA, Southgate DA, Russel J (1987). First supplement to Mccance and widowson's the component of foods. Her majesty's stationary office, London.

Perez HA, De La Rosa M, Apitz R (1994). Antimicrobial Agents. Chemother., 38: 338.

Raskin I, Ribnicky DM, Komarnytsky S, Pastor I, Fridlender B (2002). Plants and human health in the twenty – first century. Trends Biotechnol 20: 522-531

Se´verine V, Kazuhito T, Natsuo U, Kent-Olov J, Christopher JF, Didier ML (2003). Esters, Retroesters, and a Retroamide of Palmitic Acid: Pool for the First Selective Inhibitors of N-Palmitoylethanolamine-Selective Acid Amidase J. Med. Chem., 46: 4373-4376.

Sukkar SG, Edoardo R (2004). Oxidative stress and nutritutional prevention in autoimmune rheumatic diseases,Autoimmunity Rev., 3: 199 – 206.

Tsen CC (1985). Amino acid composition and biological value of cereal germs. Railed publishing co., Boston, pp. 453-466.

Vele T, Nino slav N, Anka J, Dejan D,Ljubodrag V, Ivan V and Mirjana B(2005).Volatile Components of Plum Brandies, Food Technol. Biotechnol., 43(4): 367–372.

Yusuf KO, Ekanem JT (2010). Studies of phytochemical constituents and antitrypanosomal properties of fermented wheat germ and garlic bulbs extract on Trypanosoma brucei – infected rats. J. Med. Plants Res., 4(19): 2016-2020.

# Estimation of total free fatty acid and cholesterol content in some commercial edible oils in Ethiopia, Bahir DAR

**Dimberu G. Atinafu* and Belete Bedemo**

[1]Department of Chemistry, College of Science, Bahir Dar University, Bahir Dar, Ethiopia.

**In the present investigation an attempt has been made to find out total free fatty acid and cholesterol content in some commercial edible oils in Ethiopia, Bahir Dar. Acid value, peroxide value, saponification value and cholesterol content were determined in a total of nine varieties of edible vegetable oils of which six are branded and three are non-branded. The analysis, performed using Liebermann-Burchard method, revealed varying levels of cholesterol content. Cholesterol was detected in seven of the vegetable oils while for two oil samples (Niger seed k-16 and Niger seed k-7) it was nil. Rapeseed branded vegetable oil has significantly highest cholesterol concentration (257.1 ± 0.42 mg/L) while branded palm oil has the least concentration (88.8 ± 0.85 mg/L) of cholesterol. Unlike the peroxide value, all the acid values and some saponification values show high values in comparison with the maximum permissibility level of codex standard for named vegetable oils (CODEX STAN210-1999). In conclusion, this study could be a guide line, to understand the quality of vegetable oils in Ethiopia, Bahir Dar. Consequently, these studies have been able to show that there is no cholesterol free oil in the market as opposed to what is claimed on the vegetable oil brand labels. Therefore, companies producing and marketing vegetable oils are enjoined to desist from misleading the public by labeling their products as "cholesterol free". They should indicate the amount of cholesterol present in the vegetable oil, no matter how small the quantity may be.**

**Key words:** Edible oils, free fatty acid, cholesterol, peroxide value, saponification value and liebermann-Burchard test.

## INTRODUCTION

Human beings from time immemorial have been using plants for a multitude of purposes. Oils have always been an integral part of human foods, being essential for health. Industrially, they play an important role in the development of different areas of chemical products, pharmaceutical, cosmetics, paints and most importantly, food (Atef, 2010).

Oils are naturally occurring esters of long straight-chain carboxylic acids. They belong to the saponifiable group (contain an ester groups) of lipids. Lipids are biologically produced materials that are relatively insoluble in water but soluble in polar and non-polar organic solvents. Edible oils are constituted of triacylglycerol molecules, mainly formed by unsaturated (oleic, linoleic, linolenic acids etc.) and saturated fatty acids (myristic, palmitic, stearic acids etc.) esterified to Glycerol units (Andersson et al., 2010). They can be formed from a single fatty acid that could be esterified up to three times into glycerol backbone, or at least by three different ones.

Almost every adult at present times develops some degrees of atherosclerosis, commonly known as "hardening of the arteries". Atherosclerosis leads to strokes, heart attacks and other serious health problems. High cholesterol, smoking and high blood pressure are the principal risk factors linked to heart disease.

The word "cholesterol" may quickly be associated with chronic heart disease and other heart problems. However,

*Corresponding author. E-mail: chemdimberu2009@yahoo.com.

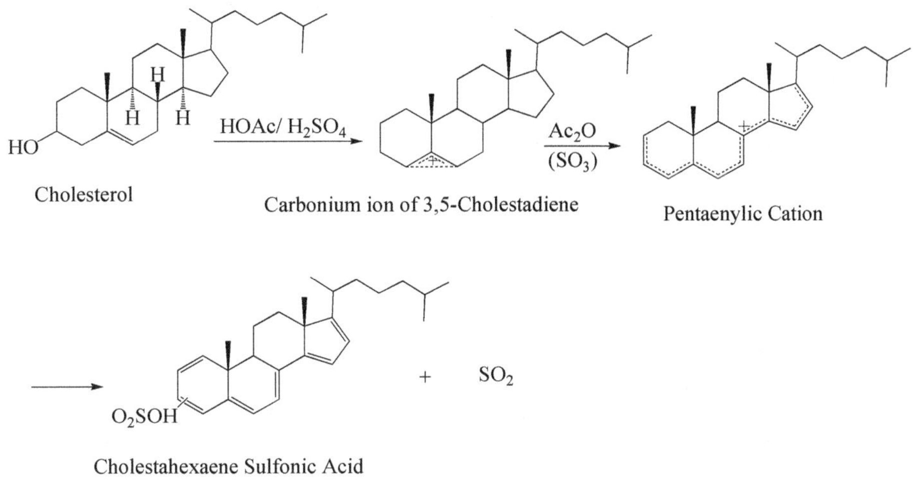


**Figure 1.** Mechanism for Liebermann-Burchard reaction.

cholesterol also has essential functions in the body such as providing essential components of membrane and serving as a precursor of bile acids, steroid hormones and vitamin D. Consuming cholesterol in our diet increases the level of low density lipoproteins (LDLs). There are so many different varieties of vegetable oil brands in our markets and all of them claim to be cholesterol free. Due to increasing awareness on the health implications of high cholesterol in the diets, most people now prefer to purchase cholesterol free vegetable oils (Attarde et al., 2010).

In general the chemistry of sterols includes a large amount of knowledge relating to the chemical properties, chemical synthesis, and analysis of sterols. The present investigations also include estimation of cholesterol in some commercial edible oil samples using UV-visible spectrophotometer by Liebermann-Burchard method (Attarde et al., 2010). Liebermann-Burchard test is used for colorimetric test to detect cholesterol, which gives a deep green color. The formation of a green or green-blue color after a few minutes mean is positive result. This color begins as a purplish, pink color and progresses through to a light green then very dark green color. The color is due to the hydroxyl group (-OH) of cholesterol reacting with the reagents and increasing the conjugation of the unsaturation in the adjacent fused ring (Figure 1). In this reaction the acetic acid in the Liebermann reagent reacts with cholesterol in the sample, which gives a green color whose absorbance, can be determined by UV-visible spectrophotometer at 640 nm (Burke et al., 1974).

## MATERIALS AND METHODS

In this study, the term branded vegetable oil (BVO) is used to indicate edible vegetable oils that are produced from registered industries in Ethiopia and imported oils with brand name labeled on the product purchased from the market, while unbranded vegetable oil (UVO) refer to the locally produced edible vegetable oils without a brand, they were obtained specifically from their raw materials processed in small scale in Bahir Dar city.

Sources of edible vegetable oils (EVOs) sold in market of Ethiopia includes both imported and derived from local oil seeds. Samples of six brands of vegetable oils produced from a variety of oil seeds such as: (oil palm, rapeseed oil, cottonseed oil, mixture of 80% cottonseed and 20% rapeseed Oil, olive oil and sunflower oil) were purchased from Bahir Dar supermarket and Modjo Edible Oil Complex, Addis Ababa, under various brand names. Samples of three non-branded vegetable oils are: two Niger seed oils from different sites and rapeseed oil were purchased from Bahir Dar in which they are produced in small scale.

Each edible oil sample was grouped into three sets: The first were produced from Ethiopian oil factory; Modjo Edible Oil Complex (cottonseed, rapeseed and 80% cottonseed + 20% rapeseed), the second were imported oils (palm oil, sunflower and olive oil) and the third were oils produced in small scale, non-branded (niger seed oils and rapeseed oils). These were purchased as packaged by plastic, highland, and stored under room temperature in the laboratory until it was required for analysis. Selected analytical methods adopted by American Oil Chemist's Society (AOCS) for determination of the composition and stability of fats and oils include peroxide value (PV), anisidine value (AV), acid value (AV), iodine value (IV), saponification value (SV) and also gas chromatography measurements, spectroscopic measurements such as UV, NMR and IR (Barbara et al., 2005; Yookyung et al., 2007; Otunola et al., 2009).

### Acid value

Each oil sample (1.0 g) was weighed and dissolved with 50 ml of ethanol in a conical flask. Two drops of phenolphthalein indicator were added and titrated to pink end point (which persisted for 15 minutes) with 0.1 N potassium hydroxide solution (KOH). Acid value was calculated (Equation 1) (Okpuzor et al., 2009):

$$\text{Acid value} = \frac{56.1 x V x C}{m} \quad (1)$$

Where 56.1 is equivalent weight of KOH, V is the volume in ml of standard volumetric KOH solution used, C is the exact

**Table 1.** Liebermann-Burchard method for cholesterol estimation.

| Reagents (ml) | $S_1$ | $S_2$ | $S_3$ | $S_4$ | $S_5$ | $S_6$ |
|---|---|---|---|---|---|---|
| Standard cholesterol solution | 0.4 | 0.6 | 0.8 | 1.0 | 1.2 | - |
| Liebermann-Burchard reagent | 2 | 2 | 2 | 2 | 2 | 2 |
| Chloroform | 7.6 | 7.4 | 7.2 | 7.0 | 6.8 | 8 |

**Table 2.** Absorbance of standard cholesterol solutions for calibration curve at different concentrations at 640 nm.

| Concentration (mg/L) | Absorbance |
|---|---|
| 80 | 0.074±0.001 |
| 120 | 0.257±0.000 |
| 160 | 0.445±0.003 |
| 200 | 0.654±0.011 |
| 240 | 0.922±0.000 |

concentration in KOH solution used (0.1 N); m is the mass in grams of the test portion (1 g).

### Saponification value

Saponification value was determined according to titremetric method discussed by Pearson (1981). Two grams of oil samples were weighed into a conical flask and 25 ml ethanolic potassium hydroxide was added. The solution was refluxed for 2 h with time to time shaking. One ml phenolphthalein was added and titrated with 0.5 N hydrochloric acid (HCl). The same process was conducted for blank determination. The value was calculated (Equation 2).

$$\text{Saponification value} = \frac{(V_o - V_1) x C x 56.1}{m} \quad (2)$$

Where 56.1 is equivalent weight of KOH, $V_0$ is the volume in ml of standard HCl solution used for the blank test, $V_1$ is the volume in ml of the standard HCl solution used for sample, C is the exact concentration of the standard HCl (0.5 N) solution and m is the mass in gram of the test portion (2 g).

### Peroxide value

Peroxide value was evaluated according to AOCS Official Method Cd 8-53 (2003). Five grams oil samples were weighed into a conical flask and 30 ml of solvent mixture of glacial acetic acid-chloroform in the ratio of 3:2, respectively, were added to the oil samples. Half ml saturated potassium iodide (KI) solution was added to the solution and allowed to stand for 1 min thereafter, 30 ml of distilled water were added and titrated with 0.01 N sodium thiosulfate solution using starch indicator until the yellow color was discharged. A blank was prepared alongside the oil samples. Peroxide value was calculated (Equation 3).

$$\text{Peroxide value} = \frac{10x(V_1 - V_2)}{m} \quad (3)$$

Where: $V_1$ volume of $Na_2S_2O_3$ for determination of test sample in ml, $V_2$ volume of $Na_2S_2O_3$ for determination of blank solution in ml and m is mass of test portion in g (5 g).

### Cholesterol estimation

Cholesterol content was estimated using Liebermann-Burchard reagent (Attarde et al., 2010). Standard cholesterol solution used was 2 mg/ml as stock solution. Liebermann-Burchard reagent was prepared with 7 ml concentrated sulfuric acid and 5 ml glacial acetic acid and was covered with black paper and kept in ice bucket in dark place.

### Preparation of sample solutions and standard cholesterol solutions

Six volumetric flasks were marked as $s_1$, $s_2$, $s_3$, $s_4$, $s_5$ and $s_6$. Standard cholesterol solution was added as 0.4, 0.6, 0.8, 1.0 and 1.2 ml in five volumetric flasks whereas, flask six was kept blank. Two ml of the Liebermann-Burchard reagent were added to all six volumetric flasks and diluted to final volume of 10 ml with chloroform (Table 1). Flasks were covered with black carbon paper and kept in dark for 15 min. Then, set zero of spectrophotometer with blank ($s_6$) at 640 nm. The absorbance of all standards (six flasks) were determined on SP65 UV/Vis spectrophotometer (Table 2) and standard graph was plotted (Figure 2).

Three ml of sample solutions were taken and their absorbances were determined on SP65 UV/Vis Spectrophotometer after adding 1 ml oil sample, 2 ml Liebermann-Burchard reagent and 7 ml chloroform. Cholesterol concentration of sample solutions was determined (Table 3) using a standard curve constructed graphically plotting the absorbance against mg/l cholesterol.

## RESULTS AND DISCUSSION

The research work was conducted to determine the extent of oxidative deterioration and cholesterol content

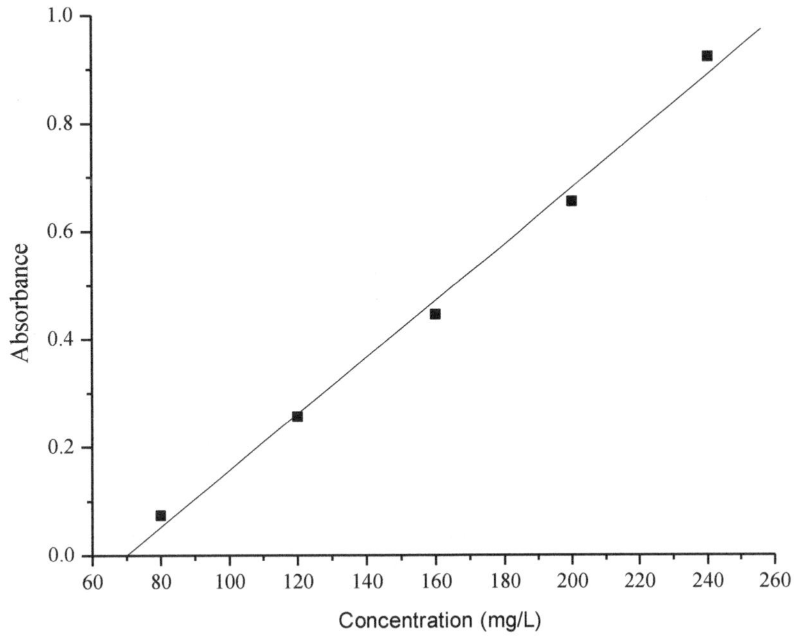


**Figure 2.** Calibration curve using cholesterol standard.

**Table 3.** Acid value, peroxide value, saponification value and cholesterol content for nine vegetable oil samples.

| Sample | Acid value (mg KOH/g) | Peroxide value (meq.peroxide/kg) | Saponification value (mg KOH/g) | Cholesterol (mg/L) |
|---|---|---|---|---|
| Cottonseed oil | $7.57 \pm 0.39^{bcd}$ | $0.60 \pm 0.85^{b}$ | $228.55 \pm 0.00^{b}$ | $131.90 \pm 0.14^{f}$ |
| 80% cottonseed + 20% rapeseed oil | $8.98 \pm 0.79^{bc}$ | $3.00 \pm 1.41^{ab}$ | $228.22 \pm 0.46^{b}$ | $190.60 \pm 0.28^{d}$ |
| Rapeseed | $7.01 \pm 1.19^{cd}$ | $2.10 \pm 0.14^{ab}$ | $32.54 \pm 0.16^{e}$ | $257.10 \pm 0.42^{a}$ |
| Rapeseed k-16 | $10.38 \pm 1.19^{ab}$ | $3.50 \pm 0.14^{ab}$ | $31.99 \pm 0.92^{e}$ | $139.20 \pm 0.28^{e}$ |
| Niger seed k-16 | $8.69 \pm 0.39^{bc}$ | $3.80 \pm 0.28^{a}$ | $162.76 \pm 0.69^{c}$ | ND |
| Niger seed k-7 | $12.06 \pm 0.40^{a}$ | $3.10 \pm 0.42^{ab}$ | $34.12 \pm 2.08^{e}$ | ND |
| Olive oil | $7.29 \pm 0.79^{cd}$ | $1.70 \pm 0.42^{ab}$ | $228.39 \pm 0.23^{b}$ | $234.80 \pm 0.00^{b}$ |
| Palm oil | $4.77 \pm 0.39^{d}$ | $2.20 \pm 0.00^{ab}$ | $260.22 \pm 1.38^{a}$ | $88.80 \pm 0.85^{g}$ |
| Sunflower oil | $5.05 \pm 0.79^{d}$ | $2.07 \pm 0.11^{ab}$ | $65.27 \pm 0.04^{d}$ | $204.60 \pm 1.13^{c}$ |

ND: not detected. Values with same letters in the same columns are not significantly different from each other according to Tukey HSD ($p < 0.05$).

from some edible oils under room temperature. Which attempts to make comparative quality assessment of industrial processed BVO against small scale processed UVOs found in Bahir Dar. This also provides supportive information for routine quality monitoring of edible vegetable oils that are used for food stuff.

The absorbance of each standard solution was measured three times at 640 nm following the procedure discussed earlier. The resulting absorbances are shown in Table 2. The equation of the calibration curve was obtained from the resulting absorbance versus concentration (mg/l) curve (Figure 2): $y = 0.005x - 0.366$: $R^2 = 0.9967$. The analytical data for acid value, peroxide value, saponification value and cholesterol content of nine oil samples are shown in Table 3.

## Acid value

Acid value is a measure of the free fatty acids in oil. Normally, fatty acids are found in the triglyceride form, however, during processing the fatty acids may get hydrolyzed into free fatty acid. The higher the acid value found, the higher the level of free fatty acids which translates into decreased oil quality. Acceptable levels for all oil samples should be below 0.6 mg KOH/g (measured in potassium hydroxide per gram) (AOCS Official Method Cd 8-53, 2003).

According to overall analysis of variance (ANOVA), acid value significantly varied among samples ($F = 18.7$, $P < 0.0001$, $R^2 = 0.94$). Niger seed k-7 and rapeseed k-16 have significantly higher acid value than the rest of oil samples (Table 3). Cottonseed, olive, rapeseed, sunflower, palm and olive oils had significantly lower acid values than the rest (Table 3). The minimum acid value was recorded for palm oil.

The acid values of UVOs were higher than BVOs, and the variations were statistically significant ($p < 0.05$). This hydrolysis is probably caused by a variety of agents presence of moisture in the oil, elevated temperature (above room temperature) and, most important of all, lipases (enzyme) coming from the source or contaminating microorganisms. This observation supports previous study that unrefined vegetable oils had higher acid value than refined oils (Rajko et al., 2010).

Among all oil samples the UVO, Niger seed oil k-7, has significantly ($p < 0.05$) highest (12.06 ± 0.40 mg KOH/g) acid value which indicate high free fatty acids and leads to a tendency to become rancid that is off-flavor (Tamzid et al., 2007). Probably, during processing the fatty acids may react with water and get hydrolyzed into the free fatty acids. In addition, niger seed oil k-7 is processed other than chemical refining methods (neutralization), which lower the free fatty acid content (Cmolik and Pokorny, 2000). Long storage of the oil seeds before or after processing may also have been responsible. Kalua et al. (2008) discussed that there were changes in oil quality during cold temperature storage of the fruit. This high content of free fatty acid levels has high probability for decreasing the ability of liver to store sugars (Gur and Harwood, 1991; Charlotte, 2004).

In contrast, palm oil has significantly ($p < 0.05$) lowest (4.77 ± 0.39 mg KOH/g) acid value. This is probably due to the higher content of saturated palmitic acid, which is less prone to oxidation than unsaturated fatty acids, linolenic acid, linoleic acid etc (Paul et al., 1992). The high degree of saturated fatty acids enables oils to be stable against oxidative rancidity and confer a longer shelf-life (Nwinuka and Barine, 2009). Interestingly, remaining seven oil samples have the acid values between these two oil samples.

All vegetable oil samples presented in this study have higher acid value than the Codex Standard for Named Vegetable Oils (CODEX-STAN210-1999) (0.6 mg KOH/g). Barring acid value, there is no significance difference ($p > 0.05$) estimated among branded local edible oils and imported edible oils. But in case of acid value branded local edible oils shows significantly high content, which implies that much free fatty acid may produced by factors.

## Peroxide value

According to the overall ANOVA, peroxide value significantly varied among samples ($F = 4.2$, $P = 0.0226$, $R^2 = 0.79$). PV is a measure of oxidation during storage and the freshness of lipid matrix. In addition, it is a useful indicator of the early stages of rancidity occurring under mild condition and it is a measure of the primary lipid oxidation products. One of the most important parameters that influence lipid oxidation is the degree of unsaturation of its fatty acids. When double bonds of unsaturated fats are oxidized, peroxides are among the oxidation products formed. High peroxide value is an indicator of oxidation level and the greater the peroxide value, the more oxidized the oil is (http://www.bioriginal.com).

Peroxide values of all the oil samples were in agreement with the maximum Codex standard peroxide value (10 meq $O_2$/Kg) for vegetable oil deterioration. Niger seed k-16 has significantly high (3.80 ± 0.28 meq $O_2$/kg) peroxide value and hence high degree of unsaturation. This observation helps to suggest that niger seed oil has high content of unsaturated fatty acids, linoleic (C18:2) and oleic acid (C18:1), which are responsible for oxidative rancidity (Mohamed and Jörg-Thomas, 2002). A significantly low value was observed in cottonseed oil (0.60 ± 0.85 meq.$O_2$/kg). More significantly, peroxide values of all edible oil samples are in agreement to the Codex Standard for named Vegetable Oils (CODEX-STAN210-1999) maximum permissibility level (10 meq $O_2$/kg). PV of all the UVOs was higher than that of the BVOs, and the variations were statistically significant ($p < 0.05$). 80% cottonseed + 20% rapeseed oil shows high (3.00 ± 1.41 meq $O_2$/kg) PV, and is significantly higher than its branded counter parts. The present studies indicate that the UVO have higher PVs, which is very likely in products that do not contain preservatives, either natural or synthetic. Acid values are also in agreement with this observation. The industrial processing of EVO includes preservatives and hence reduces the high PV over time, and present results are in line with the reported literature (Chabiri et al., 2009).

## Saponification value

Saponification value is an indication of the molecular weights of triglycerides in oil and significantly ($p < 0.05$) high proportion of saponification value in the palm oil (260.22 ± 1.38 mg KOH/g) suggests that the oil is a good raw material for soap industries. Higher Saponification value indicates high proportion of lower fatty acids since saponification value is inversely proportional to the average molecular weight or chain length of the fatty acids (Muhammad et al., 2011). Therefore, shorter the average chain length ($C_4$-$C_{12}$) the higher is the saponification number (Tamzid et al., 2007). The value obtained for palm oil during this study show that it contains high amounts of short chain fatty acids ($< C_{12}$), which is greater than the regulation of codex standard permissibility level (CODEX-STAN210-1999) (190-209 mg KOH/g).

Rapeseed k-16 has significantly ($p < 0.05$) lowest (31.99 ± 0.92 mg KOH/g) saponification value (except rapeseed and Niger seed k-7) that is also lower than codex standard permissibility level (168-181 mg KOH/g), which suggest that it contains high molecular weight long chain fatty acids hence are unsuitable for soap making and also unsuitable for human nutrition (Akinhanmi and Atasie, 2008).

Among nine oil samples studied, cottonseed oil (228.55 ± 0.00 mg KOH/g), olive oil (228.39 ± 0.23 mg KOH/g), 80% cottonseed + 20% rapeseed oil and palm oil (260.22 ± 1.38 mg KOH/g) have high saponification value than the regulation of codex standard. Typically, these oils have low molecular weight fatty acids. On the other hand, rapeseed oils (branded (32.54 ± 0.16 mg KOH/g) and unbranded (31.99 ± 0.92 mg KOH/g)), sunflower oil (65.27 ± 0.04 mg KOH/g) and niger seed oils both k-7 (34.12 ± 2.08 mg KOH/g) and k-16 (162.76 ± 0.69 mg KOH/g) have low saponification values compared to the guide line values. It indicates that they have high content of long chain fatty acids.

### Cholesterol content

Representative analytical data for cholesterol content of said oil samples are presented in Table 3. Out of these, seven oil samples contained cholesterol include cottonseed oil, 80% cottonseed + 20% rapeseed oil, rapeseed both branded and non-branded oils, olive oil, palm oil and sunflower oil. Among these rapeseed oil has significantly ($p < 0.05$) maximum (257.10 ± 0.42 mg/L) cholesterol content and palm oil has significantly ($p < 0.05$) low (88.8 ± 0.85 mg/L) cholesterol content. Surprisingly, the two Niger seed oils (niger seed k-16 and niger seed k-7), contained nil cholesterol content. However, cholesterol content of all the BVOs was higher than that of the UVOs, and the variations were statistically significant ($p < 0.05$). The peroxide value of these samples supports the observation, that is, Niger seed may have high content of the essential fatty acid; linoleic (C18:2) acid, which has the ability to decrease cholesterol levels, stimulate cholesterol excretion into the intestine and inhibit biosynthesis of cholesterol in the liver. Oils containing high level of polyunsaturated fatty acid are found to inhibit the activity of hydroxymethylglutaryl-coenzymeA-reductase (HMG-CoA-reductase) which is the regulatory enzyme in cholesterol biosynthesis (Carl et al., 2009; Seddigheh et al., 2009; Ejikeme et al., 2010).

Finding from this study supports previous work that cholesterol is present in vegetable oils, although in small proportion (Okpuzor et al., 2009; Syed et al., 2003).

### REFERENCES


Akinhanmi TF, VN Atasie (2008). Chemical Composition and Physicochemical Properties of Cashew nut (*Anacardium occidentale*) Oil and Cashew nut Shell Liquid. J. Agric. Food Environ., Sci. 2: 1-10.

Andersson B, WPD Caroline, RC Francinete, S Fabio, A Cesar, GF Antonio (2010). A simple methodology for the determination of fatty acid composition in edible oils through $^{1}$H NMR spectroscopy. Magn. Reson.Chem., 48: 642-650.

AOCS (2003). Official Method Cd 8-53. American Oil Chemists Society, Champaign, IL. Accessed on www.bioriginal.com.

Atef MA (2010). Physiological effects of some plant oils supplementation on streptozotocin-induced diabetic rats. Res. J. Med. Sci., 5: 55-71.

Attarde D, J Pawar, B Chaudhari, S Pal (2010). Estimation of sterols content in edible oil and ghee samples. Int. J. Pharm. Sci. Rev. Res., 5: 135 -137.

Barbara M, L Bernhard, M Antonio, JA Maria (2005). Direct monitoring of lipid oxidation in edible oils by Fourier transform Raman spectroscopy. Chem. Phys. Lipids, 134: 173-182.

Burke RW, BI Diamondstone, RA Velapoldi, O Menis (1974). Mechanisms of the Liebermann-Burchard and Zak Color Reactions for Cholesterol. Clin. Chem., 20: 794-801.

Carl JL, VM Richard, RM Mandeep, OV Hector (2009). Omega-3 Polyunsaturated Fatty Acids and Cardiovascular Diseases. J. Am. Coll. Cardiol., 54: 585-594.

Chabiri SA, SS Hati, GA Dimari, VO Ogugbuaja (2009). Comparative Quality Assessment of Branded and Unbranded Edible Vegetable Oils in Nigeria. Pacific J. Sci. Technol., 10: 927-934.

Charlotte G (2004). Medicine Net. Com.

Cmolik J, J Pokorny (2000). Physical refining of edible oils. Eur. J. Lipid Sci. Technol., 102: 472-486.

Ejikeme PM, LN Obasi, AC Egbuonu (2010). Physico-chemical and toxicological studies on *Afzelia africana* seed and oil. Afr. J. Biotechnol., 9: 1959-1963.

Gur MI, JL Harwood 1991. Lipid Biochemistry: An introduction. London: Chapman and Hall, p. 406. http://www.bioriginal.com. Accessed on 16/2/2011.

Kalua CM, DR Bedwood, AG Bishop, PD Pienzler (2008). Changes in virgin oil quality during low temperature fruit storage. J. Agric. Food Chem., 56: 2415-2422.

Mohamed FR, M Jörg-Thomas, (2002). Proximate Neutral Lipid Composition of Niger (*Guizotia abyssinica Cass.*) Seed. Czech. J. Food Sci. 20: 98–104.

Muhammad N, E Bamishaiye, O Bamishaiye, L Usman, M Salawu, M Nafiu, O Oloyede, (2011). Physicochemical Properties and Fatty Acid Composition of *Cyperus esculentus* (Tiger Nut) Tuber Oil. Biores. Bull., 5: 51-54.

Nwinuka NM, IN Barine, (2009). Physico-chemical Properties and Fatty Acid Composition of *Dennettia tripetala* Fruit Oil (Pepper Fruit). Niger. J. Biochem. Mol. Biol., 24: 42 - 46.

Okpuzor J, VI Okochi, HA Ogbunugafor, S Ogbonnia, T Fagbayi, C Obidiegwu, (2009). Estimation of cholesterol level in different brands of vegetable oils. Pak. J. Nutr. 8: 57-62.

Otunola GA, GB Adebayo, OG Olufemi (2009). Evaluation of some physicochemical parameters of selected brands of vegetable oils sold in Ilorin metropolis. Int. J. Phys. Sci. 4: 327-329.

Paul JN, N Manny, BG Baryan, M Rosemary, MC Peter, A Mavis (1992). Plasma lipoprotein lipid and lipid changes with substitution of elaidic acid for oleic acid in the diet. Am. J. Clin. Nutr., 55: 46-50.

Pearson D (1981). The Chemical Analysis of Food. Churchill Publishing, London, pp. 580-581.

Rajko V, V Sergeja, A Helena (2010). Biochemical parameters and oxidative resistance to thermal treatment of refined and unrefined vegetable edible oils. Czech J. Food Sci., 28: 376–384.

Seddigheh A, N Bahar, S Nizal, P Sahar, S Salbali, E Ahmad, A Leila, (2009). Evaluation of fatty acid content of some Iranian fast foods with emphasis on *trans* fatty acids. Asia Pac. J. Clin. Nutr., 18: 187-192.

Syed MS, H Imran, DAG Syed, (2003). Estimation of sterols in edible fats and oils. Pak. J. Nutr., 2: 178-181.

Tamzid HM, MT Alam, MAU Islam (2007). Physico-chemical and nutritional studies of *Terminalia belerica* roxb. Seed oil and seed kernel. J. bio-sci., 15: 117-126.

Yookyung K, SH David, EK Sandra (2007). ATR-Fourier transform mid-infrared spectroscopy for determination of *trans* fatty acids in ground cereal products without oil extraction. J. Agric. Food Chem., 55: 4327-4333.

# Response to water stress in castor (*Ricinus communis* L.) genotypes under *in vitro* conditions

**T. Radhamani[1]*, R. Ushakumari[1], R. Amudha[1] and K. Anjani[2]**

[1]Agricultural College and Research Institute, Madurai-625104, India.
[2]Directorate of Oilseeds Research, Hyderabad, India.

**A laboratory experiment was conducted to study the response of 45 castor genotypes to drought stress. Five water stress levels were used. The experiment was laid out by complete randomized design with two replications. Ten seeds of each accession were sown in sterilized sand medium and the Poly Ethylene Glycol solution was given to medium. On the tenth day, the observations were recorded. Seedling length, root length and germination percentage were observed. The highest tolerance for germination was shown by RG 2474. The values of shoot length were higher in RG 2474. The maximum root length was seen in RG 2326 and RG 3013.**

**Key words:** Castor, genotypes, water stress, *in vitro* conditions.

## INTRODUCTION

Castor is an important non edible oil seed crop grown throughout the world. Due to erratic rainfall distribution, generally crop experiences prolonged dry spells during the growth stages. Growing of drought tolerant cultivars will contribute more to stable castor production and the screening of the response of castor cultivars or breeding lines to drought stress can play a crucial role in breeding programmes. Adequate water and nutrient supply are important factors affecting optimal plant growth and successful crop production. Seed germination is first critical and the most sensitive stage in the life cycle of plants (Ashraf and Mehmood, 1990), and the seeds exposed to unfavorable environmental conditions like water stress may have to compromise the seedlings establishment (Albuquerque and Carvalho, 2003).This present study was therefore conducted with the objective to determining the response of castor genotypes to drought stress at germination and seedling stages under controlled conditions and to evaluate germination and screening criteria for drought tolerance in castor.

### MATERIALS AND METHODS

Experiment was carried out at Crop Physiology laboratory, Department of Seed Science and Technology, Agricultural College and Research Institute, Madurai. Forty five castor genotypes collected from the germplasm maintained at Directorate of Oilseeds Research, Hyderabad were tested against drought stress at germination and seedling stages under laboratory conditions. Poly Ethylene Glycol can be used to modify the osmotic potential of nutrient solution culture and thus induce plant water deficit in a relatively controlled manner, appropriate to experimental protocols. Poly Ethylene Glycol with a molecular weight of 6000 (PEG 6000) was used as a drought stimulator and five stress levels of zero (control), -0.2, -0.3, -0.4, and -0.5 MPa were developed by dissolving 11.9, 15.1, 17.8 and 20.2 g of PEG per 100 ml distilled water (Michel, 1983). Seeds were surface sterilized with 10% sodium hypochlorite solution for five minutes and then washed three times with distilled water. Ten seeds of each castor germplasm accessions were sown in germination tray containing sterilized sand medium. The experiment was laid out in a completely randomized design with two replicates for each experimental unit. On the tenth day the observations were recorded. Seedling length, root length and germination percentage were used to evaluate the genotypic response to PEG induced water stress.

## RESULTS AND DISCUSSION

Stress tolerance ability of crop plants vary with growth stages. Seed germination and seedling stages are very vulnerable. Adequate seed germination in moisture stress field condition leads to a good crop stand. There was no consistency between germination rate in the presence of osmoticum and field emergence and subsequently yield.

*Corresponding author. E-mail: radha.agri@gmail.com.

**Table 1.** Seedling germination percentage, shoot length and root length of 45 castor genotypes under water stress (-5.0 bars) condiction.

| S/N | Genotypes | Germination percentage (%) | | Shoot length (cm) | | Root length (cm) | |
|---|---|---|---|---|---|---|---|
| | | Control | Stress | Control | Stress | Control | Stress |
| 1 | RG152 | 96.32 | 44.61 | 8.52 | 4.55 | 4.12 | 2.17 |
| 2 | RG157 | 88.21 | 43.67 | 8.13 | 4.55 | 4.30 | 1.61 |
| 3 | RG551 | 93.00 | 46.29 | 11.96 | 4.75 | 4.10 | 3.42 |
| 4 | RG814 | 99.00 | 43.30 | 12.70 | 4.55 | 3.90 | 4.13 |
| 5 | RG969 | 98.20 | 54.15 | 21.00 | 9.42 | 4.23 | 4.86 |
| 6 | RG3233 | 94.72 | 62.26* | 19.20 | 7.81 | 5.30 | 3.89 |
| 7 | RG3224 | 96.37 | 60.56* | 18.86 | 6.25 | 5.10 | 2.71 |
| 8 | RG3195 | 95.50 | 59.43* | 19.60 | 11.80* | 5.21 | 2.47 |
| 9 | RG3120 | 89.37 | 54.16 | 15.40 | 8.78 | 4.30 | 2.71 |
| 10 | RG3116 | 92.36 | 50.46 | 18.50 | 10.65 | 4.72 | 2.15 |
| 11 | RG3102 | 94.00 | 62.00* | 16.20 | 7.80 | 4.98 | 3.05 |
| 12 | RG3093 | 99.00 | 56.50* | 11.74 | 5.60 | 5.10 | 4.48 |
| 13 | RG3088 | 89.00 | 61.50* | 20.50 | 9.95 | 6.21 | 4.03 |
| 14 | RG3063 | 96.34 | 68.20* | 19.70 | 15.15* | 5.72 | 4.96* |
| 15 | RG3013 | 96.51 | 61.50* | 21.00 | 14.96* | 6.20 | 6.25* |
| 16 | RG2980 | 94.22 | 59.48* | 20.23 | 15.25* | 6.50 | 5.50* |
| 17 | RG2958 | 93.72 | 62.12* | 11.90 | 6.10 | 6.42 | 4.97 |
| 18 | RG2944 | 96.58 | 58.24* | 20.00 | 10.85* | 6.32 | 6.10* |
| 19 | RG2902 | 96.00 | 59.18* | 19.50 | 11.19* | 4.32 | 3.02 |
| 10 | RG2582 | 97.60 | 61.56* | 18.00 | 8.83 | 6.12 | 4.58 |
| 21 | RG2498 | 89.00 | 57.84* | 17.25 | 9.81 | 5.72 | 4.42 |
| 22 | RG2487 | 87.00 | 56.18* | 19.32 | 11.15* | 5.69 | 4.71 |
| 23 | RG2481 | 94.43 | 56.25* | 21.00 | 14.55* | 6.40 | 5.51* |
| 24 | RG2474 | 98.81 | 68.83* | 21.55 | 16.40* | 6.42 | 5.79* |
| 25 | RG2473 | 93.36 | 58.42* | 21.53 | 13.75* | 5.90 | 4.96* |
| 26 | RG2465 | 90.84 | 50.00 | 13.45 | 12.90* | 5.42 | 3.50 |
| 27 | RG2457 | 96.81 | 52.00 | 16.70 | 7.45 | 5.70 | 3.85 |
| 28 | RG2454 | 96.38 | 61.10* | 14.20 | 8.85 | 5.12 | 4.31 |
| 29 | RG2451 | 93.65 | 58.30* | 13.10 | 10.82* | 4.46 | 3.45 |
| 30 | RG2377 | 99.00 | 50.66 | 14.00 | 10.66 | 5.72 | 4.15 |
| 31 | RG2375 | 96.68 | 53.00 | 18.90 | 11.60* | 4.31 | 3.17 |
| 32 | RG2368 | 89.97 | 56.50* | 16.00 | 15.80* | 5.49 | 5.60* |
| 33 | RG2326 | 94.58 | 51.63 | 14.10 | 10.17 | 5.19 | 6.25* |
| 34 | RG2320 | 94.46 | 43.14 | 18.20 | 12.22* | 4.80 | 5.13* |
| 35 | RG2266 | 93.68 | 45.18 | 10.10 | 9.60 | 4.20 | 2.06 |
| 36 | RG2269 | 92.55 | 48.87 | 17.60 | 7.71 | 4.81 | 2.29 |
| 37 | RG2195 | 96.00 | 49.63 | 18.20 | 5.78 | 7.42 | 4.76 |
| 38 | RG2184 | 94.33 | 43.62 | 18.30 | 6.86 | 4.31 | 2.09 |
| 39 | RG2035 | 93.28 | 47.27 | 12.25 | 8.56 | 3.73 | 1.71 |
| 40 | RG2033 | 89.00 | 52.09 | 8.76 | 4.53 | 4.16 | 3.87 |
| 41 | RG2024 | 93.64 | 56.50* | 10.70 | 4.80 | 5.12 | 4.77* |
| 42 | RG2022 | 93.22 | 44.12 | 16.10 | 5.40 | 6.22 | 5.38* |
| 43 | RG2014 | 96.12 | 50.00 | 15.40 | 5.59 | 5.17 | 4.35 |
| 44 | RG1999 | 96.29 | 58.00* | 17.70 | 4.85 | 4.13 | 3.35 |
| 45 | TMV5 | 97.18 | 55.27* | 18.80 | 4.77 | 4.25 | 3.35 |
| Grand mean | | | 54.17 | Grand mean | 9.18 | Grand mean | 3.99 |
| SE(d) | | | 0.44 | SE(d) | 0.78 | SE(d) | 0.47 |
| CD(5%) | | | 0.89 | CD(5%) | 1.58 | CD(5%) | 0.95 |

*Significant at 5% level.

Laboratory germination studies, in general, overestimate the field germination and emergence. There were no significant differences in -2.0, -3.0 and -4.0 bar (Table 1). At -6.0bar, there was no germination. It was worth noticing that increase concentration of PEG indicates precise nature of the *in vitro* screening (Kulkarni and Deshpande, 2007).

The -5.0 bar was fixed as optimum concentration to find out the performance of the genotypes. The water stress level significantly decreased all the above characters. Total of 24 genotypes were significantly higher than the grand mean for germination percentage. Similarly, 22 genotypes for seedling shoot length and 22 genotypes for root length significantly had higher values than the grand mean. The highest tolerance for germination was shown by RG2474 (68.83%). The mean seedling shoot length was maximum in RG2474 (16.40 cm) and the mean seedling root length was high in the genotype RG 3013 and RG 2326 (6.25 cm). Eleven genotypes *viz.*, RG 3088, RG3063, RG 3013, RG 2980, RG 2944, RG 2498, RG 2487, RG 2481, RG 2474, RG 2473 and RG 2368 exhibited superior mean for all the above three characters (Table 1). Shamim et al. (2009) used germination stress tolerance index (GSI), and root length stress index (RLSI) to evaluate the genotypic response to PEG-induced water stress. Hence these eleven genotypes were found to be drought tolerant under laboratory testing using PEG.

## Conclusion

Genotypes showed significant variation in all the characters studied under laboratory condition. The highest tolerance for germination was shown by RG 2474 (68.83%) followed by RG 3063 (68.20%), RG 3233 (62.26%). The values of shoot length were higher in the genotype RG 2474 (16.40 cm) followed by RG 2368 (15.80 cm), RG 2980 (15.25 cm). The maximum root length was occurred in RG 2326 and RG 3013 followed by RG 2944 (6.10 cm), RG 2474 (5.79 cm) and it was minimum in RG 157 (1.61 cm) followed by RG 2035 (1.71 cm), RG 2266 (2.06 cm). Hence, these genotypes can be performed well under drought condition.

## REFERENCES

Albuquerque FMC, de Carvalho NM (2003). Effect of type of environmental stress on the emergence of sunflower (*Helianthus annuus* L.), soybean (*Glycine max* (L.) Merril) and maize (*Zea mays* L.) seeds with different levels of vigor. Seed Sci. Technol. 31:465-467.

Ashraf M, Mehmood S (1990). Response of four *Brassica* species to drought stress. Environ. Exp. Bot. 30:93-100.

Kulkarni M, Deshpande U (2007). *In Vitro* screening of tomato genotypes for drought resistance using polyethylene glycol. Afr. J. Biotechnol. 6(6):691-696.

Michel BE (1983). Evaluation of the water potentials of solutions of polyethylene glycol 8000 both in the absence and presence of other solutes. Plant Physiol. 72:66-70.

Shamim A, Rashid A, Muhammad Y-A, Ashraf. M, Ejaz Ahmad W (2009). Sunflower (*Helianthus annuus* L.) response to drought stress at germination and seedling growth stages. Pak. J. Bot. 41(2):647-654.

# Influence of germination and fermentation on chemical composition, protein quality and physical properties of wheat flour (*Triticum aestivum*)

**Ijarotimi Oluwole Steve**

Department of Food Science and Technology (Human Nutrition Division), Federal University of Technology, Akure, Ondo State, Nigeria. E-mail: soijarotimi@gmail.com or soijarotimi@futa.edu.ng.

**The present study investigated effects of germination and fermentation on protein quality of the wheat flour. The wheat seeds were obtained locally and were divided into three portions and processed as raw wheat flour (RWF), germinated wheat flour (GWF) and fermented wheat flour (FWF) respectively. The samples were analysed for chemical and protein qualities using standard methods. Protein content (g/100g) varied between 10.77±0.66 (RWF) and 13.70±0.30 (FWF). Mineral composition of RWF, GWF and FWF had potassium as the highest while zinc (RWF and FWF) and nickel (GWF) were the least. For amino acid, glutamic acid was the most abundant; while cysteine (RWF and FWF) and methionine (GWF) were the least. Total essential amino acid plus histidine and arginine range between 26.4% (FWF) and 37.9% (GWF). For protein efficiency ratio (PER), RWF had the highest value (1.99), while GWF had the least (0.86), while for the essential amino acid index (EAAIs), RWF was higher than that of GWF and FWF. Similarly, the biological value of RWF (31.8%) was higher than those of GWF (29.4%) and FWF (29.1%) sample respectively. The anti-nutrient compositions of the wheat flour samples were low, while phytate:zinc, phytate:calcium, (Ca)(phytate):zinc and phytate:iron molar ratios were lower than the critical values. Bulk density (BD) ranged between 0.80±0.04 and 0.86±0.02, swelling capacity (SC) 0.0±0.03 and 1.23±0.21, water absorption capacity (WAC) 315 and 415% and least gellation 3.33±1.10 and 14.6±1.16%. The finding concluded that the employed processing methods, particularly fermentation, influenced the chemical composition of the wheat flour in terms of protein content and reduction of anti-nutrient composition.**

**Key words:** Germinated wheat flour, fermented wheat flour, nutritional quality.

## INTRODUCTION

Cereals such as rice, wheat and maize are members of the grass family *Graminae* and are particularly important because of their role as staple food for human nutrition, and their incorporation into various products is of great economic importance in many countries of the world (Pena et al., 2006). Cereals, particularly wheat, are important components of the daily diet, providing carbohydrates, proteins, dietary fibers and vitamins to less privilege in many countries.

The main wheat producing countries are United State of America, China, Russia, India, Pakistan, Canada, Argentina, Australia and some countries of European Union (FAO, 2009). Wheat is highly consumed in various forms like breads, biscuits, cookies, cakes, pasta, noodles (Hussain et al., 2004; Pena et al., 2006); and it is the major sources of dietary energy and protein for the survival of the people (Aslam et al., 1982; Hussain et al., 2004; Moore et al., 2006). A number of studies have indicated protective role of whole grain foods against several nutritional related diseases such as type 2 diabetes (Murtaugh et al., 2003; Pereira et al., 2002), cardiovascular diseases (Jacobs and Gallagher, 2004) and certain cancers (Larsson et al., 2005).

Scientific studies have reported that wheat contain appreciable amount of nutrient and anti-nutritional factors; and that majority of these anti-nutrients are concentrated in the aleurone layer and only 10% in the embryo (Cheryan, 1980; Muahamad et al., 2010). Evidence has shown that there are many factors, such as genetics, environmental fluctuations, type of soils, year and fertilizer application that can affect the anti-nutrient composition of cereal grains (Muahamad et al., 2010). These anti-nutritional factors reduce the nutritional quality of cereals and their removals through traditional processing techniques enhance utilization of the cereal as human food.

Processing techniques such as germination and fermentation have been found to improve the quality of cereals due to chemical changes that enhance organoleptic response (Nout, 1992), contents of free sugars, protein and vitamins, as well as bioavailability of minerals (Zamora and Fields, 1979; Chavan and Kadam, 1989a, b; Egli, 2001; Helland et al., 2002; Ochanda, et al., 2010), and results in the breakdown of some of the anti-nutritional endogenous compounds (Ahmed et al., 2006). In many instances, usage of only one method may not impart the desired removal of anti-nutritional compounds and a combination of two or more methods is required (Hassan et al., 2007).

Germination has profound effect on nutritional quality of the cereal (Chavan and Kadam, 1989b). Germination is a natural biological process of plants by which the seeds come out of latency stage (Sangronis and Machado, 2007). During germination, certain changes occur in terms of quantity and type of nutrients within the seed. These changes can vary depending on the type of vegetable, the variety of seed and the condition of germination (Bau et al., 1997; Dhaliwal and Aggarwal, 1999). An increase in bioavailability of minerals and weight has been observed during seed germination. Germinated seeds are good source of ascorbic acid, riboflavin, choline, thiamine, tocopheroles and pantothenic acid (Sangronis and Machado, 2007).

Fermentation is a metabolic process serving for some microorganisms to get energy through digestion of simple fermentable sugars, mostly glucose and fructose. It serves as a means of providing a major source of nourishment for large rural populations and contributes significantly to food security by increasing the range of raw material which can be used in the production of edible products. Fermentation enhances the nutrient of foods through biosynthesis and bioavailability of vitamins (Zamora and Fields, 1979; Gabriel and Akharaiyi, 2007), essential amino acids and improving the protein quality and fibre digestibility (Gabriel and Akharaiyi, 2007).

In view of improving food security and nutritional wellbeing of the people that relies on wheat as their staple food; thus, this present study aimed at improving the nutritional quality of wheat through germination and fermentation processing techniques. A number of studies have reported on nutritional benefits of germination and fermentation technique in terms of increasing the mineral and protein content and reducing the anti-nutrient composition of food products (Sripriya et al., 1997; Abdel Rehman et al., 2005; Hassan et al., 2007; Kouakou et al., 2008). The specific objectives of this study were to investigate protein quality, antinutritional factors and physical properties of wheat flour (*T. aestivum*). The outcomes of this study would be important for food and nutrition policy maker in terms of preventing protein energy malnutrition in Nigeria and other developing countries.

## MATERIALS AND METHOD

### Materials procurement

The wheat grains used in this study were procured in January, 2012 from Erekesan market, Akure, Ondo State, Nigeria. The chemicals used for the analyses were of analytical grade and they were purchased from Sigma Chemical Company, St. Louis, MO, USA. The chemical analyses of the wheat flour samples were carried out in the Food Science and technology laboratory, Federal University of Technology, Akure, Nigeria. The samples were divided into three batches of 1 kg each. These constituted samples for germination and fermentation while one batch that was untreated constituted the control.

### Processing of wheat flour

#### *Raw wheat flour*

One kilogram of wheat grains was weighed sorted and oven dried (Galenkamp, size 3, hotbox, London, UK) at 60°C for 20 h. The dried wheat grain was milled in attrition mill, sieved through 150 wire mesh and packed in polyethylene nylon, put in covered container and stored at room temperature prior to analysis (Figure 1)

#### *Germinated wheat flour*

One kilogram of wheat grains were weighed, sorted, steeped in distilled water for 3 h at room temperature and then completely drained of steep water using sieves. The drained wheat grain were then spread on a moistened jute sack and allowed to germinate at room temperature (4 days).The germinated grains were manually washed with distilled water, drained and oven dried at 60°C for 7 to 8 h. The dried grains were milled using attrition mill, sieved and packaged in polyethylene bags, put in an airtight container and stored under room temperature prior to analysis (Figure 1).

#### *Fermented wheat flour*

One kilogram of wheat grains were weighed, sorted, and soaked in distilled water for 2 days at room temperature. The soaked grains were washed, drained and wet milled using a Philips laboratory blender (HR2811 model), sieved with muslin cloth and fermented for 3 days at room temperature. The excess water was decanted and a moist paste that was drained to reduce the moisture content obtained. The drained paste was oven dried in hot air at 60°C for 6 to 7 h, after which it was re- milled, sieved and packaged in polyethylene bags put in a airtight container and stored at room temperature prior to analysis (Figure 1).

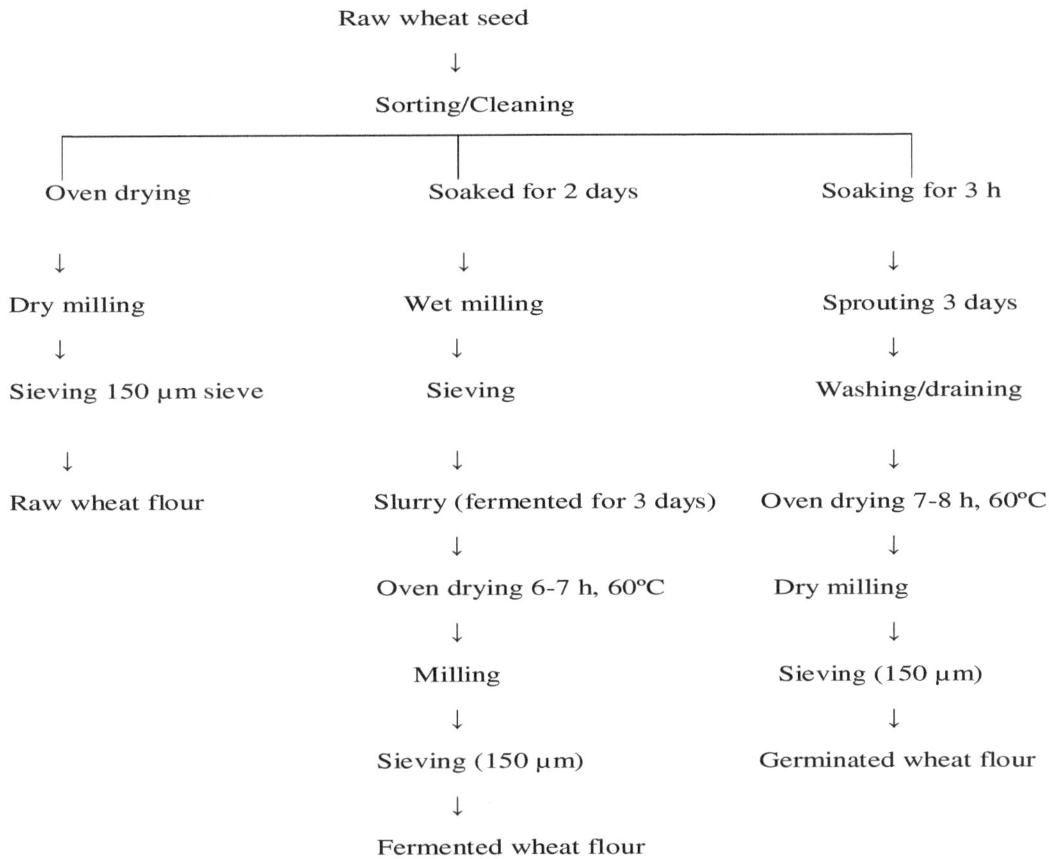


**Figure 1.** Flowchart of processing wheat seeds using different processing methods.

## Nutrient composition analysis

### *Proximate analyses*

Proximate compositions of the food sample were determined in triplicate using the standard procedures of Association of Official Analytical Chemists (AOAC) [2005]. Moisture content, crude protein (Kjeldahl method), crude fat (solvent extraction), crude fibre and ash were determined using standard methods (AOAC, 2005). The carbohydrate content was determined by difference, that is, addition of moisture, fat crude protein, ash and crude fibre, which was subtracted from 100% .This gave the amount of nitrogen free extract otherwise known as carbohydrate.

%Carbohydrate = 100 - (%Moisture + %Fat + %Ash + %Crude fibre + %Crude protein)

### *Energy value (kcal.)*

The sample calorific value was estimated [in kcal/g] by multiplying the percentages of crude protein, crude fat and carbohydrate with the recommended factors (2.44, 8.37 and 3.57, respectively) as proposed by Martin and Coolidge [1978].

### *Mineral analyses*

The standard method described by Association of Official Analytical Chemists was used for mineral content analysis of the samples (AOAC, 2005). The samples were ashed at 550°C. The ash was boiled with 10 ml of 20% hydrochloric acid in a beaker and then filtered into a 100 ml standard flask. This was made up to the mark with deionized water. The minerals were determined from the resulting solution. Sodium [Na] and Potassium [K] were determined using the standard flame emission photometer. NaCl and KCl were used as the standards (AOAC, 2005). Phosphorus was determined calorimetrically using the spectronic 20 [Gallenkamp, UK] Kirk and Sawyer [1991] with $KH_2PO_4$ as the standard. Calcium [Ca], Magnesium [Mg] and Iron [Fe] were determined using Atomic Absorption Spectrophotometer [AAS Model SP9]. All values were expressed in mg/100 g.

### *Amino acid analysis*

**Sample preparation for amino acid analysis*:*** About 2.5.0 g of each sample were weighed into the extraction thimble and the fat extracted with chloroform/methanol (2:1v/v) mixture using a Soxhlet apparatus (AOAC, 2005). The extraction lasted for 5 to 6 h.

**Hydrolysis of samples:** Thirty milligram of defatted sample was weighed into glass ampoule. A portion of 7 ml of 6 M HCl added and oxygen expelled by passing nitrogen gas into the sample. The glass ampoule sealed with a Bunsen flame and put into an oven at 105 ± 5°C for 22 h. The ampoule was allowed to cool; the content was filtered to remove the humins. The filtrate was then evaporated to dryness at 40°C under vacuum in a rotary evaporator. Each residue was dissolved with 5 ml acetate buffer (pH 2.0) and stored in a plastic specimen bottle, and kept in the deep freezer.

**Amino acid analysis:** Amino acid analysis was by ion exchange chromatography (IEC) (FAO/WHO 1991) using the Technicon Sequential Multisample (TSM) Amino Acid Analyser (Technicon Instruments Corporation, New York). The period of analysis was 76 min for each sample. The gas flow rate was 0.50 ml/min at 60°C with reproducibility consistent within ±3%. The net height of each peak produced by the chart recorder of the TSM (each representing an amino acid) was measured and calculated. The amino acid values reported were the averages of two replicates. Norleucine was the internal standard used.

**Determination nutritional quality of the samples:** Nutritional qualities of the flour samples were determined on the basis of the amino acid profiles. The Essential Amino Acid Index [EAAI] was calculated using the method of Labuda et al. (1982) according to the following equation:

$$EAAI = \sqrt[*]{\frac{[\text{Lys x Threo x Val x Meth x Isoleu x leu x Phynylal x Histi x Trypt}]a}{[\text{Lys x Threo x Val x Meth x Isoleu x leu x Phynylal x Histi x Trypt}]b}}$$

Where: [lysine, tryptophan, isoleucine, valine, threonine, leucine, phenylalanine, histidine and methionine]$_a$ in test sample and [lysine, tryptophan, isoleucine, valine, threonine, leucine, phenylalanine, histidine and the sum of methionine and cystine]$_b$ content of the same amino acids in standard protein [%] [egg or casein] respectively. Nutritional index of the food samples were calculated using the following formula:

$$\text{Nutritional index [\%]} = \frac{\text{EAAI x \%protein}}{100}$$

Biological value was calculated according to Mune-Mune et al. (2011) using the following equation:

BV = 1.09 × Essential amino acid index [EAAI] - 11.7

The Protein Efficiency Ratio [PER] was estimated according to the regression equations as given below (Mune-Mune et al., 2011):

PER = -0.468 + 0.454(LEU) - 0.105(TYR)

Amino acid scores (%) was calculated using the following formula:

$$\text{Amino acid scores (\%)} = \frac{\text{Value of essential amino acid in diet (g/100 g protein)}}{\text{FAO Ref. value for essential amino acids}} \times 100$$

**Determination of anti-nutritional composition of the samples**

***Phytic acid determination***

Phytic acid was extracted from each 3 g flour sample with 3% trichloro-acetic acid by shaking at room temperature followed by high speed centrifugation as described by Wheeler and Ferrel (1971). This method depends on an iron to phosphorus ratio of 4: 6. Five grams of the test sample was extracted with 3% tri-chloro acetic acid. The phytate was precipitated as ferric phytate and converted to ferric hydroxide and soluble sodium phytate by adding sodium hydroxide. The precipitate was dissolved in hot 3.2 N HNO and the colour read immediately at 480 nm$^3$. The standard solution was prepared from $Fe[NO_3]_3$ and the iron content was extrapolated from a $Fe(NO_3)_3$ standard curve. The phytate concentration was calculated from the iron results assuming a 4: 6 iron:phosphorus molecular ratio. The phytic acid was estimated by multiplying the amount of phytate-phosphorus by the factor 3.55 based on the empirical formula $C_6P_6O_{24}H_{18}$.

***Tannin content determination***

Tannin contents were determined by the modified vanillin-HCl methods (Burns, 1971; Price et al., 1978). A 2 g sample was extracted with 50 ml 99.9% methanol for 20 min at room temperature with constant agitation. After centrifugation for 10 min at 653 rpm, 5 ml of vanillin-HCl [2% vanilli and 1% HCl] reagent was added to 1 ml aliquots and the colour developed after 20 min at room temperature was read at 500 nm. Correction for interference light natural pigments in the sample was achieved by subjecting the extract to the conditions of the reaction, but without vanillin reagent. A standard curve was prepared using catechin [Sigma Chemical, St. Louis, MO] after correcting for blank and tannin concentration was expressed in g/100 g.

***Oxalate content determination***

Oxalate was determined by AOAC (2005) method. One gram of the sample was weighed into 100 ml conical flask. A portion of 75 ml of 3 M $H_2SO_4$ was added and the solution was carefully stirred intermittently with a magnetic stirrer for about 1 h and then filtered using Whatman No.1 filter paper. The sample filtrate [extract] (25 ml) was collected and titrated against hot [80 to 90°C] 0.1 N $KMnO_4$ solution to the point when a faint pink colour appeared that persisted for at least 30 s. The concentration of oxalate in each sample was obtained from the calculation: 1 ml 0.1 permanganate = 0.006303 g oxalate.

***Trypsin inhibition activity determination***

The trypsin inhibition activity was assayed in terms of the extent to which an extract of the defatted flour inhibited the action of bovine trypsin [EC 3.4.21.4] on the substrate benzoyl-DL-arginine-p-nitrianilide [BAPNA] hydrochloric (Kakade et al., 1969). The samples [1 g each] were extracted continuously at ambient temperature (30±2°C) for 3 h with 50 ml, 10 mM NaOH using a mechanical shaker [GallenKamp orbital shaker Surrey, UK]. The pH of the resulting slurry was adjusted to 9.4 to 9.6 with 1 M NaOH. After extraction, the suspension was shaken and diluted with distilled water such that 1 cm$^3$ of the extract produced trypsin inhibition of 40 to 60% at 37°C. The respective dilutions were noted. Consequently, TIA was calculated in terms of mg pure trypsin [Sigma type III, lot 20H0868]:

$$\text{TIA} = \frac{2.632\ \text{DA mg pure trypsin inhibited g}^{-1}\ \text{sample}}{\text{S}}$$

Where D is the dilution factor, A is the change in absorbance at 410 mm due to trypsin inhibition per cm$^3$ diluted sample extract and S is the weight of the sample.

**Functional properties**

***Water/oil absorption capacity***

Water and oil absorption capacities of the flour samples were determined by Beuchat (1977) methods. One gram of the flour was mixed with 10 ml of water or oil in a centrifuge tube and allowed to stand at room temperature (30 ± 2°C) for 1 h. It was then centrifuged at 200 × g for 30 min. The volume of water or oil on the sediment water measured. Water and oil absorption capacities were

**Table 1.** Mean (± SEM) of proximate composition of raw, germinated and fermented wheat flour.

| Nutrient/sample | Raw wheat flour | Germinated wheat flour | Fermented wheat flour |
|---|---|---|---|
| Moisture (g/100 g) | 13.20 ± 0.46$^{a}$ | 13.23 ± 1.51$^{a}$ | 12.67 ± 0.29$^{a}$ |
| Protein (g/100 g) | 10.77 ± 0.66$^{b}$ | 13.50 ± 0.49$^{a}$ | 13.70 ± 0.30$^{a}$ |
| Fat (g/100 g) | 1.93 ± 0.36$^{a}$ | 1.53 ± 0.33$^{a}$ | 1.13 ± 0.03$^{a}$ |
| Ash (g/100 g) | 0.97 ± 0.18$^{a}$ | 0.97 ± 0.36$^{a}$ | 0.77 ± 0.16$^{a}$ |
| Fiber (g/100 g) | 1.70 ± 0.34$^{a}$ | 1.93 ± 0.18$^{a}$ | 1.13 ± 0.31$^{a}$ |
| Carbohydrate (g/100 g) | 84.63 ± 0.43$^{a}$ | 82.13 ± 0.49$^{b}$ | 83.27 ± 0.66$^{ab}$ |
| Energy (Kcal.) | 398.83 ± 2.41$^{a}$ | 396.17 ± 1.05$^{a}$ | 398.13 ± 1.53$^{a}$ |

*Mean values with the same superscript in a row are not significantly different (P>0.05).

calculated as ml of water or oil absorbed per gram of flour.

#### *Bulk density*

A 50 g flour sample was put into a 100 ml measuring cylinder. The cylinder was tapped continuously until a constant volume was obtained. The bulk density (g $cm^{-3}$) was calculated as weight of flour [g] divided by flour volume ($cm^3$) (Okaka and Potter, 1979).

#### *Least gelation property*

Least gelation property was determined using the method described by Coffman and Garcia [1977]. Sample suspensions of 2 to 16% were prepared in distilled water. Each of aliquot dispersion (10 ml) was transferred into a test tube and heated in a boiling water bath for 1 h, cooled rapidly in a cold water bath, and allowed to cool further at 4°C for 2 h. The least gelation concentration was determined when the sample from the inverted test tube did not slip or fall.

#### *Swelling capacity*

This was determined with the method described by Leach et al. (1959) with modification for small samples. One gram of the flour sample was mixed with 10 ml distilled water in a centrifuge tube and heated at 80°C for 30 min. This was continually shaken during the heating period. After heating, the suspension was centrifuged at 1000×g for 15 min. The supernatant was decanted and the weight of the paste taken. The swelling power was calculated as: swelling power = weight of the paste / weight of dry flour.

#### *Determination of protein solubility*

The protein solubility of wheat flour was studied using the Were et al. (1997) method. Approximately 0.5 g each of the samples was suspended in 10 ml different salts concentrations. The solubility at natural pH was first determined, that is no acid or alkali was added and so solubility in this case was based on the normal pH of the sample in solution. The suspension was centrifuged at room temperature for 30 min at 3500rpm. The suspension obtained was filtered and the protein of filtrate was determined by biuret method with standard Bovine Serum Albumin (BSA). The Biuret method is a convenient assay for large numbers of samples of relatively soluble protein unlike the Kjeldahl method which is not a rapid and convenient assay though useful for the determination of the amount of protein in crude mixtures.

For the quantitative determination of standard protein in Biuret method, 1 g of BSA was dissolved in 100 ml distilled water in a volumetric flask. Five tubes were set up containing fractions of the BSA solution in this order: - 0.0, 0.5, 1.0, 1.5 and 2.0 ml, and they were made up to 2 ml by adding in the order 2.0, 1.5, 1.0, 0.5 and 0.00 ml of distilled water and the tubes were left to stand for 30 min. The solution from the tube containing 2.0 distilled water and 8.0 ml Biuret solution was used as the blank to standardize the UV spectrophotometer at 450nm (spectronic 20 Bausch and Lomb). The absorbance of each of the other tubes was equally taken. A standard curve was drawn for absorbance against protein concentration curve.

The determination of protein of the filtered supernatant in each sample was carried out in this way, 1.0 ml of the filtrate was pipette into a test tube and 8 ml of Biuret solution was added. The tube was let to stand for 30 min after which the absorbance was taken. The readings obtained were used to plot a graph of protein concentration for the flour samples.

#### Statistical analysis

The data were analysed using SPSS version 15.0. The mean and standard error of means (SEM) of the triplicate analyses of the samples were calculated. The analysis of variance (ANOVA) was performed to determine significant differences between the means of proximate composition, minerals, antinutritional factors, amino acid profile, fatty acids and functional properties; while the means were separated using the new Duncan multiple range test at $p<0.05$.

## RESULTS AND DISCUSSION

### Proximate composition of wheat flour

The proximate composition of wheat flour samples is presented in Table 1. The moisture content of germinated wheat flour was the highest (13.23 ± 1.51 g/100g), while that of fermented wheat flour was the lowest (12.67 ± 0.29 g/100g). These values were within the range reported by other investigators (Reihaneh and Jamuna, 2007; Rasha et al., 2011; Amagloh et al., 2012). However, investigations have shown that low moisture content of food samples is a desirable phenomenon, since the microbial activity is reduced (Oyenuga, 1968). Low moisture content in food samples increased the storage periods of the food products (Alozie et al., 2009); while high moisture content in foods encourage microbial

**Table 2.** Mineral composition of raw, germinated and fermented wheat flour (mg/100g).

| Nutrient/sample | Raw wheat flour | Germinated wheat flour | Fermented wheat flour |
|---|---|---|---|
| Calcium | $101^{c}$ | $151^{a}$ | $120^{b}$ |
| Potassium | $133.0^{c}$ | $290^{a}$ | $186^{b}$ |
| Manganese | $40.55^{a}$ | $39.95^{a}$ | $41.50^{a}$ |
| iron | $0.30^{c}$ | $0.55^{b}$ | $1.50^{a}$ |
| Zinc | $0.10^{b}$ | $0.02^{c}$ | $0.50^{a}$ |
| Sodium | $9.1^{c}$ | $19.7^{a}$ | $10.9^{b}$ |
| Magnesium | $8.50^{c}$ | $8.64^{a}$ | $8.57^{b}$ |
| Phosphorous | $9.0^{c}$ | $14.7^{a}$ | $9.8^{b}$ |
| Copper | $0.20^{b}$ | $0.00^{c}$ | $1.00^{a}$ |
| Nickel | $0.30^{b}$ | $0.01^{c}$ | $0.50^{a}$ |
| Na/K | $0.67^{b}$ | $0.69^{a}$ | $0.59^{c}$ |
| P/Ca | $0.86^{a}$ | $0.97^{a}$ | $0.82^{b}$ |

Mean with similar alphabets belong to the same homogenous subset and are not significant.

growth; hence, food spoilage (Temple et al., 1996).

The protein content of processed wheat samples varied between 10.77 ± 0.66 g/100g for raw wheat flour (RWF) and 13.70 ± 0.30 g/100g of fermented wheat flour (FWF). However, the protein content of both germinated and fermented wheat flour samples were significantly similar and higher than that of raw wheat flour (RWF) sample ($p<0.05$). This observation agreed with other scientific findings that processing techniques such as germination and fermentation improved the nutritional quality of the food products, particularly in terms of protein content (Enujiugha et al., 2003; Fasasi, 2009). In this present study, the protein content of fermented wheat flour was higher when compared with the remaining wheat samples; this finding could be attributed to the fact that during fermentation step the micro-organisms in food utilized the carbohydrate content in the food sample to synthesis amino acid needed for their growth and development (Obizoba, 1988).

Carbohydrate and energy values of the wheat flour samples range between 82.13 ± 0.49 to 84.63 ± 0.43 g/100g and 396.17 ± 1.05 to 398.83 ± 2.41 Kcal, respectively. The carbohydrate content and energy values of germinated sample were lower than those of raw and fermented wheat flour samples; this observation could be due to the utilisation of fat and carbohydrate for biochemical activities of the germinating seeds (Wang et al., 1997).

## Mineral composition of raw, germinated and fermented wheat flour

The mineral composition of raw, germinated and fermented wheat flour is shown in Table 2. The mineral composition of raw wheat flour showed that potassium had the highest value while zinc had the lowest value. For germinated wheat flour, potassium was the highest while nickel was the lowest. Also, for that of fermented wheat flour, potassium was the highest while zinc had the lowest value. In comparison, the mineral composition of germinated wheat flour sample was higher in calcium, potassium, sodium, magnesium and phosphorous than fermented and raw wheat flour sample respectively. It was observed in this study that germination and fermentation processing techniques improved the mineral composition of the flour samples except in manganese, zinc copper and nickel than that of raw wheat flour sample ($p<0.05$). This observation could be attributed to bio-synthesis and activities of micro-organism during germination and fermentation processes (Zamora and Fields, 1979; Bau et al., 1997; Dhaliwal and Aggarwal, 1999; Gabriel and Akharaiyi, 2007). Nutritionally, the ratio of Ca/P of the flour samples range between 0.82 for fermented wheat flour and 0.97 for germinated wheat flour. This finding indicates that the wheat flour samples would serve as good sources calcium and phosphorous, which are considered essential for bone and teeth formation and development in children. However, It is evident that food products containing a Ca/P ratio of >1.0 is rated good, while <0.5 is rated poor (Nieman et al., 1992). The ratio of Na/K of the food samples range between 0.59 for fermented wheat flour and 0.69 for germinated wheat flour. This indicates that consumption of wheat flour is suitable for hypertensive patients. Potassium has a beneficial effect on sodium balance. A high intake of potassium has been reported to protect against increasing blood pressure and other cardiovascular risks (Langford, 1983; Cappuccio and McGregor, 1991). Hence, the sodium to potassium (Na/K) ratio in the body is of great concern for the prevention of high blood pressure. A Na/K ratio less than one is recommended in the diets of people who are prone to high blood pressure (Langford, 1983; Cappuccio and

**Table 2.** Mineral composition of raw, germinated and fermented wheat flour (mg/100g).

| Nutrient/sample | Raw wheat flour | Germinated wheat flour | Fermented wheat flour |
|---|---|---|---|
| Calcium | $101^{c}$ | $151^{a}$ | $120^{b}$ |
| Potassium | $133.0^{c}$ | $290^{a}$ | $186^{b}$ |
| Manganese | $40.55^{a}$ | $39.95^{a}$ | $41.50^{a}$ |
| iron | $0.30^{c}$ | $0.55^{b}$ | $1.50^{a}$ |
| Zinc | $0.10^{b}$ | $0.02^{c}$ | $0.50^{a}$ |
| Sodium | $9.1^{c}$ | $19.7^{a}$ | $10.9^{b}$ |
| Magnesium | $8.50^{c}$ | $8.64^{a}$ | $8.57^{b}$ |
| Phosphorous | $9.0^{c}$ | $14.7^{a}$ | $9.8^{b}$ |
| Copper | $0.20^{b}$ | $0.00^{c}$ | $1.00^{a}$ |
| Nickel | $0.30^{b}$ | $0.01^{c}$ | $0.50^{a}$ |
| Na/K | $0.67^{b}$ | $0.69^{a}$ | $0.59^{c}$ |
| P/Ca | $0.86^{a}$ | $0.97^{a}$ | $0.82^{b}$ |

Mean with similar alphabets belong to the same homogenous subset and are not significant.

McGregor, 1991).

## Amino acids profile and nutritional quality of wheat flour samples

The amino acids profile and nutritional quality of wheat flour samples are presented in Tables 3 to 5. The result showed that glutamic acid was found to be the most abundant in raw (25.88 g/100g), germinated (20.57 g/100g) and fermented (29.23 g/100g) wheat flour; while cysteine was the least for raw (1.27 g/100g) and fermented (0.94 g/100g), and methionine (1.31 g/100g) for germinated wheat flour. The range values of total non-essential amino acid, conditionally amino acid and essential amino acid of the wheat samples were between 38.93 to 40.52 mg/100g, 14.77 to 21.91 mg/100g and 21.13 to 28.45 mg/100g, respectively. In comparison, the total amino acid composition of germinated wheat sample was higher than fermented wheat flour sample, but the value was lower than that of raw wheat sample; and this is due to the breaking down of other nutrients like carbohydrate to synthesis amino acids that the germinating seeds needed for its biochemical activities and growth. Scientific studies have documented that the protein content of sprouted or germinated plant increased due to degradation of nutrient like carbohydrate and fat in the synthesising of protein (Enujiugha et al., 2003; Fasasi, 2009).

The result of nutritional quality of the raw germinated and fermented wheat flour samples food samples is presented in Table 4. The percentage of total essential amino acid including histidine and arginine showed that germinated wheat four had the highest value (37.9%) followed by raw (35.0%) and fermented wheat flour has the least value (26.4%). The values of protein efficiency ratio (PER) of the wheat flour samples were 1.99, 0.86 and 1.61 for the raw, germinated and fermented wheat flour sample, respectively. Essential amino acid indices (EAAIs) of the raw sample had the highest value, while fermented flour sample had the lowest value. Similarly, the biological value of raw wheat flour sample (31.8%) was higher when compared with germinated (29.4%) and fermented wheat flour (29.1%) sample respectively. This observation shows that consumption of wheat flour alone without complement with other protein-based foods like legumes may not adequately meet the nutritional needs of it consumers. Scientifically, it is well known that a protein-based food material is of good nutritional quality when its biological values (BV) is high (70 to 100%) and also when the essential amino acid index (EAAI) is above 90% and to be useful as food when the values is around 80% and to be inadequate for food material when its EAAI is below 70% (Oser, 1959).

In this present study, it is observed that the PER, BV and EAAI values of germinated and fermented flour samples were low when compared with the raw wheat flour sample and also with the values reported by Oser (1959). These observations could be attributed to the complex metabolic process during germination and fermentation processing methods. For instance, it is well documented that during these processing techniques the lipids, carbohydrates and storage proteins within the seed are broken down in order to obtain the energy and amino acids necessary for the micro-organisms' and plant's development (Ferreira et al., 1995; Jachmanian et al., 1995; Podesta and Plaxton, 1994; Ziegler, 1995).

The calculated amino acid scores (Table 5) of the essential amino acids indicated that fermented wheat flour sample contained the lowest amounts of essential amino acids, while that of germinated wheat flour samples compared well with the raw flour sample. Lysine was found to be the limiting essential amino acid in the raw, germinated and fermented wheat flour samples. This

**Table 3.** Amino acid composition (mg/100 g protein) of raw, germinated and fermented popcorn flour.

| Amino acids | Raw wheat flour | Germinated wheat flour | Fermented wheat flour |
|---|---|---|---|
| **Non essential amino acids** | | | |
| Alanine | 3.48$^{b}$ | 3.72$^{a}$ | 2.25$^{c}$ |
| Aspartic acid | 5.81$^{b}$ | 9.98$^{a}$ | 4.75$^{c}$ |
| Serine | 5.35$^{a}$ | 4.66$^{b}$ | 3.18$^{c}$ |
| Glutamic acid | 25.88$^{b}$ | 20.57$^{c}$ | 29.23$^{a}$ |
| Total | 40.52 | 38.93 | 39.41 |
| **Conditionally essential amino acids** | | | |
| Proline | 8.93$^{a}$ | 4.98$^{c}$ | 6.59$^{b}$ |
| Glycine | 3.81$^{b}$ | 4.14$^{a}$ | 2.15$^{c}$ |
| Arginine | 4.14$^{b}$ | 7.92$^{a}$ | 2.99$^{c}$ |
| Cysteine | 1.27$^{b}$ | 1.32$^{a}$ | 0.94$^{c}$ |
| Tyrosine | 3.71$^{a}$ | 3.55$^{b}$ | 2.10$^{c}$ |
| Total | 21.86 | 21.91 | 14.77 |
| **Essential amino acids** | | | |
| Lysine | 3.23$^{b}$ | 3.49$^{a}$ | 2.29$^{c}$ |
| Threonine | 2.78$^{a}$ | 2.52$^{b}$ | 2.28$^{c}$ |
| Valine | 3.65$^{b}$ | 4.08$^{a}$ | 1.78$^{c}$ |
| Methionine | 2.20$^{a}$ | 1.31$^{c}$ | 1.59$^{b}$ |
| Isoleucine | 3.48$^{a}$ | 3.49$^{a}$ | 2.30$^{b}$ |
| Leucine | 6.29$^{a}$ | 6.10$^{b}$ | 5.06$^{c}$ |
| Phenylalanine | 4.24$^{a}$ | 4.69$^{a}$ | 3.56$^{b}$ |
| Histidine | 2.58$^{a}$ | 2.30$^{b}$ | 2.27$^{b}$ |
| *Tryptophan | ND | ND | ND |
| Total | 28.45 | 27.98 | 21.13 |

Mean values with the same superscript in a row are not significantly different (P>0.05).

**Table 4**. Calculated nutritional quality of raw, germinated and fermented wheat flour.

| Parameter | Raw wheat flour | Germinated wheat flour | Fermented wheat flour |
|---|---|---|---|
| TEAA+His+Arg/TAA% | 35 | 37.9 | 26.4 |
| TEAA/TAA% | 31.3 | 31.3 | 28.0 |
| TNEAA/TAA% | 68.7 | 68.7 | 71.9 |
| TSAA(Meth+Cys) | 3.45 | 2.62 | 2.52 |
| ArEAA (Phe+Tyr) | 7.93 | 8.02 | 5.64 |
| TEAA/TNEAA | 0.46 | 0.46 | 0.39 |
| PER (g/100g) | 1.99 | 0.86 | 1.61 |
| EAAI (%) | 39.9 | 37.7 | 37.4 |
| BV (%) | 31.8 | 29.4 | 29.1 |
| Nutritional index | 4.29 | 5.09 | 5.12 |

observation agreed with the report of Davidson et al. (1980), who reported that cereals are deficient in essential amino acids like lysine and tryptophan.

### Anti nutritional factor in wheat flour

The anti-nutrient composition of raw, germinated and fermented wheat flour samples is presented in Table 6. The phytic composition of the wheat flour ranged from 0.412 ± 0.01 to 0.597 ± 0.03 mg/100g, tannin content from 0.011 ± 0.00 to 0.037 ± 0.01 mg/100g, oxalate from 0.255 ± 0.01 to 0.855 ± 0.01 mg/100g; and trypsin inhibitor was between 0.707 ± 0.09 to 0.980 ± 0.02 mg/100g. The tannin, oxalate and trypsin composition of germinated wheat flour sample were higher and significant different from that of raw and fermented wheat flour samples (p<0.05), respectively. However, the values

**Table 5**. Amino acid scores of germinated and fermented wheat flour with reference to FAO/WHO standard.

| Essential amino acid for infant | FAO/WHO REF. | Amino acid scores | | |
|---|---|---|---|---|
| | | Raw wheat flour | Germinated wheat flour | Fermented wheat flour |
| Lysine | 5.8 | 55.52* | 60.17* | 39.31* |
| Threonine | 3.4 | 81.47** | 73.53** | 67.06 |
| Valine | 3.5 | 102.86 | 116.29 | 50.57** |
| Methionine | 2.2 | 99.55 | 59.09 | 72.27 |
| Isoleucine | 2.8 | 124.29 | 124.29 | 81.79 |
| Leucine | 6.6 | 95.30 | 92.27 | 76.52 |
| Phenylalanine | 2.8 | 151.07 | 160.00 | 126.79 |
| Histidine | 1.9 | 135.26 | 120.53 | 118.95 |
| Tryptophan | 1.1 | ND | ND | ND |
| Arginine | 2 | 204.00 | 395.50 | 149.00 |

Source: FAO/WHO 1991. ND (Not Detected); *First limiting amino acid; **second limiting amino acid.

**Table 6.** Anti nutritional factor of raw, germinated and fermented wheat flours (mg/g).

| Sample | Raw wheat flour | Germinated wheat flour | Fermented wheat flour |
|---|---|---|---|
| Phytic | 0.597[a] ±0.03 | 0.412[b] ±0.01 | 0.474[b] ±0.03 |
| Tannin | 0.028[b] ±0.00 | 0.037[a] ±0.01 | 0.011[c] ±0.00 |
| Oxalate | 0.585[b] ±0.01 | 0.855[a] ±0.01 | 0.255[c] ±0.01 |
| Trypsin | 0.742[b] ±0.00 | 0.980[a] ±0.02 | 0.707[c] ±0.09 |

*Mean with similar alphabet belong to the same homogenous subset and are not significantly different from each other at 5% statistical level.

**Table 7.** Relationship between phytate and bioavailability of selected minerals (zinc, iron and calcium) (mol/kg).

| Parameter | Phytate: Zinc | Phytate: Calcium | (Ca)(Phytate): Zinc | Phytate: Iron |
|---|---|---|---|---|
| Raw wheat | 2.109 | 0.128 | 0.532 | 6.055 |
| Germinated wheat | 10.045 | 0.061 | 3.792 | 2.360 |
| Fermented wheat | 0.648 | 0.166 | 0.194 | 1.861 |
| *Critical values | >15.0 | >0.24 | >200 | >1.0 |

*Sources: phytate: calcium > 0.24 (Morris and Ellis, 1985), phytate : iron > 1 (Hallberg et al., 1989), phytate : zinc >15 (Turnlund et al.,1984; Sandberg et al., 1987; Morris and Ellis, 1989), phytate : calcium/zinc > 200 (Davies et al., 1985; Bindra et al., 1986; Gibson, 2006).

were comparable to other reports (Mbithi-Mwikya et al., 2001; Ibrahim et al., 2002; Anju and Khetarpaul, 2008; Syed et al., 2011). Scientific studies have established that processing methods such as cooking, dehulling, soaking, fermetation and germination, improve the nutritional quality of food products by reducing or eliminating the anti-nutrient composition of the food products (Oboh et al., 2000; Mbithi-Mwikya et al., 2001; Ibrahim et al., 2002; Anju and Khetarpaul, 2008; Syed et al., 2011).

## Calculation of phytate and zinc, iron and calcium molar ratios to predict their bioavailability

The molar ratio of phytate and zinc, iron and calcium to predict their bioavailability is shown in Table 7. The phytate:zinc molar ratio range between 0.648 to 10.045 mol/kg, phytate:calcium 0.061 to 0.166 mol/kg, (calcium)(phytate):zinc 0.194 to 3.792 mol/kg; while that of phytate:iron was between 1.861 to 6.055 mol/kg. In comparison, the Phytate:Zinc, Phytate:Calcium,

**Table 8.** Functional properties of raw, germinated and fermented wheat flour.

| Sample | Raw wheat flour | Germinated wheat flour | Fermented wheat flour |
|---|---|---|---|
| Bulk density ($g/cm^3$) | $0.80^c \pm 0.04$ | $0.85^{ab} \pm 0.01$ | $0.86^a \pm 0.02$ |
| Swelling capacity | $0.03^b \pm 0.03$ | $1.23^a \pm 0.21$ | $0.35^b \pm 0.03$ |
| Water Absorption capacity (%) | $405^b \pm 5.0$ | $315^c \pm 5.0$ | $415^a \pm 5.0$ |
| Least Gelation (%) | $14.67^a \pm 1.20$ | $3.33^c \pm 1.16$ | $8.67^b \pm 1.16$ |

* Mean with similar alphabets belongs to the same homogenous subset and are not significantly different from each other at the 5% statistical level.

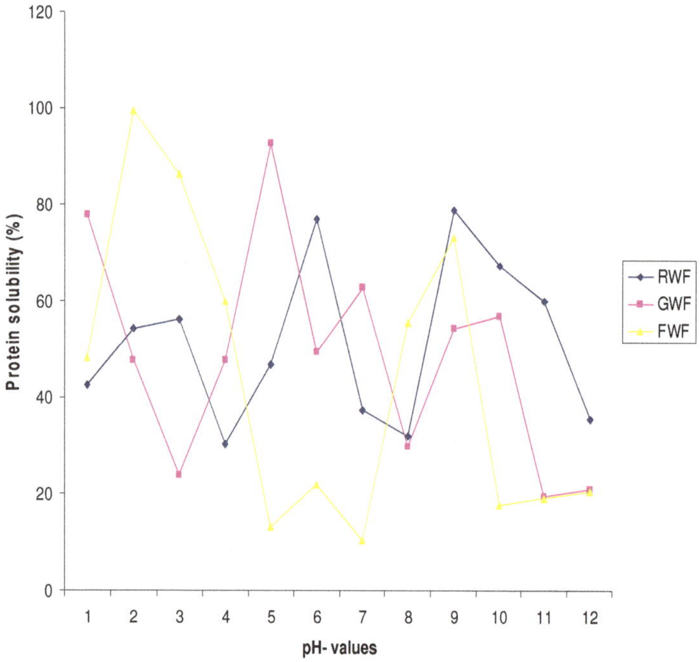


**Figure 2.** Protein solubility of the raw wheat flour (RWF), germinated wheat flour (GWF) and fermented wheat flour (FWF).

(Ca) (Phytate):Zinc and Phytate:Iron molar ratios of the wheat flour samples were lower than the critical values reported by other investigators (Morris and Ellis, 1985; Davies et al., 1985; Bindra et al., 1986; Gibson et al.,1991; Gibson, 2006). The inhibitory effect of phytate on zinc, iron and calcium absorption has been quantified by the molar ratios of phytate to zinc, iron and calcium in the diet. The ratio greater than these critical level, that is, 15.0, 0.24, 200 and 1.0 have been associated with biochemical and/or clinical evidence of zinc calcium and iron deficiency (Morris and Ellis, 1985; Davies et al., 1985; Bindra et al., 1986; Gibson et al., 1991; Gibson 2006). Evidence has shown that zinc, iron and calcium are essential trace elements for human nutrition (Kono and Yoshida, 1989); and that children are more vulnerable to sub-optimal of these trace elements with adverse effects on their growth rate and cognitive development (Hambidge et al., 1985). Phytic acid may reduce the bioavailability of dietary zinc, iron and calcium by forming insoluble mineral chelate at physiological pH.

## Functional properties of raw, germinated and fermented wheat flour

The functional properties of wheat flour samples are shown in Table 8. The Bulk density (BD) ranged between 0.80 ± 0.04 and 0.86 ± 0.02 for raw and fermented wheat flour sample respectively. Swelling capacity (SC) ranged between 0.0 ± 0.03 for raw sample and 1.23 ± 0.21 for germinated sample. For water absorption capacity (WAC) fermented wheat flour sample had the highest value (415%); while germinated had the lowest value. For least gellation, raw wheat sample had the highest value (14.6%); while fermented sample had the lowest value (31.5%).

Figure 2 shows the protein solubility of the raw, germinated and fermentation of wheat flour. The minimum protein solubility of the raw wheat flour was at pH 4, however at this pH the protein content in the food was precipitated out; while the maximum protein solubility was at pH 10. For germinated wheat flour, the minimum protein solubility was at pH 11 and maximum at pH 5; while that of fermented wheat flour sample the minimum protein solubility was at pH 7 and the maximum protein solubility was at pH 2. This observation implies that the protein content of both the germinated and fermented wheat flour samples could only be fully utilized in acidic medium. Hence, the protein content of the wheat flour samples would be fully solubilised and digested at the acidic environment (pH 2) of the stomach, which is the starting point of protein digestion in man. Also, the protein solubility was determine in order to know at what pH the protein would be more soluble, which is very important during food production system to know the state of the food if they are acidic or basic food.

Fermented wheat flour sample has the highest bulk density of 0.86 ± 0.02 $g/cm^3$, while raw wheat flour (0.80 ± 0.04 $g/cm^3$) has the least value. The value of bulk density obtained for germinated and fermented wheat flour sample in this study were higher than 0.62 $g/cm^3$ reported for tigernut flour, 0.54 $g/cm^3$ for African breadfruit kernel flour and 0.71 $g/cm^3$ reported for wheat flour (Akubor and Badifu, 2004; Oladele and Aina, 2007). The bulk density value is of importance in packaging

(Snow, 1974). The lower loose bulk density implies that less quantity of the food samples would be packaged in constant volume thereby ensuring an economical packaging. However, the packaged bulk densities would ensure more quantities of the food samples being packaged, but less economical. Nutritionally, loose bulk density promotes easy digestibility of food products, particularly among children with immature digestive system (Osundahunsi and Aworh, 2002; Gopaldas and John, 1991). Water absorption capacity (WAC) of fermented wheat flour (415%) was higher than germinated wheat flour (315%) and raw wheat flour (405%) sample. For the swelling capacity, the value for germinated wheat flour sample was 1.23, while that of raw and fermented wheat flour were 0.03 and 0.35, respectively.

The water absorption capacity values obtained for germinated and fermented wheat flour sample were higher than raw conophor flour (Odoemelam, 2003), but comparable to that of African yam bean flour (Eke and Akobundu, 1993). Water absorption capacity is an index of the maximum amount of water that a food product would absorb and retain (Mosha and Lorri, 1987; Marero et al., 1988). This result suggests that wheat flour may find application in baked products. Studies have shown that the microbial activities of food products with low water absorption capacity would be reduced (Giami and Bekeham, 1992). Hence the shelf-life of such product would be extended.

Least gelation property of the wheat flour samples range from 3.33 ± 1.16 to 14.67 ± 1.20%. It was observed that the gelation property of germinated and fermented wheat flour were comparatively lower than that of raw wheat flour sample. This implies that the flour samples have good gelating ability. Scientific studies have shown that foods that form gels at low concentrations are not ideal for infant diet, because they would require a lot of dilution in an attempt to improve digestibility in relation to volume (Ezeji and Ojimelukwe, 1993; Obatolu and Cole, 2000).

## Conclusion

This study established that germinated wheat flour sample had better nutritional parameters like mineral content, total essential amino acid, biological value and lower functional properties than fermented wheat flour sample. However, further study is required on the suitability of this processed flour in food production.

### REFERENCES


Abdel Rahman SM, Babiker EE, Abduallahi HE (2005). Effect of fermentation on antinutritional factors and HCL extractability of minerals of pearl millet cultivars. J. Food Sci. Tech. 3(4):516-522.

Ahmed MB, Hamed RA, Ali ME, Hassan AB, Babiker EE (2006). Proximate composition, antinutritional factors and protein fraction of guar gum seeds as influenced by processing treatments. Pak. J. Nutr. 5(5):481-484.

Akubor PI, Badifu GIO (2004). Chemical composition, functional properties and baking potential of African breadfruit kernel and wheat flour blends. Int. J. Food Sci. Tech. 39:223-229.

Alozie YE, Iyam MA, Lawal O, Udofia U, Ani IF (2009). Utilization of Bambara ground flour blends in bread production. J. Food Technol. 7(4):111-114.

Amagloh FK, Weber JL, Brough L, Hardacre A, Mutukumira AN, Coad J (2012).Complementary food blends and malnutrition among infants in Ghana: A review and a proposed solution Sci. Res. Essays. 9 March, 2012. 7(9):972-988,

Anju NPB, Khetarpaul N (2008). Effect of domestic processing and cooking methods on some antinutritional factors of moth bean. J. Dairying Foods. H.S. 27(2):114 - 119.

AOAC. (Association of Official Analytical Chemists) (2005). Official Methods of Analysis of the Association of Analytical Chemists International, 18th ed. Gathersburg, MD U.S.A Official methods, 2005.08.

Aslam M, Gilan AH, Qazi RA (1982). Some dimensions of rural food poverty with special emphasis on nutritional status and its improvement. Final Report German Agro Action Project, University of Agriculture, Fsb. p. 79.

Bau H, Villaume C, Nicolas J, Mejean L (1997). Effect of germination on lysine inhibiter activity in lentil and pigeon pea. J. Food Sci. Tech. 23:260-263.

Bekeham DA (1992). Proximate composition and functional properties of raw and processed full fat fluted pumpkin (*Telferia occidentalis*) seed flour. In J. Food Sci. Agric. 59:32.

Beuchat LR (1977). Functional and electrophoretic characteristics of succinylated peanut flour protein. J. Agric. Food Chem. 25:258-261.

Bindra GS, Gibson RS, Thompson LU (1986). [Phytate]/[calciumj/[zinc] ratios in Asian immigrant lacto-ovo vegetarian diets and their relationship to zinc nutriture. Nutr Res. 6:475-83.

Burns R (1971). Method for estimation of tannin in grain sorghum. Agron. J. 63:511-512.

Cappuccio FP, McGregor GA (1991). Does potassium supplementation lower blood pressure? A meta-analysis of published trials.J.Hypertens. 9:465-473.

Chavan JK, Kadam SS (1989a). Nutritional improvement of cereals by fermentation. Food Sci. Nutr. 28: 379-400.

Chavan JK, Kadam SS (1989b). Nutritional improvement of cereals by sprouting. Food Sci.Nutr.28: 401-437.

Cheryan M (1980). Phytic acid interactions in food systems. CRC Critical Rev. Food Sci. Nutr.13: 296-335.

Davidson S, Passmore S, Brock JF, Truswell AS (1980). Human Nutrition and Dietetics. Churchill Livingstone, New York :pp.213-215. www.gastrohepcom/../Arias-3e_chapter_16_Energy_Metabolism pdf.

Davies NT, Carswell AJP, Mills CF (1985).The effects ofvariation in dietary calcium intake on the phytate-zinc interaction in rats. In:Mills CF, Bremner I, Chesters JK, eds. Trace elements in man and animals-TEMA 5. Farnham Royal, UK: Commonwealth Agricultural Bureau, 125(abstr).

Dhaliwal Y, Aggarwal R (1999). Composition of fat in soyabean as affected by duration of germination and dry temperature. J. Food Sci. Tech. 36: 266-267.

Egli IM (2001).Traditional Food Processing Methods to Increase Mineral Bioavailability from Cereal and Legume Based Weaning Foods. A PhD dissertation. Swiss Federal Institute of Technol Zurich.

Eke OS, Akobundu ENT (1993). Functional properties of African yam bean (*Sphenostylis stenocarpa*) seed flour as affected by processing. Food Chem.48:337-340.

Enujiugha VN, Badejo AA, Iyiola SO, Oluwamukomi MO (2003). Effect of germination on the nutritional and functional properties of African oil bean (Pentaclethra macrophylla Benth) seed flour. Food Agric. Environ. 1:72-75.

Ezeji C, Ojimelukwe PC (1993). Effect of fermentation on the nutritional quality and functional properties of infant food formulations prepared from bambara-groundnut, fluted-pumpkin and millet seeds. Plant Foods. Hum. Nutr. 44:267-276.

FAO (2009). Food balance sheets. Food and agriculture organization, Rome.

FAO/WHO (1991). Protein quality evaluation. In Food and Agricultural Organization of the United Nations, Rome, Italy.
Fasasi OS (2009). Proximate antinutritional factors and functional properties of processed pearl millet (Pennisetum glaucum). J. Food Technol. 7(3):92-97.
Ferreira RB, Melo TS, Teixeira AN (1995). Catabolism of the seed storage proteins from Lupinus albus: fate of globulins during germination and seedling growth. Aust. J. Plant Physiol. 22:373-381.
Gabriel RAO, Akharaiyi FC (2007). Effect of spontaneous fermentation on the chemical composition of thermally treated jack beans (Canavalia ensiformis L.). Int. J. Biol. Chem. 1(2):91-97.
Giami SY, Bekeham DA (1992). Proximate composition and functional properties of raw and processed full fat fluted pumpkin (Telferia occidentalis) seed flour. J. Sci. Food Agric. 59:32.
Gibson RS (2006). Zn: the missing link in combating micronutrient malnutrition in developing countries. Proc. Nutr. Soc. 65:51-60.
Gibson RS, Smit Vaderkooy PD, Thompson L (1991). Dietary phytate x Ca/Zn millimolar ratios and Zn nutriture in some Ontario preschool children. Biol. Trace Element Res. 30:87-94.
Gopaldas T, John C (1991). Evaluation of a controlled 6 months feeding trial on intake by infants and toddlers fed highenergy-low bulk gruel versus a high energy-high bulk gruel in addition to their habitual home diet. J. Trop. Paediatr. 38:278-83.
Hallberg L, Brune M, Rossander L (1989). Iron absorption in man: Ascorbic acid and dose-dependent inhibition by phytate. Am. J. Clin. Nutr. 49:140-144.
Hambidge KM. Krebs NF, Walravens PA (1985). Growth velocity of young children receiving a dietary zinc supplement. Nutr. Res. (l):306-3l6.
Helland MH, Wicklund T, Narvhus JA (2002). Effect of germination time on *alpha-amylase* production and viscosity of maize porridge. Food Res. Int. 35:315-321.
Hassan EE. Babiker AH, Tinay EI (2007). Content of antinutritional factors and HCl-Extractability of minerals from white bean (*Phaseolus vulgaris)* cultivars: Influence of soaking and/or cooking. Food Chem.100: 362-368.
Hussain T, Abbas S, Khan MA, Scrimshaw NS (2004). Lysine of wheat flour improves selected indices of the nutritional status of predominantly cereal eating families in Pakistan. Food Nutr. Bull. 25(2):114-22.
Ibrahim SS, Habiba RA, Shatta AA, Embaby HE (2002). Effect of soaking germination, cooking and fermentation on antinutritional factors in cowpeas. Nahrung. 46:92-95.
Jachmanian I, Perifanova-Nemska M, Grompone MA, Mukherjee KD (1995). Germinating rapeseed as biocatalyst: hydrolysis of exogenous and endogenous triacylglycerols. J. Agric. Food Chem. 43:2992-2996.
Jacobs JR, Gallagher DR (2004). Whole grain intake and cardiovascular disease: a review. Current Atherosclerosis Reports 6: 415-423.
Kakade ML, Simons N, Leiner IE (1969). An evaluation of natural vs. synthetic substrates for measuring antitryptic activity of soya bean samples. Cereal Chem. 46:518-526
Kirk SR, Sawyer R (1991). Pearson's Composition and Analysis of foods, 9th edn., Longman Scientific and Technical, UK.
Kono K, Yoshida Y (1989). Current aspect of trace element research in environmental health. Bull Osaka Med. Coll. 35 (1-2):1-9
Kouakou B, Alexis KKS, Adjehi D, Marcelin DK, Dago G (2008). Biochemical changes occurring during germination and fermentation of millet and effect of technological process on starch hydrolysis by the crude enzymatic extraction of millet. J. Appl. Sci. Res. 4(11):1502-1510.
Labuda J, Kacerovský O, Kováè M, Štìrba A (1982). Výživa a krmenie hospodárských zvierat. Príroda, Bratislava. p. 164s.
Langford HG (1983). Dietary potassium and hypertension: epidemiologic data. Ann. Int. Med. 98(2):770-772
Larsson SC, Giovannucci E, Bergkvist L, Wolk A (2005). Whole grain consumption and risk of colorectal cancer: a population-based cohort of 60,000 women. Br. J. Cancer. 92:1803-1807.
Leach HW, McCowen LD, Schoch TJ (1959). Structure of the starch granules. In: Swelling and solubility patterns of various starches. Cereal Chem. 36:534-544.
Marero LM, Pajumo EM, Librando EC (1988). Technology of weaning food formulation prepared from germinated cereals and legumes. In J. Food Sci. 53:1391-1395.
Martin EA, Coolidge AA (1978). Nutrition in Action, 4th ed. Holt, R and Wilson Co. New York, USA.
Mbithi-Mwikya S, Van Camp J, Rodrı´guez R, Huyghebaert A (2001). Effects of sprouting on nutrient and antinutrient composition of kidney beans. Eur. Food Res. Technol. 212:188-191.
Moore AC, Akhter S, Aboud FE (2006). Responsive complementary feeding in rural Bangladesh. Soc. Sci. Med. 62:1917-1930.
Morris ER, Ellis R (1985). Bioavailability of dietary calcium-effect of phytate on adult men consuming non-vegetarian diets. In: ACS Symposium Series 75:Nutritional Bioavailability of Calcium. Kies C (eds), p. 63. American Chemical Society: Washington, DC, USA.
Morris ER, Ellis R (1989). Usefulness of the dietary phytic acid/zinc molar ratio as an index of zinc bioavailability to rats and humans. Biol. Trace Element Res.19:107-117.
Mosha AC, Lorri WSM (1987). High-nutrient-density weaning foods from germinated cereals.In: Alnwick D, Moses S, Schmidt OG, eds. Improving young child feeding in Eastern and Southern Africa. Nairobi, New York, Stockholm: IDRC, UNICEF, SIDA. pp. 288-299.
Muahamad N, Faqir MA, Rai MA, Moazzam RK, Shahzad H, Muhammad SJ (2010). An overview of anti-nutritional factors in cereal grains with special reference to wheat-A review Pak. J. Food Sci. 20(1-4):54-61
Mune-Mune MA, Minka SR, Mbome IL, Etoa FX (2011). Nutritional Potential of Bambara Bean Protein Concentrate. Pak. J. Nutr. 10(2):112-119
Murtaugh MA, Jacobs JR, Steffen DR, Jacob LM, Marquart L (2003). Epidemiological support for the protection of whole grains against diabetes. Pro. Nutr. Soc. 62:143-149.
Nieman DC, Butterworth DE, Nieman CN (1992). Nutrition. WmC. Brown, Dbugye, USA. pp. 237-312
Nout MJR (1992). Processed weaning foods for climates. Int. J. Food Sci. Nutr. 43:213-221.
Obizoba IC (1988). Nutritive value of malted, dry or wet milled sorghum and *corn.* Cereal Chem. 65:447-449.
Obatolu VA, Cole AH (2000). Functional property of complementary blends of Soybean and Cowpea with malted or unmalted maize. Food Chem.70: 147-153.
Oboh HA, Muzquiz M, Burbano C, Cuadrado C, Pedrosa MM, Ayet G, Osagie AU (2000). Effect of soaking, cooking and germination on the oligosaccharide content of selected Nigerian legume seeds. Plant Foods Hum. Nutr. 55:97-110.
Ochanda SO, Onyango CA, Mwasaru AM, Ochieng JK, Mathooko FM (2010). Effects of malting and fermentation treatments on group B-vitamins of red sorghum, white sorghum and pearl millets in Kenya. J. Appl. Biosci. 34:2128-2134.
Odoemelam SA (2003). Chemical composition and functional properties of conophor nut (Tetracarpidium conophorum) flour. Int. J. Food Sci. Techn. 38(6):729–734.
Okaka JC, Potter NN (1979). Physicochemical and functional properties of cowpea powders processed to reduce beany flavour. J. Food Sci. 44:1235-1240.
Oladele AK, Aina JO (2007). Chemical composition and functional properties of flour produced from two varieties of tigernut (Cyperus esculentus). Afr. J. Biotechnol. 6(21):2473-2476
Oser BL (1959). An integrated essential amino acid index for predicting the biological value of proteins. In A.A. Albanese (Ed.), Protein and amino acid nutrition NewYork: Academic Press. pp. 295-311.
Osundahunsi OF, Aworh OC (2002). A preliminary study on the use of tempeh based formula as a weaning diets in Nigeria. Plant foods Human Nutr. 57:365- 376.
Oyenuga VA (1968). Nigeria foods and feeding stuffs: their chemistry and nutritive value. Ibadan: University Press.
Pena EA, Bernardo CS, Jouve B (2006). Do tyrosine crosslinks contribute to the formation of the gluten network in common wheat dough. J. Cereal Sci. 44:144-153.
Pereira MA, Jacobs DR, Pins JJ, Raatz SK, Gross MD, Slavin JL, Seaquist ER (2002). Effect of whole grains on insulin sensitivity in overweight hyper insulinemic adults. Am. J. Clin. Nutr. 7:848-855.
Podesta FE, Plaxton WC (1994). Regulation of cytosolic carbon

metabolism in germinating Ricinus communis cotyledons I. Developmental profiles for the activity, concentration, and molecular structure of the pyrophosphate- and ATP-dependent phosphofructokinases, phosphoenolpyruvate carboxylase and pyruvate kinase. Planta 194:374-380.

Price ML, Scoyoc SV, Butler LG (1978). A critical evaluation of the vanillin reaction as an essay for tannin in sorghum grain. J. Agric. Food Chem. 26:1214-1218

Rasha MK, Esmat AA, Gibriel AY, Nagwa MHR, Ferial MA (2011). Effect of legume processing treatments individually or in combination on their phytic acid content Afr. J. Food Sci. Technol. 2(2):036-046

Reihaneh A, Jamuna PG (1997). The impact of germination and dehulling on nutrients, antinutrients, in vitro iron and calcium bioavailability and in vitro starch and protein digestibility of some legume seeds. LWT 40:1292-1299

Sandberg AS, Anderson H, Carlesson NG, Sandstrom B (1987). Degradation products of bran phytate formed during digestion in the human small intestine: effects of extrusion cooking on digestibility. In J. Nutr. 117:2061-2065.

Sangronis E, Machado CJ (2007). Influence of germination on nutritional quality of *Phaseous vulgaris* and Cajanus cajan. J. Sci. Tech. 40:116-120.

Snow JE (1974). Flexible packaging and food products compatibility in: Chemistry of food packaging. Swalon (M. ed) American Chemical Society, Washington DC. pp. 82-84.

Sripriya G, Antony U, Chandra TS (1997). Changes in carbohydrate, free amino acids, organic acids, phytate and HCl extractability of minerals during germination and fermentation of finger millet (*Eleusine coracana*). Food Chem. 58:345-350.

Syed AS, Aurang Z, Tariq M, Nadia N, Sayed JA, Muhammad S, Abdul A, Asim M (2011). Effects of sprouting time on biochemical and nutritional qualities of Mungbean varieties Afr. J. Agric. Res. 6(22):5091-5098.

Temple VJ, Badamosi EJ, Ladeji O, Solomon M (1996). Proximate chemical composition of three locally formulated complementary foods. West Afr. J. Biol.Sci. 5:134-143.

Turnlund JR, King JC, Keyes WR and Gong B, Michel MC (1984). A stable isotope study of zinc absorption in young men: effects of phytate and α-cellulose. Am. Clin. Nutr. 40:1071-1077.

Wang N, Lewis MJ, Brennan JG, Westby A (1997). Effect of processing methods on nutrients and antinutritional factors in cowpea. Food Chem. 58:59-68.

Were L, Hettiarachchy L, Kalapathy U (1997). Modified soy proteins with improved foaming and water hydration proteins. J. Food Sci. 62:21-824.

Wheeler EI, Ferrel RE (1971). A method for phytic acid determination in wheat and wheat fractions. Cereal Chem. 48:312-320.

Zamora AF, Fields ML (1979). Nutritive quality of fermented cowpeas (Vigna sinensis) and chickpeas (Cicer arietinum). J. Food Sci. 44:234-236.

Ziegler P (1995). Carbohydrate degradation during germination. In J. Kigel & G. Galili (Eds.), Seed development and germination. New York: Marcel Dekker. Inc. pp. 447-474

# 18

# Toposequential variation in soil properties and crop yield from double-cropping paddy rice in Northwest Vietnam

**Aung Zaw Oo[1], Khin Thuzar Win[1], Lam Nguyen[2] and Sonoko Dorothea Bellingrath Kimura[1]**

[1]Tokyo University of Agriculture and Technology, Saiwaicho 3-5-8, Fuchu, Tokyo, 183-8509, Japan.
[2]Hanoi University of Agriculture, Trau Quy, Gia Lam, Hanoi, Vietnam.

**To access the spatial differences in soil properties and crop yield at cascade level affected by either sediment induced or farmers' fertility practice, field experiments were conducted in Cheing Khoi watershed area during spring and summer seasons with two different cascades. The cascades had fertilized and unfertilized parts and all the measurements were conducted at the top, middle and bottom field of each part. In both rice cascades, middle and bottom fields showed high silt and clay content while sand was the dominant type in the top field. Total nitrogen (TN) and carbon (TC) contents were significantly higher in lower lying fields than top field. Seasonal average surface water $NH_4$ concentrations for top, middle and bottom fields were 2.7, 3.4 and 3.2 mg $L^{-1}$ in spring and 2.8, 3.5 and 3.1 mg $L^{-1}$ in summer, respectively. Differences were substantial in yield components parameters and grain yield depending on toposequence position. Grain yields in the middle fields of both rice cascades were higher than other field positions in both fertilized and unfertilized fields. The grain yields were significantly related with surface water $NH_4$ concentration, TN content, sand and silt content of soil. The larger toposequential differences in crop yield require different field specific management practices for each position in order to improve rice production in this watershed area.**

**Key words:** Crop performance, double-cropping, paddy rice, spatial variability, toposequence.

## INTRODUCTION

In Northern mountainous region of Vietnam, an important agro-ecosystem is a composite swidden agriculture which integrates annual food crops, such as maize, cassava and upland rice, and fallow in the uplands and permanent wet rice fields in valley bottoms of the catchment (Lam et al., 2005) to form a single household resource system (Rambo, 1998). More recently, traditional agriculture methods involving fallows are more and more replaced by market oriented land use annual mono-cropping systems, that have a low soil cover during their establishment phase (e.g. maize), inducing severe erosion on steep slopes. Such land use alterations have dramatic environmental effects (Wezel et al., 2002) and high precipitation will lead to accelerate soil degradation due to erosion of the steep slopes used for agriculture.

Soil erosion is considered to have serious impacts on the current productivity and sustainability of the land. Upstream erosion will lead to sedimentation and siltation of downstream water bodies and paddy fields as well as nutrient transportation within sediments and irrigation water. During erosion events, nutrients are removed and attached to eroded sediments, reallocated in the watershed

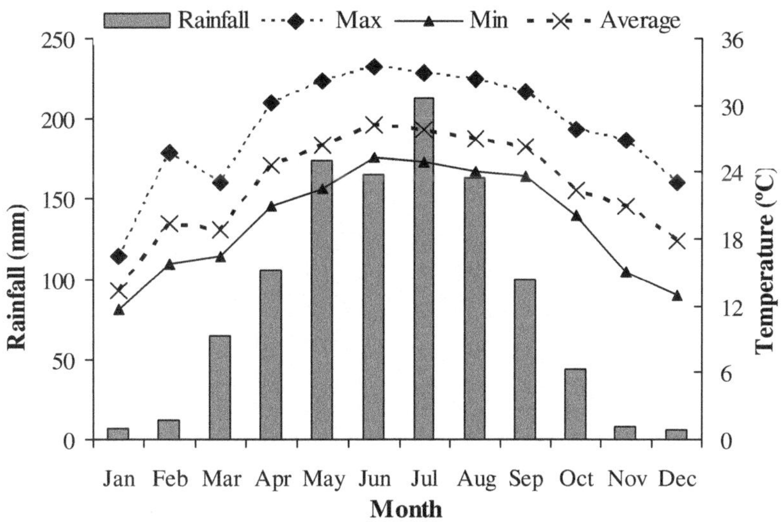


**Figure 1.** Monthly average rainfall and temperature of Yen Chau city recorded during the year 2011.

(Dung et al., 2008; Pansak et al., 2008). Sediment rich water will flow into the paddy fields at the upper side and flow out at the lower side so that the distribution of the sediments is unequal throughout the rice fields. The deposited sediments create patterns of spatial variability in soil fertility of downstream watershed (Gao et al., 2007; Mingzhou et al., 2007) and will influence the crop productivity of the terrace rice fields.

Rice fields are located on the gently sloping land which differs in elevation for a few meters in an undulating topography. These differences in toposequence position may lead to differentiation in soil properties and hydrological conditions (Hseu and Chen, 2001; Tsubo et al., 2006) and therefore crop yield. Topography directly affects soil-forming processes through erosion and deposition, and variation has been observed in soil texture (Yamauchi, 1992), nitrogen (N), phosphorus (P) and potassium (K) content (Moormann et al., 1977; Eshett et al., 1989; Posner and Crawford, 1992; Yamauchi, 1992) among the toposequence positions. Furthermore, organic compounds present in the water also will influence soil fertility in paddy rice fields. However, redistribution of nutrients through erosion-sedimentation processes in upland-lowland areas and its impact on soil fertility in the lowland are too often neglected (Mochizuki et al., 2006; Rüth and Lennartz, 2008).

In general, large field to field variations in soil fertility and rice growth along the toposequence are not desirable for rice production. For developing the site-specific fertilizer management strategies of crop, it is essential to know the spatial variability of soil factors and to assess their influence on the variability of crop growth and yield. Accessing spatial variation among the toposequence field positions is necessary for identifying and quantifying the limiting factors for rice growth, and addressing the spatial variability of rice yield. There have been studies about spatial variability of yield, crop growth performance and soil (Wezel et al., 2002; Homma et al., 2003, Tsubo et al., 2006; Rüth and Lennartz, 2008) but the knowledge about sediment inducing spatial variation in soil properties and crop yield among the toposequence position due to upland soil erosion is still limited.

Thus, the objectives of this study were (1) to access the sediment inducing toposequential variability of soil properties and crop yield at cascade level and (2) to distinguish between the inherent spatial variability in soil fertility induced by sediment deposits and soil fertility influenced by farmers' fertilization practices.

## MATERIALS AND METHODS

The field experiments were carried out from February until November 2011, during two rice cropping seasons in the Chieng Khoi commune (350 m.a.s.l., 21° 7′60″N, 105°40′0″E), Yen Chau district, Northwest Vietnam. The studied area is located in the tropical monsoon area with very hot and rainy summers with 858 mm (May to October, 2011) and cool and dry winters with 201 mm rainfall (November to April, 2011) (Figure 1). The annual precipitation was 1059 mm in 2011. Open channel irrigation system and a lake reservoir allow for two rice cropping seasons a year; spring crop from February to June and summer crop from July to November.

Two rice cascades (Cascade 1 and 2) were selected for this experiment. The length and altitude differences were 83 and 7 m for cascade 1 and 87 and 5 m for cascade 2, respectively. Both cascades contained 5 to 6 successive paddy fields, covering a total area of 0.8 ha. The uppermost field of cascades received water directly from the irrigation channel. All other fields received water from a single inlet from above lying field and drain via single outlet to the lower situated field. At the beginning of the first crop season, sediments entering the rice fields are mainly provided through irrigation system. The irrigation water had an average concentration of 2 mg $l^{-1}$ organic C and 1.5 mg $l^{-1}$ total N when no rain occurred (Schmitter et al., 2010). During rainy season, additional sediments

**Table 1.** Physico-chemical properties of the experimental soils among the field positions before transplanting of spring rice, 2011.

| Field position | Sand (%) | Silt (%) | Clay (%) | TN (g $kg^{-1}$) | TC (g $kg^{-1}$) | pH | EC ($msm^{-1}$) |
|---|---|---|---|---|---|---|---|
| | | | | **Cascade 1** | | | |
| Top | 54.0 | 32.6 | 13.4 | 0.17 | 3.2 | 8.1 | 24.2 |
| Middle | 23.3 | 51.9 | 24.8 | 0.32 | 5.6 | 8.0 | 25.3 |
| Bottom | 31.9 | 49.7 | 18.5 | 0.29 | 5.3 | 8.3 | 21.7 |
| $LSD_{(0.05)}$ | 8.5 | 8.4 | 4.6 | 0.05 | 1.0 | 0.1 | 3.5 |
| | | | | **Cascade 2** | | | |
| Top | 58.9 | 34.1 | 7.0 | 0.18 | 2.5 | 8.0 | 25.3 |
| Middle | 25.5 | 43.4 | 31.2 | 0.29 | 4.4 | 8.2 | 23.4 |
| Bottom | 40.3 | 37.3 | 22.4 | 0.26 | 4.8 | 8.4 | 17.6 |
| $LSD_{(0.05)}$ | 12.5 | 4.1 | 8.6 | 0.04 | 0.8 | 0.1 | 3.7 |

from upland are delivered to some of the rice fields besides the irrigation system.

The experiment was laid out in a split plot design with three replications at each site. All fields in both cascades were divided into two parts resulting in two strips per cascade. Two sets of factors included in this experiment were as follows: different toposequence positions (top, middle and bottom) as the main plot and with (+F) and without (-F) fertilizer application as the subplot. The soil type was Gleysols (silty loam in the different horizons) (UNESCO, 1974). The applied chemical fertilizers were 213 kg N $ha^{-1}$, 150 kg P $ha^{-1}$ and 93 kg K $ha^{-1}$ per rice season with split applications according to the local recommendations by extension service. The first dressing, at transplanting, contained 56% N, 100% P and 34% K of the total amount of fertilizer applied in the form of NPK and urea. Second and third dressings contained 22% N and 33% K of the total amount of fertilizer in the form of urea and Kali ($K_2SO_4$ – 40% $K_2O$) which were applied at active tillering and heading stage.

The sticky rice variety (*Oryza sativa*, Var. Nep 87) was used in both cropping season for both cascades. Seedlings of 18 days old were transplanted in the well-puddle fields. Rice seedlings were transplanted on February 28 and harvested on June 30, 2011 for the spring rice and the corresponding dates for the summer rice are July 28 and November 7, 2011, respectively. After harvesting the spring rice, all the fresh crop residues leftover were incorporated into the soil by plowing and harrowing. All management practices were followed by farmer practices in both cropping seasons. Previous to this experiment, all fields were used similarly for continuous double-cropping paddy rice for several years.

Top soil samples at 0 to 5 cm depth were taken before transplanting and after harvest for sand, silt, clay, total nitrogen (TN) and total carbon (TC) content. Soil particle analysis was done by pipette method (Gee and Bauder, 1986). (TN) and total (TC) contents were analyzed by using a NC analyzer (Sumigraph NC-80; Sumika Chemical Analysis Service Co., Japan). The soil pH was measured in the supernatant suspension of a 1 : 2.5 soil: water mixture using a portable pH meter equipped with a combined electrode (glass : Ag/AgCl, Horiba, Japan). Electrical conductivity of the soil water was measured in the supernatant suspension of a 1 : 5 soil : water mixture using EC meter (OM-51, Horiba, Japan). Water samples were collected at 10 days interval for $NH_4$ concentration analysis. The analysis was done by using ion selective pack test for $NH_4$, simplified chemical analysis (Kyoritsu Chemical-Check Lab., Corp).

Seven samples hills were collected from each plot for collection of data on plant characters and yield components. Grain yield was determined from 1 $m^2$ sampling area at harvest and was expressed as rough (unhulled) rice at 14% moisture content. All the data were evaluated by an analysis of variance (ANOVA) by using CropStat 7.2 statistical software. Treatment means were compared by least significant differences (LSD) at p=0.05.

## RESULTS

### Soil properties

Soil physical properties varied significantly among the field positions in both rice cascades (Table 1). Sand was the dominant particle type in the top fields containing 50.0% and 58.9% in cascade 1 and 2, respectively. The middle and bottom fields showed higher silt content in both cascades; 51.9 and 49.7% in cascade 1 and 43.4 and 37.3% in cascade 2, respectively. In cascade 1, clay content was significantly ($p<0.01$) higher in middle field but there were no statistical differences in clay content between top and bottom field. Significantly ($p<0.01$) higher clay content was found in the middle and bottom fields of cascade 2 than in the top field.

Soil TN and TC contents were significantly ($p<0.01$) different among the field positions (Table 1). In cascade 1 and 2, the average values of TN from different field positions were 0.25 and 0.24 g $kg^{-1}$ in the top soil layer, respectively. TC content for cascade 1 and 2 were 4.75 and 3.89 g $kg^{-1}$, respectively. When comparing different toposequence positions, TN content was significantly ($p<0.01$) higher in the middle field with 0.32 and 0.29 g $kg^{-1}$, followed by bottom field with 0.29 and 0.26 g $kg^{-1}$ and the lowest was observed in the top field with 0.17 and 0.18 g $kg^{-1}$ in cascade 1 and 2, respectively. In both rice cascades, similar trend like TN was also found for TC content. The highest TC content was observed in the middle fields followed by bottom field and the lowest was in the top field. There were no significant differences in TN and TC content of the soil taken before transplanting and after harvest (data not shown). Soil pH ranged from

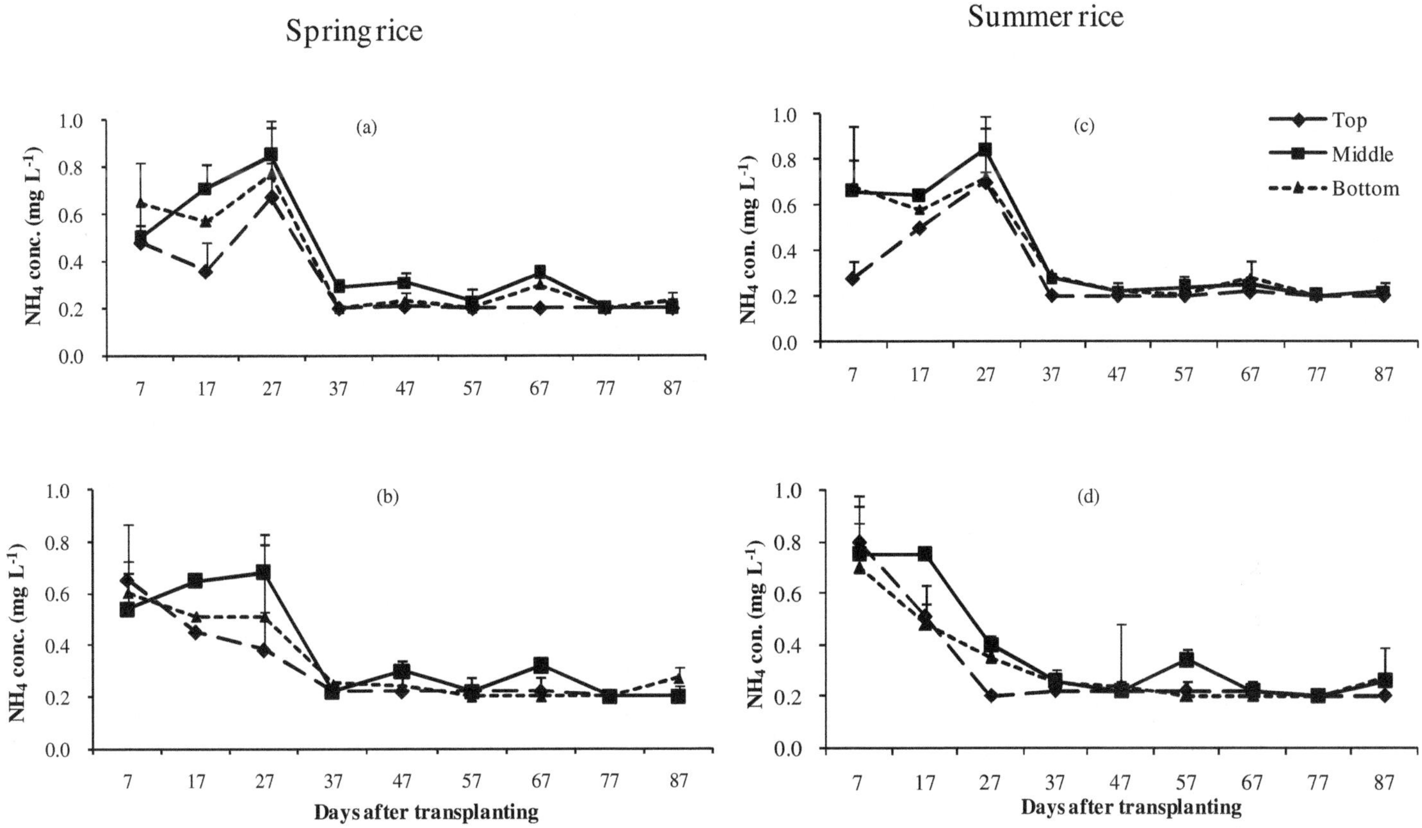


**Figure 2.** Seasonal variations in $NH_4$ concentration of surface water; (a and c -cascade 1) and (b and d – cascade 2) among the positions of toposequence rice field (bar – standard deviation).

8.0 to 8.4 in both rice cascades and significant ($p<0.05$) higher soil pH was found in bottom fields than the others (Table 1). Electrical conductivity (EC) value was found significantly higher in top field than bottom fields in both rice cascades (Table 1).

## $NH_4$ concentration in surface water

Seasonal variations in $NH_4$ concentrations were observed among the field positions in both cascades in spring rice (Figures 2a and b) and also in summer rice (Figures 2c and d). In spring rice, the concentration of $NH_4$ in surface water ranged from 0.2 to 0.85 mg $L^{-1}$ for cascade 1 and 0.2 to 0.68 mg $L^{-1}$ for cascade 2 with the highest peaks at 27 DAT. Seasonal $NH_4$ concentrations for top, middle and bottom fields in average of both cascades were 2.7, 3.5 and 3.1 mg $L^{-1}$, respectively. In summer rice, the concentrations of $NH_4$ in flood water were observed in the range of 0.2 to 0.84 mg $L^{-1}$ in cascade 1 and 0.2 to 0.8 mg $L^{-1}$ in cascade 2 with the highest peaks occurred between 17 and 27 DAT during rice growing season. Top, middle and bottom fields contained average $NH_4$ concentration in surface water of 2.8, 3.5 and 3.1 mg L-1, respectively, for both rice cascades.

## Number of effective tillers

Significant differences ($p<0.05$) in effective tiller per hill were observed among the different toposequence positions in both cascades (Table 3 and Figure 3). In spring rice, for the effect of field position on average of both cascades, number of effective tillers per hill was found the highest in the middle field than the other fields. The lowest number of effective tillers was observed at top field in cascade 1 and at bottom field in cascade 2. The fertilized part produced more effective tillers than unfertilized part in both rice cascades.

In summer rice, the middle field also showed the highest number of effective tillers in average of both cascades followed by bottom field and the lowest was found at top field. Application of fertilizer produced significantly ($p<0.05$) higher number of effective tillers than unfertilized part in both cascades.

## Yield component parameters

In spring rice, fertilizer treatment did not show any significant variation with respect to panicle length and filled grain proportion in cascade 1 and 2 (Tables 2 and

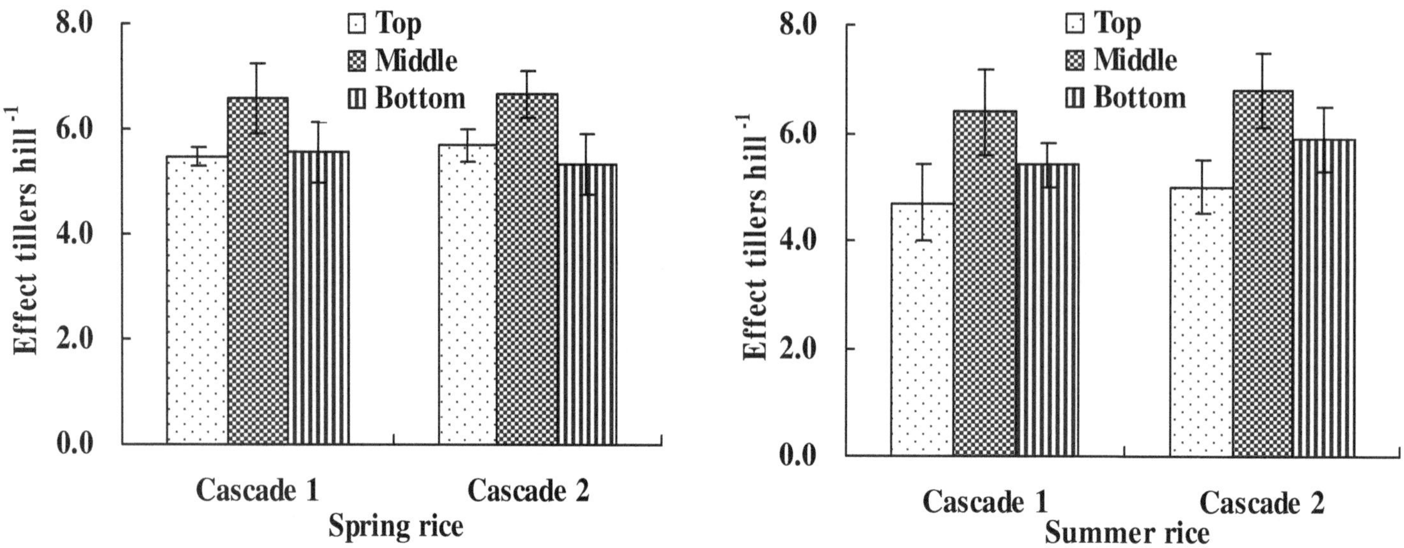


**Figure 3.** Number of effective tillers at different field position of cascade 1 and 2 in spring and summer rice seasons (Error bar indicates standard deviation).

3). The overall mean values were higher in fertilized part than unfertilized part. Regarding the effect of different toposequence positions, significant ($p<0.05$) differences in panicle length were observed with the longest in the top field followed by the middle and the lowest was observed in the bottom field in both cascades (Table 2). Middle field showed the highest filled grain proportion followed by the bottom field and the lowest was found in the top field in both cascades.

In summer rice season, fertilizer application did not show any significant variation in respect of filled grain proportion but showed significant ($p<0.05$) differences in panicle length in both cascades (Table 3). Among the field positions, middle field showed significantly higher value in panicle length and filled grain proportion than other fields in both cascades (Table 2). The variation in panicle length and filled grain proportion was influenced by interaction of fertilizer and field position only in cascade 2 of spring rice. In summer rice, only panicle length was significantly ($p<0.05$) influenced by fertilized and field position in both cascades.

In both rice seasons, significant ($p<0.05$) differences in grains per panicle and 1000 grain weight were observed due to fertilizer and different toposequence positions (Table 3). Fertilization increased grains per panicle and 1000 grain weight than unfertilized part in both rice seasons (Table 2). In spring rice, the middle field showed the highest value of grains per panicle and 1000 grain weight in both cascades. The variation in grain per panicle was also influenced by fertilizer and field position in cascade 2 and 1000 grain weight in cascade 1. In summer rice, the highest values of grains per panicle and 1000 grain weight were also observed in middle field followed by the bottom field and the lowest was found at the top field in both cascades. There were no interaction effect of fertilizer and field position on grain per panicle and 1000 grain weight in summer rice in both cascades.

### Spatial variability in grain yield

The grain yield of rice differently responded to the different toposequence positions of the fields and to the different fertilizer treatments (Figure 4 and Table 3). Grain yield was also influenced by interaction of fertilizer and field position in both rice cascades except in summer rice of cascade 1 (Table 3). In spring rice, a significant ($p<0.05$) difference of grain yield was observed between the fertilizer treatments in both rice cascades with fertilizer application showing a higher value of grain yield (+F: 644.3 g $m^{-2}$) than unfertilized plot (–F: 483.3 $m^{-2}$) in the average of cascade 1 and 2. In summer rice, significantly ($p<0.05$) higher grain yield was observed in fertilized part (+F: 599.8) than unfertilized part (-F: 480.8) in the average of both cascades.

Significant ($p<0.05$) yield differences were observed among the toposequence positions in both cascades of spring rice (Table 3) and the highest yields were observed in the middle field in both cascades. The middle field produced the highest grain yield of 644.8 g $m^{-2}$ followed by the first field; 518.2 g $m^{-2}$ and the lowest value 512.1 g $m^{-2}$ was found in the last field during spring rice season. The obtained grain yields for spring rice in cascade 1 and 2 were significantly ($p<0.01$) and positively related with TN content in the soil by a simple linear regression model (Figure 5). Grain yields were also found to be negatively related ($p<0.05$) with sand content in cascade 2 but not related in cascade 1 (Figure 6). Observed grain yields in spring rice season were significantly ($p<0.05$) related with silt content only in

**Table 2.** Effect of fertilizer and toposequence position on yield components of spring and summer rice.

| Parameter | Spring rice | | | | Summer rice | | | |
|---|---|---|---|---|---|---|---|---|
| | Panicle length (cm) | Grains panicle$^{-1}$ | Filled grain % | 1000- grain wt (g) | Panicle length (cm) | Grains panicle$^{-1}$ | Filled grain % | 1000-grain wt (g) |
| **Cascade 1** | | | | | | | | |
| **Treatment** | | | | | | | | |
| -F | 21.7 | 136.1 | 78.9 | 24.1 | 18.9 | 80.5 | 76.2 | 24.2 |
| +F | 22.5 | 154.4 | 80.6 | 25.0 | 20.3 | 101.0 | 77.6 | 25.4 |
| $Lsd_{(0.05)}$ | 1.2 | 13.2 | 4.5 | 0.8 | 0.6 | 12.1 | 6.7 | 1.0 |
| **Position** | | | | | | | | |
| Top | 23.3 | 167.0 | 76.5 | 22.4 | 19.3 | 77.9 | 76.2 | 23.6 |
| Middle | 21.5 | 185.8 | 82.7 | 25.3 | 20.4 | 101.8 | 77.9 | 25.8 |
| Bottom | 20.1 | 95.9 | 80.0 | 24.0 | 19.9 | 92.4 | 77.4 | 24.9 |
| $Lsd_{(0.05)}$ | 1.4 | 13.2 | 5.5 | 1.12 | 0.8 | 14.0 | 12.1 | 0.1 |
| **Cascade 2** | | | | | | | | |
| **Treatment** | | | | | | | | |
| -F | 21.5 | 132.1 | 72.4 | 24.2 | 19.2 | 86.0 | 72.8 | 24.0 |
| +F | 22.1 | 153.7 | 76.8 | 25.3 | 20.4 | 90.9 | 76.8 | 25.5 |
| $Lsd_{(0.05)}$ | 0.8 | 16.1 | 4.7 | 0.8 | 0.7 | 10.8 | 6.0 | 0.2 |
| **Position** | | | | | | | | |
| Top | 23.3 | 129.3 | 68.3 | 25.5 | 19.9 | 85.5 | 67.1 | 24.0 |
| Middle | 22.3 | 189.3 | 80.6 | 25.7 | 20.4 | 107.7 | 77.7 | 25.6 |
| Bottom | 19.7 | 110.0 | 74.9 | 22.3 | 19.1 | 95.6 | 73.3 | 25.1 |
| $Lsd_{(0.05)}$ | 0.9 | 19.8 | 5.8 | 1.0 | 0.8 | 13.2 | 12.1 | 0.2 |

-F and +F stand for unfertilized and fertilized part, respectively.

cascade 2 (Figure 7). In spring rice, grain yields were significantly ($p<0.05$, $r=0.812$) correlated with $NH_4$ concentration in surface water in both rice cascades.

In summer rice, the highest value 559.2 g $m^{-2}$ of grain yield was also observed in middle field followed by the bottom field; 517.7 g $m^{-2}$ and the last was in the top field; 400.2 g $m^{-2}$ in average of both cascades. The obtained grain yields for summer rice in both cascades were significantly ($p<0.01$) and positively related with TN content (Figure 5) and negatively ($p<0.01$) related with sand fraction by a simple linear regression model (Figure 6). Observed grain yields in summer rice season were significantly ($p<0.01$) related with silt content in both cascades (Figure 7). According to spearman rank order correlation analysis, grain yields were significantly ($p<0.05$, $r=0.88$) related with average concentration of $NH_4$ in surface water for both rice cascades in summer rice.

## DISCUSSION

### Influence of toposequence position on soil properties

There were substantial differences in soil texture among the different toposequence positions (Table 1). The downward movement of finer nutrient rich soil material along with irrigation and runoff water from upland were deposited to the

**Table 3.** Results of the ANOVA test for growth and yield components of rice during spring and summer rice growing period in 2011.

| | Effective tiller no. | Panicle length | Grain panicle$^{-1}$ | Filled grain (%) | 1000 grain weight | Grain yield |
|---|---|---|---|---|---|---|
| **Cascade 1** | | | | | | |
| **Spring** | | | | | | |
| Fertilizer (F) | ** | ns | * | ns | * | * |
| Field position(FP) | ** | ** | ** | ns | ** | ** |
| F x FP | * | ns | ns | * | * | * |
| **Summer** | | | | | | |
| Fertilizer (F) | * | ** | ** | ns | ** | ** |
| Field position(FP) | * | * | * | * | ** | ** |
| F x FP | ns | * | ns | ns | ns | ns |
| **Cascade 2** | | | | | | |
| **Spring** | | | | | | |
| Fertilizer (F) | * | ns | * | ns | * | * |
| Field position(FP) | ** | ** | ** | ** | ** | ** |
| F x FP | ** | ** | * | ns | ns | * |
| **Summer** | | | | | | |
| Fertilizer (F) | * | ** | * | ns | * | ** |
| Field position(FP) | ** | * | ** | * | ** | ** |
| F x FP | * | * | ns | ns | ns | * |

Ns, Not significant, ** and * indicate significant at 0.01 and 0.05 levels, respectively.

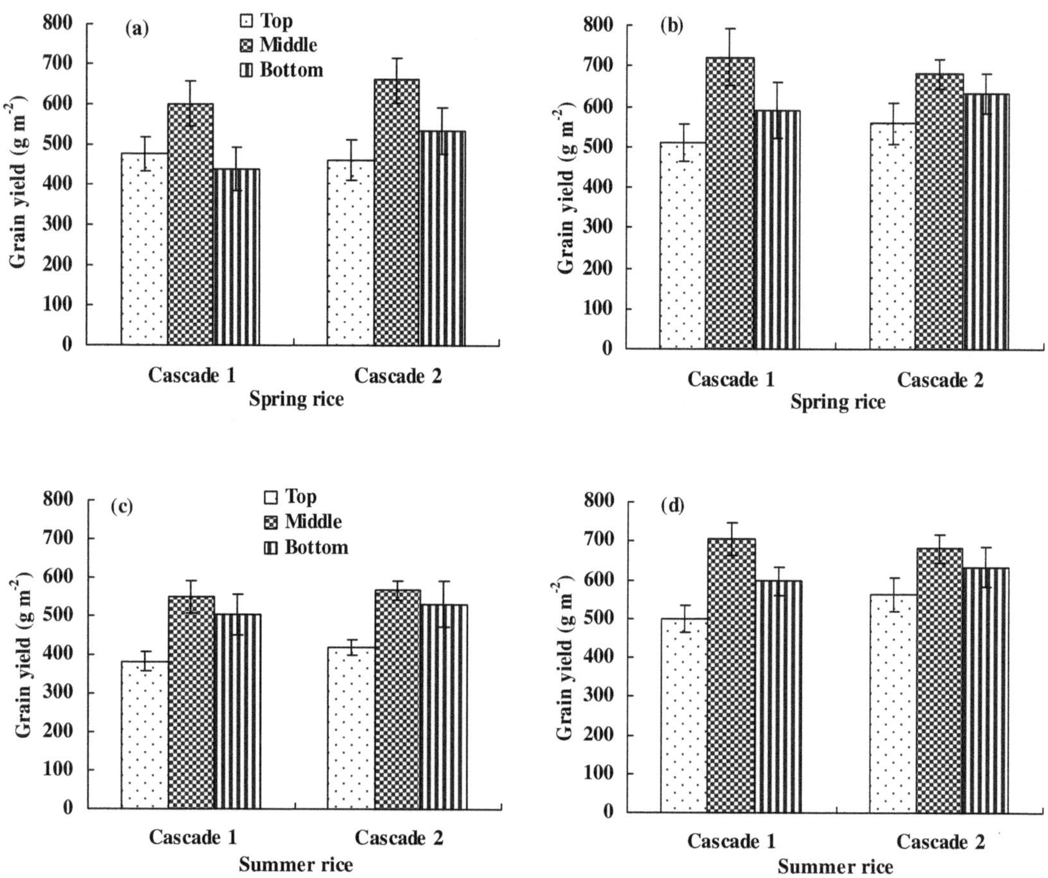


**Figure 4.** Grain yield in non-fertilized (a and c) and fertilized plots (b and d) at different toposequence positions of cascade 1 and 2 in spring and summer rice seasons (Error bar indicates standard deviation).

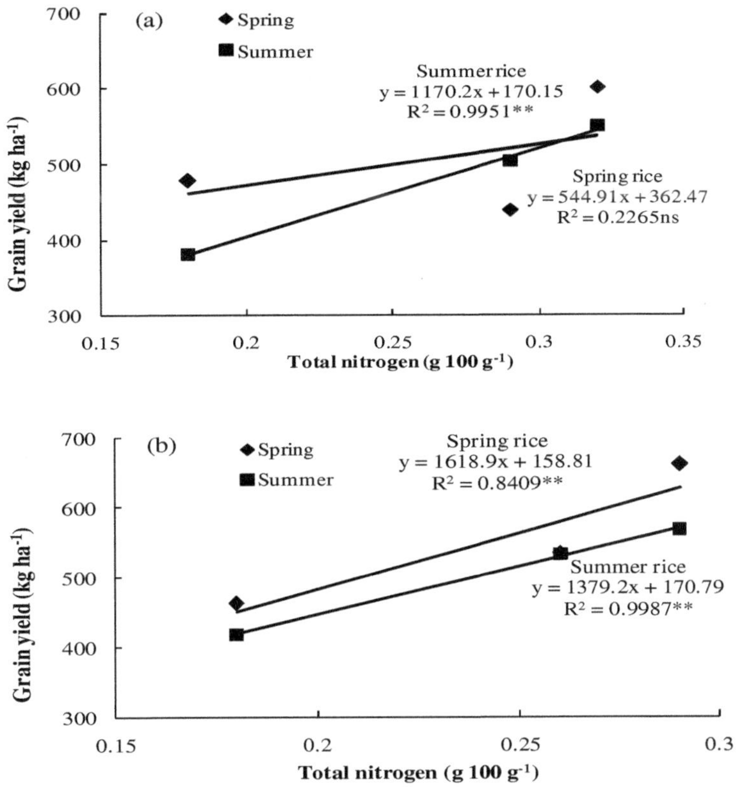


**Figure 5.** Relationship between grain yield and total nitrogen content of soil (a) Cascade 1 and (b) Cascade 2.

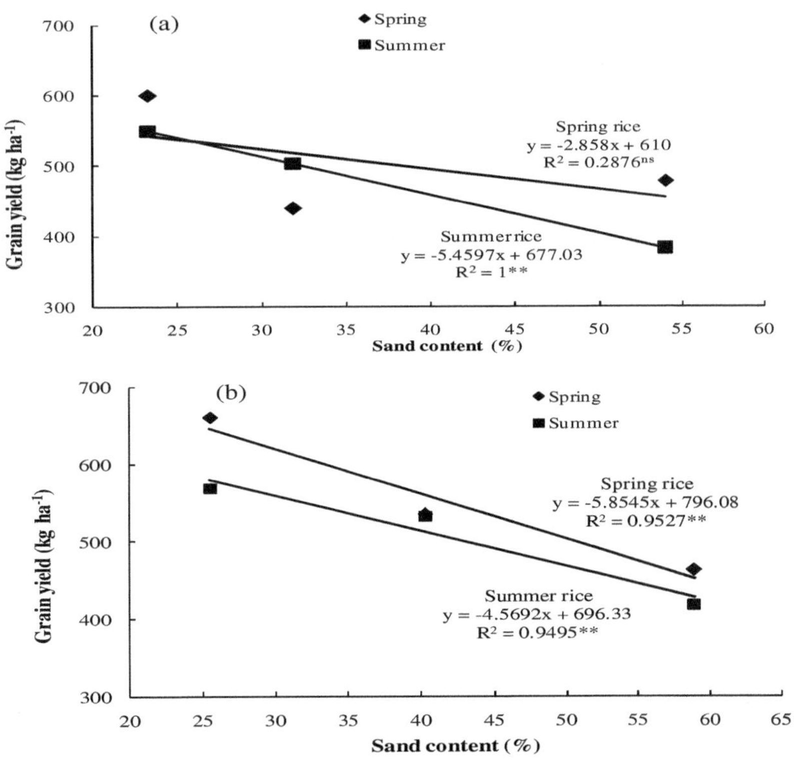


**Figure 6.** Relationship between grain yield and sand content (a) Cascade 1 and (b) Cascade 2.

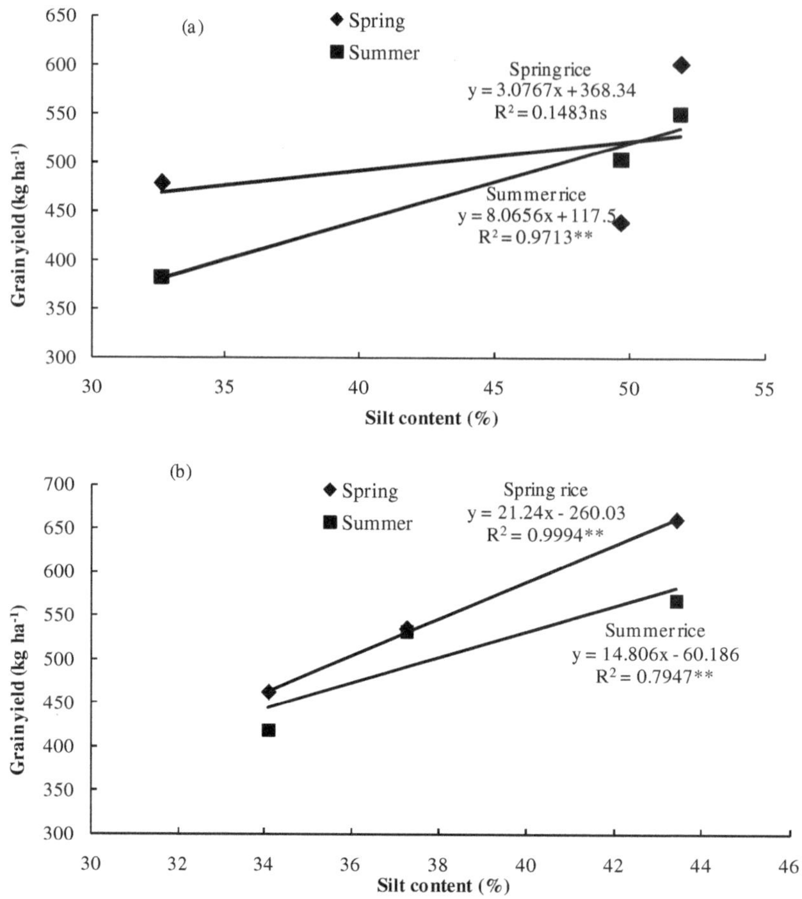


**Figure 7.** Relationship between grain yield and silt content (a) Cascade 1 and (b) Cascade 2.

lower lying fields of the cascades (Schmitter et al., 2010). Therefore, silt and clay contents were higher in the middle and bottom fields than top field. The increasing trend in clay content from top to bottom along the toposequence was also observed by Eshett et al. (1989) in Nigeria and Posner and Crawford (1992) in Senegal and Boling et al. (2008) in Thailand and Indonesia.

TN and TC content also varied among the different toposequence positions (Table 1). The first field in the cascade had lower TN and TC content with higher sand fraction than the lower fields. Increase in soil TN and TC were related to the different toposequence positions, which were influenced by the distance from the irrigation channel and possibility of material transportation between fields due to runoff water from the upper fields. Tsubo et al. (2007) also reported the downward movement of finer nutrient in soil in rainfed rice terraces. Increase in TN and TC in the middle and bottom fields were related to increase in clay and silt fraction in these fields. The results of analysis of $NH_4$ concentration in surface water showed that the middle and bottom fields of both rice cascades were higher in concentration than top fields (Figure 2). This indicates transportation of organic rich sediment material from the irrigation channel (Poch et al., 2006). In irrigated rice fields, irrigation water has been identified as a considerable source of additional particulate N and organic C to be transported into irrigated fields depending on its origin (King et al., 2009). Besides irrigation water, runoff water from the surrounding steep slopes could be identified as an additional pathway of sediment delivery into the paddy fields influencing strongly the SOC content and textural changes of the paddy soil (Schmitter et al., 2010).

## Influence of management and toposequence position on yield and yield component parameters

Cascade irrigation water can lead to an enrichment or depletion of soil fertility, which can explain the spatial variability patterns in landscape level (Schmitter et al., 2010). This study showed that the number of effective tillers was related to higher content of TN in both cascades (Figure 3). Furthermore, both cascades showed generally higher number of effective tillers in the fertilized fields than in the unfertilized ones. This corresponds to the findings of Salahuddin et al. (2009) in Mymensingh, Bangladesh. They reported that the number of effective tillers increased with increasing N levels in soil. Significant higher grains per panicle, proportion of filled grain and 1000 grain weight were observed in the middle field which might be related to higher silt and clay content, TN and TC content and $NH_4$ concentration in surface water of the middle field in both rice seasons. Only the panicle length was the highest in the first field in spring rice.

The spatial variation in grain yield observed in both cascades showed a clear difference depending on the different topsequence positions within each season (Figure 4). In both rice cascades, the grain yield increased along the cascade with increasing distance from the irrigation channel, which resembled the distribution of soil particles, and TN and TC content within the cascade. Schmitter et al. (2011) stated that the increase in grain yield towards the bottom fields of toposequence was related to an increase in soil organic C and decrease in sand content in Chieng Khoi area. Rice performance increased when finer nutrient rich sediments were deposited (Schmitter et al. 2011). In this experiment, the higher grain yield in middle fields were significantly related with the higher $NH_4$ concentration in surface water in both rice seasons. Mochizuki et al. (2006) also pointed out that the incorporation of SOC enriched sediments in lowland rice paddy fields had positive effect on grain yield. For both rice seasons, the TN and TC content and sand fraction significantly affected the obtained rice yields with higher yield at the lower situated rice fields containing higher silt, clay, TN and TC content.

## Seasonal differences

In both cascades, among the toposequence positions, the highest yield was observed in the middle fertilized fields in spring and summer rice seasons (Figure 4). Grains per panicle, effective tillers, proportion of filled grain and 1000 grain weight were also highest in the middle field. Grain yield was higher at bottom field than top field position in fertilized part of both cascades and unfertilized part of cascade 2 in spring rice. However, that was not the case at cascade 1 of unfertilized part in spring rice. This might be due to flood damage in the fields lower on the toposequence of cascade 1 which was located next to the river. Schmitter et al. (2010) pointed out that cascade 1 was affected by flooding during typhoons and discussed that flooding events in relation to sediment deposits altered the linear trend of soil fertility changes along a cascade.

In summer rice, higher grain yield was observed in bottom field than top field position in all parts of both cascades. The increase in grain yield in the bottom fields in both rice seasons might be related to an increase in TN and TC content and a decrease in sand content. Schmitter et al. (2011) reported that the grain yield increase along the cascade with increasing distance from irrigation channel resembled the soil fertility pattern observed in the top soil (that is, an increase in SOC and decrease in sand content in this watershed area). A similar spatial variability of grain yield has been reported by Homma et al. (2003), Rüth and Lennartz (2008) and Haefele and Konboon (2009) who observed increasing grain yields due to deposition of nutrient rich fine sediments at the end of a toposequence.

## Influence of fertilization

The effect of sedimentation can be seen clearly in the unfertilized part where the middle field produced the highest grain yields than other fields in both crop seasons (Figure 4). Application of the recommended amount of fertilizers increased yields over unfertilized plots but fertilizer management did not mitigate spatial variation trend that middle field also produced the highest yield in both spring and summer rice. Mochizuki et al. (2006) reported that combining chemical fertilizer with the incorporation of the sediment soil into the paddy soil increased the grain yield significantly, while without fertilizer the sediments had no effect on yield. This result was in contrast to the findings of our research, where fertilized plots produced higher yield, but still showed the effect of toposequence positions. The study by Schmitter et al. (2011) showed that an average, the fertilized fields yielded continuously more than the fields without fertilizer in Chieng Khoi area. In unfertilized part, there was a significant difference in grain yield among toposequence positions which showed the significant influence of sediment deposition from the upland to the lowland rice fields. Grain yield increased in lower lying fields when finer nutrient rich sediments were deposited depending on quality and quantity of the sediments and the deposition pathways.

## Conclusion

This study showed clearly that the spatial variations in soil properties and crop production in the cascades could be linked to sediment deposition onto the top soil. The downward movement and deposition of sediment along the toposequence positions created spatial variability patterns and influenced the crop productivity. Although recommended types of fertilizer management practice increase grain yield over unfertilized parts, toposequence specific management practice such as the same amount of fertilization in all toposequence positions should not be practiced in this watershed area. The results highlighted that spatial variations in soil properties and crop yields required field specific and field adapted management practices for each position rather than toposequence specific management practice for sustainable land use in this watershed area.

## ACKNOWLEDGEMENTS

This research was done under the Uplands Program (SFB 564) "Research for Sustainable Land Use and Rural Development in Mountainous Regions of Southeast Asia". A special expression of gratitude for their support goes to the Centre for Agricultural Research and Ecological Studies (CARES) in Vietnam. The first author would like to thank all the colleagues of the Uplands Program for their support during research work.

### REFERENCES

Boling AA, Tuong TP, Suganda H, Konboon Y, Harnpichitvitaya D, Bouman BAM, Franco DT (2008). The effect of toposequence position on soil properties, hydrology, and yield of rainfed lowland rice in Southeast Asia. Field Crop Res. 106:22-33.

Dung NV, Vien TD, Lam NT, Tuong TM, Cadish G (2008). Analysis of the sustainability within the composite swidden agroecosystem in Northern Vietnam. 1. Partial nutrient balance and recovery times of upland fields. Agric. Ecosys. Environ. 128(1-2):37-51.

Eshett ET, Omueti JAI, Juo ASR (1989). Soil properties and mineralogy in relation to land use on a sedimentary toposequence in South-Eastern Nigeria. J. Agric. Sci. 112:377-386.

Gao P, Pasternack GB, Bali KM, Wallender WW (2007). Suspended-sediment transport in an intensively cultivated watershed in Southeastern California. Catena 69(3):239-252.

Gee GW, Bauder JW (1986). Particle-size Analysis. In: A. Klute (Editor), Methods of Soil Analysis. Part 1. 2nd ed. Agron. J. 9:383-399.

Haefele SM, Konboon Y (2009). Nutrient management for rainfed lowland rice in Northeast Thailand. Field Crop Res. 114:374-385.

Homma K, Horie T, Shiraiwa T, Supapoj N, Matsumoto N, Kabaki N (2003). Toposequential variation in soil fertility and rice productivity of rainfed lowland paddy fields in mini-watershed (Nong) in Northeast Thailand. Plant Prod. Sci. 6:147-153.

Hseu ZY, Chen ZS (2001). Quantifying soil hydromorphology of a rice-growing ultisol toposequence in Taiwan. Soil Sci. Soc. Am. J. 65:270-278.

King AP, Evatt KJ, Six J, Poch RM, Rolston DE, Hopmans JW (2009). Annual carbon and nitrogen loadings for a furrow-irrigated field. Agric. Water Manage. 96:925-930.

Lam NT, Patanothai A, Limpinuntana V, Vityakon P (2005). Land-use sustainability of composite swiddening in the uplands of Northern Vietnam: nutrient balances of swidden fields during the cropping period and changes of soil nutrients over the swidden cycle. Int. J. Agric. Sustain. 3(1):1-12.

Mingzhou Q, Jackson RH, Zhongjin Y, Jackon MW, Bo S (2007). The effects of sediment-laden waters on irrigated lands along the lower Yellow River in China. J. Environ. Manage. 85(4):858-865.

Mochizuki A, Homma K, Horie T, Shiraiwa T, Watatsu E, Supapoj N, Thongthai C (2006). Increased productivity of rainfed lowland rice by incorporation of pond sediments in Northeast Thailand. Field Crop Res. 96:422-427.

Moormann FR, Veldkamp WJ, Ballaux JC (1977). The growth of rice on a toposequence- a methodology. Plant Soil 48(3):565-580.

Pansak W, Hilger TH, Dercon G, Kongkaew T, Cadisch G (2008). Changes in the relationship between soil erosion and N loss pathways after establishing soil conservation systems in uplands of Northeast Thailand. Agric. Ecosys. Environ. 128:167–176.

Poch RM, Hopmans JW, Six JW, Rolston DE, McIntyre JL (2006). Considerations of a field-scale soil carbon budget for furrow irrigation. Agric. Ecosys. Environ. 113(1-4):391-398.

Posner JL, Crawford EW (1992). Improving fertilizer recommendations for subsistence farmers in West Africa: the use of agro-economic analysis of on-farm trials. Fert. Res. 32:333-342.

Rambo AT (1998). The composite swiddening agroecosystem of the Tay Ethnic Minority of the Northwestern Mountains of Vietnam. In: Land Degradation and Agricultural Sustainability: Case Studies from Southeast and East Asia, ed. by A. Patanothai. KhonKaen, Thailand: Regional Secretariat, the Southeast Asian Universities Agroecosystem Network.

Rüth B, Lennartz B (2008). Spatial variability of soil properties and rice yield along two catenas in Southeast China. Pedosphere 18:409-420.

Salahuddin KM, Chowhdury SH, Munira S, Islam MM, Parvin S (2009). Response of nitrogen and plant spacing of transplanted aman rice. Bangladesh. J. Agric. Res. 34(2):279-285.

Schmitter P, Dercon G, Hilger T, Thi Le Ha T, HuuThanh N, Lam N, DucVien T, Cadisch G (2010). Sediment induced soil spatial variation in paddy fields of Northwest Vietnam. Geodama 155:298-307.

Schmitter P, Dercon G, Hilger T, Hertel M, Treffner J, Lam N, Duc Vien T, Cadisch G (2011). Linking spatio-temporal variation of crop response with sediment deposition along paddy rice terraces. Agric. Ecosys. Environ. 140:34-45.

Tsubo M, Basnayake J, Fukai S, Sihathep V, Siyavong P, Sipaseuth, Chanphengsay M (2006). Toposequential effects on water balance and productivity in rainfed lowland rice ecosystem in Southern Laos. Field Crop Res. 97:209-220.

Tsubo M, Fukai S, Basnayake J, TO PT, Bouman B, Harnpichitvitaya D (2007). Effects of soil clay content on water balance and productivity in rainfed lowland rice ecosystem in Northeast Thailand. Plant Prod. Sci. 10 (2):232-241.

UNESCO (United Nations Educational, Scientific and Cultural Organization) (1974). FAO/UNESCO Soil map of the world, 1:5,000,000 Vol. 1 Paris: UNESCO.

Wezel A, Steinmüller N, Friederishen JR (2002). Slope position effects on soil fertility and crop productivity and implications for soil conservation in upland Northwest Vietnam. Agric. Ecosys. Environ. 91:113-126.

Yamauchi M (1992). Growth of rice plants in soils of toposequence in Nigeria. Jpn. J. Trop. Agric. 36:94-98.

# Nutrient composition of *Allanblackia Paviflora* seed kernels and oil compared with some plant fats and oils and application of the oil in soap preparation

**Joseph Adubofuor[1]*, Wilfred Sefah[2] and James Henry Oldham[3]**

[1]Department of Food Science and Technology, Kwame Nkrumah University of Science and Technology, Kumasi, Ghana.
[2]Department of Hotel, Catering and Institutional Management, Kumasi Polytechnic, Kumasi, Ghana.
[3]Department of Biochemistry and Biotechnology, Kwame Nkrumah University of Science and Technology, Kumasi, Ghana.

**The Allanblackia tree is found in parts of West, Central and East Africa and grows primarily in tropical rainforests. The oil from the seeds has been used by local communities but not on a commercial scale. The study was conducted to determine the nutrient composition of *Allanblackia paviflora* seeds and compare with some plant fats as well as assess the use of the oil in soap preparation. The oil was extracted by a screw press and also by solvent extraction using petroleum ether. The proximate composition showed that the dried seed kernels have a high fat content of 67.59%, low protein and ash contents of 4.27 and 1.98%, respectively. The major minerals in the seeds were potassium and phosphorus. The predominant fatty acids in the extracted oils were stearic acid and oleic acid with contents of about 52 and 45% respectively and the lowest amount was palmitic acid. The extremely high stearic acid content and very low palmitic acid content distinguished Allanblackia oil from palm and palm kernel oils. Sensory evaluation of soap samples prepared from the oil indicated that the soap had good lathering ability and washing efficiency.**

**Key words:** *Allanblackia paviflora*, nutrient composition, plant fats and oils, sensory evaluation.

## INTRODUCTION

*Allanblackia* belongs to the Guttiferae family, subfamily Clusiodeae and is a non-timber forest product (NTFP) which can be found as wild and undomesticated tree species in the rainforests of African countries such as Nigeria, Kenya, Congo, Angola, Cameroon, Tanzania, Uganda and Ghana (http://www.abs-initiative.info/uploads/media/Addis_0909_Muchugi_01.pdf; Bürkle and Palenberg, 2009).

*Allanblackia* is an evergreen forest tree confined to tropical Africa, to a height of 30 m tall. *Allanblackia* trees of which there are only nine similar species, occur from Sierra Leone to Tanzania. Most attention is currently focused on *Allanblackia paviflora* (Upper Guinea and, Ghana westwards), *Alanblackia floribunda* (from Benin and Nigeria east to eastern DR Congo and Angola), *Allanblackia stuhlmannii and Allanblackia ulugurensis* which occur in East Africa in the mountains of Eastern Tanzania. The tree is common in the wild and is frequently used as a shade tree for cocoa. *A. parviflora* is sometimes considered a synonym of *A. floribunda*. However, their areas of distribution are disjunct. The two species are very similar, but *A. floribunda* can be distinguished by the folded disk glands of the male flowers and longer pedicel. The fats from the seeds of *A. floribunda* and *A. stuhlmannii* are very similar in comparison to that of *A. parviflora*. *A. Parviflora* is an oil-rich tree crop that can also be used as shade tree and for medicinal purposes. They are mostly found in the wild forests in the Western, Ashanti and Eastern regions of

*Corresponding author. E-mail: jadubofuor@gmail.com.

Ghana (Ofori et al., 2008; Orwa and Oyen, 2007; Stucki, 2005).

In Ghana and from the uplands of the Eastern Arc Mountains in Tanzania to the lowlands of Sierra Leone in West Africa, the oil-rich seeds of the Allanblackia tree have been harvested by local communities for generations for food, to make cooking oil, and in more recent times for soap, similar to Elaeis palm oil, but have not been used on a commercial scale. Allanblackia oil has a higher melting point and in cooking it can congeal before the food is eaten. Originally collected mainly from the wild, the domestication of this tree and increased demand for the oil is an opportunity to boost incomes for African farmers. The increase in demand has come from potential uses of the oil identified by the international company Unilever for use in food products like spreads; vegetable based dairy products and ice cream. The seeds are valued because of the highly nutritive oil which can be used in the food and cosmetic industries (Buss and Tissari, 2010; Bürkle and Palenberg, 2009). Seeds and fruits are edible. The seeds yield vegetable fat while the bitter seedcake is used as an animal feed (Orwa et al., 2009). The seed oil of *A. paviflora* is of prime importance as a foreign exchange earner and is being developed as a rural based enterprise in many African countries notably Ghana, Nigeria, Cameroon and Tanzania (Shrestha and Akangaamkum, 2008).

Allanblackia oil is very similar to current oils in that it consists of familiar triglycerides in commonly used vegetable oils. It contains minor constituents like tocopherol and has good storage stability. The oil can also be mixed with other fats and oils to modify physical properties. Its composition and relatively high melting point makes the fat a valuable raw material that can be used without transformation to improve the consistency of margarines, cocoa butter substitutes and similar products. On a commercial scale, the use of the oil has been at a low level. In most parts of Africa, the traditional use has decreased over the last 50 years and been in favour of other liquid oils. Nevertheless, Allanblackia has been identified by FAO as a crop of high potential interest to the development of rural communities and the seed oil has the potential to become a new profitable and sustainable raw material which could be recognized internationally (http://www.allanblackiaghana.com/pages/FAG.html).

The objectives of this research work was to determine the nutrient composition of Allanblackia seed oil and compare with some plant fats and also to assess the use of the oil in soap preparation.

## MATERIALS AND METHODS

### Source and preparation of *A. paviflora* seeds

The seeds of *A. paviflora* used for the study were manually removed from the fruits obtained from a village in New-Edubease in the Ashanti Region of Ghana. Twenty kilograms (20 kg) of the seeds were washed and sun dried for three days. The seeds were then stored in polyethylene bags under room temperature for two weeks after which they were dehulled and milled for the analyses. Figures 1 and 2 shows the whole dried dehulled and milled seed kernels of *A. paviflora*.

### Proximate analyses of Allanblackia Kernels

The methods for the determination of the proximate composition of the kernels were based on AOAC (1990). Analyses were carried out in triplicates. Reagents were obtained from the Department of Biochemistry and Biotechnology and Department of Chemistry, KNUST.

### Determination of the mineral composition

In the mineral analyses, 2 g of the dried sample was ashed in a muffle furnace at 600°C for 2 h. The ash was then dissolved in 10 ml of 5 M HCl. The dissolved sample was filtered into a 50 ml volumetric flask and diluted to the 50 ml mark. The sample was then directly analyzed for the minerals. Using the Atomic Absorption Spectrophotometer (AAS) (Model: UNICAM 929AA Spectrophotometer, UK), the mineral analyzed were Sodium (Na), Copper (Cu), Potassium (K), Calcium (Ca), Phosphorus (P) and Manganese (Mn) (James, 1995).

### Extraction of Allanblackia oil

Two oil samples were extracted by mechanical and solvent extraction. The mechanical extracted oil (designated as SPO) by means of a screw press was based on the optimum conditions established by Sefah et al. (2010) by heating the sieved milled samples with a moisture content of 13% to a temperature of 100°C for 2 h. Figure 3 shows the manual screw press used for the mechanical extraction of the oil. Solvent extracted oil (SEO) was by means of the Soxhlet apparatus and extraction was carried out using petroleum ether as the solvent for 16 h.

### Qualitative test of phytochemicals in the Allanblackia oil

Terpenes and carotenoids were determined qualitatively by methods described by Asante (1993) and analyses on saponins and tannins were based on the method outlined by Harbone (1984).

### Fatty acid profile of Allanblackia oil

#### *Preparation of fatty acid methyl esters*

0.1 g of the oil was dissolved in 20 ml of diethyl ether followed by the addition of 2 ml of 2 M methanolic KOH. The mixture was well corked and then swirled for thorough mixing. It was then centrifuged for 15 min at 250 rpm and aliquots of the organic layer were taken for analyses.

Gas chromatographic method was used for the determination of the fatty acids. The fatty acid analyses was carried out by using a Perkin Elmer, Autosystem XL, Gas Chromatogragh (USA) equipped with a flame ionization detector (FID), coupled to a Varian integrator and chart recorder. A column of diethylene glycol succinate (DEGS) of 15% packed on Chromosorb W-AN-I20 mesh was used. The flow rate of the carrier gas, $N_2$, was 22.6 ml/min. and the F.I.D. hydrogen flow rate was 300 ml/min. The GC was set on attenuation 4, column temperature of 160°C and detector temperature of 200°C. Volume injected was 1 µl. Identification and quantification of fatty acids were

**Figure 1.** Dehulled seed kernels of *Allanblackia parviflora*.

**Figure 2.** Milled *A. paviflora* seed kernels.

based mainly on direct comparison of the gas chromatogram response to available fatty acid methyl ester standards.

### Determination of the energy content

The energy content of the kernels was determined by calculation based on Atwater factors by using 17 kJ/g for protein and carbohydrate and 37 kJ/g for fat.

### Preparation of soap from Allanblackia oil

100 g of the Allanblackia fat was weighed into a 1litre beaker. The oil was then heated and saponification initiated by adding 20 ml of 23.5% NaOH. Sixty (60.0) g NaOH pellets was then added gradually while stirring until saponification was completed (the mixture was no longer oily). Eight grams of salt (NaCl) in the form of saturated solution was then used to grain out the soap. The soap obtained was washed with water and filtered through a linen cloth. It was air-dried and a small amount of water was added to soften it whilst heating. The molten soap was then poured in casts and allowed to dry (Asante, 1993).

### Sensory evaluation prepared soap from Allanblackia oil

The lathering ability, washing efficiency, colour, hardness and odour were physical properties determined by 53 panelists using a seven-point hedonic scale. The score was one (1) for very good

**Figure 3.** Manual screw press used for oil extraction.

and seven (7) for very bad. The colour was assessed by visual inspection whilst the odour was detected by smelling. The hardness was determined by pressing the soap in between the thumb and the third finger. Lather was formed by vigorously shaking the soap in about 200 ml of water between 2 and 5 min. The soap was used to wash an oil-stained cloth to measure its washing efficiency (Asante, 1993).

### Chemical properties of prepared soap from Allanblackia oil

#### *Total fatty matter*

Five grams of the finished soap was dissolved in 30 ml distilled water. The fatty acid was precipitated from the salt (neutral soap) present in the solution with 0.5 M HCl. The solution was filtered and diethyl-ether was added to the precipitated fatty acid to aid drying. The dried fatty acid obtained was then weighed and the percentage fatty acid determined (Asante, 1993; Jolly, 1963).

#### *Free caustic alkali*

Five grams of a sample of the soap was weighed and dissolved in 30 ml of 95% ethanol. A few drops of phenolphthalein indicator and 10 ml of 20% $BaCl_2$ were added. The resulting solution was then titrated against 0.05 M $H_2SO_4$. The volume (v) of the acid obtained was multiplied by the factor 0.31 and divided by the weight of sample taken to obtain the percent $Na_2O$ present (Asante, 1993; Jolly, 1963).

### Statistical analyses

The data obtained was subjected to statistical analyses using analysis of variance (ANOVA). Differences among the mean values were determined using Duncan's multiple range test.

## RESULTS AND DISCUSSION

Table 1 shows the proximate composition and energy content of *A. paviflora* kernels compared with shea kernels and cocoa beans. The seeds contain a fat that is solid at ambient temperatures. The kernel, which makes up about 60 to 80% of the whole seed weight, contains about 72% fat. In some areas in Africa the seeds are eaten by children as a high-energy snack (Orwa et al., 2009; Orwa and Munjuga, 2007). Allanblackia kernels have comparatively very high content of fat and energy.

**Table 1.** Proximate composition of Allanblackia seed kernels compared with cocoa and shea kernels.

| Component | Allanblackia | Shea kernels[1] | Cocoa beans[2] |
|---|---|---|---|
| Moisture (%) | 3.40±0.03 | 3.3 | 5 |
| Fat (%) | 67.59± 0.02 | 31.7 | 58 |
| Protein (%) | 4.27±0.01 | 7.7 | 13 |
| Ash (%) | 1.98±0.11 | 4.6 | 3 |
| Crude Fibre (%) | 5.70± 0.02 | 6.2 | 10 |
| Carbohydrate (%) | 17.06 | 46.6 | 11 |
| **Energy (kJ/100 g)** | **2863.44** | **2085.80** | **2554** |

Source: [1]Ugese et al. (2010), [2]Bockish (1993).

**Table 2.** Mineral content of Allanblackia seed and shea kernels (mg/kg).

| Type of mineral | Allanblackia | Shea kernels[2a] |
|---|---|---|
| Sodium(Na ) | 5.24 | 12.17± 0.83 |
| Copper (Cu) | 0.04 | - |
| Potassium (K) | 8.41 | 1.55±0.03 |
| Manganese (Mn) | 0.13 | 4.15±0.78 |
| Calcium (Ca) | 0.10 | 0.28±0.343 |
| **Phosphorus (P)** | **8.34** | - |

Source: [2a]Alhassan et al. (2011).

The seeds are comparatively low in protein, ash and fibre content. The results of six minerals analyzed and compared to shea kernels are shown in Table 2. With the exception of potassium, the minerals in shea kernels are higher than the amounts in Allanblackia kernels. The main minerals essential for human body include sodium, potassium, magnesium, calcium and phosphorus and these are required in amounts greater than 50 mg/day. The values obtained for the main minerals were lower than the daily required amounts. The minimum daily requirement of potassium is estimated to be 782 mg. Trace elements such as manganese and copper are essential in concentrations of less than 50 mg/day. The main and trace elements have varied functions in the body and play vital roles in metabolism and the functions of nerves, muscles and enzymatic systems in the human body. The concentration of copper was less than the maximum level of 0.4 mg/kg specified by Codex Allimentarius Commission. Copper is required in trace amounts and is poisonous at high concentrations. Manganese is an important component in a number of systems and is present in all living tissues. The concentration in the seed kernels was lower than the recommended daily intake of 2 to 5 mg (Ronsivalli and Viera, 1992; Belitz et al., 2004; CODEX STAN 211 -1999)

Table 3 shows the results of qualitative tests of the presence of phytochemicals in the Allanblackia oil samples obtained by solvent extraction (SEO) and mechanical extraction (SPO) using the screw press. Phytochemicals are naturally occurring, biologically active chemical compounds in plants which provide colour, aroma, flavour and also act as a natural defense system for host plants. Carotenoids are plant pigments which are responsible for colour (CFNI, 2005). Tannins contribute to taste in terms of bitterness and astringency. Terpenes impart pleasant smell and exhibit special pharmacological activities. Saponins have potent antifungal properties and have bitter taste (Ajayi et al., 2011; http://www.wiley-vch.de/books/sample/3527317864_c01.pdf). The tests indicated the absence of carotenoids, terpenes, saponins and tannins in both samples. It could however be that some of these phytochemicals were present in very small quantities below detection by the methods used in the analyses.

## Fatty acid profile of Allanblackia oil and its comparison with other vegetable fats

The fatty acid profile of Allanblackia oil and its comparison with other vegetable fats is shown in Tables 4 and 5. The Allanblackia seed oil is low in palmitic acid and is solid at room temperature. The fatty acid composition of stearic acid and oleic acid has been reported to range from 45 to 58 and 40 to 51% respectively (Orwa and Munjuga, 2007). The values obtained for the two samples in Table 3 fell within the reported ranges. In Table 5, which shows the average

**Table 3.** Qualitative analyses of lipid soluble phytochemicals in Allanblackia oil.

| Sample | Carotenoids | Terpenes | Saponins | Tannins |
|---|---|---|---|---|
| SPO | Absent | Absent | Absent | Absent |
| SEO | Absent | Absent | Absent | Absent |

SPO, Screw-pressed oil; SEO, solvent extracted oil.

**Table 4.** Fatty acid characteristics of Allanblackia oil, borneo (tallow fat), shea butter and cocoa butter.

| Fatty acids (%) | Allanblackia | Shea Butter[4] | Cocoa Butter[3] | Borneo Tallow fat[3] |
|---|---|---|---|---|
| Palmitic | 2.93 | 4 | 25 | 20 |
| Stearic | 52.25 | 42 | 37 | 42 |
| Oleic | 44.82 | 45 | 34 | 36 |

Source: Belitz[3] et al. (2004); [4]Alander and Anderson (2002).

**Table 6.** Fatty acid characteristics of Allanblackia oil, palm oil and palm kernel oil.

| Fatty acids (%) | Oleic acid (C 18:1) | Stearic acid (C 18:0) | Palmitic acid (C 16:0) | Myristic acid (C 14:0) | Lauric acid (C 12:0) |
|---|---|---|---|---|---|
| Allanblackia oil[4] | 40-51 | 45-58 | 0-2 | 0-1 | 0-1 |
| Palm oil[2] | 26-52 | 1-8 | 32-59 | 0-7 | 0-2 |
| Palm kernel oil[2] | 10-23 | 1-3 | 7-11 | 14-20 | 41-55 |

[2]Bockish (1993); [4]EFSA (2007).

fatty acid compositions of both SEO and SPO, stearic acid content of Allanblackia is relatively higher than the other fats. The oleic acid composition compares very well with shea butter and is comparatively higher than the amounts found in cocoa butter and Borneo tallow fat. Cocoa butter and fats (solid at room temperature) belong to the oils and fats rich in palmitic and stearic acids with the latter referred to as cocoa butter substitutes (or cocoa butter interchangeable fats). They are relatively hard and can crystalline in several polymorphic forms. Their melting points are between 30 and 40°C. When cocoa butter melts in the mouth a pleasant cooling sensation is experienced. This is characteristic of only a few types of triacylglycerols present in fats which contain predominantly palmitic, oleic and stearic acids. The fatty acid composition is also reflected by the resistance of these fats to auto-oxidation and microbiological deterioration. These fats are utilized preferentially in the manufacture of candy, chocolates and confections (Belitz et al., 2004). Although the major fatty acids represented in Table 4 show some degree of variation, the four plant fats are characterized by high levels of stearic and oleic acids and can similar be used together with cocoa butter as raw materials in the cosmetic and pharmaceutical industries.

Table 5 shows the fatty acid profile of *A. paviflora* oil, palm oil and palm kernel oil. The predominant fatty acids in Allanblackia oil being oleic acid and stearic acid are very high compared to the amounts found in palm oil and palm kernel oil. Allanblackia oil differs uniquely and distinctively from the other two oils by the extremely high stearic acid content ranging between 45 and 58% as against the respective amounts of 1 to 8 and 1 to 3% in palm oil and palm kernel oil, respectively.

Palmitic acid is the predominant saturated fatty acid in palm oil and in this sense, clearly distinguishes itself from the other two oils by the higher amounts of the fatty acids ranging from 32 to 59%. Palm kernel oil also differs completely from the other two oils in terms of the relatively high percentage composition of myristic and lauric acids. Altogether the saturated fatty acid composition of palm kernel oil ranges between 55 and 75%.

In terms of the monounsaturated fatty acid (MUFA), the three oils have considerable amounts with *A. paviflora* oil having the highest percentage composition of 40 to 51% followed by palm oil with a percentage composition of 26 to 52%. Among the saturated fatty acids (SFAs), stearic acid is comparatively very high. Major food sources of stearic acid are meat, poultry, fish, grain products and milk/milk products. Fats rich in stearic acid include cocoa butter (typically consumed as chocolate), beef tallow, lard and butter (Kris-Etherton et al., 2005). Contrary to the perception that all SFAs are alike in terms of their cardiovascular disease risk, research demonstrates that

**Table 6.** Percentage composition of triglycerides in Allanblackia oil, cocoa butter and shea butter.

| Type of fat | Tryglyceride **fraction** | Composition (%) |
|---|---|---|
| Shear butter[5] | SOS | 40 |
| | SOO | 27 |
| | POS | 6 |
| Allanblackia oil[4] | SOS | 69.1 |
| | SOO | 23 |
| Cocoa butter[6] | SOS | 26.4 |
| | POS | 42.1 |
| | **POP** | **18.3** |

Source: [5]Alander (2004), [4]EFSA (2007), [6]Lipp et al. (2001).

individual SFAs differ in their effects on blood cholesterol levels. Although it is classified as a saturated fatty acid, stearic acid is unique among the SFAs, in that investigations (in both human and animal studies) have consistently shown that stearic acid, unlike other long chain SFAs (those longer than 10 carbon chains) has no effect on blood total, LDL cholesterol (that is, 'bad' cholesterols) and high density lipoprotein (HDL) (that is, the 'good' cholesterol) cholesterol levels in adults. Unlike the other predominant long chain SFAs which are palmitic (C16:0), myristic (C14:0) and lauric (C12:0) acids (which increase blood cholesterol levels), stearic acid has been shown to have a neutral effect on blood total and low density lipoprotein cholesterol levels. This implies that this long chain SFA may not increase the risk for cardiovascular disease. Several possible explanations have been proposed to explain why stearic acid does not raise blood cholesterol levels and it has been a subject of considerable research. As to what biological mechanism(s) underlines stearic acid's regulation of cholesterol metabolism remains a question to be resolved (Kris-Etherton et al., 2005; Haumann, 1998; Grundy, 1994; Pearson, 1994; Mensink, 2005).

## Triglyceride fraction

Table 6 shows the tryglyceride composition of Allanblackia oil, cocoa butter and shea butter. In shea butter the most common combinations of fatty acid is SOS (S-stearic, O-oleic) making up to 40% of the total trigylcerides followed by SOO (27%) and POS (6%) (P-palmitic). The study of Di Vincenzo et al. (2005), found that SOO, OOO and SOS are the major triglycerides in shea butter with regional variation. The distinctive triglyceride composition of shea butter and Allanblackia dominated by SOS and SOO makes it easy to separate high melting stearin fraction from the low melting olein fraction by fractionation and the resultant stearin could be further used in the mixture with cocoa butter equivalents. Shea butter is highly regarded in the cosmetic field because of its high emolliency and moisturization capacities. The cosmetic and soap making industries can use shea butter to formulate cosmetics, soaps, shampoos, creams and balsams for hair and other skin cleansing and rejuvenating products (Alander, 2004). The similar characteristics of Allanblackia with shea butter, makes it potentially and comparably a useful raw material also in the cosmetic industry. Cocoa butter triglycerides are composed predominantly of POS, SOS and POP. Others include SOO and SLS (Lipp et al., 2001).

## Sensory evaluation and properties of *Allanblackia* soap

The sensory properties of the Allanblackia soap samples evaluated by 53 panelists who used the soap to wash an oil-stained cloth are represented in Table 7. Figure 4 shows a sample of soap prepared from Allanblackia oil. The control sample used was Key bar soap, which is popularly sold in the market and usually used by many households for washing. There was no significant difference ($p > 0.05$) in the physical properties of the soap produced from the screw-pressed oil (SPO) and the solvent extracted oil (SEO) in terms of texture, lathering ability, washing efficiency and odour. Both soap samples showed good lathering ability and washing efficiency. The texture of both soap samples was found to be very hard and had a mild (that is, neither strong nor weak) odour).

Comparatively, the two soap samples produced from *Allanblackia* fat were harder than the Key bar soap but had similar washing efficiency. Also, there was a significant difference ($p < 0.05$) between *Allanblackia* soap and the Key bar soap in terms of lathering ability and odour. The lather formed from the Key bar soap was very good compared to the *Allanblackia* soap. In terms of overall acceptability, the Key bar soap was preferred compared to the *Allanblackia* soap. This might be due to its very good lathering ability, colour and odour. The *Allanblackia* soap samples were produced without addition of colourants and aroma inducing materials but due to the creamy colour of the soap produced from the screw-pressed oil, it was more preferred, compared to the soap produced from the solvent extracted oil.

The chemical properties of the soap samples were significantly different ($p<0.05$). Comparatively, the solvent extracted fat had the highest total fatty matter and less moisture content than the screw-pressed oil. Both samples had no excess caustic alkali in the soap. The total fatty matter, free caustic alkali and moisture content of the soap prepared from all the two oil samples were found to satisfy regulations set by the Ghana Standards Authority.

## Conclusion

*A. paviflora* kernels are high yielding oilseed which can

**Table 7.** Physical and chemical properties of *Allanblackia* soap.

| Soap sample | Physical properties | | | | |
|---|---|---|---|---|---|
| | Texture | Lathering ability | Washing efficiency | Odour | Colour |
| SPO | Very hard | Good | Good | Neither strong nor weak | Cream |
| SEO | Very hard | Good | Good | Neither strong nor weak | White |
| Key soap | Neither hard nor soft | Very good | Good | Strong | Pale Yellow |

| | Chemical properties | | |
|---|---|---|---|
| | Moisture (%) | Total fatty matter (%) | Free caustic alkali |
| SPO | $27.70 \pm 0.13^{a}$ | $59.96 \pm 0.02^{a}$ | Below detection |
| SEO | $26.52 \pm 0.11^{b}$ | $61.89 \pm 0.04^{b}$ | Below detection |
| GSA | < 28 | > 59 | < 0.25 |
| LSD | 1.094 | 0.522 | - |
| SED | 0.254 | 0.1213 | - |

Values with different superscripts (a and b) in the same column are significantly different ($p<0.05$); LSD, least significant differences of means (at 5%); SED, standard error of differences of means. GSA, Ghana Standards Authority.

**Figure 4.** A prepared soap from Allanblackia oil.

serve commercially as a rich source of oil. Proximate composition showed a high fat content of 67.59% in *A. paviflora* kernels and low amounts of protein and minerals. Comparatively, the mineral content in Allanblackia are lower than the amounts in shea kernels. The fatty acid profile of solvent and screw-pressed oils indicated stearic acid and oleic acid as the predominant acids with contents of about 52 and 45% respectively. The extremely high stearic acid content and very low palmitic acid content distinguished Allanblackia from palm and palm kernel oils. The percentage composition of the triglycerides shows that *A. paviflora* can be used as an alternative vegetable oil to palm oil, cocoa butter and shea butter in the production of food and non-food products. Sensory evaluation of the properties of soap samples prepared from the oil indicated that the soap had good lathering ability and washing efficiency and was comparatively harder than Key soap which was used as the control sample. There was a significant difference ($p< 0.05$) between the Allanblackia soap and the Key Bar soap in terms of lathering ability and odour.

## REFERENCES


Ajayi IA, Ajibade O, Oderinde RA (2011). Preliminary phytochemical analysis of some plants seeds. J. Chem Sci. 1(3):58-62

Alhassan E, Agbemava SE, Adoo NA, Agbodemegbe VY, Bansah CY, Della R, Appiah GI, Kombat EO, Nyarko BJB (2011). Determination of trace elements in Ghanaian shea butter and shea nuts by neutron activation analysis. Res. J. Appl. Sci. Eng. Technol. 3(1):22-25

Alander J (2004). Shea Butter- a Multi Functional Ingredient for Food and Cosmetics. Lipid Technol. 16:202-205.

Alander J, Andersson AC (2002). The Shea Butter Family – the Complete Emollient Range for Skin Care Formulations. Cosmetics and Toiletries Manufacture Worldwide 2002:28–32.

AOAC (1990) Official Methods of Analysis. 15th Edition. Association of Official Analytical Chemists, AOAC Pub, Virginia, U.S.A. 69-84:951-979

Asante FA (1993). The Chemistry of three indigenous seeds and their extraction, nutritional and industrial potentials of their oils. MPhil thesis. Department of Biochemistry, KNUST, Kumasi. pp. 42-73

Bockish J (1993). Handbook of Food Technology-Food fats and oils. Eugen Ulmer Limited and Co. p. 240

Buss C, Tissari J (2010). Allanblackia-an ingredient for poverty reduction www.rural21.com/uploads/media/R21_Allanblackia_0310_01.pdf visited on November 12, 2012

Belitz HD, Grosch W, Schieberle P (2004). Food Chemistry third Edition, pp.427-431, 643, 651-652

Bürkle E, Palenberg M (2009). Allanblackia Supply chain as a Strategic Alliance with Unilever R&D Netherlands. Global Public Policy Institute, GPPi. Pp. 19-20.

CODEX STAN 211-1999, www.Codexallimentarius.org/input/.../standards/337/cxs_211e.pdf visited on December 03, 2012.

CFNI (2005). Phytochemicals. Carribean Food and Nutrition Institute, UWI Campus, st. Augustine Trinidad, W.I.

http://www.paho.org/english/cfni/nyamnewsdec1-205.pdf
Di Vincenzo D, Maranz S, Serraiocco A, Vito R, Wiesman Z, Bianchi G (2005). Regional Variation in Shea Butter Lipid and Triterpene Composition in Four African Countries. J. Agric. Food Chem. 53:7473-7479.
European Food Safety Authority, EFSA (2007). EFSA J. 580:1-10
Grundy SM (1994). Influence of stearic acid on cholesterol metabolism relative to other long-chain fatty acids. Am. J. Clin. Nutr. 60:986-990.
Harbone JB (1984). Phytochemical Methods. A guide to modern techniques of plant analysis. 2nd ed Chapman and Hall New York. pp. 120-126.
Haumann BF (1998). Stearic acid: a 'different' saturated fatty acid. INFORM (American Oil Chemists' Society) 9(3):202-208.
http://www.allanblackiaghana.com/pages/FAG.html).visited on July, 2011
http://www.abs-initiative.info/uploads/media/Addis_0909_Muchugi_01.pdf, visted on February 01, 2013
http://www.wiley-vch.de/books/sample/3527317864_c01.pdf visited on December 03, 2012
James CS (1995). Analytical Chemistry of foods. Chapman and Hall.,Glasgow, UK 46:76-77.
Jolly SC (1963). Official Standardized and Recommended Methods of Analysis. Analytical methods committee, W. Heffer & Son Ltd, Cambridge, pp. 144-148.
Kris-Etherton PM, Griel AE, Psota TL (2005). Dietary stearic acid and risk of cardiovascular disease: intake, sources, digestion, and absorption. Lipids 40:1193-1200,
Lipp M, Simoneau C, Ulberth F, Anklam E, Crews C, Brereton P, de Grey W, Schwack W, Wwiedmaier C (2001). Composition of Genuine Cocoa Butter and Cocoa Butter Equivalents. J Food Compos. Anal. 14:399-408.
Mensink RP (2005). Effects of stearic acid on plasma lipid and lipoproteins in humans. Lipids 40:1201-1205,
Ofori DA, Peprah T, Ilenneh S, Bosompem, SA (2008) *Allanblckia* Domestication Project; Report on training workshop on vegetative propagation of *Allanblckia parviflora,* CSIR-Forestry Research Institute of Ghana, Kumasi 25th to 29th February, 2008.
Orwa C, Mutua A, Kindt R, Jamnadass R, Simons A (2009). Agroforest tree Database: a tree reference and selection guide version 4.0 (http://www.worldagroforestry.org/af/treedb/).
Orwa C, Munjuga M (2007). Allanblackia floribunda Oliv. In: van der Vossen, H.A.M. and Mkamilo, G.S. (Editors). PROTA 14: Vegetable oils/Oléagineux. [CD-Rom]. PROTA, Wageningen, Netherlands. Prota
Orwa C, Oyen LPA (2007). Allanblackia parviflora A.Chev. In: van der Vossen, H.A.M. & Mkamilo, G.S. (Editors). PROTA 14: Vegetable oils/Oléagineux. [CD-Rom]. PROTA, Wageningen, NetherlandsPearson TA (1994). Stearic acid: a unique saturated fatty acid. Am. J. Clin. Nutr. 60(suppl):983-1072
Ronsivalli LJ, Viera ER (1992). Elementary food science 3rd Edition, Van Nostrand Reinhold, NY. USA. pp. 26-27
Sefah W, Adubofuor J, Oldham JH (2010). Optimum conditions for expression of oil from Allanblackia floribunda seeds and assessing the quality and stability of pressed and solvent extracted oil. Afr. J. Food Sci. 4(9):563-570.
Shrestha RB, Akangaamlmm AD (2008). Novella partnership, a partnership for poverty reduction through sustainable enterprise development based on Allanblackia, SNV Ghana,Pp 20: In utility of grafting in tree domestication programme with special reference *toallanjjlac/(fa parv/flora* a. ciiev.D. A. Ofori; T. Peprah1; S Ilenneh JB. Yon Berg Z, Tchoundjeu R, Jamnadass 5 and AJ. Simons (2008) Gh. J. Forest. 23(24):42-48
Stucki V (2005). Allanblackia oil to boost rural income, The Forest conservation Programme, IUCN-The World Conservation Union, Rue Mauverney 28, Gland 1196, Switzerland. www.iucn.org/forest
Ugese FD, Baiyeri KP, Mbah BN (2010). Proximate traits of the seed and seed cake of shea buttertree (Vitellaria paradoxa C. F. Gaertn.) in Nigeria's savanna ecozone. J. Appl. Biosci. 31:1935-1941

# Comparative assessment of metalaxyl enhanced protection of pearl millet varieties in the control of downy mildew

**B. Aliyu[1], S. S. Hati[2]*, G. A. Dimari[3] and P. O. Donli[4]**

[1]Department of Biological Sciences, Gombe State University, Gombe, Nigeria.
[2]Department of Chemistry, Gombe State University, Gombe, Nigeria.
[3]Department of Chemistry, University of Maiduguri, Maiduguri, Nigeria.
[4]Department of Biological Sciences, University of Maiduguri, Maiduguri, Nigeria.

**A comparative assessment was carried out to determine the enhanced protection by metalaxyl in the control of downy mildew on three varieties of pearl millet (SOSAT-C-88, Ex-Borno and GB 8735). Field trials were carried out in 2007 and 2008 growing seasons. The split-plot design was used with the chemical treatments (metalaxyl, and control) in the main plots while the varieties were assessed in the sub-plots. The treatments were replicated three times. The result of this study showed varying enhanced protections of the pearl millet varieties against downy mildew incidence on the vegetative shoots. SOSAT-C-88 showed the most significantly enhanced ($p<0.05$) protection against the incidence of downy mildew and consequently on percentage yield. The order: SOSAT-C-88 > Ex-Borno > GB 8735 of enhanced protection was recorded. This trend was consistent for the year 2007 and 2008 under study. Gene effects have been attributed to the resistance potentials of pear millet varieties against downy mildew incidence as indicated also from the results of this study. However, with enhanced protection by suitable fungicides such as metalaxyl, a significant protection against downy mildew incidences can be achieved and yield increased.**

**Key words:** Metalaxyl, pearl millet, downy mildew, enhanced protection.

## INTRODUCTION

Metalaxyl, $C_{15}H_{21}NO_4$, methyl *N*-(2, 6-dimethylphenyl)-*N*-(methoxyacetyl) DL alanine, 1) a fungicide with a unique combination of residual and systematic properties, is highly active both *in vitro* and *in vivo* against several fungi especially those causing downy mildews. Its application consists of mixtures as foliar spray for tropical and subtropical crops, as a soil treatment for control of soil-borne pathogens and as seed treatment (Kenneth, 1981; Park et al., 2002; Fourie, 2004). The mode of action of metalaxyl includes the inhibition of protein and ergosterol synthesis, by interference with the synthesis of ribosomal RNA. This action is systemic with protective and curative action, absorbed through the leaves, stems and roots (Tadeo et al., 2008). Its chemical and structural properties have been attributed to its action. A single crystal X-ray diffraction study by Park et al. (2002) presents the structure of metalaxyl (Figure 1). The toxicological and environmental assessment of metalaxyl has been well documented in USEPA (1994). Studies of its biodegradation by bacteria and mould strains have also been reported (Tykva and Morais, 2002).

Downy mildew caused by *Slerospora graminicola* is the most damaging disease of pearl millet (*Pennisetum glaucum)* in the world and is particularly devastating in agricultural monocultures. Reduction in yield results in losses in effort and funds used in the production (Williams and Singh, 1981). Williams (1984) reported that downy mildew is still one of the greatest threats to pearl

*Corresponding author. E-mail: stevehati@yahoo.com.

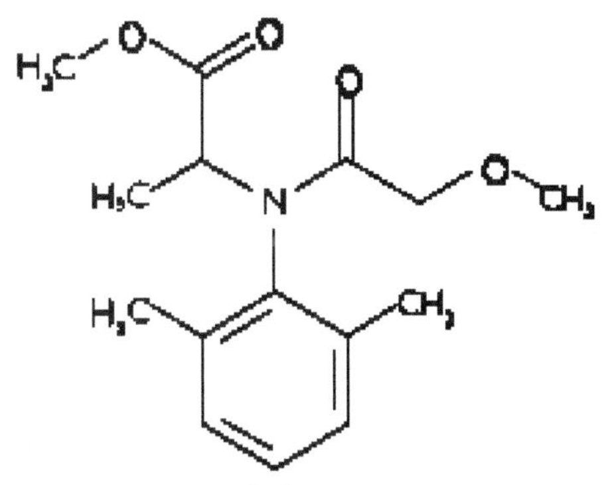

(a)

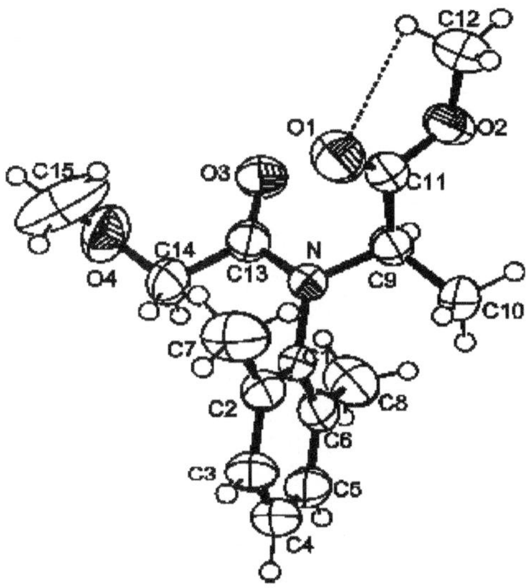


(b)

**Figure 1.** Metalaxyl (a) Structural formula, (b) Crystallographic structure.

millet cultivation in Africa and India. The disease is caused by the Oomycetes fungus *S. graminicola* (Porter, 1926), under conditions of high (>95%) relative humidity and moderate temperature (20-22°C) (Williams, 1984).

An annual loss due to downy mildew in Nigeria was put at 10% (Subramanyam, 1963; King and Webster, 1970; Williams and Singh, 1981). Results from seed dressing experiment in Africa, recorded losses ranging between 10% in Mozambique (Decarvalho, 1949), and 60% Nigeria (King and Webster, 1970). In control plots where seeds were not treated with metalaxyl in Nigeria, annual yield loss was estimated at 10% and losses of more than 50% have been seen in some fields (King and Webster, 1970, Anaso; 1996). The variable nature of pearl millet makes it react differently to downy mildew infection, hence influenced yield.

Millet (*P. glaucum* (L) R. Br.) is one of the most important staple food crops in Africa and India (Williams, 1984). The grain is nutritious and contains 5 to 7% oil. It has higher protein level and more mineral constituent than maize and rice (NRC, 1996). Pearl millet is rich in genetic diversity possibly to counteract the widely differing weather conditions in the production areas (Chahal et al., 1975). The genetically different cultivars ensure bumper harvest regardless of seasonal hazards (Odvody et al., 1995). It is however, the main cultivated host of downy mildew (Williams, 1984); although it is attacked by various diseases and pests on the field (Porter, 1926).

In northern Nigeria, pearl millet is regarded as second only to sorghum in importance. The grain is used in a variety of traditional foods such as "dambu" "shadaka" "fura", "biski", "masa", "yartsala", "tuwo", "kunu", "burukutu" and "pito". Its production supersedes that of sorghum because of its outstanding ability to withstand drought, its early maturing advantage over other crops and it is the first grain crop grown by farmers in Northern Nigeria in anticipation of early grain for consumption when other reserved grains become exhausted. In an earlier work (Aliyu et al., 2008) the combine effect of metalaxyl and neem seed oil was investigated on downy mildew.

It was found that the combined effect provided statistically significant protection against downy mildew. However, data seems to be scanty in defining the interaction effect of metalaxyl and variety in the control of the disease. It was therefore considered essential to carry out a study that would assess the interaction of metalaxyl and some millet (SOSAT-C-88, Ex-Borno and GB8735) varieties on the incidence of pearl millet downy mildew for the control of this disease.

## MATERIALS AND METHODS

### Metalaxyl

Metalaxyl fungicide in Apron 42DS (Syngenta Crop Protection Canada, Inc.) was obtained and used in this study. This brand formulation is about 33.3% metalaxyl-M (mefenoxam) formulated as a powder for use by seed treatment.

### Pearl millet seeds

Millet varieties, SOSAT-C-88, Ex-Borno and GB 8735 used in this study were obtained from the germplasm bank of the Lake Chad Research Institute, Maiduguri Nigeria.

### Experimental site

Field trials were carried out in 2007 and 2008 cropping seasons in Maiduguri (Latitude 11°51' N; 13° 15'E) at the Teaching and Research Farm of the Department of Crop Science, Faculty of Agriculture, University of Maiduguri Nigeria. The experimental site is a sandy loam and has been classified as typic ustipsamment. The mean annual rainfall range between 40 and 250 mm with an annual average temperature ranges from 20 to 43°C. The area has been cropped with millet for the several years ensuring build up of disease inoculum. Natural epiphytotic in field was therefore relied upon as the source of inoculum in each season. The split - plot design was employed in this study (Anaso, 1996).

### Chemical treatment of seeds

The three pear millet varieties were treated with Metalaxyl according to methods described by Smiley et al. (2002). Metalaxyl (0.4 g) was used per 50 g of all pearl millet varieties separately. Control (50 g of each of the variety remain untreated).

### Cultural practices and management

The site was harrowed, disked and leveled before marking out the plots. The main plots were 12 X 4.0 m (48 $m^2$) each and separated by 1 m from each other were assigned to the chemical treatments (Metalaxyl). The sub-plot of 4 X 4 m (16 $m^2$) and spaced at 0.5 m comprised the three millet varieties. The seeds were sown when the rains established at the rate of ten seeds per hole and spaced at 75 X 30 cm between rows and within rows, respectively. At three weeks after sowing during the first weeding plants were thinned to two plants per stand.

Fertilizers were applied at the recommended rates of 60 kg/N/ha, 30 kg $P_2O_5$/ha and 30 kg $K_2O$/ha (Anaso, 1989). At planting, basal dose of NPK (15:15:15) at the rate of 48 g N, 48 g, $P_2O_5$ and 48 g $K_2O$ were applied respectively to each plot. The remaining dose of 48 gN was applied at six weeks after sowing by the side placement method. Weeds were controlled manually using hoes on the third and the sixth weeks after sowing. Harvest was done manually at maturity using cutlass. The plants were cut at their bases and placed in their respective plots for further drying after which the panicles were cut and threshed.

### Downy mildew incidence on vegetative shoot

Infected main stems and basal tillers plants were counted at 7 days intervals for 56 days after sowing (DAS) and downy mildew incidence was computed as the number of diseased main stand or basal tillers expressed as percentage of the total number of main stands or basal tillers assessed according to James (1983) as follows: Downy mildew incidence (%) = Number of diseased plants/plot ÷ Total number of plants/plot x 100

### Parameters measured and data analysis

Millet stands in each plot were counted and recorded at 56 days after sowing. Grain yields for all treatments were obtained after harvesting, sun-drying, threshing and winnowing. They were then weighed on a Mettler balance (0 to10 kg) and yields per plot determined. The yields were later converted to kg/ha. The number and weight of panicles per plot were determined. Data collected were subjected to analysis of variance (ANOVA), using coupled Microsoft Excel + Analyse-it v. 2.12 (Analyse-it® 2007). Variations of results were considered significant at $p<0.05$.

## RESULTS AND DISCUSSION

The interaction effect of metalaxyl and varieties (Table 1) revealed high downy mildew (7.26) and (23.24) incidence on metalaxyl treated and control seeds of GB8735 respectively on 28 (DAS) in 2007 compared to other varieties. In contrast the interaction effect on Ex-Borno was lower (7.04) and (17.00) on metalaxyl treated seeds and control respectively compared to the incidence on GB8735 in both 2007 and 2008. The incidence was lower (2.97) and (7.26) on Ex-Borno in 2008 on metalaxyl treated seeds and control respectively, compared to result recorded on the same variety in 2007. The lowest (6.79), (10.58), (1.77) and (2.76) incidence was recorded on main stands of SOSAT-C-8735 of both metalaxyl treated seeds and control compared to all the varieties in 2007 and 2008, with lowest (1.77) downy mildew incidence on millet stands grown from SOSAT-C-88 seeds treated with metalaxyl than all the varieties under the same conditions in both `years. The trend is similar on 35 (DAS) and 42 (DAS). Table 2 indicated lower (0.43), (1.19), (0.83) and (1.70) downy mildew incidence on SOSAT-C-88 in both metalaxyl treated seeds and control plots, on 21 (DAS) in 2007 and 2008, compared to other varieties on the same condition. Similar trend was recorded in plots that received metalaxyl and control on 28, 35, 42, 49 and 56 (DAS). Figure 2 generally indicated higher grain yield in 2008 than in 2007 of all three pearl millet varieties studied. The order of yield is as presented on Figure 2: SOSAT-C-88 > Ex-Borno > GB8735. The downy mildew incidence was significantly ($p<0.05$) lower on SOSAT-C-88, in both treated and control plots followed by Ex-Borno and highest on GB8735 stands, that received metalaxyl and its control. The interaction effect between seed treatment and the variety in the control of pearl millet downy mildew was significantly higher on SOSAT-C-88, which indicated greater protection against downy mildew, among the varieties and yielded significantly higher compared to Ex-Borno and GB8735. Incidence of the disease on SOSAT-C-88 basal tillers not treated or the control was significantly lower than on Ex-Borno which was also lower than GB8735 that were not treated with metalaxyl. The control crop of GB8735 had significant higher disease incidence than all varieties in both (2007 and 2008) years.

Downy mildew incidence was significantly higher on GB8735 among the three varieties, followed by Ex-Borno. The result depicted significantly higher grain yield from millet stands grown from seeds treated with metalaxyl than the untreated of the same variety. Yield also varied from varietal stands grown from seeds treated with metalaxyl, with SOSAT-C-88 significantly producing grains and was followed by Ex-Borno. The yield in 2008 was significantly higher than in 2007, the disease incidence increase with the increase in the number of days after sowing. With low inoculum the main stands may escape the disease. However, downy mildew being a polycyclic disease with compound interest mode of infection (Singh and Williams, 1980). It may not be surprising, therefore, the higher incidence of the disease on the secondary (basal) tillers which develop later and receive their inoculums from the main tillers. In the present study metalaxyl was tested against downy mildew.

**Table 1.** Interaction of variety x metalaxyl on incidence of downy mildew on main plant in 2007 and 2008.

| | | % Downy mildew incidence | | | |
|---|---|---|---|---|---|
| | | 2007 | | 2008 | |
| Days after owing (DAS) | Variety | Metalaxyl | Control | Metalaxyl | Control |
| | SOSATC88 | 6.79 | 10.58 | 1.77 | 2.76 |
| 28 | Ex-Borno | 7.04 | 17.00 | 2.97 | 7.26 |
| | GB8735 | 7.26 | 23.24 | 8.19 | 26.27 |
| | SOSATC88 | 7.27 | 11.68 | 16.13 | 24.50 |
| 35 | Ex-Borno | 7.79 | 22.67 | 21.34 | 34.69 |
| | GB8735 | 10.99 | 31.31 | 20.19 | 49.26 |
| | SOSATC88 | | | 25.46 | 36.45 |
| 42 | Ex-Borno | | | 27.41 | 40.60 |
| | GB8735 | | | 33.04 | 60.95 |

**Table 2.** Interaction of variety x metalaxyl on incidence of downy mildew on basal tillers in 2007 and 2008.

| | | % Downy mildew incidence | | | |
|---|---|---|---|---|---|
| | | 2007 | | 2008 | |
| Days after sowing (DAS) | Variety | Metalaxyl | Control | Metalaxyl | Control |
| | SOSATC88 | 0.43 | 1.19 | 0.83 | 1.70 |
| 21 | Ex-Borno | 0.65 | 1.53 | 1.53 | 2.58 |
| | GB8735 | 1.0 | 3.51 | 1.59 | 4.51 |
| | SOSATC88 | 0.43 | 8.32 | 4.21 | 10.94 |
| 28 | Ex-Borno | 0.65 | 12.99 | 5.89 | 18.96 |
| | GB8735 | 1.0 | 14.56 | 11.61 | 28.28 |
| | SOSATC88 | | | 24.02 | 44.11 |
| 42 | Ex-Borno | | | 43.87 | 30.35 |
| | GB8735 | | | 33.14 | 70.81 |
| | SOSATC88 | | | 24.02 | 45.13 |
| 49 | Ex-Borno | | | 44.80 | 37.20 |
| | GB8735 | | | 43.62 | 74.65 |
| | SOSATC88 | | | 24.66 | 49.80 |
| 56 | Ex-Borno | | | 51.30 | 54.28 |
| | GB8735 | | | 47.58 | 77.67 |

Metalaxyl is very soluble in water: 7.1 g/L (7100 mg/L), moderately stable under normal environmental conditions. It is photolytically stable in water when exposed to sunlight, with a half-life of 400 days. Monitoring data demonstrate that metalaxyl has the potential to reach groundwater (USEPA 1994). These factors may likely also be responsible for the effects imparted on the millet varieties.

Waller and Ball (1982) reported consistent disease incidence of 10 to 15%. Both the pathogen and disease have been reported from more than twenty countries in Asia, Africa, Europe and USA (Safeeula, 1976). Yield losses of up to 50% due to downy mildew have been reported from West Africa (Singh et al., 1987). Result indicated that continuous seed treatment with metalaxyl for years was the most effective when used for the control of pearl millet downy mildew. This work agrees with the findings of Williams and Singh (1981) who observed

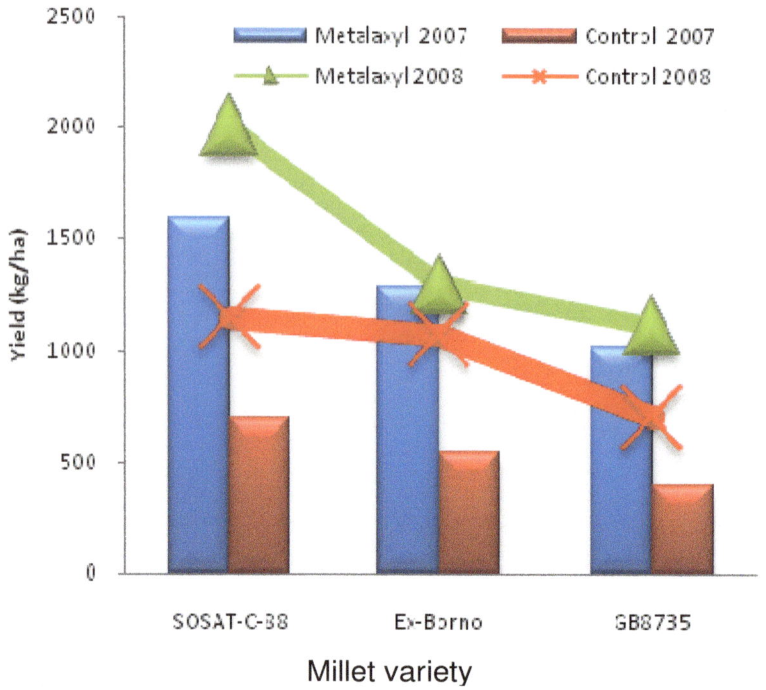


**Figure 2.** Effects of metalaxyl protection on grain yield of pearl millet in Maiduguri, 2007 and 2008.

that metalaxyl effectively controlled downy mildew in pearl millet, and Anaso et al. (1989), who quantitatively assayed the impact of the disease on grain yield and demonstrated the negative effects of disease incidence and grain yield and found significant reduction in grain yield, directly proportional to the incidence of infection.

In conclusion therefore, metalaxyl treatment lowers incidence of the disease and resulted in higher grain yield. The yield of SOSAT-C-88 with metalaxyl treatment increased in the following year which may likely be as a result of subsequent used of metalaxyl which resulted in the reduction of inoculum or Oospore in the area. It is expected that a consistent use of metalaxyl will greatly reduce Oospore or inculum therefore resulting in a better grain yield of pearl millet especially SOSAT-C-88 that stood out as the most suitable variety.

## REFERENCES


Aliyu B, Hati SS, Donli PO, Anaso AB (2008). Effect of metalaxyl and neem seed oil on the incidence of downy mildew of pearl millet. Cont. J. Agron. 2: 1-7.

Analyse-it® (2007). General and Clinical Laboratory Analyses Software Version 2.10.Analyse-it Software Ltd. P.O. Box 77, Leeds, LS12 5XA, England.

Anaso AB (1996). Pearl millet downy mildew in Maiduguri Nigeria disease incidence and yield loss estimates, Niger. J. Plant Protection 16: 82-92.

Anaso AB (1989). Variety selection and planting strategies of maize for control of sorghum downy mildew in Nigeria Guinea Savanna. Appl. Agric. Res., 3: 196-200

Anaso AB, Emechebe AM, Tyagi PD, Manzo, SK (1989). Assessment in yield due to sorghum downy mildew of maize in Nigeria Guinea Savanna tropical Pest Management pp. 301-303.

Chahal SS, Gill KSK, Phul PS, Anjula SS (1975). Screening of Pearl Millet germiplasm for Sclerospora graminicola (Sacc.) Shroeter. Crop Improvement 2: 65-70.

Decarvalho T (1949). Omildio de Mildo (*Sclerospora* sp) Gacta do Agricultor, Mozambique, 1: 61-63.

Fourie PH (2004). Metalaxyl Sensitivity Status of Downy Mildew Populations in Western Cape Vineyards S. Afr. J. Enol. Vitic, 25(1): 19-22

James WC (1983). Crop loss assessment in plant pathologist pocketbook 2nd edition (A. Johnston and C. Booth Eds) Commonwealth Mycological Institute Kew 130-143

Kenneth RG (1981). Downy Mildews of graminaceous crops in the downy mildew Ed. by D.M Spencer Academic Press London, pp.367-394.

King SB, Webster OJ (1970). Downy mildew of sorghum in Nigeria India Phytopathology, 212-319.

National Research Council (NRC) 1996. Lost crops of Africa Grains National Academy Press Washington D.C., 1 (36): 1-12.

Nwasike CC, Baker EFI, Egharevba PN (1983) The Potential for Improving Millet (*Pennisetum typhoides*) in farming systems of the Semi Arid Areas of Nigeria. Agric. Environ., 7: 15-21.

Odvody B, Ejeta GM, Hanna WW, Kumar AK (1995). Diversity Among Africa pearl millet land race population. Crop Sci. Darker, 35(3): 919-924.

Park KII, Kim YK, Cho SII, Yoo MH (2002). Molecular and Crystal Structure of Metalaxyl, $C_{15}H_{21}NO_4$ Korean J. Crystallogr., 13(3/4): 148-151.

Safeeula KM (1976). Biology and control of downy mildew of pearl millet Sorghum and Funger millet, Wesley press, Mysore, India, pp.285-304.

Singh SD, Balls SL, Thakur DP (1987). Problems and strategies in the control of downy mildew (a review article), Proceedings of the International Pearl millet Workshop International Crops Research Institute for the Semi-Arid Tropics, Patancheru AP (India) pp. 161-172.

Singh SD, Williams RJ (1980). The role of sporangia in the epidemiology of pearl millet downy mildew. Phytopathol., 70: 1187-1190.

Smiley R, Cook RJ, Paulitz T (2002). Seed Treatments for Small Grain Cereals, Agricultural Research Center, Oregon State University EM, 8797: 1-5

Subramanyan AD (1963). Infectivity of Oospore Material of Sclerospora graminicola (Sacc) Schroet. The Bajra Green Ear Pathoge. India Phytopathol., 15: 247-249.

Tadeo JL, Sánchez-Brunete C, González L (2008). Pesticides: Classification and Properties CRC Press, pp.1-35

Tykva R, Morais C (2002). Metalaxyl Biodegradation by Bacterial and Mold Isolates From Portuguese Vineyard Soils Institute of Organic Chemistry and Biochemistry, Academy of Sciences of the Czech Republic, CZ – 166 10 Prague 6, Czech Republic, pp.1-6

USEPA: United States Environmental Protection Agency (1994). Reregistration Eligibility Decision (RED) Metalaxyl Pesticides Prevention And Toxic Substances EPA 738-R-94-017 (7508W).

Waller JM, Ball SL (1982). Pearl millet downy mildew interactions in Durable Resistance in crops, Plenum Press London. pp. 80-125.

Williams RJ (1984). Downy mildew of tropical cereals, Adv. Plant Pathol., 2: 1-03.

Williams RJ, Singh SD (1981). Control of pearl millet downy mildew by seed treatment with metalaxyl, Annals Appl. Biol., 97: 263-268.

# Comparative productivity under special crop programme in Benue State, Nigeria: A case of participant and non-participant soybean growers

**G. A. Abu**

Department of Agricultural Economics, College of Agricultural Economics, Extension and Management Technology, University of Agriculture, Makurdi, Benue, Nigeria. E-mail: goddy_master@yahoo.co.uk.

**This study set out to analyze the scarce resource allocation in the special crop programme between farmers who participated in this programme and who did not. Data collected were analyzed using descriptive statistics, gross margin analysis, Z-test and regression analysis. The results of the study showed that most of the participant soybean farmers and non participant farmers were within the age bracket of 31 to 40 years. Most of the respondents were males for both soybeans. A significant difference in output was found between participant and non-participant farmers. The per hectare average cost of production for soybean participant farmers was ₦33,624. The gross margins per hectare N 26,734 soybean were found to be profitable. The results of the multiple regression analysis showed that 83 and 67% of the variations in soybean yield were explained by the combined effect of herbicide, fertilizer, seed and labor for participant and non-participant farmers respectively. Soybean farmers (both participants and non participants) were producing in stage two, the rational stage of production. The ratios of** Marginal Value Product (MVP) **to marginal factor cost (MFC) were greater than unity, hence, they were economically inefficient in resource use. The major problems encountered by the farmers were inadequate storage facilities and inadequate quantity of fertilizer. The study concluded that participant and non-participant farmers in the special crop programme were inefficient in resource use. The study recommended that production inputs such as fertilizer and herbicide should be provided through institutional sources at the required time and quantity and they should be made affordable for the end users.**

**Key words:** Comparative, productivity, crop, programme, Nigeria, participant, non-participant.

## INTRODUCTION

The Special Crop Programme was launched in 2001 production season and therefore had a short gestation period in Benue State. The programme objectives were to rapidly increase productivity and output of soybean, maize, rice, cassava and sesame (mandated crops) on an economically and environmentally sustainable basis and also to reduce year-to-year variability in the production of these mandated crops. These objectives were to be achieved through the implementation of specified activities. Soybean was selected for this study because it is one of the crops that the programme successfully implemented in the three senatorial zones at inception in 2001. Also, the demand for soybean for food and raw materials was high from government and private investors. In one instance, large quantities of soybean were imported to meet the Taraku soybean factory needs. Three sites, one in each senatorial district, were selected in the state and concrete action plan developed. In each of the sites, farmers were selected and given production inputs and technical assistance at concessional rates. Such categories of farmers form the participant group. The participant farmers cultivate soybean as sole crop. Other farmers in the community that were not selected to participate in the programme constitute the non-participant farmers. The non-participant farmers also carry out specified activities intended for the participant farmers in anticipation of being recognized and incorporated into the programme as the activity expands. The non-participant farmers also cultivate soybean as sole crop.

The State Government undertook huge investment in procuring production inputs and technical assistance to participating farmers in the state. After five years of successful programme implementation, the objective of equilibrating the demand of soybean to supply was not achieved. This is evidenced by severe shortages experienced by government and private sector consumers of this crop. In one instance, the State Government imported large quantities of the crop to meet local demand. The performance of the programme as demonstrated by its failure to meet the demand for food and raw materials has been of much concern to government and agricultural researchers. Given this backdrop, this study set out to analyze how efficient farmers in the programme were allocating scarce resources and compare that to the farmers that are not participating in the programme. A pertinent question to ask is "what are the prospects of this new programme?" This question is pertinent in view of the many problems that led to the failure of previous programmes. Any agricultural programme should be based on clear understanding of the factors that prevail at the farm level. Soybean production is hindered by scarcity of productive resources and low productivity of available resources. Production relationship in traditional agriculture is subject to wide variation over time and space. This is because such natural and economic factors such as weather and prices, which influence these relationships, are not fixed. Thus it might be wrong to base present decisions and policies on past results. This study will address the following research questions: (1) what are the socioeconomic characteristics of soybean farmers in the study area? (2) Is the production of soybean under the special crop programme profitable? (3) Are the farmers under the special programme more efficient in their resource allocation than those outside the programme?

The objectives of this study were to:

1. Examine the socio-economic characteristics of soybean farmers.
2. Determine and compare costs and returns in the production of soybean under the Special Crop Programme.
3. Determine and compare the efficiency of the resource-use in the production of soybean under the special programme and those outside the programme.

## RESEARCH METHODOLOGY

This study was conducted in Benue State, Nigeria. The state is situated in the middle belt region of Nigeria between longitude 7°44 E and 9°55' E and between latitude 6° 29' N and 8°7' N of the' equator (MANR, 1984). It is to the east of the rivers Niger-Benue confluence. The state shares boundaries with Nasarawa state to the North, Kogi state to the west, Cross River state to the south and Enugu state to the southwest. The state also shares an international boundary with the Republic of Cameroon in the southeast. The predominant vegetation in the state is the southern Guinea Savannah. The state covers total landmass of 33,706 $km^2$ with a population of 2,780,398 people (NPC, 2006). The predominant ethnic groups are Tiv and Idoma other indigenous ethnic groups are the Igedes and Etulos.

This study used a purposive random sampling technique in the selection of respondents from the three geo-political zones. The first stage is to get the list of the participating and non-participating farmers in each zone. It was found that there were fifty participant soybean farmers in each zone. From the list of the fifty participant farmers from each of the zone, a random sample of twenty (20) soybean farmers participating in the programme was drawn. The non-participant farmers are more in number but for the sake of comparative analysis, twenty (20) soybean growers were selected. In total there were 60 growers in each group (participant and non-participant) in three zones. Data were collected with the assistance of the Divisional Agricultural officers of the Ministry of Agriculture and staff of the Benue Agricultural and Rural Development Authority (BNARDA).

Primary data were collected from sampled farmers for the three cropping seasons of 2002, 2003, and 2004. The instruments for the data collection include sets of questionnaire and interview schedules. Simple descriptive statistics, Gross margin, regression analysis and Z-test were employed in this study. The functional forms of the regression considered were the linear production function, quadratic production function and the Cobb Douglass functions.

The linear production function with five variable inputs namely land, herbicide, fertilizer, seed and labour is expressed as:

$$Y = b_o + b_1X_1 + b_2X_2 + b_3X_3 + b_4X_4 + b_5X_5 + e$$

where, Y = crop output (kg); $X_1$ = total land (ha); $X_2$ = total quantity of herbicide (lit); $X_3$ = total quantity of fertilizer (kg); $X_4$ =. Total quantity of seed input (kg); $X_5$ = total labour (mandays); e = error term; $b_1, b_2$- - -$b_5$ are regression coefficients to be estimated

The linear production function assumes a linear relationship between the output and the inputs as well as constant marginal productivities of resources used. Various studies have fitted them into all kinds of agricultural production data, each subject to its assumptions and limitations.

The quadratic production, for a five-input case is expressed as:

$$Y = b_o + b_1X_1 + b_2X_2 + b_3X_3 + b_4X_4 + b_5X_5 + b_6X_1^2 + b_7X_2^2 + b_8X_3^2 + b_9X_4^2 + b_{10}X_5^2 + b_{11}X_1X_2 + b_{12}X_1X_3 + b_{13}X_1X_4 + b_{14}X_1X_5 + b_{15}X_2X_3 + b_{16}X_2X_4 + b_{17}X_2X_5 + b_{18}X_3X_4 + b_{19}X_3X_5 + b_{20}X_4X_5 + b_{21}X_1X_2X_3X_4X_5 + e$$

Where $X_1$ - $X_5$ is defined as in the linear production function.
$X_iX_j$ = interaction terms of variables i and j ,$b_0$= constant term
$b_1 - b_{21}$ = Regression coefficients.

The quadratic function can have a distinct peak, denoting maximum output for a single combination of factors and the elasticity of production is not constant, but declines with input magnitude. The advantages of using this model were mainly to ease the estimation and the fact that it shows whether there is increasing or decreasing return to factors of production. Among its disadvantages, it cannot show both diminishing and increasing return to factor in a single curve. At very high or low levels of inputs, the function may predict a negative total product. This form usually requires a large number of regression coefficients to be estimated for a given number of variable inputs. Thus, it generally leads to loss of many degrees of freedom and often results in too many regression coefficients not being significant in a profitable sense. The Cobb Douglass production which is amongst the several forms fitted is specified as:

$$\text{Log } Y = \text{Log } a + b_1\text{Log}X_1 + b_2\text{Log}X_2 + b_3\text{Log}X_3 + b_4\text{Log}X_4 + b_5\text{Log}X_5 + \text{Log } E$$

where, Y = crop output (kg); $X_1$ = total land (ha); $X_2$ = total quantity of herbicide (lit); $X_3$ = total quantity of fertilizer (kg); $X_4$ = Total quantity of seed input (kg); $X_5$ = total labour (mandays); E = error term; $b_1, b_2 - - b_5$ are regression coefficients to be estimated.

All the above explanatory variables were included based on the assumptions that they were the only common resources and that they all contributed positively to crop output. Therefore, all their coefficients were expected to have positive signs.

From the estimated regression coefficients $\{b_1\}$, beta coefficients $\{b_1\}$ were computed for each of all the resources. The beta coefficient $\{b_1\}$ is a standardized measure of the relative importance of individual explanatory variables, irrespective of the units in which the regression coefficients were measured. The marginal value product {MVP} of each resource was calculated as:

$$MVP_x = MPPx.Py$$

where $MPPx = b_i\ Y/X_i$

This value {MVP} was compared with the cost of one unit of the particular resource to make inference on resource-use efficiency.

To achieve the efficiency of resource use in a factor-product relationship, the optimal level of variable input must be determined as well as its optimum level output equivalent theoretically, this is obtained when the MVP equates the unit price of the input or when the ratio of MVP to MFC equates, that is MVP/MFC-1 (Olukosi and Ogungbile, 1982).

The choice of the better functional form to use for result interpretation was based on the relative magnitude of the adjusted coefficient of multiple determinations ($R^2$) in relation to the statistical significance of the regression coefficients, the appropriateness of the signs of the regression coefficients and the relative conforming of the various parameters estimates to reality.

The Cobb Douglas production function was selected as the lead equation for further analysis. The choice of the Cobb Douglas was informed by the fact that it gave the highest adjusted coefficient of multiple determinations ($R^2$) for both soybean and maize (participant and non-participant) farmers, all the coefficients were positive and most of them were statistically significant for all the category of farmers.

The linear production function had relatively low values of adjusted coefficient of multiple determinations ($R^2$). In the non-participant farmer group, $b_0$ (constant) was found to be negative.

The quadratic production function had insignificant t-values for most of the coefficient for soybean and maize non-participant farmers.

## RESULTS AND DISCUSSION

### Socio-economic characteristics of soybean farmers

The various elements of the farmers' environment often have pronounced influence on farming decisions and output. Some of these elements include farmers' socio-economic characteristics, their resource endowment, climatic conditions and government policies. It is for the purpose of understanding the prevalent farming conditions during the survey years that this section discusses the farmers' socio-economic characteristics and their production problems.

### Age distribution of respondents

Most of the farmers in these two groups (participant and non-participants to special crop programme) were within the age brackets of 31 to 40 years with an average age of 32 years and 34 years respectively for soybean participants and non-participant farmers respectively. Table 1 clearly illustrates the age distribution of respondents. In the non-participant farmer category, farmers below the age of twenty years were found. Age is one of the influencing socio-demographic attributes that affects the adoption of innovations responsible for higher farm outputs. In this study area, because of very nature of labour intensive farming, the relatively younger person will have more strength and work output to crop more area of land relatively to an old age group. In both participant and non-participant groups, farmers above the age of seventy were found.

### Sex distribution

Male domination in farming in both groups has been noticed. The commonly held belief in this study area shows that soybean was perceived as male crops. This perception stems from the fact that the crops were cultivated mainly for cash. It is only recently that soybean was being processed for household consumption. Sex distribution, as a social construct, had so much to do with roles and task to carry out in the course of soybean production. Notwithstanding that the majority of the farms were owned by men, women have major contributions in terms of weeding and other harvest and post harvest activities. Tables 2 show the sex distribution of soybean farmers.

### Household size

The household size influences labour availability. Given that the mode of soybean production rely so much on family labour, basic cultural practices such as weeding depends so much on family labour. The average household size ranged from six to ten residents with an average of seven and nine, six and ten residents respectively for the participant and non-participant farmers. This figures agrees with the findings of Edache et al. (1998) who found the average households to range between six to ten members. Tables 3 show the distribution of soybean farmers according to household size.

### Level of education

Education positively affects adoption of innovation. The

**Table 1.** The age distribution of soybean farmers.

| Age (years) | Participant soybean farmers | % | Non-Participant soybean farmers | % |
|---|---|---|---|---|
| | Number of respondents | | Number of respondents | |
| < 20 | - | - | 10 | 16.67 |
| 21- 30 | 5 | 8.33 | 8 | 13.33 |
| 31- 40 | 37 | 61.67 | 22 | 36.67 |
| 41-50 | 8 | 13.33 | 12 | 20.00 |
| 51- 60 | 7 | 11.67 | 6 | 10.00 |
| > 70 | 3 | 5.00 | 2 | 3.33 |
| Total | 60 | 100.00 | 60 | 100.00 |

**Table 2.** Sex distribution of soybean participant and non-participant farmers.

| Sex | Participant soybean farmers | % | Non-Participant soybean farmers | % |
|---|---|---|---|---|
| | Number of respondents | | Number of respondents | |
| Male | 45 | 75.00 | 39 | 65.00 |
| Female | 15 | 25.00 | 21 | 35.00 |
| Total | 60 | 100.00 | 60 | 100.00 |

**Table 3.** Distribution of soybean participant and non-participant farmers according to household size.

| Household size | Participant soybean farmers | % | Non-Participant soybean farmers | % |
|---|---|---|---|---|
| | Number of respondents | | Number of respondents | |
| 1-5 | 11 | 18.33 | 9 | 15.00 |
| 6-10 | 34 | 56.67 | 42 | 70.00 |
| 11-15 | 7 | 11.67 | 7 | 11.67 |
| 16-20 | 5 | 8.33 | 2 | 3.33 |
| 21-25 | 3 | 5.00 | - | - |
| Total | 60 | 100.00 | 60 | 100.00 |

use of recommended farming practices are important managerial concern. Skills are acquired to implement specific instructions. Education facilitates the understanding and skill acquisition. Education makes the farmer favorably disposed to receive advice from extension agents. The most of the respondents had primary, secondary, tertiary or adult education. Only negligible number of farmers in both participant and non-participant categories had no formal education. Table 4 shows the distribution of soybean farmers according to level of education.

### Farming experience

Majority of the farmers had between 11 to 15 years of experience with an average experience of 13 years for participant soybean farmers, 12 years for non-participant soybean farmers. There is no difference between participant and non-participant farmers in terms of farming experience. Farming experience counts given that the individual farmer possesses value, makes choices and takes action. This implies that the individual farmer takes his own decisions regarding his resources and output guided by the desire to maximize his utility in time and space. Farming experience therefore facilitates good decision-making. Table 5 shows the distribution of soybean farmers according to farming experience.

### Land acquisition

Land is an important means of production. Most of the farmers acquired their farmlands through inheritance and same was concluded by Allen and Lenk (1998). Most of the respondents have one field for the production of soybean. The average land under soybean cultivation was 5.0 ha for participant farmers and 4.13 ha for non-participant farmers in 2002. In 2003, the average land under soybean cultivation was 5.03 ha for participant farmer and 4.13 for non-participant farmers. In 2004, the average land under soybean cultivation was 6.75 ha for

**Table 4.** Distribution of soybean participant and non-participant farmers according to level of education.

| Level of education | Participant soybean farmers Number of respondents | % | Non-Participant soybean farmers Number of respondents | % |
|---|---|---|---|---|
| Primary education | 22 | 36.67 | 31 | 51.67 |
| Secondary education | 14 | 23.33 | 9 | 15.00 |
| Tertiary education | 8 | 13.33 | 7 | 11.67 |
| Adult education | 6 | 10.00 | 3 | 5.00 |
| No formal education | 10 | 16.67 | 10 | 16.67 |
| Total | 60 | 100.00 | 60 | 100.00 |

**Table 5.** Distribution of soybean participant and non-participant farmers according to farming experience.

| Farming experience in years | Participant soybean farmers Number of respondents | % | Non-Participant soybean farmers Number of respondents | % |
|---|---|---|---|---|
| <5 | 3 | 5.00 | - | - |
| 6-10 | 11 | 18.33 | 8 | 13.33 |
| 11-15 | 32 | 53.33 | 28 | 46.67 |
| 16-20 | 8 | 13.33 | 12 | 20.00 |
| 21-25 | 4 | 6.67 | 10 | 16.67 |
| 26-30 | 2 | 3.34 | 2 | 3.33 |
| Total | 60 | 100.00 | 60 | 100.00 |

participant farmers and 5.63 ha for non-participant farmers. On account of increase in population, the Tiv communities overlap into neighboring Taraba and Nassarawa States. Agriculture is the main economic activity in these areas. The mode of agricultural practice in this study area is the type that requires more land to be cultivated in order to get more output. Due to the increase in population density, the number of persons per square kilometer increases and the attendant consequence is a decrease in size of farmland. Table 6 shows the distribution of soybean farmers.

## Cost of production and returns for soybean

The total revenue and cost per hectare were calculated for soybean farmers with the special crop programme and those not participating in the programme. These were then compared to determine the profitability of soybean under the programme. The Z-statistic showed a significant difference at 5% level between the gross margin per hectare and the total cost per hectare for soybean production under the programme. The average cost and revenue and the gross margin per hectare for soybean participant farmers were ₦33,830, ₦59,844 and ₦26,782 respectively. The average cost, the average revenue and the gross margin per hectare for soybean non-participant farmers were ₦39,251.70, ₦67,891.70 and ₦28,640, respectively. Tables 7 and 8 show detailed breakdown of cost and return for soybean, hence, soybean under the special crop programme was found profitable.

The hypothesis that soybean under the special crop programme was not profitable is thus rejected

## Efficiency of resource use in soybean production by participant and non participant farmers

Data collected from the respondents were analyzed to determine the relationship between inputs (land, herbicides, fertilizer, seeds and total labour) and output of soybean. Several functional forms were fitted but the Cobb Douglas gave the best fit. It is reported in Table 9.

$$Y=58.47X_1^{0.19}X_2^{0.34}X_3^{0.17}X_4^{0.27}$$

The production function for the soybean production by participant farmers is given by:

$$\text{Log } Y=\log 58.47+ 0.19 \log X_1+ 0.34\log X_2 +0.17\log X_3+ 0.27\text{og } X_4$$

The regression results indicate that 83% of the variations in soybean yield among the sampled farmers were explained by the factor inputs included in the model. The signs of the coefficients estimated were positive for all inputs and statistically significant at 5% level of probability in the case of the participant farmers. The Average Physical Product (APP) of herbicide, fertilizer, seed and

**Table 6.** Distribution of soybean participant and non-participant farmers according to mode of land acquisition.

| Tenure system | Participant soybean farmers | % | Non-Participant soybean farmers | % |
|---|---|---|---|---|
| | Number of respondents | | Number of respondents | |
| Inheritance | 38 | 63.33 | 44 | 73.33 |
| Rents | 10 | 16.67 | 5 | 8.33 |
| Purchase | 8 | 13.33 | 4 | 6.67 |
| Allocation | 4 | 6.67 | 7 | 11.67 |
| Total | 60 | 100.00 | 60 | 100.00 |

**Table 7.** Costs and returns for soybean under the special crop programme.

| Variable | Total revenue(N/Ha) | Total costs (N/Ha) | Gross margin (N/Ha) |
|---|---|---|---|
| Maximum | 79.350 | 48.000 | 45.350 |
| Minimum | 15.633 | 25.000 | - 17.000 |
| Average | 59.844 | 33.830 | 26.782 |
| Std. dev. | 11.016 | 2.210 | 10.113 |
| Z-Statistic | 24.68* | | |

*N = Naira. 1 US$ = N150

**Table 8.** Costs and returns per hectare for soybean participant and non-participant farmers.

| Variable | Participant farmer ( ₦ ) | Non-participant farmer ( ₦ ) |
|---|---|---|
| Herbicide | 2,716.00 | 3,654.00 |
| Fertilizer | 8,625.00 | 10,902.50 |
| Seeds | 2745.60 | 2,695.20 |
| Labour | 19,743.40 | 22,000.00 |
| Total variable cost | 33,830.00 | 39,251.70 |
| Total revenue | 59,844.00 | 67,891.70 |
| Gross margin | 26,782.00 | 28,640.00 |

**Table 9.** Regression coefficients for soybean production: Participant and non- participant farmers.

| Variable | Participant farmers | Non-participant farmers |
|---|---|---|
| | Regression coefficients | Regression coefficients |
| Constant | 58.47 (6.54)* | 38.24 (4.32)* |
| Herbicide ($x_1$) | 0.19 (2.11)* | 0.16 (1.72)* |
| Fertilizer ($x_2$) | 0.34 (4.36)* | 0.29 (2.80)* |
| Seed ($x_3$) | 0.17 (1.96)* | 0.21 (2.01)* |
| Labour ($x_4$) | 0.27 (2.53) | 0.33 (4.20)* |
| $R^2$ | 0.83 | 0.67 |
| F | 195.83* | 137.26* |

Figures in brackets are t-values; *Significant at 5%.

labour were computed as 405.54, 5.82, 36.45 and 18.40, respectively. The Marginal Physical Product (MPP) of Herbicide, fertilizer, seed and labour were estimated as to be 77, 1.98, 6.20 and 4.96, respectively.

The Marginal Physical Product of inputs was valued at the prevailing product price of soybean, that is, ₦35/kg weight of Soybean. Thus, the MVPs of herbicide, fertilizer, seed and labour were estimated as ₦2695, ₦69.26, ₦217 and ₦173.60, respectively. Table 10 gives the details. The MFC for herbicide, fertilizer, seeds and

**Table 10.** Average and marginal productivity values and efficiency ratios for soybean farmers with the programme.

| Variable | APP | MPP | MVP | MFC | MVP/MFC |
|---|---|---|---|---|---|
| Herbicide | 405.54 | 77.00 | 2695.00 | 700 | 3.85 |
| Fertilizer | 5.82 | 1.98 | 69.26 | 30 | 2.31 |
| Seed | 36.45 | 6.20 | 217.00 | 35 | 6.20 |
| Labour | 18.40 | 4.96 | 173.60 | 850 | 0.02 |

**Table 11.** Average and marginal productivity values and efficiency ratios for non participating soybean farmers.

| Variable | APP | MPP | MVP | MFC | MVP/MFC |
|---|---|---|---|---|---|
| Herbicide | 317.72 | 50.84 | 1779.40 | 700 | 2.54 |
| Fertilizer | 5.87 | 1.70 | 59.50 | 30 | 1.98 |
| Seeds | 31.58 | 6.63 | 232.05 | 35 | 6.63 |
| Labour | 22.57 | 7.44 | 260.40 | 850 | 3.06 |

Source: Field survey (2004).

labour were estimated as N700 /L, ₦30 /kg, ₦35 /kg and ₦850 /man-day, respectively. The return to scale which is (0.97) the sum of elasticity in a Cobb-Douglass production suggests diminishing return to scale, implying that soybean farmers were producing in stage two. The comparison of APP and MPP values also confirms that the farmers were producing in stage two. That is, all values of APP were greater than those of MPP while MPP values are all positive ($0 \leq MPP \leq APP$). Gordon and Dadidovan,2004, Doran, 1985, Rahman et al., 1998 and Helfand and Levine, 2004 employed similar approaches in their studies.

The Marginal Values Product were compared with unit input price to determine how close the farmers were to the theoretical optimum performance of MVP = P = MFC. The MVPs of herbicide, fertilizer and seeds computed as ₦2695, ₦69.26 and ₦217 respectively were greater than the cost of obtaining additional units of these inputs (N700, N30 and N35) respectively.

However, the ratio of MVP to MFC for labour is far less than one (0.02). The result indicates under utilization for land, herbicide, fertilizer, and seeds and over utilization for labour.

In the case of the non participant farmers, data from this group were analyzed to determine the relationship between inputs (herbicide, fertilizer, seeds and total labour and the output) of soybean. For this group, the general equation is specified as:

$$Y = 38.24X_1^{0.16} X_2^{0.29} X_3^{0.21} X_4^{0.33}$$

Linearizing the equation to get:

$$LogY = Log\ 38.24 + 0.16\ Log\ X_1 + 0.29\ Log\ X_2 + 0.21 Log\ X_3 + 0.33\ Log\ X_4$$

The regression result indicates that 67% of the variations in soybean yield among the sampled fields were explained by the factor inputs included in the model. The signs of the coefficients estimated were positive for all inputs and statistically significant at 5% level of probability.

Average Physical Products (APP) of herbicides, fertilizers, seeds and labour were computed as 317.72, 5.87, 31.58 and 22.57, respectively. The Marginal Physical Product (MPP) of herbicides, fertilizers, seeds and labour were estimated to be 50.84, 1.70, 6.63 and 7.74, respectively. The Marginal Physical Product of inputs was valued at prevailing product price of ₦35/kg weight of soybean. The Marginal Value Product (MVP) of herbicides, fertilizer, seed and labour were estimated as ₦1779.40, ₦59.50, ₦232.05, and ₦ 260 .40, respectively as shown in Table 11

The sum of elasticity which is the return to scale is 0.99, suggesting a diminishing return to scale. This further implies that the farmers were operating in stage two. The comparison of the APP and MPP values also confirms that the farmers were operating in stage two, that is, all values of APP were greater than those of MPP. In other words, this implies that the soybean farmers were technically efficient in their use of resources.

The marginal value product when compared with unit input price to determine how close the farmer is to the theoretical optimum performance of MVP= P = MFC. The MVPs of herbicides, fertilizer, seeds and labour were substantially greater than the cost of obtaining additional units of these inputs respectively. Both categories of farmers (participants and non participants) producing soybean were found to be technically efficient. This was evidenced from the fact that both were operating in stage two. However, the non participants' farmers were more

**Table 12.** Efficiency ratios for soybean participant and non-participant farmers.

| Variable | Participants farmers MVP/MFC | Non-participants farmers MVP/MFC |
|---|---|---|
| Herbicide | 3.85 | 2.54 |
| Fertilizer | 2.31 | 1.98 |
| Seeds | 6.20 | 6.63 |
| Labour | 0.02 | 3.06 |

efficient as their ratios of MVP to MFC for all inputs except seed and total labour were lower than those of the participant's farmers.

Table 12 clearly shows that the non-participant farmers were more efficient than the participating farmers. This is contrary to expectation. It was expected that the participant farmers who have received production inputs at subsidized rates and taught modern methods should be more efficient than the non-participating farmers. The participant farmers relied on the state to provide all the inputs. However, these inputs apparently were not provided on time, and when provided, the quantities were inadequate. The shortage was not augmented. Also, there were occasional cases of inputs supplied being diverted and resold. From the foregoing, participant farmers did not use their inputs timely and neither managed their farms in terms of timely weeding and application of fertilizer and herbicide as the non participating farmers.

## Conclusion

The study concludes that soybean production under the special crop programme is profitable. The farmers, both participant and non participants were found to be efficient in their resource allocation. This was evident from the ratios of APP greater than MPP. The non participant farmers were more efficient comparatively to the participant group. This study recommends that more farmers should be networked into the special crop programme.

## REFERENCES


Allen, D,. and Lenk, D (1998). "The nature of the farm". J. Law Econ. 1: 343-386.

Doran HE (1985). "Small" or "Large" farm: some methodological considerations". Am. J. Agric Econ.1: 130-132.

Edache EO, Abdulsalam Z, Edache EO, Mahmud MA (1998). Economic Analysis of Factors Affecting Small-Scale Farming in Zaria Area of Kaduna State in Ramalan, A.A. (ED) Irrigation in Sustainable Agriculture Proceeding of the 12th National Irrigation and Drainage Seminar, 14-16th April 1998.

Gordon M, Dadidova S (2004). Farm Productivity and Efficiency in the CEE Applicant countries: a Synthesis of results. Agric. Econ. 31: (2004) pp 241-249.

Helfand SM, Levine ES (2004). Farm size and the determinants of Productive efficiency in the Brazilian center west. Agric. Econ. 31:241-249.

Ministry of Agriculture and Natural resources (MANR) (1984). Annual report of crop, area yield survey in Benue in Benue State, Nigeria, Nigeria.

National Population Commission (NPC) (2006). Report of the 2006 population census.

Olukosi JO, Erhabor PO (1982). Introduction to Farm Management Economics: Principles and Applications. AGITAB Ltd., Samaru, Zaria, Nigeria.

Rahman S, Maiangwa M, Agada JE (1998). Resource productivity and Returns in Rice Production under Irrigation system in Toto LGA of Nassarawa State in Rahmalan AA (Ed) Irrigation in Sustainable Agriculture Proceeding of the 12th National Irrigation and Drainage Seminar. pp. , 14-16.

# Impact of various rice based cropping systems on soil fertility

**S. Porpavai[1]*, P. Devasenapathy[2], K. Siddeswaran[3] and T. Jayaraj[2]**

[1]Agricultural Engineering College and Research Institute, Kumulur, Trichy 621 712, India.
[2]Tamil Nadu Rice Research Institute, Aduthurai, 612 101, India.
[3]Agricultural College and Research Institute, Coimbatore, 641003, India.

**Field experiments with ten cropping systems were evaluated during 2002 to 2006 at Soil and Water Management Research Institute, Thanjavur, Tamil Nadu, India during Kharif (June to Sep), rabi (Oct to Jan) and summer seasons (Feb to Apr). The cropping systems were evaluated for their productivity, and to assess their effect on the soil organic carbon content and soil available nitrogen. Inclusion of legumes in the cropping system improved the organic carbon status of the soil. The cropping systems rice (*Oryza sativa*) – rice – blackgram (*Vigna mungo*), onion (*Allium cepa*) – rice – blackgram, groundnut (*Arachis hypogea*)– rice – blackgram and rice – rice – greengram, (*Vigna radiata)* improved the soil organic carbon content and soil available N status. Inclusion of blackgram and greengram in rice based cropping system increased the yield of succeeding crop of rice.**

**Key words:** Cropping systems, soil organic carbon, soil available N.

## INTRODUCTION

Rice (*Oryza sativa*) is a component of widely varying cropping systems. Rice – based cropping systems form an integral part of agriculture in Tamil Nadu. Several intensive rice based cropping systems have been identified and are being practiced by the farmers. While intensive agriculture, involving exhaustive high yielding varieties of rice and other crops, has led to heavy withdrawal of nutrients from the soil, imbalanced and discriminate use of chemical fertilizers has resulted in deterioration of soil health (John et al., 2001). Suitable rice based cropping has to be evaluated, to assess the stability in production. Inclusion of pulses vegetable in the cropping system is more beneficial than cereals after cereals (Kumpawat, 2001). Rice-rice-pulse and rice-rice-sesame are the existing dominant cropping system of Thanjavur district. Continuous cropping results in rapid decline in soil fertility and thus require a special attention. Thus, the existing rice based cropping system has to be diversified with the inclusion of vegetables, pulses, maize and oilseed crops in kharif and summer seasons, to meet the problem of water and labour scarcity and to sustain the soil health. Experiments on cropping systems are ultimate solution to overcome the drawbacks of mono cropping system. Hence, the present study was conducted to assess the impact of various rice based cropping systems on the soil health, in terms of the soil organic carbon content and soil available nitrogen.

## MATERIALS AND METHODS

Field experiments were conducted at Soil and Water Management Research Institute, Kattuthottam, Thanjavur during the kharif (June to Sep), rabi (Oct to Jan) and summer (Feb to Apr) seasons of 2002 to 2006. The geographical location of the research institute has the reference to 79° east longitude,10°45 North latitude and 50 m above mean sea level. The average annual rainfall is 930 mm. The mean maximum and minimum monthly temperature vary from 39.9℃ in May to 29.4℃ in December and from 27.1℃ in May to 20.8℃ in January respectively. The soil was sandy clay loam, neutral in reaction (pH 6.5), low in available nitrogen (217 kg/ha), high in available phosphorus (265 kg/ha) and medium in available potassium (153 kg/ha) with organic carbon content 0.390%.

The experiment was laid out in randomized complete block design with four replications in the same location throughout the study period. Treatments comprised of ten rice based cropping

*Corresponding author. E-mail: porpavaiswmri@yahoo.co.in.

**Table 1.** Details of fertilizer program.

| Crop | Fertilizer recommendation (kg / ha) | | |
|---|---|---|---|
| | N | P | K |
| Rice | 125 | 50 | 50 |
| Rice | 150 | 60 | 60 |
| Black gram | 25 | 50 | - |
| Green gram | 25 | 50 | - |
| Sesame | 35 | 23 | 23 |
| Maize | 135 | 62.5 | 50 |
| Onion | 100 | 50 | 100 |
| Bhendi | 50 | 50 | 30 |
| Radish | 50 | 100 | 50 |
| Groundnut | 17 | 34 | 54 |
| *Lab-Lab* | *25* | *50* | -- |

systems, namely, rice (*O. sativa* )– rice – blackgram *(Vigna mungo)*, rice – rice – sesame (*Sesamum indicum)*, rice – rice – bhendi *(Abelmoschus esculentus)*, lab-lab *(Dolichus lablab)* – rice – maize, onion (*Allium cepa* )– rice – black gram, rice – rice – onion, bhendi – rice – radish *(Raphanus sativus)*, maize (*Zea mays)* – rice – sesame, groundnut (*Arachis hypogea* )– rice – black gram and rice – rice – green gram were evaluated. The crops were raised under irrigated condition with recommended package of practices. The grain yield was recorded from the net plot and computed to kg / ha at 14% moisture for rice and at 8% moisture for other component crops.

#### Details of fertilizer program

The soil was analyzed for organic carbon (wet digestion method, Walkley and Black, 1934) and available nitrogen (Subbaiah and Asija, 1956) after the completion of cropping system in the first, second, third and fourth years of study. The data for the characters studied under various treatments were analysed statistically using the Agres Software Version 3.01 (1994) (Table 1).

## RESULTS AND DISCUSSION

### Organic carbon content of soil

The organic carbon status of the soil increased after all the crop sequences tested at the end of each year. The maximum organic carbon build up was observed with the inclusion of leguminous crops. Among the ten cropping systems tested rice – rice – black gram, onion – rice – black gram, groundnut – rice – black gram and rice – rice – green gram cropping sequences recorded higher organic carbon content in the first, second, third and fourth year respectively. These cropping systems contributed to an increase of 0.04% organic carbon content at the end of the fourth year. This increase in organic carbon content was found to be superior than the other cropping systems.

Such an increase in the organic carbon content is attributed to the accumulation of root residues and shedding of leaves by the leguminous crops (Thakur and Sharma, 1988). Kumar et al. (2001) also reported that, inclusion of leguminous crops in the system increased the organic carbon and available nitrogen phosphorus, potassium and sulphur content of the soil. This may be due to the addition of nutrients by biological N fixation of these crops. The minimum increase in the organic carbon content was noticed in rice – rice – sesame (0.006%), rice – rice – onion (0.001%) bhendi – rice – radish (0.008%) and rice-rice-onion (0.005%) cropping system over four years.

### Post harvest soil available N

The initial available nitrogen status of the soil was 217 kg/ha. Progressive increase in available N status after each year was observed with the inclusion of leguminous crop in the rotation. Among the cropping systems tested rice – rice – black gram, onion – rice – black gram, groundnut – rice – black gram and rice – rice – green gram cropping systems significantly improved the soil available nitrogen status in the first, second, third and fourth year respectively. The fixation of atmospheric nitrogen by the leguminous crops might have contributed for the increased soil N status (Mongia et al., 1989). When compared to first year, the soil available N status was improved in groundnut – rice – blackgram (14 kg/ha), rice – rice – greengram (12 kg/ha), rice – rice – blackgram (9 kg/ha) onion – rice – blackgram (8 kg/ha) at the end of the fourth year. Though the quantity of available nitrogen contributed was not huge, positive trend was observed with the maintenance of soil available nitrogen status. The soil available N status was depleted in maize – rice – sesame (20 kg/ha) and rice – rice – sesame (19 kg/ha) due to the exhaustive nature of maize and sesame. This was followed by bhendi-rice-radish (13 kg/ha), rice-rice-onion (8 kg/ha) cropping systems. Rice – rice – blackgram, onion – rice – blackgram, groundnut – rice – blackgram and rice – rice – greengram cropping systems improved the post harvest soil available N status. Increase in available nitrogen, phosphorus, potassium and sulphur content in cropping sequences involving vegetable, pea, green gram were reported by Gangwar and Ram (2005). Cultivation of legume crop is viewed more as a soil fertility improver than as independent crops grown for their grain output. This is because legume crops are self sufficient in N supply (Kanwarkamla, 2000). (Table 2).

The preceding crop species can have a beneficial or detrimental effect on the performance of the succeeding crop. The well known beneficial effects of preceding legume crop on cereal are also found in multiple cropping systems, but the magnitude of the effects varies with management practices and the legume species used. Progressive increase in the grain yield of kharif crops preceded by pulses was recorded in rice-rice- blackgram and rice-rice- green gram cropping systems in the

**Table 2.** Influence of cropping sequence on soil organic carbon and post harvest available N.

| Trs. | Crop sequence | Organic carbon content of soil (%) after completion of cropping system | | | | Soil available N (kg/ha) | | | |
|---|---|---|---|---|---|---|---|---|---|
| | | I Year | II Year | III Year | IV Year | I Year | II Year | III Year | IV Year |
| $T_1$ | Rice – rice – blackgram | 0.416 | 0.426 | 0.430 | 0.453 | 215 | 218 | 220 | 224 |
| $T_2$ | Rice – rice – sesame | 0.395 | 0.398 | 0.399 | 0.401 | 196 | 190 | 186 | 177 |
| $T_3$ | Rice – rice – bhendi | 0.397 | 0.399 | 0.400 | 0.402 | 195 | 190 | 189 | 186 |
| $T_4$ | Lab-lab – rice – maize | 0.396 | 0398 | 0.399 | 0.401 | 190 | 182 | 180 | 178 |
| $T_5$ | Onion – rice – blackgram | 0.413 | 0.423 | 0.431 | 0.453 | 217 | 219 | 221 | 225 |
| $T_6$ | Rice – rice – onion | 0.392 | 0.395 | 0.398 | 0.402 | 193 | 190 | 188 | 185 |
| $T_7$ | Bhendi – rice – radish | 0.393 | 0.396 | 0.399 | 0.401 | 197 | 192 | 188 | 184 |
| $T_8$ | Maize – rice – sesame | 0.395 | 0.398 | 0.402 | 0.404 | 192 | 187 | 179 | 172 |
| $T_9$ | Groundnut – rice – blackgram | 0.415 | 0.425 | 0.433 | 0.450 | 215 | 218 | 224 | 229 |
| $T_{10}$ | Rice – rice – greengram | 0.416 | 0.426 | 0.435 | 0.454 | 216 | 219 | 223 | 228 |
| | Initial | 0.390 | | | | 217 | | | |
| | CD (P=0.05) | 0.010 | 0.014 | 0.016 | 0.015 | 7.26 | 8.30 | 9.65 | 10.40 |

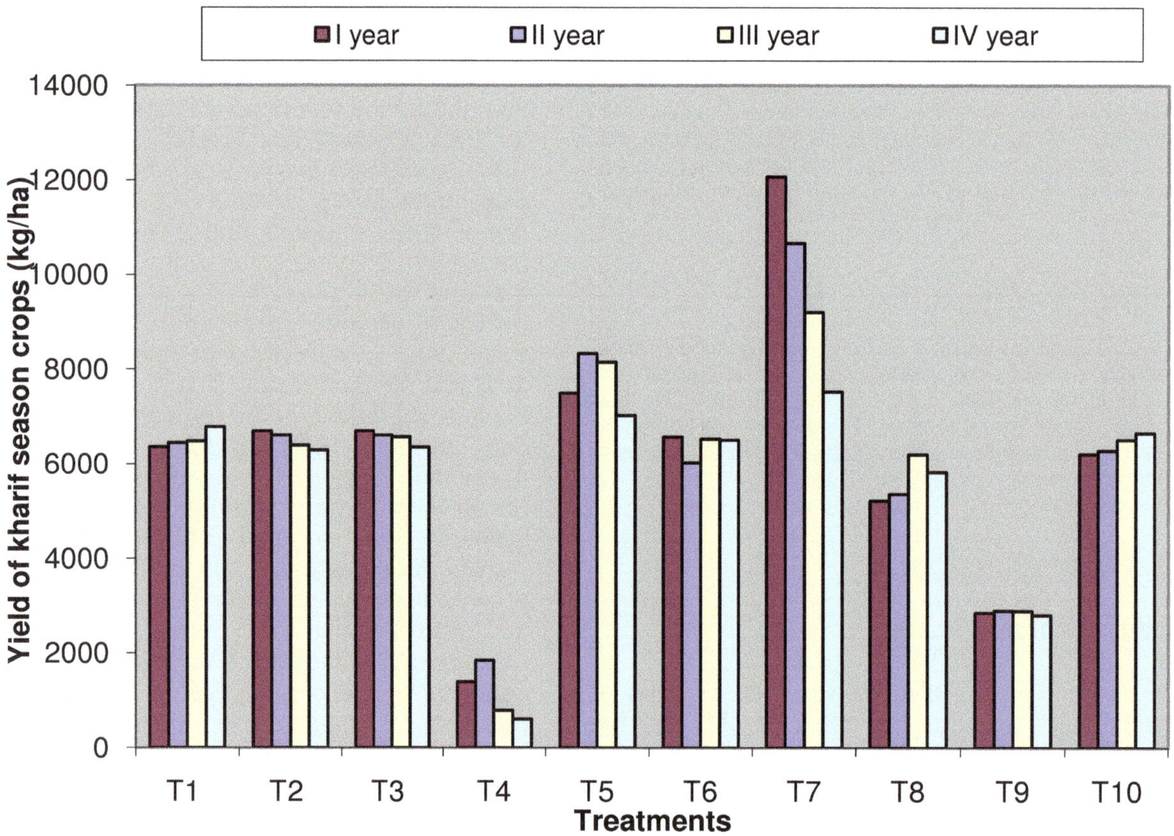


**Figure 1.** Influence of summer crop on the yield of Kharif season crop.

second, third and fourth year. When compared to the first year, the yield increase of 430 to 446 kg was recorded in the fourth year (Figure 1). Kumar et al. (1993) reported similar result, pointing out the superiority of leguminous crops in increasing the yield of the succeeding crops of rice. They also observed a reduction in the yield of rice grown after sesame. The grain yield of kharif season crops decreased in rice-rice-sesame, rice-rice-bhendi, lab-lab-rice- maize, bhendi-rice-radish, maize-rice-sesame cropping systems. The grain yield of kharif season crops are maintained without any reduction in onion-rice-black gram, rice-rice-onion and groundnut-rice-black gram cropping systems. Legumes were potentially important to diversify cereal based mono cropping into cereal-legume sequences which had nutrient cycling advantages.

## ACKNOWLEDGEMENT

The author sincerely thanks the Project Directorate for Farming System Research, Modipuram for funding.

### REFERENCES

Agres statistical software (1994). Version 3.01. Pascal International Software Solution, USA.

Gangwar B, Ram B (2005). Effect of crop diversification on productivity and profitability of rice (Oryza sativa)-wheat (Triticum aestivum) system. Indian J .Agric. Sci., 75(7): 435-438.

John PS, George M, Jacob R (2001). Nutrient mining in agro – climatic zones of Kerala. Fertilizer News, 46: 45 – 52, 55 – 57.

Kanwarkamla (2000). Legumes – the soil fertility improver. Indian Farming, 50(5): 9

Kumar V, Jayakrishna UK, Nair CS, Pushpakumar R, Nair VR (1993). Economics of high intensity crop sequences in a rice based cropping system. Oryza, 30: 302-304.

Kumar AL, Yadav DS, Singh RM, Achal (2001). Productivity and stability of rice (Oryza sativa) based cropping systems in eastern Uttar Pradesh. Indian J. Agron., 46(4): 573-577.

Kumpawat BS (2001). Production potential and economics of different crop sequences. Indian J. Agron., 46(3): 421 – 424.

Mongia AD, Gangwar B, Shyamsingh V, Kumar BAK (1989). Long term effect of fertilizers on important crops in Andaman and Nicobar Islands, Fertil. News, 34(4): 81 – 85.

Subbaiah BV, Asija CL (1956). A rapid procedure for the determination of available nitrogen in soil. Curr.Sci., 25: 259-260.

Thakur HC, Sharma NN (1988). Effect of various cropping patterns including cereals, pulses and oilseeds on Chemical properties of the soil. Indian J. Agric. Sci., 58(9): 708 – 709.

Walkley A, Black CA (1934). An examination of the method for determining soil organic matter and proposed modification of the chromic acid titration method. Soil Sci., 37: 29-34.

# Studies on the compatibility of neem oil with predator, *Chrysoperla carnea* for the management of aphids (Homoptera: Aphididae) in canola (*Brassica napus* L.)

**Muhammad Hamayoon Khan , Nazir Ahmad, S. M. Masoom Shah Rashdi, Muhammad Ismail, Imran. Rauf and Muhammad Tofique**

Nuclear Institute of Agriculture, Tandojam-70060, Pakistan.

**The present studies were conducted to find out the compatibility of neem oil with the predator, *Chrysoperla carnea* for the management of aphids in canola. Among the different treatments tested, module consisting of neem oil 2% + *C. carnea* proved very effective in reducing the aphid population with an average of 18.6/plant. Neem oil 1% and *C. carnea* alone also produced significant results compared to untreated check where mean per plant population of aphid was 39.3 and 41.3, respectively. Maximum seed yield (3295.1 kg/ha) was recorded from neem oil 2% + *C. carnea* followed by neem oil 1% + *C. carnea* (3219.1 kg/ha) and neem oil 2% (2809.4 kg/ha), respectively. Over all mean population of *C. carnea* was highest (0.48/plant) in plots treated with predators alone. The same was second most abundant (0.40/plant) in plots treated with neem oil 1% + *C. carnea* followed by plots treated with neem oil 2% + *C. carnea* (0.36/plant). The study manifested neem oil concentrations relatively safe to beneficial insects and suitable for use in integrated pest management of aphids in canola.**

**Key words:** Neem, *Chrysoperla carnea*, Aphids, canola, botanical pesticides, infestation.

## INTRODUCTION

Canola (*Brassica napus* L.) is the traditional oilseed crop and second most important source of edible oil after cottonseed in Pakistan. Domestic rapeseed/canola production has gradually declined since 2008 and accounts for about 6% of total oilseed production in Pakistan (Rehman, 2010). There are many reasons responsible for this declined trend like competition with other crops, lack of utilization of production technology, higher prices of agricultural inputs, marketing problems but insect pests infestation seem to be the main constraint.

About 43 insect species have been recorded attacking these plants, but the most important are aphids which attack canola. The three species of aphids, that is, cabbage aphid (*Brevicoryne brassicae* L.), mustard aphid (*Lipaphis erysimi* Kalt.) and green peach aphid (*Myzus persicae* Sulz.) are more abundant and widely distributed. They suck sap from plants and can be found massed on young, rapidly growing leaves depending on the aphid species. The colonies are mainly established on the top inflorescence of the plant during flowering stage. It usually causes damage by, wilting, flower abortion and reduced pod set. They may also excrete honey dew and cover the plants, which encourage the growth of black sooty mould, thereby reducing the plants' ability to photosynthesize and generally decreases plant vigour

(Berlandier et al., 2010). The aphids are outstanding in their distribution, in host plant range and able to transfer over 100 virus diseases of plants on about thirty different families including many major crops such as Brassica (Devi and Singh, 2007). Many factors are involved in the widespread of aphid populations. Besides indiscriminate use of insecticides without regard for economic thresholds and development of resistance, there are many interactions that affect its population trends. The farmers solely depend on pesticides for the control of this insect pest. It is a known fact, that often only 1% of the active ingredients reach the target pests, while 99% of these substances, some of which are highly toxic, trouble the environment (Hassan, 1992). Non-selective use of pesticides is responsible for water pollution, soil degradation, insect resistance and resurgence, destruction of native flora and fauna (Baloch and Haseeb, 1996). Keeping in view the ill effects of pesticides, emphasis should be focused to replace pesticides with alternative means of control that are less toxic, safe, low in cost, local in production and also environmentally friendly (Mohammad et al., 2010).

These alternatives include integrated pest management, which is bringing together and utilizing all the control measures concurrently or successively in an integrated way against a particular pest and thereby minimizing the need for chemical control (Shahid, 2003). The use of botanical insecticides and biological control for the management of aphids in canola are good alternatives for pesticides. One of the important botanicals is neem whose products are derived from the neem tree, *Azadirachta indica* A. Juss. (L.) that grows in arid tropical and subtropical regions on several continents. The principal active compound in neem is azadirachtin, a bitter, complex chemical. The neem products have been found very effective in causing aphid mortality and hence can be used for the management of aphids in different crops (Tess Henn and Weinzeiri, 1989). Similarly the bio-control agent, *Chrysoperla* spp. also play a significant role in the management of aphids while some other natural enemies feeding on aphids are chrysopids, coccinellids and syrphids (Kannan, 1999). Holland and Thomas (1997) concluded that polyphagous predators can efficiently manage aphid infestation late in the season when population increase rapidly. Messina and Sorenson (2001) reported that lacewings reduced the aphid population on some plants and their effectiveness was 84%. Neem extracts are usually safe for beneficial organisms, such as bees, predators and parasitoids, mammals, and for the environment (Tang et al., 2002). Neem oil are comparatively less toxic to bio-control agents compared to synthetic pesticides and hence can be raked as harmless (Sattar et al., 2011). Neem-derived insecticides are regarded as generally compatible with insect natural enemy conservation (Schmutterer, 1997). Devi et al. (2002) reported that neem, endosulfan and phosalone could be used along with the biological control agents for the control of mustard aphid.

These reports emphasize the importance of testing neem-derived insecticides against not only the target pest insects but also key biological control agents. Keeping this backdrop in view, the present research studies were executed to deal with the population of aphids using neem oil and augmenting *C. carnea* and to find out the compatibility of neem oil with the predator, *C. carnea* under field conditions.

## MATERIALS AND METHODS

The experiment was conducted on field located at the experimental farm of Nuclear Institute of Agriculture, Tandojam, Sindh, Pakistan during winter season 2011. The test cultivar Wester was used for this study. Crop sowing was completed in the third week of November, after rice had been harvested. The field was properly prepared, weeds and stubbles were removed, and the land was finally leveled and prepared by laddering. The pure non contaminated canola seed was planted with a hand drill at appropriate soil moisture. Standard agronomic practices were followed for raising the crop and the field was hand weeded periodically. The plants were maintained at 10 cm spacing by thinning of plants and 30 cm between rows. The experiment was laid out in a randomized complete block design with six treatment (neem oil 1%, neem oil 2%, neem oil 1% + *C. carnea*, neem oil 2% + *C. carnea*, *C. carnea* alone and untreated check) replicated three times. The plot size was 1 m x 3.5 m (3.5 $m^2$). The bio-control agent *C. carnea* obtained from insect rearing laboratory of Nuclear Institute of Agriculture, Tandojam, were released regularly at fortnightly interval to the field. Neem oil was purchased from the local market and used at concentration of 1 and 2%. The neem oil concentrations, in the respective treatments, were applied on the crop in the form of spray with the help of knapsack hand sprayer having 20 liters capacity fitted with hollow cone nozzle. Pest sampling started with the appearance of aphids in the field and continued till harvesting of the crop. The prevalence of aphid per treatment was assessed based on numbers recorded on 5 randomly selected plants, at least one of which had been colonized by the aphids in each replicate from the start of infestation. The population of *C. carnea* was also regularly recorded by observing five randomly selected plants from each replication. Yield per plot was also taken to evaluate the loss of plants by the attack of aphid and subsequent effects of treatments. Data on yield were taken from three replicates of each treatment which was as g/3.5 $m^2$ plot and further transformed into kg per hectare. The aphid densities were transformed to mean values for analyses. Computer software Statistix was used to analyze the data and significance of difference in mean population of aphids was sorted out with LSD (5% significance level).

## RESULTS AND DISCUSSION

Results revealed (Table 1) significant differences in the aphid densities among different treatments at all the post treatment observations. Among the different treatments tested, module consisting of neem oil 2% + *C. carnea* proved the most effective in reducing the aphid population with an average 18.6/plant followed by module neem oil 1% + *C. carnea* (22.6) and treatment of neem oil 2% (29.0). Neem oil 1% and *C. carnea* alone also produced

**Table 1.** Mean population of aphids and seed yield recorded from different treatments.

| Treatments | Aphids/plant | Yield (g/3.5 m²) | Yield (kg/ha) |
|---|---|---|---|
| Neem oil 1% | 39.3$^{B}$ | 956.7$^{B}$ | 2733.4$^{B}$ |
| Neem oil 2% | 29.0$^{C}$ | 983.3$^{B}$ | 2809.4$^{B}$ |
| Neem oil 1% + *C. carnea* | 22.6$^{CD}$ | 1126.7$^{A}$ | 3219.1$^{A}$ |
| Neem oil 2% + *C. carnea* | 18.6$^{D}$ | 1153.3$^{A}$ | 3295.1$^{A}$ |
| *C. Carnea* alone | 41.3$^{B}$ | 856.7$^{C}$ | 2447.7$^{C}$ |
| Untreated control | 98.4$^{A}$ | 683.3$^{D}$ | 1952.2$^{D}$ |
| LSD | 9.45 | 67.31 | 192.31 |

Means within a column followed by different letters are significantly different ($P<0.05$).

**Table 2.** Population of different stages of bio-control agent *C. Carnea.*

| Treatments | Population of *C. carnea*/plant | | |
|---|---|---|---|
| | Eggs | Larvae | Adult |
| Neem oil 1% | 0.11$^{B}$ | 0.11$^{C}$ | 0.14$^{B}$ |
| Neem oil 2% | 0.11$^{B}$ | 0.11$^{C}$ | 0.11$^{B}$ |
| Neem oil 1% + *C. carnea* | 0.44$^{A}$ | 0.33$^{AB}$ | 0.44$^{A}$ |
| Neem oil 2% + *C. carnea* | 0.37$^{A}$ | 0.29$^{B}$ | 0.44$^{A}$ |
| *C. Carnea* alone | 0.49$^{A}$ | 0.44$^{A}$ | 0.51$^{A}$ |
| Untreated control | 0.14$^{B}$ | 0.14$^{C}$ | 0.11$^{B}$ |
| LSD | 0.13 | 0.12 | 0.20 |

Means within a column followed by different letters are significantly different ($P<0.05$).

significant results compared to untreated check where mean per plant population of aphid was 39.3 and 41.3, respectively. The highest prevalence of insect pest was found in the control treatment (98.4/plant) which was significantly higher than all treatments. Yield data also confirmed the same trend as maximum 1153.33.5 g/3.5 m² (3295.1 kg/ha) was recorded from neem oil 2% + *C. carnea* followed by 1126.7 g/3.5 m² (3219.1 kg/ha) and 983.3 g/3.5 m² (2809.4 kg/ha) from neem oil 1% + *C. carnea* and neem oil 2%, respectively. But there was no statistically significant difference observed in the yield data obtained from neem oil 2% + *C. carnea* and neem oil 1% + *C. carnea* during the course of study. One factor of particular interest in this study was to find out the negative effect of neem oil on the survival of *C. carnea*. Data recorded on the establishment of released predators revealed that neem oil had no serious effect on the survival of predator, *C. carnea* (Table 2). Highest mean population of eggs, larvae and adults of *C. carnea* per plant (0.49, 0.44 and 0.51 respectively) was investigated in treatment where *C. carnea* alone were released. However it was found non significantly different with treatments of neem oil 1% + *C. carnea*, and neem oil 2% + *C. carnea*, except for the population of larvae in the latter treatment where it was significantly lower. Over all mean population of *C. carnea* was maximum (0.48/plant) in plots where predators alone were released. The same was second most abundant (0.40/plant) in plots treated with neem oil 1% + *C. carnea* followed by plots treated with neem oil 2% + *C. carnea* (0.36/plant). These two treatments also exerted a negative effect on aphids showing drastic reduction in their population as discussed above. Furthermore, the encouraging population of predators in the neem treated plots pointed out that the predatory insects are not affected to a great extent reflecting the undamaging effects of neem oil on the bio-control agents. These findings confirmed that the strategy of using both (neem oil and predators) does not affect key biological control agents. These results are in conformity with those of Schuster and Stansly (2000), who tested Azatin EC on two species of green lacewings and found Neem product non toxic to eggs, larvae and adults. Similarly Hermann et al. (1997) reported no negative effect of NeemAzal T/s and NeemAzal-F on *C. carnea* efficacy. The initial application of NSKE 5% followed by the release of *C. carnea* effectively managed the mustard aphid population (Pandey and Narendra, 2008). Unlike these findings NeemAzal-T/S was found harmful to larvae of the lacewing *C. carnea* (Stephen) causing mortality (Srinivasan and Babu, 2000). Medina et al. (2004) revealed no effect on mortality or fertility of the adult green lacewings, *C. carnea* (Stephens), when provided with azadirachtin treated water. The treatments where neem oil alone (1 and 2%) was applied also proved effective for the management of aphids (Dhaliwal et al., 1998). It is important to mention that the compatibility of bio-control agents to botanical pesticides is based on the type of botanical, natural enemy and stage of development. Therefore, further studies are needed to screen out different other botanicals for compatibility with the natural enemies in order to avoid disrupting successful biological control programs.

## Conclusion

The present research findings elucidated neem oil very effective and compatible with predator, *C. carnea* for the management of aphids in canola. Neem oil concentrations appeared to be relatively benign to beneficial insects and are suitable for inclusion in integrated pest management programs. However, a short delay between the treatment and the release of the bio-control agents is suggested for the successful combination of the use of neem and *C. carnea* in conjunction.

### REFERENCES

Baloch UK, Haseeb M (1996). Xenobiotics in Third World Agriculture Environment. In Environmental xenobiotics ML Richerdosn, Taylor and Francis, Caller Macmillian Publisher London, U.K. P. 286.

Berlandier F, Severtson D, and Mangano P (2010). Aphid management in canola crops. Farm note, Department of Agriculture and Food, Govt of Western Australia.

Devi LC, Singh TK, Varatharajan R (2002). Management of mustard aphid with natural enemies, plant product and chemical insecticides. Indian J. Entomol. 64(3):373-376.

Devi PB, Singh TK (2007). Studies on the morphometry of the green peach aphid, *Myzus persicae* (Sulzer) (Homoptera: Aphididae). Entomol. Res. 37:81-85.

Dhaliwal GS, Gill RS, Dilwarl VK, Dhawan AK (1998). Management of insect-pest complex of cabbage with neem based insecticides. In: "Ecological Agric. Sustain. Develop. II:306–314.

Hassan SA (1992). Guideline of the side-effects of plant protection product on *Trichogramma chilonis*. In Guideline for testing the effect of pesticides on beneficial organisms, (ed) Hassan, S.A. IOBC/WPRS Bulletin 15(3):18-39.

Hermann P, Zebitz CPW, Kienzle J (1997). Effects of different NeemAzal- formulations on larvae of *Chrysoperla carnea* in laboratory and semi-field. In Kleeberg, Zebitz, editors. Practice oriented results on use and production of neem-ingredients and pheromones. Proceedings of the 5th Workshop, Wetzlar, Germany, 1996 22 – 25 January, pp. 183-188.

Holland JM, Thomas SR (1997). Assessing the role of beneficial invertebrate in conventional and integrated farming system during an outbreak of *Sitobion avenae*. Biol. Agric. Hort. 15(1-4):73-82.

Kannan HO (1999). Population dynamics of the wheat aphid, *Schizaphis graminum*, (Rondani) (Homoptera: Aphididae) and its natural enemies in the field. Sudan J. Agric. Res. 2:65-68.

Medina P, Budia F, Del Estal P and Nuela VIE (2004). Influence of azadirachtin a botanical insecticide on *Chrysoperla carnea* (Stephens) reproduction: toxicity and ultrastructural approach. J. Econ. Entomol. 97:43–50.

Messina FJ, Sorenson SM (2001). Effectiveness of lacewing larvae in reducing Russian wheat aphid population on susceptible and resistant wheat. Biol. Cont. 21(1):19-26.

Mohammad GTK, El-Shereif , Saehn (2010). Toxic Effect of Capsicum and Garlic Xylene Extracts in Toxicity of Boiled Linseed Oil formulations against Some Piercing Sucking Cotton Pests. American-Eurasian J. Agric. Environ. Sci. 8(4):390-396.

Pandey R, Narendra NS (2008). Effectiveness of biocontrol based ipm modules against *lipaphis erysimi* kaltenbach (Hemiptera: Aphididae). J. Plant Protect. Res. 48 (1):1-7.

Rehman MS (2010). Pakistan oilseeds and annual products. Global Agricultural Information Network, USDA foreign agricultural service.

Sattar S, Farmanullah, Saljoqi AR, Arif M, Sattarand H, Qazi JI (2011). Toxicity of Some New Insecticides Against *Trichogramma chilonis* (Hymenoptera: Trichogrammatidae) Under Laboratory and Extended Laboratory Conditions. Pak. J. Zool. 43(6):1117-1125.

Schmutterer H (1997). Side-effects of neem (*Azadirachta indica*) products on insect pathogens and natural enemies of spider mites and insects. J. App. Entomol. 121:121–128.

Schuster DJ, Stansly PA (2000). Response of two lacewing species to biorational and broad-spectrum insecticides. Phytoparasitica 28:297-304.

Shahid M (2003). Principles of insect pests management. Higher Education Commission Islamabad Pak. 30:161-166.

Srinivasan G, Babu PCS (2000). Effect of Neem products on predatory green lacewing, *Chrysoperla carnea*. Pest Res J. 12:123– 126.

Tang YQ, Weathersbee A.A, Mayer RT (2002). Effect of neem extract on the brown citrus aphid (Homoptera: Aphididae) and its parasitoid *Lysiphlebus testaceipes* (Hymenoptera: Aphididae). Environmental Entomol., 31:172-176.

Tess H, Weinzeiri (1989). Use of botanical insecticides for pests control. J. Bot. 10(9):21-28.

# Effect of bio-control agents on yield, yield components and root rot control in two wheat cultivars at New Valley region, Egypt

**Moubarak M. Y.[1] and Abdel-Monaim M. F.[2]***

[1]Wheat Research Department, Field Crops Research Institute, A.R.C. Giza, Egypt.
[2]Plant Pathology Research Institute, A.R.C. Giza, Egypt.

**The main aim of this study is to evaluate efficiency of four bioagents, *Bacillus subtilis* (isolate BSM1), *Bacillus megaterium* (isolate BMM5), *Trichoderma viride* (isolate TVM2) and *Trichoderma harzianum* (isolate THM4) for the control of rot pathogens *Fusarium graminearum, Drechslera halodes and Rhizoctonia solani* on two wheat cultivars (Sakha 93 and Bani Suif 5) under greenhouse conditions. Moreover, their effect on wheat growth and yield were also studied under field conditions. *In vitro*, all tested bioagents significantly reduced redial growth of the pathogenic fungi. *T. viride* was active more than the other tested bioagents followed by *T. harzianum*, while *B. subtilis* was the least ones. Under greenhouse conditions, all tested bioagents were able to reduce significantly damping-off and root rot caused by the tested pathogens compared with control and increased fresh and dry weight of the survival plants when applied as soil or grain treatments however, there was variation among bioagent isolates effect on reduction of disease severity both application methods. *T. viride* and *B. megaterium* were recorded the highest effective in this respect compared with other tested bioagents. Under field conditions, analysis of variance and mean performance were estimated for four characters: grain yield 4.8 $m^{-2}$, No. of spikes $m^{-2}$, No. of kernels $spike^{-1}$ and 1000-kernel weight. Significant mean squares were obtained for all studied characters between the seasons (S), methods (M), (S) × (M), treatments (T), (S) × (T), (M) (T) and interaction (S) × (M) × (T) for Sakha 93 cultivar and Bani Suif 5 cultivars except application methods and (S) × (M) for number of kernels $spike^{-1}$ and 1000-kernel weight, respectively. While the treatments *T. harzianum* and *B. megaterium* were best treatments to increase grain yield, the treatments *B. subtilis* and *T. harzianum* were best treatments to increase number of spikes and the treatments *B. subtilis* and *B. megaterium* were best treatments to increase number of kernels for soil and grain application methods, respectively.**

**Key words:** Bio-agents, root rot disease, wheat growth, yield and yield components.

## INTRODUCTION

Wheat is one of the most important feeding crops in Egypt and many other countries in the world. It is primarily grown as a food crop but the straw is also used for industrial products as feed for livestock. Wheat is subjected to relatively large number of diseases during its growing season which attack all plant parts causing serious losses in crop productivity (Bakr, 1997).

Root-rot diseases caused by soil-borne fungi are the most important wheat diseases. Several fungi were recorded as causal pathogens of root-rot diseases such as *Fusarium graminearum, Fusarium equiseti, Fusarium solani, Drechslera halodes and Rhizoctonia solani* (Hashem and Hamada, 2002; Atef, 2008; Asran and Eraky, 2011). These diseases remain prevalent because of the trends toward a higher frequency of cereals in the rotation, including, commonly, continuous cereals, and the use of less or no tillage.

Agricultural practices for management of soil borne pathogens in the field includes cultural practices, crop rotation, fungicide applications, methyl bromide fumigation,

*Corresponding author. E-mail: fowzy_2008@yahoo.com.

soil solarization and use of resistant or tolerant varieties. At present, no single method provides adequate control of soil borne diseases (Hausbek and Lamour, 2004). Using of chemicals to control soil borne pathogens causes several negative effects, such as: i) development of pathogen resistance, ii) hazards to humans, iii) damage to beneficial organisms and iv) environmental pollution. Moreover, many chemicals will be banned in near future. However, for sustainable production, pathogens still need to be controlled in order to ensure healthy plant establishment and growth (Gerhardson, 2002). Therefore, developments of various beneficial micro-organisms or biological control agents' methods are urgently needed in order to provide an alternative to chemical control. Among different biological approaches, use of the microbial antagonists like yeasts, fungi and bacteria could be promised, effectively, safely and eco-friendly in controlling many of soil borne pathogens (Gravel et al., 2004). Many biological control agents such as *Trichoderma* spp. and *Bacillus* spp. could be effectively used in suppressing diseases caused by *Fusarium* spp. *Drechslera halodes* and *Rhizoctonia solani* as reported by many workers (Hashem and Hamada, 2002; Soleimani et al., 2005; Nourozian et al., 2006; Abdel- Monaim, 2010). Modes of action for beneficial micro-organisms include direct parasitism of plant pathogens, competition for space or nutrients, or production of antibiotics, enzymes or plant hormones (Lugtenberg et al., 2003). This led to promote plant growth during the growing season as reported by Mercier and Manker (2005). However, up to date, only a few antagonist microorganisms have been identified as potential, effective bio-control agents against soil borne pathogens (Spadaro and Gullino, 2005). Also, these bioagents increased significantly due to seed germination and increased of plant growth in wheat and other many crops (Riungu et al., 2007; Zafari et al., 2008; El-Mohamedy et al., 2011). The differences among genotypes, among methods and among treatments within all methods were significant for number of kernels spike$^{-1}$, 1000-kernel weight and grain yield on the wheat (Moubarak, 2007). There were significant differences among the genotypes (Mahdy et al., 1996 and Pawar et al., 1997).

The main objective of the present study was to evaluate the effectiveness of isolated micro-organisms from soil under New Valley governorate conditions as bio-control agents against the incidence of root rot disease caused by *Fusarium graminearum, Drechslera halodes* and *Rhizoctonia solani* and its impact on vegetative growth, yield and yield components of two wheat cultivars under field conditions.

## MATERIALS AND METHODS

During 2008/2009 growing seasons, samples of wheat plants exhibiting root rot were collected from different fields in New Valley Governorate. Samples were washed thoroughly with tap water. Small portions of the root rot diseased samples were surface sterilized in 1% sodium hypochlorite solution for 5 min, rinsed in sterilized water and dried between folds of sterilized filter papers. The portions were placed on potato dextrose agar (PDA) medium and incubated at 25±1°C. The obtained fungal colonies were purified using single spore or hyphal tip techniques suggested by Dhingra and Sinclair (1984). Identification of the fungi detected was conducted according to Booth (1985) and Gilman (1998). Stock cultures were maintained on PDA slants and kept in a refrigerator at 5°C for further studies.

### Pathogenicity tests

Groups of formalin-sterilized soil were separately infested with fungal inocula of four pathogenic fungi (15-days-old) grown on barley sand medium (3%, w/w), then slightly watered every other day for a week. Soil provided with the same amount of barley sand medium and free from fungal inocula were used as control. The infested and uninfested soils were packed in formalin-sterilized pots (30-cm-diam.). Wheat grain (cvs. Sakha 93 and Bani Suif 5) superficially sterilized with 1% sodium hypochlorite were sowed at the rate of 10 grains/pot. A set of 4 replicates were used for each fungus or check treatments. Percentages of damping-off determined after 30 days from planting. After 90 days from planting, wheat plants were removed from the soil, washed thoroughly to remove soil debris, root rot were recorded as follows:

0 = roots without discoloration (no infection), 1= 1-20%, 2= 21-40%, 3= 41-75%, 4= 75-100% discoloration root mass and 5= completely dead plants include pre-or post emergence damping-off or old plants for each replicate. The mean of disease index (DI) and disease severity index (DSI) for each replicate was calculated by the formula suggested by Liu et al. (1995) was calculates:

$$DSI = \frac{\Sigma d}{d\,\max \times n} \text{X}100$$

Where as: d is the disease rating of each plant, d max is the maximum disease rating and n is the total number of plants examined for each replicate.

### Efficacy of antagonistic bioagents against pathogenic fungi *in vitro*

*Bacillus subtilis* (isolate BSM1), *B. megaterium* (isolate BMM5), *Trichoderma viride* (isolate TVM2) and *T. harzianum* (isolate THM4), were used in these experiments as bioagents against root rot diseases on wheat were obtained from Plant Pathology Research Department, New Valley Agricultural Research Stations.

Antagonism of *Trichoderma viride, T. harzianum* on *Fusarium graminearum, Drechslera halodes and Rhizoctonia solani*, the causal pathogens of root rot on wheat plants, was studied by dual culture technique (Rama et al., 2000). A mycelial disc (7 mm diameter), obtained from the peripheral region of 5 days old cultures tested pathogenic and *bioagent* fungi were placed simultaneously on the periphery, about 1 cm from the edges of the Petri-dishes (9 cm diameter) content PDA medium at opposite sides. While in each isolate of *B. megaterium* and *B. subtilis* was streaked individually at one side on PDA medium in plates and incubated for 24 h, then one disc (7 mm in diameter) of any pathogenic isolate was placed on the opposite side (Kaur et al., 2007). The Petri dishes containing the PDA medium inoculated with the tested pathogens alone served as control. All plates were

**Table 1.** The name origin and pedigree of two wheat cultivars.

| No. | Cultivars | Origin | Scientific name | Pedigree |
|---|---|---|---|---|
| 1 | Sakha 93 | Egypt | *Triticum aestivum* L. | Sakha 92 / TR 810328. |
| 2 | Bani Suif 5 | Egypt | *Triticum durum* L. | Dipper-2 / Bushen-3. |

incubated at 28°C and measurements were taken after 5 days. At the end of incubation period, radial growth was measured. The percentage inhibition growth of tested pathogens in presence of bioagents was calculated. The growth inhibition was calculated by using the formula: Mycelial growth inhibition= 100 (C- T / C), Where C= growth in control and T = growth in treatment.

#### Efficacy of grain and soil treatments with some bioagents under greenhouse and field conditions

##### *Production of bioagents inocula used for grain and soil treatments*

Antagonistic bacterial inocula were produced as described by Landa et al. (2001). Bacterial concentration in the suspension was adjusted to proximately $5 \times 10^8$ cells $ml^{-1}$ by measuring absorbance at 600 nm in a spectrophotometer and using standard curves for each bacterial isolate. Inocula of antagonistic fungi were prepared as described by Sallam et al. (2008). The fungal suspension was adjusted to $3 \times 10^7$ cfu $ml^{-1}$, using haemocytometre for each isolate. The bacterial and fungal suspension was used for grain coating and soil drenching.

##### *Under greenhouse conditions*

Formalin-sterilized soil were inoculated with the highly pathogenic isolated fungi that is *R. solani*, *F. graminearum* and *D. halodes* individually at rate 3% then irrigated and left for 7 days. The infested pots were divided into two groups, the first group planted with wheat grains soaked in suspension of bioagent isolates for 6 h. before sowing and the second groups were drenching with 50 ml of bio agents isolates then planting with untreated wheat grains. Ten grains were sown in pots which were infested with the pathogens. Also untreated grains were sown in pots infested with the tested pathogens to serve as a control. All treatments were in 4 replicates and arranged in complete randomized design.

##### *Under field conditions*

The present study was carried out at the New Valley Agricultural Research Station by Wheat Research Department, during two successive growing seasons, 2009/2010 and 2010/2011. Two wheat (*Tritium* spp.) cultivars were used to study the bio-control agents.

The name and pedigree of these cultivars are presented in Table 1 In the growing seasons; the two wheat cultivars were grown for evaluation in a randomized complete blocks design (RCBD) with four replications. Each plot was planted in eight rows; four meter long, 20 cm. apart. The recommended cultural practices were applied from sowing to harvesting. Data for the studied characters recorded on six internal rows.

The studied characters were grain yield 4.8 $m^{-2}$ (Kg.), number of spikes $m^{-2}$, number of kernels $spike^{-1}$ and 1000-kernel weight (g.). The obtained data were analyzed to estimate better methods and treatments of using bio-control agents.

#### Statistical analysis

The least significant differences (LSD.) test at 5% level of probability, according to Steel and Torrie (1980) was used to compare among means.

## RESULTS AND DISCUSSION

Isolation trials from diseased roots of wheat plants resulted in five pathogenic fungi, that is *Fusarium graminearum, F. equiseti, F. solani, Drechslera halodes and Rhizoctonia solani.*

### Pathogenicity tests

Pathogenicity of isolated fungi was tested on two wheat cultivars (Sakha 93 and Bani Suif 5). Data in Table 2 indicate that all the tested fungi significantly caused damping-off and root rot in both wheat cultivars subsequently, decreased the percentages of healthy survival plants compared with the check treatment. *Rhizoctonia solani* was the most pathogenic fungi as they recorded percentage damping-off (60 and 50%) and 26.8 and 23% root rot in both cultivars, respectively followed by *D. halodes* where caused 50 and 40% damping-off and 25.8 and 20.4% On the contrary, *F. solani and F. equiseti* were the least pathogenic ones recording the lowest percentages of these criteria. It was also observed that the percentages of damping-off in all tested fungi were higher than those of root rot. These results are in harmony with those reported by Hashem and Hamada (2002), Atef (2008) and Asran and Eraky (2011).

### Efficacy of antagonistic bioagents against pathogenic fungi *in vitro*

*Bacillus subtilis, B. megaterium, T. viride* and *T. harzianum,* strains were evaluated for antagonistic effect against *F. graminearum, D. halodes* and *R. solani* on Petri dishes containing PDA medium. Table 3 show that the bioagent strains succeeded in reducing the radial growth of the tested pathogenic fungi. *T. viride* was active more than the other tested bioagents for reducing the radial growth of fungi. The percent inhibition of radial growth of tested fungi viz., *F. graminearum* (67.7%), *D. halodes* (74.0%), and *R. solani* (59.5%) were reduced by *T. viride* which grown over the colonies of all three

**Table 2.** Pathogenic ability of four soil borne fungi on damping-off and root rot severity of wheat plants under greenhouse conditions.

| Treatment | Cultivars | | | | | |
|---|---|---|---|---|---|---|
| | Sakha 93 | | | Bani Suif 5 | | |
| | Damping-off (%) | Root rot (%) | Survival plants (%) | Damping-off (%) | Root rot (%) | Survival plants (%) |
| *F. graminearum* | 40 | 30.0 | 30.0 | 35 | 22.4 | 42.6 |
| *F. equiseti* | 25 | 18.2 | 56.8 | 20 | 14.8 | 65.2 |
| *F. solani* | 20 | 15.4 | 64.6 | 20 | 10.0 | 70.0 |
| *D. halodes* | 50 | 25.8 | 24.2 | 40 | 20.4 | 39.6 |
| *R. solani* | 60 | 26.8 | 13.2 | 50 | 23.0 | 27 .0 |

| LSD at 0.05 | Damping-off | Root rot | Root rot | Survival plants |
|---|---|---|---|---|
| Treatments (A) | 3.70 | 3.20 | 3.20 | 5.78 |
| Cultivars (B) | 2.33 | 2.03 | 2.03 | 3.65 |
| Interaction (AxB) | 5.23 | ns | ns | ns |

**Table 3.** Inhibitory effect of some bio-against isolates on mycelial growth of wheat pathogenic fungi.

| Treatment | Mycelial growth inhibition (%) | | | |
|---|---|---|---|---|
| | *F. graminearum* | *D. halodes* | *R. solani* | Mean |
| *B. subtilis* | 39.2 | 53.1 | 41.8 | 44.70 |
| *B. megaterium* | 52.0 | 58.5 | 48.1 | 52.87 |
| *T. viride* | 67.7 | 74.0 | 59.5 | 67.07 |
| *T. harzianum* | 60.2 | 66.7 | 50.9 | 59.27 |
| Mean | 54.77 | 63.07 | 50.07 | - |

LSD at 0.05 for: treatments (A) =3.55; pathogenic fungi (B) = 2.97; introduction (AxB) = 6.85.

pathogens. On the other hand Trichoderma isolates were effective more than Bacillus isolates in inhibition of redial growth of all the tested fungi and the greatest reduction occurring in *D. halodes* (63.07%) followed by *F. graminearum* (54.77%), while *R. solani* less affected ones (50.07%). The growth inhibition of wheat root rot fungi by dual culture in this study could be due to its fast growing nature, secretions of armful extra-cellular compounds like antibiotics, cell wall degrading enzymes such as β-1, 3 gluconase, endochitinase and chitinase enzymes which degrade the cell wall leading to lyses of mycelium of the pathogen and mycoparasitism in dual culture as found with other fungi (Lugtenberg et al., 2003; Riungu, 2008 and Rahman, et al., 2010).

## Efficacy of grain and soil treatments with some bioagent isolates

### *Greenhouse experiments*

In the seed coating and soil treatment tests, all the tested bioagents were able to reduce significantly damping-off and root rot caused by *F. graminearum, D. halodes* and *R. solani* compared with control, however, there was variation among bioagent isolates effect on reduction of disease severity both soil and seed application methods and both cultivates (Table 4A, B and C). *B. megaterium* and *T. viride* recorded the highest reduction in damping-off and root rot severity caused by the tested fungi, while soil and/or grain treatments with *B. subtilis* and *T. harzianum* gave the lowest reduction for these diseases in both cultivars.

Concerning the method of applications, it is cleared that there were a significant differences among method of applications. The highest significant values for suppressing fusarium, drechslera and rhizoctonia root rots incidence on wheat plants were recorded with soil treatment in all cases.

Also, there were significant variation among both cultivars; however cv. Sakha 93 was more susceptible than cv. Bani Suif 5 in all cases. On the other hand, the obtained data show that all of the soil and grain application treatments increased significantly the fresh and dry weight of the survival wheat plants compared with control, but there was difference among their effect

**Table 4A.** Effect of soil drenching, wheat grain treatment with different bioagents on damping-off and root rot diseases, fresh and dry weight plant$^{-1}$ under artificial infection with *Fusarium graminearum* under greenhouse conditions.

| Treatment | Application methods | Sakha 93 | | | | Bani Suif 5 | | | |
|---|---|---|---|---|---|---|---|---|---|
| | | Damping-off (%) | Root rot (%) | Fresh weight (g plant$^{-1}$) | Dry weight (g plant$^{-1}$) | Damping-off (%) | Root rot (%) | Fresh weight (g plant$^{-1}$) | Dry weight (g plant$^{-1}$) |
| *B. subtilis* | Soil | 20 | 14.0 | 3.161 | 0.691 | 15 | 10.0 | 3.422 | 0.725 |
| | Grain | 30 | 16.4 | 2.481 | 0.524 | 25 | 12.8 | 2.235 | 0.473 |
| *B. megaterium* | Soil | 5 | 6.6 | 4.583 | 0.972 | 10 | 5.2 | 4.25 | 0.907 |
| | Grain | 10 | 8.0 | 4.028 | 0.872 | 15 | 6.8 | 3.657 | 0.796 |
| *T. viride* | Soil | 15 | 7.6 | 4.209 | 0.923 | 5 | 4.5 | 3.969 | 0.795 |
| | Grain | 15 | 9.4 | 3.412 | 0.731 | 10 | 6.0 | 3.414 | 0.699 |
| *T. harzianum* | Soil | 20 | 12.4 | 3.800 | 0.786 | 15 | 9.0 | 3.55 | 0.77 |
| | Grain | 25 | 13.6 | 3.078 | 0.622 | 25 | 10.6 | 2.914 | 0.621 |
| Control | | 45 | 33.6 | 1.333 | 0.277 | 35 | 25.4 | 1.182 | 0.251 |

| **LSD at 0.05** | **Damping-off** | **Root rot** | **Fresh weight** | **Dry weight** |
|---|---|---|---|---|
| Treatments (A) | 1.71 | 1.20 | 0.177 | 0.057 |
| Application methods (B) | * | * | * | * |
| Cultivars (C) | * | * | * | * |
| Interaction (AxBxC) | ns | 2.55 | ns | 0.111 |

**Table 4B.** Effect of soil drenching, wheat grain treatment with different bioagents on damping-off and root rot diseases, fresh and dry weight plant$^{-1}$ under artificial infection with *Drechslera halodes* under greenhouse conditions.

| Treatment | Application methods | Sakha 93 | | | | Bani Suif 5 | | | |
|---|---|---|---|---|---|---|---|---|---|
| | | Damping-off (%) | Root rot (%) | Fresh weight (g plant$^{-1}$) | Dry weight (g plant$^{-1}$) | Damping-off (%) | Root rot (%) | Fresh weight (g plant$^{-1}$) | Dry weight (g plant$^{-1}$) |
| *B. subtilis* | Soil | 20 | 12 | 3.142 | 0.674 | 20 | 10.8 | 2.916 | 0.634 |
| | Grain | 30 | 13.6 | 2.627 | 0.557 | 25 | 13.0 | 2.014 | 0.431 |
| *B. megaterium* | Soil | 5 | 6.8 | 5.092 | 1.088 | 5 | 5.8 | 4.071 | 0.833 |
| | Grain | 15 | 8.2 | 4.022 | 0.863 | 10 | 6.2 | 3.214 | 0.643 |
| *T. viride* | Soil | 5 | 10.6 | 3.613 | 0.738 | 5 | 7.6 | 3.208 | 0.628 |
| | Grain | 10 | 12.0 | 3.429 | 0.702 | 5 | 9.4 | 2.814 | 0.603 |

**Table 4B contd.**

| | | | | | | | | | |
|---|---|---|---|---|---|---|---|---|---|
| *T. harzianum* | Soil | 25 | 19.8 | 3.633 | 0.731 | 15 | 16.6 | 3.029 | 0.617 |
| | Grain | 35 | 22.2 | 3.240 | 0.672 | 25 | 19.8 | 2.492 | 0.541 |
| Control | | 50 | 28.4 | 1.037 | 0.209 | 40 | 24.6 | 0.805 | 0.171 |

| **LSD at 0.05** | **Damping-off** | **Root rot** | **Fresh weight** | **Dry weight** |
|---|---|---|---|---|
| Treatments (A) | 2.45 | 1.28 | 0.177 | 0.048 |
| Application methods (B) | * | * | * | * |
| Cultivars (C) | * | * | * | * |
| Interaction (AxBxC) | 4.89 | 2.56 | 0.352 | 0.097 |

in this regard. The maximum fresh and dry weight obtained in *B. megaterium* followed by *T. viride*. While the lowest increased of fresh and dry weight of survival wheat plants were recorded in case of *B. subtilis* application.

On the other hand, soil application treatment improved the fresh and dry weights of wheat were more than grain application method in all cases.

Also, cv. Bani Suif 5 gave fresh and dry weights high than cv. Sakha 93 in most cases.

Generally, *B. megaterium* and *T. viride* recorded the highest reduction of damping-off, root rot and increased fresh and dry weight of survival wheat plants especially in soil application method. The obtained results are in good accordance with previous studies which have been concluded that *Trichoderma* spp. and *Bacillus* spp. can effectively protect many plant species against root rot diseases (Hashem and Hamada, 2002; Soleimani et al., 2005; Nourozian, et al., 2006; Atef, 2008; Abdel- Monaim, 2010).

According to Harman (2001) natural factors limiting the number of soil borne pathogens occur through a combination of antagonism by other soil fungi and bacteria, natural release of antibiotics from other bacteria and fungi, and by competitive exclusion of habitat in the root zone or rhizosphere.

The mechanism of Trichoderma and Bacillus action on pathogens may be by attacking and binding the pathogenic organisms by sugar linkage and begins to secrete extracellular protease and lipase (Soleimani et al., 2005; Zaghloul et al., 2007), produce siderophores and hydrogen cyanide (Soleimani et al., 2005), production of secondary metabolites such as Phenazine -1-Carboxilic acid (PCA), 2,4-Pyrrolnitrin, Oomycin (Knudsen, 1995).

## Under field conditions

Mean squares for all studied characters in (Table 5), showed significant differences between the seasons (S), methods (M), (S) x (M), treatments (T), (S) x (T), (M) x (T) and (S) x (M) x (T) for Sakha 93 cultivar, except methods and (S) x (M) for number of kernels spike$^{-1}$. These results are confirmed with those obtained by Mahdy et al. (1996), Pawar et al. (1997) and Moubarak (2007). Mean squares for all studied characters (Table 6), showed significant differences between the seasons (S), methods (M), (S) × (M), treatments (T), (S) × (T), (M) × (T) and (S) × (M) × (T) for Bani Suif 5 cultivar, except methods and (S) × (M) for 1000-kernel weight. These results are confirmed with those obtained by Mahdy et al. (1996), Pawar et al. (1997) and Moubarak (2007).

The results illustrated (Tables 7 and 8) showed that, in soil application method, the treatments *T. harzianum*, *B. subtilis*, *B. subtilis* and *B. megaterium* had highest values for grain yield, number of spikes m$^{-2}$, number of kernels spike$^{-1}$ and 1000-kernel weight in the two seasons, respectively, while in grain application method, the treatments *B. megaterium*, *T. harzianum*, *B. megaterium* and *B. megaterium* had highest values for grain yield, number of spikes m$^{-2}$, number of kernels spike$^{-1}$ and 1000-kernel weight in the two seasons, respectively. From the previous results, it could be concluded that, the treatments *T. harzianum* and *B. megaterium* were best treatments to increase grain yield, the treatments *B. subtilis* and *T. harzianum* were best treatments to increase number of spikes and the treatments *B. subtilis* and *B. megaterium* were best treatments to increase number of kernels for soil application method and grain application

**Table 4C.** Effect of soil drenching, wheat grain treatment with different bioagents on damping-off and root rot diseases, fresh and dry weight plant$^{-1}$ under artificial infection with *Rhizoctonia solani* under greenhouse conditions.

| Treatment | Application methods | Sakha 93 | | | | Bani Suif 5 | | | |
|---|---|---|---|---|---|---|---|---|---|
| | | Damping-off (%) | Root rot (%) | Fresh weight (g plant$^{-1}$) | Dry weight (g plant$^{-1}$) | Damping-off (%) | Rootrot (%) | Fresh weight (g plant$^{-1}$) | Dry weight (g plant$^{-1}$) |
| *B. subtilis* | Soil | 20 | 14.2 | 2.345 | 0.512 | 15 | 10.0 | 2.267 | 0.501 |
| | Grain | 25 | 17.4 | 1.958 | 0.405 | 15 | 13.8 | 2.001 | 0.402 |
| *B. megaterium* | Soil | 10 | 7.6 | 3.900 | 0.809 | 10 | 6.0 | 3.100 | 0.668 |
| | Grain | 15 | 8.2 | 3.409 | 0.735 | 10 | 8.0 | 2.069 | 0.443 |
| *T. viride* | Soil | 5 | 5.6 | 2.967 | 0.602 | 5 | 4.6 | 2.933 | 0.625 |
| | Grain | 10 | 6.4 | 2.491 | 0.542 | 5 | 6.0 | 2.725 | 0.579 |
| *T. harzianum* | Soil | 25 | 16.4 | 2.817 | 0.600 | 20 | 13.6 | 2.521 | 0.512 |
| | Grain | 35 | 22.6 | 2.183 | 0.452 | 30 | 15.6 | 2.055 | 0.439 |
| Control | | 55 | 30.4 | 1.038 | 0.219 | 45 | 23.2 | 1.058 | 0.229 |

| LSD at 0.05 | Damping-off | Root rot | Fresh weight | Dry weight |
|---|---|---|---|---|
| Treatments (A) | 2.00 | 1.50 | 0.229 | 0.051 |
| Application methods (B) | * | * | * | * |
| Cultivars (C) | * | * | * | * |
| Interaction (AxBxC) | 3.98 | Ns | 0.460 | ns |

**Table 5**. Mean square analysis of four characters for Sakha 93 bread wheat cultivar using bio-control agents by soil and grain methods in 2009/2010 and 2010/2011 seasons.

| S.O.V | d.f. | Mean squares | | | |
|---|---|---|---|---|---|
| | | Grain yield 4.8m$^{-2}$ | No. of spikes m$^{-2}$ | No. of kernels spike$^{-1}$ | 1000-kernel weight |
| Replications | 3 | 0.004 | 34.533 | 13.933 | 0.635 |
| Seasons (S) | 1 | 9.716* | 12152.450* | 2121.800* | 1570.878* |
| Methods (M) | 1 | 0.021* | 6265.800* | 0.200 | 24.753* |
| SM | 1 | 0.036* | 2184.050 * | 7.200 | 110.215* |
| Treatments (T) | 4 | 0.388* | 4976.513* | 122.856* | 298.386* |
| ST | 4 | 0.120* | 892.825* | 42.581* | 11.924* |
| MT | 4 | 0.201* | 5423.737* | 308.606* | 40.884* |
| SMT | 4 | 0.023* | 926.800* | 14.231* | 17.766* |
| Error | 57 | 0.004 | 46.130 | 3.065 | 2.101 |

* Significant at 5% probability.

**Table 6**. Mean square analysis of four characters for Bani Suif 5 durum wheat cultivar using bio-control agents by soil and grain methods in 2009/2010 and 2010/2011 seasons.

| | | Mean squares | | | |
|---|---|---|---|---|---|
| **S.O.V** | **d.f.** | **Grain yield 4.8 $m^{-2}$** | **No. of spikes $m^{-2}$** | **No. of kernels spike$^{-1}$** | **1000-kernel weight** |
| Replications | 3 | 0.007 | 41.233 | 1.512 | 2.204 |
| Seasons (S) | 1 | 3.113* | 9159.200* | 1453.513* | 1553.203* |
| Methods (M) | 1 | 0.021* | 8946.450* | 32.513* | 0.001 |
| SM | 1 | 0.031* | 530.450* | 35.112* | 0.435 |
| Treatments (T) | 4 | 0.520* | 4649.019* | 371.294* | 119.963* |
| ST | 4 | 0.032* | 1000.544* | 15.919* | 7.977* |
| MT | 4 | 0.315* | 5239.794* | 383.106* | 3.414* |
| SMT | 4 | 0.018* | 1127.294* | 64.206* | 23.208* |
| Error | 57 | 0.004 | 45.996 | 3.153 | 1.616 |

* Significant at 5% probability.

**Table 7.** Effect of bio-control against on grain yield (Kg 4.8 $m^{-2}$), No. of spikes $m^{-2}$, kernels $spikes^{-1}$ and 1000-kernel weight for Sakha 93 bread wheat cultivar in 2009/2010 and 2010/2011 seasons.

| Season | Treatment | Application methods | Grain yield (Kg 4.8 $m^{-2}$) | No. of spikes $m^{-2}$ | No. of kernels $spike^{-1}$ | 1000-kernel weight |
|---|---|---|---|---|---|---|
| 2009/ 2010 | *B. subtilis* | Soil | 1.73 | 349.50 | 62.25 | 50.63 |
| | | Grain | 1.66 | 330.50 | 56.25 | 45.08 |
| | *B. megaterium* | Soil | 1.62 | 237.50 | 46.50 | 57.10 |
| | | Grain | 1.89 | 347.75 | 61.00 | 49.45 |
| | *T. viride* | Soil | 1.68 | 309.00 | 55.75 | 42.93 |
| | | Grain | 1.70 | 316.25 | 48.25 | 45.95 |
| | *T. harzianum* | Soil | 1.86 | 326.00 | 57.75 | 53.53 |
| | | Grain | 1.69 | 368.25 | 54.25 | 46.40 |
| | Control | | 1.59 | 305.00 | 59.25 | 44.95 |
| 2010/ 2011 | *B. subtilis* | Soil | 2.44 | 359.00 | 74.25 | 56.88 |
| | | Grain | 2.12 | 336.75 | 67.50 | 55.15 |
| | *B. megaterium* | Soil | 2.46 | 322.75 | 60.25 | 64.03 |
| | | Grain | 2.86 | 358.75 | 76.25 | 64.03 |
| | *T. viride* | Soil | 2.41 | 342.50 | 63.25 | 48.30 |
| | | Grain | 2.23 | 337.50 | 63.25 | 53.08 |
| | *T. harzianum* | Soil | 2.76 | 351.25 | 66.75 | 57.85 |
| | | Grain | 2.49 | 378.75 | 61.00 | 60.98 |
| | Control | | 2.10 | 327.00 | 65.50 | 54.65 |

| LSD at 0.05 | Grain yield 4.8 $m^{-2}$ | No. of spikes $m^{-2}$ | No. of kernels $spike^{-1}$ | 1000-kernel weight |
|---|---|---|---|---|
| Seasons (S) and methods (M) | 0.057 | 6.075 | 1.566 | 1.297 |
| (S) (M) | 0.08 | 8.591 | 2.215 | 1.833 |
| Treatments (T) | 0.057 | 6.075 | 1.566 | 1.296 |
| (S) (T) and (M) (T) | 0.081 | 8.592 | 2.215 | 1.833 |
| (S) (M) (T) | 0.114 | 12.15 | 3.132 | 2.593 |

method, respectively. The obtained results are in harmony with that obtained by Riungu et al. (2007). Zafari et al., 2008 on wheat. All these workers reported that using beneficial micro-organisms as biocontrol agents led to enhancement of plant growth parameters.

Such enhancement may be due to induce plant resistance (De Meyer et al., 1998), produce extracellular enzymes and antifungal or antibiotics, which decrease biotic stress on plant, and produce growth promoters substances (Szczech and Shoda, 2004).

In addition, Egamberdiyeva (2007) hypothesized that there are several mechanisms by which rhizosphere bacteria may stimulate plant growth, such as production of plant growth substances, nitrogen fixation, phytohormones, vitamins, solublizing minerals besides, their role in

**Table 8.** Effect of bio-control against on grain yield (Kg 4.8 $m^{-2}$), No. of spikes $m^{-2}$, kernels $spikes^{-1}$ and 1000-kernel weight for Bani Suif 5 durum wheat cultivar in 2009/2010 and 2010/2011 seasons.

| Season | Treatments | Application methods | Grain yield (Kg 4.8 $m^{-2}$) | No. of spikes $m^{-2}$ | No. of kernels $spike^{-1}$ | 1000-kernel weight |
|---|---|---|---|---|---|---|
| 2009/ 2010 | *B. subtilis* | Soil | 2.99 | 325.50 | 62.25 | 53.23 |
| | | Grain | 2.83 | 287.75 | 62.00 | 52.60 |
| | *B. megaterium* | Soil | 2.81 | 289.75 | 47.25 | 60.20 |
| | | Grain | 3.08 | 280.00 | 67.50 | 58.43 |
| | *T. viride* | Soil | 2.88 | 310.50 | 59.00 | 51.00 |
| | | Grain | 2.77 | 269.00 | 49.25 | 53.88 |
| | *T. harzianum* | Soil | 3.34 | 286.00 | 65.75 | 55.68 |
| | | Grain | 2.99 | 295.00 | 55.25 | 55.90 |
| | Control | | 2.72 | 250.50 | 50.25 | 56.65 |
| 2010/ 2011 | *B. subtilis* | Soil | 3.28 | 350.00 | 73.25 | 61.15 |
| | | Grain | 3.10 | 256.25 | 69.25 | 64.60 |
| | *B. megaterium* | Soil | 3.18 | 287.50 | 59.25 | 65.83 |
| | | Grain | 3.76 | 311.75 | 75.25 | 67.30 |
| | *T. viride* | Soil | 3.26 | 357.50 | 62.25 | 62.88 |
| | | Grain | 3.19 | 288.75 | 67.25 | 57.65 |
| | *T. harzianum* | Soil | 3.72 | 317.50 | 68.25 | 64.48 |
| | | Grain | 3.42 | 324.25 | 64.25 | 64.00 |
| | Control | | 3.08 | 282.50 | 57.50 | 67.23 |

| LSD at 0.05 | Grain yield 4.8$m^{-2}$ | No. of spikes $m^{-2}$ | No. of kernels $spike^{-1}$ | 1000-kernel weight |
|---|---|---|---|---|
| Seasons (S) and methods (M) | 0.054 | 6.066 | 1.588 | 1.137 |
| (S) (M) | 0.077 | 8.579 | 2.246 | 1.608 |
| Treatments (T) | 0.054 | 6.066 | 1.588 | 1.137 |
| (S) (T) and (M) (T) | 0.077 | 8.579 | 2.246 | 1.608 |
| (S) (M) (T) | 0.108 | 12.133 | 3.177 | 2.274 |

direct inhibition of pathogen growth and suppression of diseases caused micro-organisms and increased of plant growth and yield.

## Conclusion

Specific rhizobacterial and rhizofungal agents can influence disease suppression and could be considered as part of disease control strategy under an integrated pest management which offers a successful approach for the deployment of both agro-chemicals and biocontrol agents.

Also, this study suggest that simultaneous screening of rhizobacteria and rhizofungi for growth and yield promotion under field experiment is a good tool to select effective rhizobacteria and rhizofungi for biofertilizer development biotechnology.

The treatments *T. harzianum* and *B. megaterium* were best treatments to increase grain yield, the treatments *B. subtilis* and *T. harzianum* were best treatments to increase number of spikes and the treatments *B. subtilis* and *B. megaterium* were best treatments to increase number of kernels for soil application method and grain application method, respectively.

## REFERENCES

Abdel-Monaim MF (2010). Integrated management of damping- off, root and/or stem rot diseases of chickpea with sowing date, host resistance and bioagents. Egypt. J. Phytopathol., 38(1-2): 45-61.

Asran MR, Eraky Amal MI (2011). Aggressiveness of certain *Fusarium graminearum* isolates on wheat seedling and relation with their Trichothecence production. Plant Pathol. J., 10(1): 36-41.

Atef Nagwa M (2008). *Bacillus subtilits* and *Trichoderma harzianum* as wheat inoculants for biocontrol of *Rhizoctonia solani*. Australian J. Basic Appl. Sci., 2(4): 1411-1417.

Bakr DW (1997). Studies on some root rots of wheat . M. Sc. Thesis, Faculty Agriculture Assuit University, Assiut, Egypt.

Booth C (1985).The genus *Fusarium*. Kew, Surrey Commonwealth Mycological Institute, 2nd Ed., p. 237.

De Meyer G, Bigirimana J, Elad Y, Höfte M (1998). Induced systemic resistance in *Trichoderma harzianum* T39 bio-control of *Botrytis cinerea*. Eur. J. Plant Pathol., 104: 279-286.

Dhingra OD, Sinclair JB (1985). Basic Plant Pathology Methods. CRC, Boca Raton, Florida, USA.

Egamberdiyeva D (2007). The effect of plant growth promoting bacteria on growth and nutrient uptake of maize in two different soils. Appl. Soil Eco., 36: 184-189.

El-Mohamedy RSR, Abd El-Samad EH, Habib Hoda AM, Fath El-Bab TSH (2001). Effect of using bio-control agents on growth, yield, head quality and root rot control in broccoli plants grown under greenhouse conditions. Int. J. Acad. Res., 3(2): 71-80.

Gerhardson B (2002). Biological substitute for pesticides.Trends Biotech., 20: 338-343.

Gilman JC (1998). A manual of soil fungi: Iowa. State. Univ. Press. Ames. Iowe. U.S.A.

Gravel V, Martinez C, Antoun H, Tweddell RJ (2004). Evaluation of antagonistic microorganisms as bio-control agents of root rot (*Pythium ultimum*) of greenhouse tomatoes in rock wool. Can. J. Plant Pathol. 26: 152-159.

Harman GE (2001). *Trichoderma* spp., including *T. harzianum, T. viride, T. koningii, T. hamatum* and other spp. Deuteromycetes, Moniliales (asexual classification system) In: "Biological Control: A Guide to Natural Enemies in North America". (C.R. Weeden, A.M. Shelton, M. P. Hoffman eds.), Cornell University, USA.

Hashem M, Hamada Afaf M (2002). Evaluation of two biologically active compounds for control of wheat root rot and its causal pathogens. Microbiology, 30(4): 233-239.

Hausbek MK, Lamour KH (2004). *Phytophthora capsici* on vegetable crops: research progress and management challenges. Plant Disease 88: 1292-1303.

Kaur R, Singh RS, Alabouvette C (2007). Antagonistic activity of selected isolates of fluorescent *Pseudomonas* against *Fusarium oxysporum* f. sp. *ciceri*. Asian J. of Plant Sci. 6(3): 446-454.

Knudsen IMB, Hockenhull J, Jensen DF (1995). Biocontrol of seedling diseases of barley and wheat caused by *Fusarium culmorum* and *Bipolaris sorokiniana*:Effects of selected fungal antagonists on growth and yield components. Plant Pathol., 44: 467-477.

Landa BB, avas-Cortés, JA Hervás, Jiménez-Díaz RM 2001. Influence of temperature and inoculum density of *Fusarium oxysporum* f. sp. *ciceris* on suppression of Fusarium wilt of chickpea by rhizosphere bacteria. Phytopathol., 91: 807-816.

Liu L, Kloepper JW, Tuzun S (1995). Introduction of systemic resistance in cucumber against Fusarium wilt by plant growth-promoting rhizobacteria. Phytopathol., 85: 695-698.

Lugtenberg BJJ, Bloemberg GV, Woeng CA, Thomas FC (2003). Phenazines and their role in bio-control by *Pseudomonas* bacteria. New Phytologist, 153: 503-523.

Mahdy EE, Ismail AA, Kheiralla KA (1996). The relative merits of selection index and pedigree selection in improving grain yield of spring wheat. Assiut J. Agric. Sci., 27: 17–33.

Mercier J, Manker DC (2005). Bio-control of soil borne diseases and plant growth enhancement in greenhouse soilless mix by the volatile-producing fungus *Muscodor albus*. Crop Protection, 24: 355-362.

Moubarak MYGH (2007). Efficiency of Some Breeding Methodologies on Some Bread Wheat Populations under New Valley Conditions. Ph. D. Thesis, Agron. Dep. Fac. Agric. Minia Univ. Egypt.

Nourozian J, Etebarian HR, Khodakaramian G (2006). Biological control of *Fusarium graminearum* on wheat by antagonistic bacteria. Nutraceutical and Functional Food, 28(1): 29-38.

Pawar IS, Yunus M, Karwasra SS, Prabalee S (1997). Use of single spike selection procedure in wheat improvement. Haryana Agric. Univ. J. Res., 27: 167–169.

Rahman T, Ahmed AU, Islam MR, Hosen MI (2010). Physiological study and both *in vitro* and *in vivo* antifungal activities against *Stemphylium botrysum* caused stemphylium blight disease in lentil (*Lens culinaris*). Plant Pathol. J. 9(4): 179-187.

Rama B, Raju M, Krishna Murthy KVM (2000). Efficacy of *Trichoderma* spp. In the management of collar rot of groundnut caused by *Aspergillus niger* Van Teighem. Indian J. Plant Prot., 28: 197-199.

Riungu GM, Muthomi JW, Narla RD, Wagacha JM, Gathumbi JK (2008). Management of Fusarium head blight of wheat and deoxynivalenol accumulation using antagonistic microorganisms. Plant Pathol. J., 7: 13-19.

Sallam Nashwa MA, Abo-Elyousr KAM, Hassan MAE (2008). Evaluation of *Trichoderma* species as biocontrol agent for damping-off and wilt diseases of *Phaseolus vulgaris* L. and efficacy of suggested formula. Egypt. J. Phytopathol., 36(1-2): 81-93.

Soleimani MJ, Shamsbakhsh M, Taghavi M, Kazemi SH (2005). Biological control of stem and root rot of wheat caused by *Bipolaris* spp. by using antagonistic bacteria, fluorescent Pseudomonas and Bacillus sp. J. Biological Sci., 5(3): 347-353.

Spadaro D, Gullino ML (2005). Improving the efficacy of bio-control agents against soil borne pathogens. Crop Protection 24: 601-613.

Steel RGD, Torri JH (1980). Principles and procedures of statistical biometrical approaches 2nd ED Mc Graw Hill Book company – New York. USA.

Szczech M, Shoda M (2004). Bio-control of *Rhizoctonia* damping-off of tomato by *Bacillus subtilis* combined with *Burkholderia cepacia*. J. Phytopathol., 152: 549-556.

Zafari D, Koushki MM, Bazgir E (2008). Biocontrol evaluation of take-all disease by *Trichoderma* screened isolates. African J. Biotech., 7: 3653–3659.

Zaghloul RA, Hanafy Ehsan A, Neweigy NA, Khalifa Neamat A (2007). Application of biofertilization and biological control for tomato production. 12th Conference of Microbiology; Cairo, Egypt (18-22) March, pp. 198-212.

# 25

# Induction of dormancy in non-dormant varieties of groundnut

**M. Sudha Rani[1] , P. S. Rao[2], K. Kanaka Durga[1], V. Bharathi[3], K. Bharathi[4] and A. Vishnuvardhan Reddy[4]**

[1]Plant Breeding, Seed Research and Technology Centre, Rajendranagar, Hyderabad-30, India.
[2]Plant Physiology, Institute of Biotechnology, Rajendranagar, Hyderabad-30, India.
[3]Plant Pathology, Seed Research and Technology Centre, Rajendranagar, Hyderabad-30, India.
[4]Seed Research and Technology Centre, Rajendranagar, Hyderabad -30, India.

**A field experiment was conducted at Seed Research and Technology Centre, Hyderabad to induce dormancy in groundnut using maleic hydrazide (MH) during *rainy season*, 2010 and 2011. A non - dormant, bunch type of groundnut variety Narayani was utilized. Two foliar sprays of MH at different concentrations *viz*., 500, 750, 1000 and 1250 ppm were imposed at 60 and 90 days after sowing. Significantly higher germination of 80.0 and 81.73% immediately after harvest were recorded by untreated control seed lots during 2010 and 2011 seasons, respectively. The treatment of 1250 ppm of MH showed the lowest germination percent of 32.33 during 2010, while it was 36.33 during 2011, followed by 1000 ppm of MH at 60 and 90 days after sowing. Similarly, the same treatment had the least seedling vigour index of 1201 while that of untreated plot was 1905 during 2010 and 2012 during 2011. It was found that two foliar sprays of MH at 1250 ppm at 60 and 90 days after sowing would induce dormancy in non-dormant bunch type groundnut cultivar up to 30 days and protects the seed from *in situ* germination due to untimely rains at the time of harvest.**

**Key words:** Pre-harvest sprouting, chemicals methods, induction of dormancy, groundnut.

## INTRODUCTION

Groundnut (*Arachis hypogea* L.) is an important crop among oil seeds, as it can be consumed and utilized in diverse ways. Groundnut seeds are rich in oil and protein. Groundnut is grown on nearly 23.95 million ha worldwide with the total production of 36.45 million tons with an average yield of 1520 kg / ha in 2009 (FAO, 2011). In India, it occupies an area of 6.41 million ha with a production of 9.36 million tones, which accounts for a productivity of 1460 kg/ha during 2007 to 2008 (Anonymous, 2008). In India, 70% of the groundnut area and 75% of the production are concentrated in the four states namely, Gujarat, Tamilnadu, Andhra Pradesh and Karnataka. In groundnut, the sub species *fastigiata,* which includes the *Spanish* and *Valencia* are the most preferred groundnut sub species in the semi-arid tropics, which accounts for about 60% of the world's groundnut production area. Most of the popularly cultivated varieties in southern India are non-dormant and harvesting these cultivars in *rainy season* poses the problem of sprouting in the field and subsequently deterioration of the quality of the kernel to a tune of 20 to 50% (Varman and Raveendran, 1991). This problem becomes more obvious if harvesting is further delayed. Yield losses due to *in situ* germination in bunch type of groundnut varieties have been estimated to range between 20 and 40% by Reddy et al. (1987) and Nagarjun and Radder (1983). Thus, a short period of seed dormancy is necessary to reduce these losses. Kumar et al. (1991) reported that the cultivation of sub species *fastigiata* generally lacked dormancy while those of sub species *hypogea* were characterized by long dormancy periods.

Maleic hydrazide (MH) (Diethanolamine salt of 1, 2-dihydroxy-3,6 pyridazine-dione), a growth inhibitor has been successfully used to induce dormancy and thus to reduce sprouting losses in potato, sugarbeet, onion, carrot and rice. Vaithialingam and Rao (1973) reported that induction of dormancy in TMV-2, a bunch groundnut variety by the foliar application of MH-30, in a field trial conducted at Coimbatore. Nagarjun and Radder (1983) and Gupta et al. (1985) also reported that foliar application of MH could induce dormancy in bunch type of groundnut varieties in the field trials. Spraying of MH at 250 ppm at 60 days after sowing (DAS) was as effective as higher concentration sprayed at later stages in inducing seed dormancy in bunch type groundnut (Nagarjun and Radder, 1983). Similarly, Naitiyal (2004) observed that foliar application of MH at 1000 ppm at 60 days after crop emergence was superior in induction of dormancy in groundnut. However, the information available on the effect of MH in inducing seed dormancy in groundnut is meager and inconclusive. Therefore, the present study was taken up to observe the effect of MH to induce dormancy in a bunch groundnut cultivar and also to standardize its dosage and time of foliar application.

## MATERIALS AND METHODS

The experiment was conducted at field plots of Seed Research and Technology Centre, Rajendranagar, Hyderabad, during rainy season, 2010 and 2011. The non dormant, bunch type of groundnut variety namely, Narayani was utilized in the current study. Two foliar sprays of maleic hydrazide at different concentrations *viz*., 500, 750, 1000 and 1250 ppm were given to the crop at 60 and 90 days after sowing. The experiment was laid out in a factorial randomized block design with three replications. The gross plot size was 4 × 3 m and the net plot size was 3.8 × 2.6 m. The row to row spacing was 30 cm, while the plant to plant spacing was 10 cm. Observations were recorded on germination, root length, shoot length, total seedling length and seedling vigour index (SVI) based on seedling length at regular intervals till the germination of the treated seed lots becomes equal to the untreated control with the harvested produce. The germination percent was tested using between paper method of testing suggested by International Seed Testing Association (1985). The seedling vigour index was calculated by multiplying the germination per cent with mean seedling length. The induction of dormancy was tested by conducting the germination test immediately after harvest and was continued till the germination percent of both control as well as the treated lots declines to 5% below the minimum seed certification standards (MSCS) and the interval of testing was five days. The dormancy duration was computed as the period from harvest till the germination percent of the treated seed lots becomes equal to the untreated control. The mean values of the data were statistically analyzed following factorial Randomized Block Designs (RBD) using Windostat software version 8.6.

## RESULTS AND DISCUSSION

The data on germination percentage indicated significant differences between control and MH treated seed lots. Significantly higher germination of 80.0 and 81.73% immediately after harvest were recorded by untreated control seed lots during 2010 and 2011 seasons, respectively, as shown in the Table 1. However, the treatment 1250 ppm of MH showed the lowest germination percent of 32.33 during 2010, while it was 36.33 during 2011, followed by 1000 ppm of MH during both the seasons. The germination percent recorded at 5 days interval showed that spray of MH at 1250 ppm could restrict the germination of groundnut cultivar Narayani for 30 days (Table 1). From the current study, it has been further noticed that there was an increased trend with respect to germination percentage as the number of days of harvest increased. The lower germination percent observed in the MH treated plots for a period of 30 days may be due to the blockage of any of the sequential processes involved in the germination. The research work of Bewley and Black (1982) revealed that the application of an inhibitor (MH) could bring about certain changes in the physiological and biochemical processes like alteration in promoter to inhibitor ratio, moisture content of the seed, water absorption capacity of the seeds, protein content and oil content of the seed, which are responsible to make the seed dormant by way of arresting the growth of the embryo. Another important conception was that, the dormant and non-dormant states of the seed were dependant on relative levels of inhibitors and promoters present in the seed. The results obtained in the present investigation are in confirmation with the earlier reports of Naitiyal (2004) and Nirmala (2006), who reported that MH spray at 1000 ppm induced the dormancy in non-dormant groundnut genotypes. Similarly, Jayadeva (2008) also reported that foliar application of MH at 1000 ppm at 60 and 90 days after sowing could induce dormancy up to 45 days in groundnut varieties, SB-X1 and TG-26 while only five days dormancy could be induce in TAG-24.

The seed without foliar spray of MH that is, control seed lot recorded higher seedling vigour index (Based on seedling length) during all the periods of testing (Table 2). During *rainy season*, 2010, immediately after harvest the seed harvested from control plots (No foliar spray of MH) recorded significantly superior seedling vigour index (SVI) of 1905. However, MH spray @ 1250 ppm recorded the least SVI of 1201 followed by 1000 ppm concentration (1324). Similar trend of higher SVI was shown by seed lots which were harvested from control plots with out foliar spray of MH during 2011 also. The SVI of untreated seed of untreated plot was 2012 immediately after harvest and it reached to a tune of 2235 after 35 days of harvest. As seen in case of germination percent, it has been noticed that there was an increased trend with respect to SVI also as the number of days after harvest

**Table 1.** Effect of pre harvest foliar spray of maleic hydrazide (MH) on seed germination per cent at different days after harvest in groundnut.

| | Days after harvest (2010) | | | | | | | | |
|---|---|---|---|---|---|---|---|---|---|
| | 0 | 5 | 10 | 15 | 20 | 25 | 30 | 35 | Mean |
| Control | 80.00(63.44) | 81.77(64.75) | 82.33(65.12) | 83.66(66.19) | 86.33(68.28) | 86.33(68.28) | 88.31(70.00) | 90.66(72.24) | 84.92 |
| 500 ppm MH | 58.33(49.78) | 65.00(53.73) | 81.77(64.75) | 84.41(66.74) | 88.31(70.00) | 87.11(68.95) | 88.31(70.00) | 92.33(73.89) | 80.69 |
| 750 ppm MH | 49.33(44.60) | 60.32(50.94) | 80.00(63.44) | 81.77(64.75) | 84.41(66.74) | 86.33(68.28) | 86.33(68.28) | 88.31(70.00) | 77.1 |
| 1000 ppm MH | 42.33(40.57) | 51.77(46.03) | 65.00(53.73) | 66.22(54.45) | 80.00(63.44) | 83.66(66.19) | 86.33(68.28) | 86.33(68.28) | 70.2 |
| 1250 ppm MH | 32.33(35.24) | 36.33(37.05) | 51.77(46.03) | 60.32(50.94) | 66.22(54.45) | 65.68(54.15) | 80.66(63.94) | 81.77(64.75) | 59.38 |
| Mean | 52.46 | 59.04 | 72.17 | 75.28 | 81.05 | 81.82 | 85.99 | 87.88 | 74.46 |
| | S.E.D± | C.D (5%) | | | | | | | |
| Treatment | 0.58 | 0.25 | | | | | | | |
| Days | 0.76 | 0.66 | | | | | | | |
| Interaction | 1.32 | 1.14 | | | | | | | |

| | Days after harvest (2011) | | | | | | | | |
|---|---|---|---|---|---|---|---|---|---|
| | 0 | 5 | 10 | 15 | 20 | 25 | 30 | 35 | Mean |
| Control | 81.73(64.77) | 83.33(65.12) | 81.77(64.75) | 84.66(66.97) | 84.33(66.66) | 84.15(66.50) | 85.77(67.86) | 86.33(68.28) | 84.01 |
| 500 ppm MH | 5.33(48.04) | 65.44(53.97) | 80.66(6.94) | 81.66(64.67) | 85.33(67.45) | 86.33(68.28) | 87.11(68.95) | 86.33(68.28) | 78.52 |
| 750 ppm MH | 59.66(50.59) | 63.33(52.71) | 79.00(62.72) | 82.33(65.12) | 84.41(66.74) | 83.66(66.19) | 84.41(66.74) | 84.00(66.42) | 77.60 |
| 1000 ppm MH | 41.66(40.22) | 49.33(44.60) | 52.66(47.12) | 63.66(52.95) | 75.00(60.00) | 80.66(63.94) | 83.33(65.88) | 86.33(68.28) | 66.58 |
| 1250 ppm MH | 36.33(37.05) | 42.00(40.40) | 55.66(48.27) | 60.66(51.18) | 64.11(53.66) | 62.66(65.42) | 70.00(60.00) | 83.44(65.96) | 59.35 |
| Mean | 54.94 | 60.69 | 69.95 | 74.59 | 78.64 | 79.49 | 82.12 | 85.27 | 73.21 |
| | S.E.D± | C.D(5%) | | | | | | | |
| Treatment | 0.25 | 0.50 | | | | | | | |
| Days | 0.66 | 1.30 | | | | | | | |
| Interaction | 6.14 | 2.26 | | | | | | | |

(Figures in the paranthesis are the arc sine values).

**Table 2.** Effects of pre harvest foliar spray of maleic hydrazide (MH) on seedling vigour index (SVI) at different days after harvest (cv. Narayani).

| | Days after harvest(2010) | | | | | | | | |
|---|---|---|---|---|---|---|---|---|---|
| | 0 | 5 | 10 | 15 | 20 | 25 | 30 | 35 | Mean |
| Control | 1905 | 1926 | 1976 | 2093 | 2145 | 2173 | 2208 | 2220 | 2080.7 |
| 500 ppm MH | 1691 | 1730 | 1821 | 1803 | 1926 | 1979 | 2014 | 2076 | 1880.0 |
| 750 ppm MH | 1556 | 1596 | 1623 | 1674 | 1721 | 1750 | 1822 | 1851 | 1699.1 |
| 1000 ppm MH | 1324 | 1344 | 1453 | 1501 | 1602 | 1679 | 1721 | 1746 | 1546.2 |
| 1250 ppm MH | 1201 | 1223 | 1272 | 1343 | 1379 | 1427 | 1476 | 1501 | 1352.7 |

**Table 2. contd.**

| | | | | | | | | | |
|---|---|---|---|---|---|---|---|---|---|
| Mean | 989 | 1011 | 1093 | 1123 | 1122 | 1322 | 1421 | 1422 | 1187.8 |
| | S.E.M± | C.D (5%) | | | | | | | |
| Treatment | 9.65 | 121 | | | | | | | |
| Days | 12.47 | 214 | | | | | | | |
| Interaction | 24.69 | 326 | | | | | | | |

| | Days after harvest (2011) | | | | | | | | |
|---|---|---|---|---|---|---|---|---|---|
| | **0** | **5** | **10** | **15** | **20** | **25** | **30** | **35** | **Mean** |
| Control | 2012 | 2064 | 2122 | 2012 | 2173 | 2202 | 2203 | 2235 | 2127.8 |
| 500 ppm MH | 1763 | 1803 | 1872 | 1901 | 1893 | 1946 | 1993 | 2053 | 1903.0 |
| 750 ppm MH | 1623 | 1647 | 1734 | 1812 | 1872 | 1864 | 1903 | 1946 | 1800.1 |
| 1000 ppm MH | 1372 | 1346 | 1421 | 1503 | 1564 | 1623 | 1691 | 1725 | 1530.6 |
| 1250 ppm MH | 1254 | 1247 | 1293 | 1335 | 1391 | 1422 | 1476 | 1493 | 1363.8 |
| Mean | 1604.8 | 1621.4 | 1688.4 | 1712.6 | 1778.6 | 1811.4 | 1853.2 | 1890.4 | 1745.1 |
| | S.E.M± | C.D (5%) | | | | | | | |
| Treatment | 7.54 | 114 | | | | | | | |
| Days | 16.54 | 136 | | | | | | | |
| Interaction | 19.23 | 248 | | | | | | | |

increased. Among the various concentrations studied, 1250 ppm of MH was found to be effective as growth retardant to reduce SVI which could be considered as a sign of induction of dormancy. The decline in seedling vigour index could be due to lower germination percent as well as poor seedling growth as a result of MH spray. The current findings are in accordance with the results of Jayadeva (2008).

## Conclusion

Two foliar sprays of maleic hydrazide at 1250 ppm at 60 and 90 days after sowing would induce dormancy in non-dormant bunch type groundnut cultivar up to 30 days (4 weeks) and protects the seed from *in situ* germination due to untimely rains at the time of harvest of the crop.


## ACKNOWLEDMENT

The authors are thankful to Acharya N.G. Ranga Agricultural University, Rajendranagar, Hyderabad for providing the financial support to carry out the study.


### REFERENCES


Anonymous (2008). Agricultural Statistics at a Glance-2007-2008. Directorate of Economics and Statistics, Ministry of Agriculture, GOI, Krishi Bhavan, New Delhi.

Bewley JD, Black M (1982). Physiology and Biochemistry of Seeds. II. Development, Germination and Growth. Spriger-Verlag Berlin Heidelberg, Newyork. pp. 88-101.

FAO (2011). The state of the world's land water resourses for food and agriculture. Managing systems at risk. Food and Agriculture Organisation of United Nations, Rome, 2011.

Gupta RK, Singh SS, Verma MM (1985). Induction of dormancy in groundnut (*Arachis hypogea* L.) variety T-64 by maleic hydrazide. J. Agric. Res. 19(2):82-86.

International Rules for Seed Testing (ISTA) (1985). Seed Sci. Technol. 13:299-335.

Jayadeva B (2008). Induction of dormancy in summer groundnut (*Arachis hypogea* L.). M.Sc (Agric) Thesis, Mahatma Phule Krishi Vidyapeeth, Rahuri, Maharastra, India.

Kumar TSA, Gowda MVC, Nadef HL (1991). Seed dormancy in erect bunch genotypes of groundnut (*Arachis hypogea* L.). J. Oil seeds Res. 8(2):166-172.

Nagarjun P, Radder GD (1983). Studies on induction of seed dormancy in bunch type groundnut. Seed Res.11(1):24.

Naitiyal PC (2004). Issues related to maintenance of seed viability and regulation of dormancy in groundnut. Groundnut Research in India, pp. 321-338..

Nirmala MS (2006). Field germinabiliity and storability of rice seeds treated with maleic hydrazide spray. Mad. Agric.

J. 93(7-12):271-272.
Reddy PS, Basu MS, Tiwari SP, Devidayal, Radhakrishnan T (1987). Spanish groundnut strains with fresh seed dormancy. Cur. Sci. 56(8):368-369.
Vaithialingam R, Rao JS (1973). Induction of dormancy in groundnut by pre-harvest foliar spray of M.H. Mad. Agric. J. 6(9-12):1862-1863.
Varman PV, Raveendran TS (1991). New sources of seed dormancy in bunch groundnut. Cur. Res. 20(11):237-238.

# *In vitro* and *In vivo* responses of different treating agents against wilt disease of safflower

**M. Govindappa[1]*, V. Ravishankar Rai[2] and S. Lokesh[3]**

[1]Department of Biotechnology, Shridevi Institute of Engineering and Technology, Sira Road, Tumkur-572 106, India.
[2]Department of Studies in Microbiology, University of Mysore, Manasa Gangothri, Mysore-570 006, India.
[3]Department of Studies in Applied Botany, Seed Pathology and Biotechnology, University of Mysore, Manasa Gangothri, Mysore-570 006, India.

**In safflower Anucop, captan and Mancozeb M-45 were found extremely effective in reducing *Fusarium oxysporum* f.sp. *carthami* wilt. The seed treatments improved seed germination, seedling vigour and plant stand. Due to these treatments many of the seed-borne fungi failed to express in the normal way. Bioagents formulations viz., *Trichoderma harzianum, Bacillus subtilis* and *Pseudomonas fluorescens* reduced the wilt incidence both under greenhouse and field conditions, thereby enhancing the growth of the seedlings. These antagonists significantly reduced the population of *Fusarium oxysporum* f.sp. *carthami*, increased the seed germination and seedling vigour. Leaf extracts of *Becopa monniera* and *Adathoda vasica* were found effective in the control of safflower wilt. Comparatively *B. monniera* enhanced the seed germination and quality parameters of plants both under greenhouse and field conditions and also effectively suppressed the wilt up to flowering.**

**Key words:** Fungicides, bioagents, plant extracts, *Fusarium oxysporum* f.sp. *carthami*, disease management.

## INTRODUCTION

In safflower, seed-borne fungal diseases play a major role in the yield loss, among which is wilt caused by *Fusarium oxysporum* f.sp. *carthami* Klisiewicz and Houston, a serious disease in almost all major growing areas. Main symptoms of the disease are unilateral yellowing of the foliage followed by wilting, one sided infection on plant parts, golden yellow leaf discoloration, curving of the midrib towards yellow side and dark brown discoloration of vascular tissue of root and stem (Klisiewicz, 1963; Holdeman and McCartney, 1964). In severe infection, flowering is delayed and size of the head remains small and partially blossomed. A large number of ovaries fail to develop seeds or form blackish, small, distorted, chaffy seeds which are abortive (Chakrabarti, 1976). The severity of symptoms varies depending on the time of infection and the genotype with a grain yield loss ranging from 7.2 to 100% (Sastry and Ramachandran, 1993).

Lower dose of fungicides and several micro organisms have been successfully used to control the disease (Raghuchander et al., 1997; Vidyashekaran et al., 1997; Prasad and Rangeswaran, 1999; Krishnamurthy et al., 2001; Kumar and Dubey, 2001). In addition to this many workers have used plant extracts against pathogenic fungi (Tewari and Kurochevu, 1999; Varshnay, 2001). Though many attempts have been made to control the severity of the fungal diseases many lacunae have been left out with respect to dose, type of fungicides as well as bioagents. Since the effectiveness of the chemicals and potential of the organisms varies depending upon the agroclimatic conditions, in the present study few synthetic fungicides, biological agents and plant extracts were used for seed treatment to test their efficacy against safflower wilt disease causing fungus borne *in vitro* and *in vivo*.

*Corresponding author. E-mail: dravidateja07@yahoo.co.in.

## MATERIALS AND METHODS

### Collection of seed samples and incidence of the wilt pathogen and their effect on seed quality

Seed samples of safflower cultivar Manjira were collected from Annigeri, Dharwad district of Karnataka, India during February 2002 and stored in polyethylene bags for further use. In order to assess the incidence of *F. oxysporum* f.sp. *carthami*, the seeds were subjected to standard blotter method. Seeds were plated on 3 layers of wet blotters in plastic plates and incubated following the procedures of ISTA (Anonymous, 1996). On 8th day of incubation, seeds were observed for the occurrence of *F. oxysporum* f.sp.*carthami* and its incidence was recorded and tabulated. To study effect of *F. oxysporum* f.sp. *carthami* on growing seedlings, 200 seeds were surface sterilized with 2% NaOCl solution and individually plated on 1% water agar in the test tubes, which were. The entire set up was incubated under alternate cycles of 12/12-h light and darkness for a period of 2 weeks. The seeds were examined carefully for the occurrence of colonies of *F. oxysporum* f.sp. *carthami* and seedlings showing symptoms were recorded. The symptoms developed on the seedlings were compared with the other set in which the seeds were inoculated with the spore suspension of *F. oxysporum* f.sp.*carthami*.

### Incidence of the *F. oxysporum* f.sp. *carthami* in water agar method

In addition to the water agar seedling symptom test, both uninoculated and *F. oxysporum* f.sp.*carthami* inoculated seeds were sown in thoroughly washed sand bed and were incubated for a prolonged period of 30 days. During incubation, sand bed was kept moist to prevent desiccation. The seedlings which showed symptoms were identified and such seedlings were finally excised and cut into pieces of 1 cm and incubated on wet blotters. Such components were observed for the occurrence of *F. oxysporum* f.sp. *carthami* and its incidence was recorded.

### Fungicide treatment

In order to find out an effective treating agent for the management of wilt disease a few chemical fungicides viz., Anucop, Bavistin, Captan, Mancozeb M-45, Milzim, Vitavax and Volzim were used at the concentration of 0.1, 0.2 and 0.3%. In this case each fungicide was separately dusted on the seeds and was smeared in a clean polyethylene pouch.

### Isolation, mass culture and formulation preparation of biocontrol agents for seed treatment

For the purpose of utilizing bioagents, a few microbes that included a fungus *Trichoderma harzianum*, bacteria like *Bacillus subtilis* and *Pseudomonas fluorescens* were isolated from safflower rhizosphere soil obtained from the experimental plots of Safflower Research Station, Annigeri, Dharwad district of Karnataka by serial dilution technique (Crossan, 1967).

*T. harzianum* was grown on potato dextrose agar (PDA) in Petri plates and spores were harvested from 12 days old sporulating colonies. 10 ml of sterilized distilled water was added to each of the culture plate and was gently agitated with the help of a fine sterilized camel hairbrush. The spore suspension was collected and filtered through muslin cloth and centrifuged at 10,000 rpm. The sediment was collected, further air-dried and mixed with purified talc powder in the ratio of 1:10 (w/w).

Bacterial biocontrol agents, *B. subtilis* and *P. fluorescens* were cultured on nutrient broth and King's broth, respectively. Two-day-old cultures of *B. subtilis* and *P. fluorescens* were centrifuged at 10,000 rpm for 10 min. The supernatant was discarded and the pellets were washed in sterilized distilled water repeatedly thrice and finally suspended in sterile distilled water. Using spectrophotometer, the optical density of the solution was adjusted to 0.45 at 610 nm, which gave the concentration of $4 \times 10^8$ and $9 \times 10^8$ cfu/ml, respectively. Each of the bacterial formulation was prepared under aseptic condition by mixing 400 ml of bacterial suspension with one kg of purified talc powder. It was air dried and stored in polyethylene bags at 4°C. The bacterial formulations were separately dusted to seeds at three different concentrations of 5, 10 and 15 g/kg of seeds.

### Preparation of the plant extracts

Plant materials such as fresh leaves of *Adathoda vasica* and *Becopa monniera* were harvested and thoroughly washed in tap water. 100g of leaves of each plant were macerated to thick paste with the help of a mortar and pestle. It was extracted with 100 ml of distilled water and filtered through the muslin cloth. The extract was then centrifuged at 5,000 rpm for 15 min. The supernatant obtained was collected and stored at $5^0$C for further use. Seeds were separately soaked in 25, 50 and 75 % leaf extract for 24 h at $28\pm2^0$ C. The treated seeds were then air-dried and used for the assessment of *F. oxysporum* f.sp. *carthami* both in laboratory and field conditions.

### Efficacy of different treating agents against *F.oxysporum* f.sp. *carthami* under laboratory conditions

In all the cases treated seeds were plated on wet blotters and incubated following standard procedures (Anonymous, 1996). The incubated seeds were further carefully evaluated for the occurrence of mycoflora. Similarly treated seeds were subjected to paper towel method, and their percent germination was recorded. The root-shoot length of the seedlings was also measured after 7 days; vigour index was calculated based on mean root length × mean shoot length/ no. of germinated seeds and tabulated. For comparison, untreated seeds maintained under similar conditions served as corresponding control.

### Effect of different treating agents on the incidence of *Fusarium* wilt under greenhouse condition

To evaluate the efficacy of different treatments, treated seeds were sown in soil and separately maintained under greenhouse and field conditions. In greenhouse conditions seeds were sown in the earthen pots (22-cm diameter at 10 seeds/pot containing soil, sand and manure in the ratio of 1:1:1). The germination and the disease incidence were recorded on 7th and 60th days of sowing, respectively.

### Efficacy of different treating agents against *Fusarium* wilt under field conditions

In the experimental plots, field trails were conducted during rabi (2002-2003) season, in randomized block design. The seeds of different treatments were separately sown in the plots of size 5×6 $m^2$ with an inter row spacing of 30 cm. Untreated seeds sown under similar conditions served as control. Along with this, seeds were dressed with target pathogen *F. oxysporum* f.sp.*carthami* and sown for comparison. The pots and the field were applied with known

**Table 1.** Incidence of *F. oxysporum* f.sp. *carthami* in safflower seedlings Under different laboratory conditions.

| Seedling symptom test | Seed germination (%) | % incidence of *Fusarium oxysporum* f.sp. *carthami* on seeds/ seedlings |
|---|---|---|
| Water agar method | 66 | 18.33±0.33[b] |
| Growing-on test | 66 | 21.33±0.33[a] |

Values are mean of four replicates of 100 seeds each. According to Duncan's Multiple Range Test (DMRT), values followed by different superscripts are significantly different at $P \leq 0.05$.

**Table 2.** Effect of *F. oxysporum* f.sp. *cathami* on seed germination , seedling vigour, plant height of safflower in paper towel method.

| Seed treatment | Variation in the seedling quality of safflower due to *Fusarium oxysporum* f.sp. *carthami* | | | | |
|---|---|---|---|---|---|
| | Seed germination (%) | MRL±SE | MSL±SE | VI | Fresh weight (gm) |
| Control | 64 | 8.5±0.03[a] | 4.1±0.03[a] | 806 | 0.31±0.29[a] |
| Control+ pathogen | 58 | 7.4±0.03[b] | 4.0±0.08[a] | 661 | 0.27±0.20[b] |

Values are mean of four replicates of 100 seeds each. Control=seed without *F.oxysporum* inoculation, pathogen=*F. oxysporum* f.sp.*carthami*, MRL-mean root length, MSL-mean shoot length, VI-vigour index. According to Duncan's Multiple Range Test (DMRT), values followed by different superscripts are significantly different at $P \leq 0.05$.

quantity of fertilizers (180-300 kg N+P+K $ha^{-1}$). Apart from the rate of seedling emergence, plant height, number of branches, number of capitula, number of leaves, average leaf area and size of the stem were also recoded at flowering stage. Sequentially disease incidence was also recorded from emergence till seed setting. Percent disease incidence, variation in plant growth and yield were calculated using following formula described by Mukherjee et al. (2001).

% disease incidence = [(Wilt in treated plots - wilt in check plots) / Wilt in check plots] x 100

% Increase in seedling emergence= [(Seedling emergence in treated plots - Seedling emergence in check plots) / Seedling emergence in check plots] x 100

% Increase in yield = [(Yield in treated plots - Yield in check plots) / Yield in check plots] x 100

% Increase in plant height = [(Plant height in treated plots - Plant height in check plots) / Plant height in check plots] x 100

% Increase in branches = [(No. of branches in treated plots - No. of branches in check plots) / No. of branches in check plots] x 100

% Increase in capitula = [(No. of capitula in treated plots - No. of capitula in check plots) / No. of capitula in check plots] x 100

% Increase in leaves = [(No. of leaves in treated plots - No. of leaves in in check plots) / No. of leaves in check plots] x 100

% Increase in size of stem = [(Average stem girth in treated plots - Average stem girth in check plots) / Average stem girth in check plots] x 100

The resultant data from repeated experiments were combined and analyzed statistically based on ANOVA followed by Duncan's Multiple Range Test.

## RESULTS AND DISCUSSION

*F. oxysporum* f.sp. *carthami* infected seeds of safflower and seedlings showed less germination correspondingly in water agar and growing on test was high incidence of the fungus (recast statement) (Table 1). This pathogen also affected the seedling quality in paper towel method (Table 2).

The results depict that there was reduction in the incidence of *F. oxysporum* f.sp. *carthami* at different concentrations of fungicides. Compared to control among the fungicides tested, incidence of the pathogen was inhibited significantly with Captan (0.2%) followed by Anucop, Bavistin, Blitox, Vitavax, Catafol Milzim (0.3%) Mancozeb M-45 and Volzim (Figure 1). Besides, the fungicide treated seeds enhanced the germination compared to control. Sugha et al. (1995) and Sugha (2001) have reported the effectiveness of Bavistin, Captan and Vitavax on population dynamics of white rot fungus (*Fusarium* sp.) of pea. According to Gupta and Aneja (2001) have reported considerable decline in the pathogen population due to treatment with Captan and Mancozeb M-45 in soybean. Singh et al. (2002) have made a comparative *in vitro* study with some fungicides like Captan, Dithane M-45, Vitavax, Bavistin for their efficacy in controlling *Fusarium* species on mungbean. Dubey and Patel (2001) also tested different concentrations of fungicides against various pathogens. Dithane M-45 and Bavistin were reported to be effective in reducing seed-borne infection of *Fusarium* sp. on maize seeds (Kumar and Agarwal, 1998). These fungicides might have paralyzed the fungus at varied level depending upon their dose and type.

Among the bioagents, seeds treated with *T. harzianum* (1%) showed maximum reduction in the incidence of wilt pathogen. Application of *B. subtilis* (1%) and *P. fluorescens* (1.5%) also reduced the incidence of the pathogen compared to control (Figure 1). Bioagents treated seeds also enhanced the germination compared to control. These results are in support of the findings of

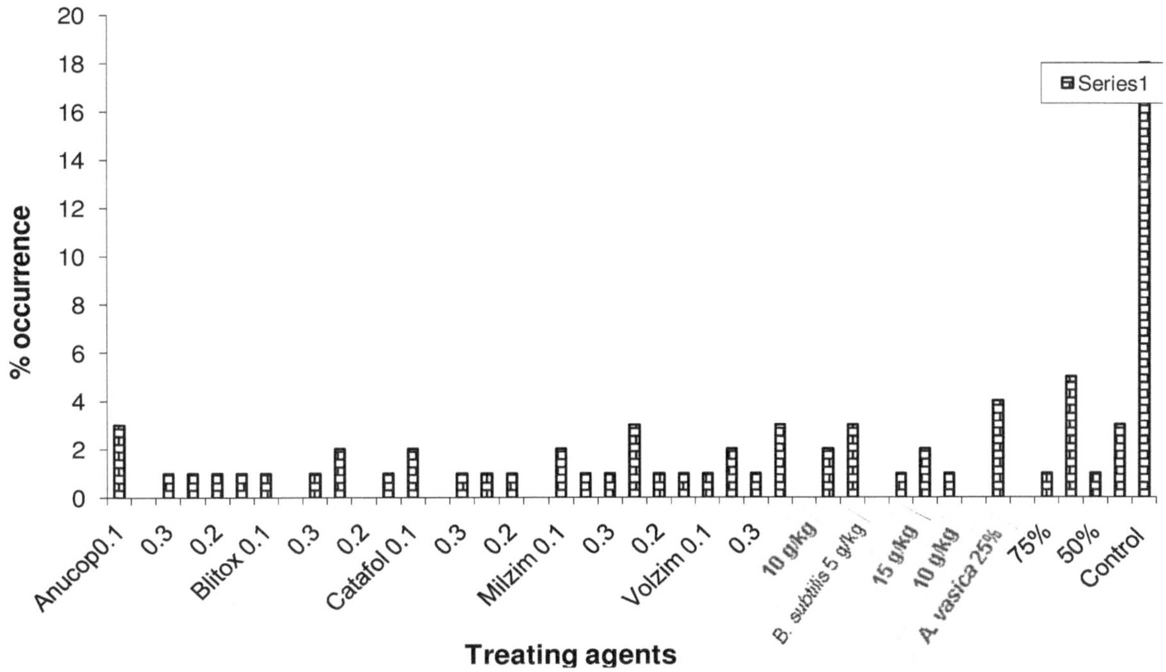


**Figure 1.** Effect of different treatments on the incidence of wilt fungus *Fusarium oxysporum* f.sp. *carthami* in safflower seeds.

Jeyalakshmi et al. (1998) who reported the biological control potential of *T. harzianum*. *Trichoderma* spp. And bacterial bioagents produce mycolytic enzymes, thus playing an important role in the degradation of target pathogens (Elad et al., 1982; Aziz et al., 1993; Baker and Dickman, 1993; Sivakumar and Narayanaswamy, 1998). Among plant extract treatments, 50% extract of *B. monniera* increased the germination, root-shoot length and seedling vigour in paper towel method and decreased the incidence of *F. oxysporum* f.sp. *carthami* which was followed by *A. vasica* compared to control (Figure 1 and Table 3). Many plant extracts are reported to increase seed germination through decreasing *F. oxysporum* incidence (Wegrazyka, 1984; Bansal and Gupta, 2000).

Nine fungicide enhanced seed germination, root-shoot length and seedling vigour. Among them Captan (0.2%) and Mancozeb M-45 (0.3%) followed by Vitavax, Bavistin, Catafol, Milzim (0.2%) and Volzim (0.3%) were more effective (Table 3). Present findings are in support of the reports of several workers (Dey et al., 1989; Chakrabarty and Rao, 1992; Laxminaryan et al., 1996). Bioagents treated seeds showed beneficial effects on germination which resulted in increased root-shoot length and seedling vigour. *T. harzianum* treated seeds were more efficient than *B. subtilis* and *P. fluorescens* treated samples (Table 3). Bunker and Mathur (2000) also reported the similar results with usage of *T. harzianum*.

In greenhouse experiments, it was observed that all fungicides reduced the disease incidence from 18 to 4% (Table 4). Comparatively it was noticed that the reduction in disease incidence was maximum in Anucop, Bavistin, Captan (0.2%) and Mancozeb M-45 (0.3%). Captan and Mancozeb M-45 stood superior over Anucop (0.2%), Bavistin (0.2%), Blitox (0.2%), Milzim (0.2%). Fungicides acted possibly by inducing metabolic changes leading to development of toxic factors, resulting in the internal environment unfavorable for pathogens growth and activity, ultimately inducing the resistance and protection against infection. These results are confirmatory with the observations of Prakash and Ganesan (1998), Parimala et al. (1998), Padmavathi et al. (1998), Sandrou et al. (1998) and Kumar and Dubey (2001) in cowpea, blackgram, brinjal and sunflower, respectively. Reports of De and Chaudhary (1999) are also in confirmation with the present findings, who observed the minimization of wilt disease due to Bavistin, Mancozeb M-45 and Vitavax. Bhatti et al. (1985), Pedgoanker and Mayee (1989) and Cicero et al. (1992) have reported the effectiveness of Thiram, Cabendazim against safflower wilt pathogen *F. oxysporum* f.sp. *carthami*. All the fungicidal seed treatments increased germination, plant height which correlated with the findings of Pathak et al. (2001). Bioagents formulations gave significant results controlling the wilt disease, and also enhanced seed germination and plant height (Table 4). *T. harzianum* was more effective than other bioagents which minimized the wilt followed by *B. subtilis* and *P. fluorescens*. Application of biocontrol agents as a seed treatment may induce the accumulation of lytic enzymes in safflower. The increased

**Table 3.** Effect of treatments on seed germination, seedling growth and vigour of safflower var. manjira.

| Treatments | Dose (%) | Germination (%) | MRL±S.E. | MSL±S.E | Vigour index |
|---|---|---|---|---|---|
| **Fungicides** | | | | | |
| Anucop | 0.1 | 70 | 8.8±0.03[b] | 6.2±0.03[cd] | 1057 |
| | 0.2 | 88 | 9.5±0.08[a] | 6.8±0.05[c] | 1447 |
| | 0.3 | 72 | 8.8±0.05[b] | 6.4±0.04[c] | 1094 |
| Bavistin | 0.1 | 72 | 8.7±0.03[b] | 6.8±0.03[c] | 1140 |
| | 0.2 | 81 | 9.4±0.03[a] | 6.7±0.03[c] | 1309 |
| | 0.3 | 69 | 8.5±0.03[bc] | 6.6±0.12[c] | 1049 |
| Blitox | 0.1 | 70 | 8.1±0.03[c] | 6.2±0.03[cd] | 1001 |
| | 0.2 | 84 | 8.8±0.03[b] | 6.3±0.04[cd] | 1274 |
| | 0.3 | 78 | 8.2±0.03[c] | 6.2±0.05[cd] | 1123 |
| Captan | 0.1 | 70 | 9.1±0.03[a] | 8.6±0.05[a] | 1128 |
| | 0.2 | 92 | 9.8±0.03[a] | 8.8±0.05[a] | 1629 |
| | 0.3 | 77 | 8.9±0.03[b] | 6.8±0.02[c] | 1217 |
| Catafol | 0.1 | 68 | 9.0±0.08[a] | 5.8±0.01[e] | 1000 |
| | 0.2 | 72 | 8.5±0.05[bc] | 5.3±0.03[d] | 1066 |
| | 0.3 | 69 | 8.8±0.05[b] | 5.6±0.03[d] | 1001 |
| Dithane M-45 | 0.1 | 86 | 8.6±0.03[b] | 6.6±0.05[c] | 1309 |
| | 0.2 | 84 | 8.8±0.03[b] | 7.3±0.06[b] | 1361 |
| | 0.3 | 92 | 9.0±0.03[a] | 8.4±0.00[ab] | 1595 |
| Milzim | 0.1 | 68 | 8.1±0.03[c] | 5.7±0.03[e] | 0945 |
| | 0.2 | 90 | 8.7±0.03[b] | 7.3±0.06[b] | 1440 |
| | 0.3 | 80 | 8.7±0.03[b] | 6.7±0.04[c] | 1224 |
| Vitavax | 0.1 | 80 | 8.5±0.03[bc] | 6.6±0.01[c] | 1208 |
| | 0.2 | 96 | 8.6±0.03[bc] | 6.8±0.03[c] | 1498 |
| | 0.3 | 90 | 8.4±0.08[bc] | 7.1±0.03[b] | 1395 |
| Volzim | 0.1 | 80 | 8.4±0.03[bc] | 7.3±0.05[b] | 1353 |
| | 0.2 | 78 | 8.6±0.03[bc] | 6.7±0.03[c] | 1201 |
| | 0.3 | 88 | 9.1±0.03[a] | 6.9±0.03[c] | 1408 |
| **Bioagents formulations** | | | | | |
| *Trichoderma harzianum* | 0.5 | 78 | 8.8±0.35[b] | 7.4±0.03[c] | 1272 |
| | 1.0 | 90 | 9.2±0.01[a] | 6.8±0.03[d] | 1443 |
| | 1.5 | 79 | 8.7±0.05[b] | 6.6±0.03[d] | 1209 |
| *Bacillus subtilis* | 0.5 | 70 | 8.9±0.06[b] | 7.9±0.05[ab] | 1176 |
| | 1.0 | 81 | 9.2±0.01[a] | 8.0±0.03[a] | 1401 |
| | 1.5 | 70 | 8.8±0.04[b] | 7.8±0.03[b] | 1162 |
| *Pseudomonas fluorescens* | 0.5 | 62 | 7.8±0.03[d] | 7.7±0.03[b] | 1023 |
| | 1.0 | 68 | 8.1±0.02[c] | 7.8±0.03[b] | 1081 |
| | 1.5 | 78 | 8.8±0.04[b] | 8.3±0.03[a] | |

**Table 3.** Contd.

| | | | | | |
|---|---|---|---|---|---|
| **Plant extracts** | | | | | |
| *Becopa monniera* | 25 | 76 | $9.1 \pm 0.03^{b}$ | $7.3 \pm 0.03^{a}$ | 1246 |
| | 50 | 83 | $9.7 \pm 0.03^{a}$ | $7.6 \pm 0.01^{a}$ | 1436 |
| | 75 | 80 | $8.7 \pm 0.03^{c}$ | $7.5 \pm 0.03^{a}$ | 1304 |
| *Adathoda vasica* | 25 | 76 | $9.1 \pm 0.01^{b}$ | $7.3 \pm 0.05^{a}$ | 1239 |
| | 50 | 79 | $9.4 \pm 0.02^{a}$ | $7.4 \pm 0.03^{a}$ | 1339 |
| | 75 | 77 | $9.1 \pm 0.03^{b}$ | $7.3 \pm 0.03^{a}$ | 1299 |
| Control | | 64 | $8.5 \pm 0.08^{c}$ | $4.1 \pm 0.03^{b}$ | 0813 |
| Control + Pathogen | | 58 | $7.4+0.16^{d}$ | $4.0 \pm 0.05^{b}$ | 0661 |

MRL-mean root length, MSL-mean shoot length, SE-standard error. Values are mean of four replications. According to Duncan's Multiple Range Test (DMRT), values followed by different superscripts are significantly different at $P \leq 0.05$.

**Table 4.** Effect of Seed treatment on seedling emergence, plant height, disease incidence under greenhouse conditions.

| Treatments | Dose (%) | Germination (%) | Plant height (cm) | Incidence of the wilt (%) |
|---|---|---|---|---|
| | | **Fungicides** | | |
| | 0.1 | 70 | $58.1 \pm 00.8^{j}$ | $4+0.57^{cd}$ |
| Anucop | 0.2 | 88 | $61.3 \pm 0.11^{c}$ | - |
| | 0.3 | 72 | $59.1 \pm 0.57^{ef}$ | $3+0.57^{de}$ |
| | 0.1 | 72 | $57.8 \pm 0.57^{jk}$ | $1+0.57^{fg}$ |
| Bavistin | 0.2 | 81 | $62.1 \pm 00.8^{bc}$ | - |
| | 0.3 | 69 | $56.7 \pm 0.11^{lmn}$ | $2+0.57^{ef}$ |
| | 0.1 | 70 | $57.4 \pm 0.23^{jkl}$ | $3+0.57^{de}$ |
| Blitox | 0.2 | 84 | $62.2 \pm 0.57^{b}$ | $1+0.57^{fg}$ |
| | 0.3 | 78 | $58.9 \pm 0.57^{ghi}$ | $3+0.57^{de}$ |
| | 0.1 | 70 | $57.2 \pm 1.66^{pq}$ | $1+0.57^{fg}$ |
| Captan | 0.2 | 92 | $64.1 \pm 0.57^{a}$ | - |
| | 0.3 | 77 | $59.3 \pm 0.57^{ef}$ | $1+0.57^{g}$ |
| | 0.1 | 68 | $51.1 \pm 0.57^{s}$ | $2+0.57^{fg}$ |
| Catafol | 0.2 | 72 | $54.3 \pm 0.57^{p}$ | $1+0.57^{ef}$ |
| | 0.3 | 69 | $51.4 \pm 1.00^{pqr}$ | $1+0.57^{fg}$ |
| | 0.1 | 86 | $59.7 \pm 0.57^{q}$ | $1+0.57^{fg}$ |
| Dithane M-45 | 0.2 | 84 | $60.3 \pm 0.57^{d}$ | $1+0.57^{fg}$ |
| | 0.3 | 92 | $64.0 \pm 0.57^{a}$ | - |
| | 0.1 | 68 | $53.1 \pm 0.57^{qr}$ | $2+0.57^{ef}$ |
| Milzim | 0.2 | 90 | $56.2 \pm 0.57^{o}$ | $1+0.57^{fg}$ |
| | 0.3 | 80 | $52.9 \pm 0.57^{r}$ | $2+0.88^{ef}$ |
| | 0.1 | 80 | $57.9 \pm 0.57^{j}$ | $4+1.15^{cd}$ |
| Vitavax | 0.2 | 96 | $60.0 \pm 0.57^{de}$ | $4+0.57^{cd}$ |
| | 0.3 | 90 | $59.7 \pm 0.57^{def}$ | $2+0.57^{ef}$ |
| Volzim | 0.1 | 80 | $56.3 \pm 0.57^{o}$ | $2+0.57^{ef}$ |

**Table 4.** Contd.

| | | | | |
|---|---|---|---|---|
| | 0.2 | 78 | 54.1±0.57 p | 4+0.57 ef |
| | 0.3 | 88 | 59.2±0.57 gh | 3+0.57 cd |
| | **Bioagents formulations** | | | |
| | 0.5 | 78 | 58.4±0.11 hij | 3+0.57 de |
| *Trichoderma harzianum* | 1.0 | 90 | 59.9±0.57 def | - |
| | 1.5 | 79 | 59.1±0.57 f | 2+0.57 g |
| | 0.5 | 70 | 56.9±0.57 klm | 3+0.57 de |
| *Bacillus subtilis* | 1.0 | 81 | 60.1±0.88 de | - |
| | 1.5 | 70 | 57.8±0.57 jk | 1+0.57 fg |
| | 0.5 | 62 | 55.8±0.57 h | 2+0.57 ef |
| *Pseudomonas fluorescens* | 1.0 | 68 | 56.4±0.57 mn | 1+0.57 fg |
| | 1.5 | 78 | 57.9±0.57 j | - |
| | **Plant extracts** | | | |
| | 25 | 76 | 56.1±0.00 mn | 4+0.57 cd |
| *Becopa monniera* | 50 | 83 | 59.3±0.57 gh | - |
| | 75 | 80 | 57.9±0.57 j | 1+0.57 kg |
| | 25 | 76 | 56.8±0.57 lm | 5+0.57 c |
| *Adathoda vasica* | 50 | 79 | 58.1±0.57 ij | 1+0.57 fg |
| | 75 | 77 | 55.4±0.57 o | 3+0.57 de |
| Control | | 64 | 40.4±0.11 t | 18±0.57 b |
| Control + Pathogen | | 58 | 38.8±0.57 u | 58 ±0.57 a |

Values are mean of four replications of 100 seeds each, According to Duncan's Multiple Range Test (DMRT), values followed by different superscripts are significantly different at P≤ 0.05.

activities of PAL, lytic enzymes and accumulation of phenolics in safflower in response to seed treatment with *T. harzianum* and *P. fluorescens* might have also contributed to increase resistance against pathogen. Effect of these bioagents in plant disease management have been reported by Chang et al. (1986), De et al. (1996), De and Chaudhary (1999), Prasad and Rangeswaran (1999). But *B. subtilis* enhanced the plant height compared to *T. harzianum* and *P. fluorescens*. In case of plant extracts, *B. monniera* (50%) reduced the disease incidence and increased the seed germination, plant height followed by *A. vasica* compared to control. Various workers (Natarajan and Lalithakumari, 1987) have also recorded similar results using plant extracts on rice.

Five fungicides, which were effective under laboratory conditions, were also tested under field conditions. They enhanced seed germination, plant height, number of branches, mean leaf area, number of leaves, girth of stem, number of capitula and number of seeds/capitula compared with untreated seeds (Table 5). Among them, Captan (0.2%) showed maximum increase in all the growth parameters, Mancozeb M-45 (0.3%), Anucop, Bavistin and Blitox (0.2%) were the next effective treatments, which increased the plant growth parameters over control. These fungicides also decreased the wilt incidence during different growth stages of the plants. The early occurrence of wilt was recorded at 30 days after sowing. The highest wilt incidence was in control and even in Blitox (0.2%) treated plants. Seed treatment improved field standing of the plants against the disease through increased plant growth. Seed treatment with fungicides ultimately enhanced the yield through the reduction of wilt disease. In various crops, Taylor and Harman (1990), Mishra and Dharam (1992), Mukhopadyay et al. (1992), De and Chaudhary (1999) and Kumar and Dubey (2001) have reported the similar findings with respect to yield and diseases.

Van Loon et al. (1998) and Laha and Venkataraman (2001) have suggested different fungicides treatment for

**Table 5.** Effect of seed treatments on different growth parameters of safflower against *Fusarium oxysporum* f.sp.*carthami* under field conditions.

| Seed treatments | Dose (%) | Responses of different treatments on quality of safflower plants | | | | | | | |
|---|---|---|---|---|---|---|---|---|---|
| | | Seed germin. (%) | Plant height (cm) | Mean branches/ plant | Leaf area/ (cm) | Mean no.of leaves/ plant | Mean girth of stem/plant (mm) | Total no. of heads/plant | Mean no. of seeds/plant |
| **Fungicides** | | | | | | | | | |
| Anucop | 0.2 | 90 | $62.0\pm0.57^{c}$ (53.5) | $7.3\pm0.57^{ab}$ (34.2) | $67\pm0.57^{b}$ (26.4) | $74.4\pm0.57^{d}$ (59.7) | $6.36\pm0.57^{ab}$ (22.1) | $10.3\pm0.57^{c}$ (212.1) | $19+0.57^{c}$ |
| Bavistin | 0.2 | 88 | $63.2\pm0.57^{bc}$ (56.4) | $6.0\pm0.57^{bc}$ (10.3) | $65\pm0.57^{c}$ (22.6) | $68.1\pm0.57^{e}$(46.1) | $5.48\pm0.57^{b}$ (5.2) | $7.8\pm0.57^{d}$ (135.2) | $15+0.57^{p}$ |
| Blitox | 0.2 | 84 | $62.4\pm0.57^{c}$ (54.4) | $5.6\pm0.57^{bc}$ (3.0) | $62\pm0.57^{d}$ (17.0) | $77.8\pm0.57^{b}$ (67.0) | $6.46\pm0.57^{b}$ (24.0) | $7.1\pm0.57^{de}$ (114.8) | $13+0.57^{ef}$ |
| Captan | 0.2 | 92 | $74.0\pm0.57^{a}$ (83.2) | $7.7\pm0.57^{a}$ (41.6) | $75\pm0.57^{a}$ (41.5) | $80.2\pm0.57^{a}$ (72.1) | $7.43\pm0.57^{a}$ (42.6) | $14.7\pm0.57^{a}$ (345.5) | $24+0.57^{a}$ |
| Mancozeb M-45 | 0.3 | 92 | $64.4\pm0.57^{b}$ (59.4) | $7.5\pm0.57^{a}$ (37.9) | $74\pm0.57^{a}$ (39.6) | $75.4\pm0.57^{c}$(61.8) | $6.48\pm0.57^{ab}$ (24.4) | $11.6\pm0.57^{b}$ (252.7) | $21+0.57^{b}$ |
| **Bioagents formulations** | | | | | | | | | |
| *Trichoderma harzianum* | 1.0 | 90 | $63.8\pm0.57^{a}$ (57.8) | $7.1\pm0.57^{ab}$ (30.5) | $64\pm0.57^{c}$ (20.8) | $72.2\pm0.57^{d}$ (54.9) | $6.23\pm0.57^{ab}$ (19.6) | $6.2\pm0.57^{ef}$ (86.4) | $14+0.57^{e}$ |
| *Bacillis subtilis* | 1.0 | 86 | $61.8\pm0.57^{c}$ (52.5) | $5.8\pm0.57^{c}$ (6.6) | $57\pm0.57^{e}$ (7.5) | $66.1\pm0.57^{f}$ (41.8) | $5.39\pm0.57^{b}$ (3.5) | $5.8\pm0.57^{ef}$(77.6) | $12+0.57^{f}$ |
| *Psuedomonas fluorescens* | 1.5 | 84 | $57.2\pm0.57^{e}$ (41.6) | $5.7\pm0.57^{c}$ (4.8) | $55\pm0.57^{e}$ (3.8) | $65.2\pm0.57^{f}$ (39.9) | $5.31\pm0.57^{a}$ (2.0) | $5.7\pm0.57^{ef}$ (72.2) | $10+0.57^{g}$ |
| **Plant extracts** | | | | | | | | | |
| *Becopa monniera* | 50 | 83 | $59.3\pm0.57^{d}$ (46.8) | $5.9\pm0.57^{c}$ (8.4) | $56\pm0.57^{f}$ (5.7) | $75.4\pm0.57^{c}$ (61.8) | $6.39\pm0.57^{ab}$ (22.6) | $5.4\pm0.57^{f}$ (64.2) | $9+0.57^{gh}$ |
| *Adathoda vasica* | 50 | 80 | $58.1\pm0.57^{de}$ (43.8) | $5.7\pm0.57^{c}$ (4.77) | $54\pm0.57^{g}$ (1.88) | $64.8\pm0.57^{f}$ (39.1) | $5.84\pm0.57^{ab}$ (12.1) | $5.1\pm0.57^{f}$ (52.4) | $8+0.57^{h}$ |
| Control | | 64 | $40.4+0.57^{f}$ | $5.44\pm0.57^{c}$ | $53\pm0.57^{g}$ | $46.6\pm0.57^{g}$ | $5.21\pm0.57^{b}$ | $3.3\pm0.57^{g}$ | $4\pm0.57^{i}$ |
| Control + pathogen | | 58 | $38.8+0.57^{f}$ | $5.20\pm0.57^{c}$ | $50.4\pm0.57^{h}$ | $42.1\pm0.57^{h}$ | $2.13\pm0.57^{c}$ | $1.8\pm0.57^{c}$ | $2\pm0.57^{j}$ |

Data given in parenthesis is refers to per cent increment over control. According to Duncan's Multiple Range Test (DMRT), values followed by different superscripts are significantly different at $P\leq 0.05$. Values are mean of four replications of 100 seeds each.

better suppression of various diseases. Seed treatments with fungicides were found effective against wilt pathogen (*F. oxysporum* f.sp. *phaseoli*) of French bean under field conditions. They also increased seed germination, plant height and grain yield (Mukherjee et al., 2001). Bioagents showed significant effect on seed germination and improved the quality of safflower plants such as plant height, branches/plant, leaf area and number of leaves, girth of stem, number of heads/plant and number of seeds/plant (Table 5). Among bioagents treatment *T. harzianum* showed maximum beneficial effects followed by *B. subtilis* and *P. fluorescens*. These bioagents also decreased the wilt incidence during different growth stages. The highest wilt incidence was recorded in *F. oxysporum* f.sp. *carthami* treated seeds and untreated control plots (18%). Bioagent formulations improved field standing of the plant against the disease and alsoimproved the plant growth parameters. Improvement in quality of safflower plants due to treatment with *Trichoderma* species, *B. subtilis* and *P. fluorescens* against *Fusarium* species has also been reported by Gehlot and Purohit (2002). Results with respect to efficacy of the plant extracts tested for the control of safflower wilt under field conditions revealed less incidence of wilt (2%) due to *B. monniera* over by *A. vasica* (3%). Both plants extracts showed moderate effect on germination, as well as increased the shoot length, number of leaves, plant height, number of heads and other growth parameters (Tables 5 and 6). Plant extracts with their antifungal properties may affectrespiration through the presence of inhibitors, carboxin, which

**Table 6.** Effect of different seed treatments on the incidence of wilt of safflower under field conditions.

| Treatment | Dose | % variation in the development of wilt due to various treatments in safflower during different growth stages | | | | | | |
|---|---|---|---|---|---|---|---|---|
| | | 30 DAS | | | 60 DAS | | 90DAS | |
| | (%) | Seed germination (%) | DI (%) | DIR (%) | DI (%) | DIR (%) | DI | DIR (%) |
| **Fungicides** | | | | | | | | |
| Anucop | 0.2 | 90 | - | $90\pm0.57^{a}$ | $1\pm0.57^{c}$ | $84\pm0.57^{c}$ | $1\pm0.57^{de}$ | $84+0.57^{c}$ |
| Bavistin | 0.2 | 88 | - | $88\pm0.57^{c}$ | $1\pm0.57^{c}$ | $83\pm0.57^{d}$ | $1\pm0.57^{de}$ | $83+0.57^{c}$ |
| Blitox | 0.2 | 84 | $1\pm0.57^{c}$ | $77\pm0.57^{f}$ | $1\pm0.57^{c}$ | $79\pm0.57^{e}$ | $2\pm0.57^{cd}$ | $75+0.57^{f}$ |
| Captan | 0.2 | 92 | - | $92\pm0.57^{a}$ | - | $92\pm0.57^{a}$ | - | $92+0.57^{a}$ |
| Mancozeb M-45 | 0.3 | 92 | - | $92\pm0.57^{a}$ | - | $92\pm0.57^{a}$ | - | $92+0.57^{a}$ |
| **Bioagents formulations** | | | | | | | | |
| *Trichoderma harzianum* | 1.0 | 90 | - | $90\pm0.57^{b}$ | - | $90\pm0.57^{b}$ | - | $90+0.57^{b}$ |
| *Bacillus subtilis* | 1.0 | 86 | - | $86\pm0.57^{d}$ | $1\pm0.57^{c}$ | $81\pm0.57^{d}$ | $1\pm0.57^{de}$ | $81+0.57^{d}$ |
| *Pseudomonas fluorescens* | 1.5 | 84 | $1\pm0.57^{c}$ | $78\pm0.57^{f}$ | $1\pm0.57^{c}$ | $79\pm0.57^{e}$ | $1\pm0.57^{de}$ | $79+0.57^{e}$ |
| **Plant extracts** | | | | | | | | |
| *Becopa monnieri* | 50 | 83 | - | $83\pm0.57^{e}$ | $1\pm0.57^{c}$ | $78\pm0.57^{e}$ | $2\pm0.57^{cd}$ | $74+0.57^{f}$ |
| *Adathoda vasica* | 50 | 80 | $1\pm0.57^{c}$ | $74\pm0.57^{g}$ | $1\pm0.57^{c}$ | $75\pm0.57^{f}$ | $3\pm0.57^{c}$ | $67+0.88^{g}$ |
| Control | | 64 | $14\pm0.57^{b}$ | | $16\pm0.57^{b}$ | | $18\pm0.57^{b}$ | |
| Control + pathogen | | 58 | $58\pm0.57^{a}$ | | $58\pm0.57^{a}$ | | $58\pm0.57^{a}$ | |

DI-disease incidence, DIR-disease incidence reduction. According to Duncan's Multiple Range Test (DMRT), values followed by different superscripts are significantly different at $P\leq 0.05$. Values are mean of four replications of 100 seeds each.

interfered with synthesis of protein, DNA and RNA in the rapidly metabolizing cells of the pathogens (Raysdale and Silser, 1970; Natarajan and Lalithakumari, 1987). Varshey (2001) evaluated the effect of some plant extracts on important diseases of plants on rice crop. Banasal and Gupta (2001) have used many plant extracts against *F. oxysporum* for its effective control.

Efficacy of fungicides, bioagents and plant extracts showed similar results in *in vitro* and *in vivo* in the management of *Fusarium* wilt of safflower. Among fungicides Captan (0.2%), Mancozeb M-45 (0.3%) and Anucop (0.2%), in bioagents *T. harzianum* (1.0%), plant extract *B. monniera* (50%) were found highly effective in controlling safflower diseases in turn enhancing the quality of plants under both greenhouse and field conditions. Thus, the present study highlight the importance of seed treatment with most effective fungicides, bioagents and plant extracts in suppression of safflower wilt disease.

## ACKNOWLEDGEMENTS

We are highly thankful to Prof. H. Shekar Shetty, Emeritus Professor, DOS in Applied Botany, Seed Pathology and Biotechnology, University of Mysore and Dr MR Hulinaykar, Managing Trustee, Shridevi Charitable Trust (R.), Tumkur for their encouragement during the course of the study.

## REFERENCES

Anonymous (1996). International Rules for Seed Testing. Seed Sci. Technol., 24 supplement.

Aziz AY, Foster HA, Fairhurst CP (1993). Extra cellular enzymes of *Trichoderma harzianum, T. polysporum* and *Scytadium lignicola* in relation to biological control of Dutch elm disease. Arboric J., 17: 159-170.

Baker R, Dickman MB (1993). Biocontrol with fungi. *In:* Soil Microbial Ecology Application in Agricultural and Environmental Management (Blain Meeting Jr.F.Ed.). Marcel Dekker Inc. New York, pp. 275-306.

Bansal AK, Gupta RK (2000). Evaluation of plant extracts against *Fusarium oxysporum*, wilt pathogen of fenugreek. Indian Phytopathol., 53(1): 107-108.

Bhatti MAR, Randhwa MA, Ali S (1985). Chemical control of fungi occurring on safflower seeds. Pakistan J. Scientific Industrial Res. 28:31-32.

Bunker RN, Mathur K (2000). Integration of biocontrol agents and fungicides for suppression of dry rot of *Capsicum frutescens*. J. Mycol. Plant Pathol., 31(3): 330-334.

Chakrabarti DK (1976). Intensive Agriculture, 14(2): 16.

Chakrabarty SK, Rao RDJP (1992). Eradication of *Fusarium moniliforme* from maize seeds. Indian J. Plant Protection. 20: 105-107.

Chang YL, Vhang YC, Baker R, Kleifeld O, Chet I (1986). Increased growth of plants in presence of biological agent *T. harzianum*. Plant Dis., 70: 145-148.

Cicero SM, Chamma HMCP, Novembre ADLC, Marracs MDH (1992). Physiological sanitary quality of corn seeds submitted to different fungicide treatments. Seed Sci. Technol., 20: 295-302.

Crossan DF (1967). Selective isolation for soil microorganisms by means of differential media. In: source book of laboratory exercises in plant pathology (Freeman W.H. Ed.). American Phytopathological Society. San Francisco and London, p. 387.

De KR, Chaudhary RG, Naimudin (1996). Comparative efficacy of biocontrol agents and fungicides on chickpea wilt caused by *Fusarium oxysporum* f.sp. *ciceri*. Indian J. Agric. Sci., 66(6): 37-373.

De KR, Chaudhary RG (1999). Biological and chemical seed treatment against lentil wilt. LENS Newsletters, 26(1&2): 28-31.

Dey SK, Gill KS, Aulakh SS, Sokhi RS, Khetira, Grewel RK (1989). Effect of seed dressing chemicals on germination and seedling vigour of maize. Plant Dis. Report, 4: 36-42.

Dubey SC, Patil B (2001). Determination of tolerence in *Thanatephores cucumeris*, *Trichoderma viride*, *Gliocladium virens* and *Rhizobium* sp. To fungicides. Indian Phytopathol., 54(1): 98-101.

Elad YI, Chet, Henis Y (1982). Degradation of plant fungi by *Trichoderma harzianum*. Canadian J. Microbiol., 28: 729-725.

Gehlot P, Purohit DK (2002). Biocontrol of *Fusarium* wilt of chilli through seed bacterization with *Pseudomonas fluorescens*. J. Mycol. Plant Pathol., 32(1): 133.

Gupta A, Aneja KR (2001). Field efficacy of seed dressing chemicals on seedling emergence, seed yield and seed weight in soybean. Indian Phytopathol., 28(1): 54-58.

Holdeman OL, Mc Cartnery WO (1964). In disease of safflower bureau-plant pathology. Department of Agriculture. California, USA, pp. 26.

Jeyalakshmi C, Durairaj P, Seetharam K, Sivaprakasam K (1998). Biocontrol of fruit rot and die back of chilli using antagonistic microorganisms. Indian Phytopathol., 51(2): 180-183.

Klisiewicz JM (1963). Phytopathology, 53: 1046-1048.

Krishnamurthy CD, Lokesh S, Shetty HS (2001). Occurrence, transmission and remedial aspects of *Drechslera oryzae* in paddy (*Oryza sativa* L.). Seed Res., 29(1): 63-70.

Kumar M, Agarwal VK (1998). Effect of fungicidal seed treatment on seed-borne fungi, germination and seedling vigour of maize. Seed Res., 26(2): 147-161.

Kumar D, Dubey SC (2001). Management of collar rot of pea by the integration biological and chemical methods. Indian Phytopathol., 54(1): 62-66.

Laha GS, Vekataraman S (2001). Sheath blight management in rice with biocontrol agents. Indian Phytopathol., 54(4): 461-464.

Laxminarayan C, Sinha SK, Aujla SK, Consul, Sinha SN, Payak MM, Nene YL, Renfereo BC (1996). Fungicidal seed treatment of maize in India. Indian Phytopathol. Bull., 3: 18-26.

Mishra AK, Dharam Vir (1992). Field evaluation of fungicides against fungi causing discoloration of paddy seeds. Indian Phytopathol. 45(1): 49-54.

Mukherjee S, Tripathi HS, Rathi YPS (2001). Integrated management of wilt complex frenchbean (*Phaseolus vulgaris* L.). J. Mycol. Plant Pathol., 31(2): 213-215.

Mukhopadhyay AN, Shrestha SM, Mukherjee PK (1992). Biological seed treatment for control of soil-borne plant pathogens. FAO Plant Protection Bull., 40(1): 21-30.

Natarajan MR, Lalitha Kumari D (1987). Antifungal activity of leaf extracts of *Lawsonia inermis* on *Drechslera oryza*. Indian Phytopathol., 40: 393-395.

Padmavathi M, Prakasan B, Sunilkumar A, Anbu R, Ganesan J (1998). Effect of chemicals and growth regulators on the seed germination and vigour of brinjal. In: proceedings of National symposium of future goal of physiological research for the improvement of plant research, 2: 42-47.

Pathak AK, Godika S, Jain JP, Murali S (2001). Effect of antagonistic fungi and fungicides on the incidence of stem rot of mustard. J. Mycol. Plant Pathol., 31(3): 327-329.

Parimala K, Thangavel P, Saravanan K, Anbuselvan A, Ganesan J (1998). Effect of lant growth regulators and chemicals on seed germination in blackgram. In: proceedings of National symposium of future goal of physiological research for the improvement of plant research, 2: 42-47.

Pedgoankar SM, Mayee CD (1989). Screening of safflower genotypes and efficacy of seed dressing fungicides against safflower wilt. Second International Conference, 9-13 January, Hyderabad, pp. 265-270.

Prasad RD, Rangeswaran R (1999). Granular formulation of *Trichoderma* and *Gliocladium* spp. In biocontrol of *Rhizictonia solani* of chickpea. J. Mycol. Plant Pathol., 29(20): 222-226.

Prakash M, Ganesan J (1998). Effect of plant growth regulators and chemicals on seed germination in jack (*Artocarpus heterophylla* L.) in the proceedings of National symposium of future of Physiological research for the improvement of plant research, 2: 42-47.

Raghuchendar T, Rajappan K, Samiappan R (1997). Evaluation methods of application of biocontrol agents in the control of mungbean root rot. Indian Phytopathol., 50:229-234.

Raysadale NN, Sisler HD (1970). Metabolic effects to fungitoxicity of carboxin. Phytopathol., 60:1422-1427.

Sandrou S, Wahab K, Kavitha R (1998). Effect of hardening on physiological growth and yield parameters of sunflower cv. Modern. In: proceedings of In: proceedings of National symposium of future goal of physiological research for the improvement of plant research. 2;70-75

Sastry KR (1996). Symptoms of wilt disease-clue for use in resistance breeding. In: proceedings of Training programme on breeding approaches for improving productivity of safflower and group meeting on heterosis breeding in safflower (Hegde D.M., Raghvaiah, C.V. and D. Patil, Eds.). Directorate of oilseed research. Hyderabad, pp. 25-32.

Sastry KR, Ramachnadran R(1992). Differential genotypic response to progressive development of wilt in safflower. J. Oilseeds Res., 9: 297-305.

Singh SD, Rawal P, Shekawat NS, Lodha PC (2002). Management of mungbean (*Vigna radiata* (L.) Wikzek) seed mycoflora by seed dressing fungicides. J. Mycol. Plant Pathol., 23(1): 149.

Sivakumar G, Narayanaswamy NT (1998). Biological control of sheath blight of rice with *Pseudomonas fluorescens*. Oryzae, 35(1): 57-60.

Sugha SK (2001). Management of white rot of pea by fungicides. J. Mycol. Plant Pathol., 31(3): 335-338.

Sugha SK, Kapoor SK, Singh BM (1995). Management of chickpea wilt with fungicides. Indian Phytopathol., 48(1): 27-35.

Taylor AG, Harman GE (1990). Concepts and technologies of selected seed treatments. Annual Rev. Phytopathol., 28: 321-339.

Tewari SN, Kuruchevu V (1999). Fungicidal activity of plant and animal products. Annals Agric. Res., 20(10): 113-115.

Varshney V (2001). Effect of plant extracts on *Drechslera graminae*, the causal agent of stripe disease of barley. Indian Phytopathol., 54(1): 88-90.

Van Loon LC, Bakker PAHM, Pieterse CMJ (1998). Systemic resistance induced byrhizosphere bacteria. Annual Rev. Phytopathol., 36: 435-483.

Vidyasekaran P, Rabidran M, Muthamila K, Rajappan N, Subramanian, Vasumathi K (1997). Development of a powder formulation of *Pseudomonas fluorescens* for control of rice blast. Indian Phytopathol., 46: 291-297.

Wegrazyka FH (1984). Czechoslovakia academy technique. Ralnictowo, 39: 137-15

# Effect of different doses of herbicides on yield attributes and grain yield of dry seeded rice (*Oryza sativa* L.)

**Fayaz Ahmed Bahar and Zahida Rashid**

Department of Agronomy, G. B. Pant University of Agriculture and Technology, Pantnager-263145 (Uttaranchal), India.

**A field experiment was conducted during *kharif* season of 2002 to find out the effective dose and stage of application of cyhalofopbutyl at 80, 100 and 120 g ha$^{-1}$, quinclorac at 125, 187.5 and 250 g ha$^{-1}$, clefoxydim at 50, 75 and 100 g ha$^{-1}$ applied at 15 DAS and pretilachlor at 0.5, 0.75 and 1.0 kg ha$^{-1}$, pendimethalin at 1.0 kg ha$^{-1}$ applied as pre-emergence for studying their effect on yield attributes and grain yield of dry seeded rice (*Oryza sativa* L.). Highest yield attributes and grain yield (62.8 q ha$^{-1}$) were observed under weed free condition followed by pendimethalin at 1.0 Kg ha$^{-1}$ (54.0 q ha$^{-1}$). The lowest grain yield (0.85 q ha$^{-1}$) was recorded in weedy check. Higher doses of cyhalofopbutyl, quinclorac, clefoxydim and pretilachlor recorded significantly higher yield attributes and grain yield than their respective lower doses. Uncontrolled weed growth caused 98.64% reduction in grain yield of rice. None of the treatments could produce grain yield at par with weed free treatment and pendimethalin at 1.0 Kg ha$^{-1}$ proved to be more promising than all other herbicides.**

**Key words:** Dry seed, herbicide, rice, weed.

## INTRODUCTION

Weed infestation in dry seeded rice (*Oryza sativa* L.) is one of the major constraints responsible for low yield and the depression of yield due to weed competition may be as high as 40 to 100% (Choubey et al., 2001) depending upon the intensity and type of weed flora. Prolonged use of herbicides with same mode of action can result in development of herbicide resistance in weeds (Malik and Singh, 1995). Repeated use of any single herbicide in a crop also generates a shift in the composition of weed flora with the result that secondary weeds may become of primary concern. In view of the above facts, the present investigation was undertaken to find out the effective dose and stage of application of herbicides.

## MATERIALS AND METHODS

A field expression was conducted in the D2 block at Crop Research Center, GBPUA & T, Pantnagar, district U.S. Nagar, Uttaranchal, during *kharif* season of 2002. The soil of the experimental site was loam in texture with pH 7.6. The treatments consisted of cyhalofopbutyl at 80, 100, 120 g ha$^{-1}$, quinclorac at 125, 187.5, 250 g ha$^{-1}$, clefoxydim at 50, 75, 100 g ha$^{-1}$ ,all applied at 15 DAS; pretilachlor at 0.5, 0.75, 1.0 Kg ha$^{-1}$ and pendimethalin at 1.0 Kg ha$^{-1}$ ,both applied at 2 DAS, compared with weedy and weed-free checks. The experiment was laid out in randomized block design replicated three times. Herbicides were applied at spray volume of 600 1 ha$^{-1}$. Rice variety Sarjoo-52 at 50 Kg seed ha$^{-1}$ was sown in rows 20 cm apart during the third week of June.

## RESULTS AND DISCUSSION

All the treatments gave significantly higher grain yield then weedy check except clefoxydim at 50 g ha$^{-1}$. The highest grain yield (62.81 q ha$^{-1}$) was recorded in weed free treatment followed by pendimethalin at 1.0 kg ha$^{-1}$

**Table 1.** Effect of treatments on the yield attributes and grain yield of rice.

| Treatment | Dose ($g\ ha^{-1}$) | No.of panicles ($m^{-2}$) | No.of grains $panicle^{-1}$ | 1000-grain weight (g) | Grain yield ($q\ ha^{-1}$) | Straw yield ($q\ ha^{-1}$) | Grain/straw ratio | Crop weed competition index(WI) |
|---|---|---|---|---|---|---|---|---|
| Cyhalofopbutyl | 80 | 168 | 85 | 22.36 | 12.3 | 35.6 | 0.34 | 80.5 |
| Cyhalofopbutyl | 100 | 238 | 108 | 22.86 | 17.1 | 57.3 | 0.29 | 72.6 |
| Cyhalofopbutyl | 120 | 299 | 131 | 22.90 | 35.1 | 75.3 | 0.46 | 44.2 |
| Quinclorac | 125 | 231 | 104 | 22.40 | 8.9 | 24.6 | 0.35 | 85.7 |
| Quinclorac | 187.5 | 235 | 107 | 22.70 | 23.3 | 70.5 | 0.32 | 62.8 |
| Quinclorac | 250 | 263 | 114 | 22.78 | 31.6 | 78.1 | 0.39 | 49.7 |
| Clefoxydim | 50 | 145 | 76 | 22.56 | 8.3 | 24.8 | 0.33 | 86.8 |
| Clefoxydim | 75 | 209 | 86 | 22.75 | 8.7 | 33.0 | 0.26 | 86.1 |
| Clefoxydim | 100 | 242 | 107 | 22.76 | 23.3 | 66.8 | 0.34 | 63.0 |
| Pretilachlor | 500 | 253 | 106 | 22.83 | 22.4 | 52.0 | 0.42 | 64.4 |
| Pretilachlor | 750 | 252 | 115 | 22.90 | 23.6 | 56.6 | 0.41 | 62.4 |
| Pretilachlor | 1000 | 265 | 122 | 23.03 | 31.3 | 73.7 | 0.42 | 49.8 |
| Pendimethalin | 1000 | 310 | 143 | 23.53 | 54.0 | 116.9 | 0.46 | 13.7 |
| Weed free | - | 325 | 146 | 22.73 | 62.8 | 125.0 | 0.50 | 0.00 |
| Weedy check | - | 102 | 73 | 21.73 | 0.85 | 9.1 | 0.09 | 98.6 |
| C.D at 5% | - | 29 | 17 | 1.1 | 7.50 | 15.20 | 0.07 | 4.2 |

(54.0 $q\ ha^{-1}$) which was significantly lower than weed free. The lowest grain yield (0.85 $q\ ha^{-1}$) was recorded in weedy check. Uncontrolled weed growth caused 98.6% reduction in grain yield of rice. Similar findings have also been reported by Kolhe and Tripathi (1998) and Choubey et al. (2001). In weed free treatment, the highest number of panicles (325 $m^{-2}$), number of grain per panicle (146) and 1000 grain weight (23.73 g) were recorded. However, all these yield components were significantly reduced in weedy check. The decrease in yield components in weedy treatment and increase in yield attributes in weed free and pendimethalin at 1.0 Kg $ha^{-1}$ may be attributed to the more crop weed competition in weedy check and very low degree of crop weed competition in other two treatments (Table 1). The highest grain yield in weed free treatment has also been reported by Singh et al. (2003).

Cyhalofopbutyl at 120 g $ha^{-1}$ and clefoxydim at 100 g $ha^{-1}$ produced significantly more grain yield than cyhalofopbutyl at 80, 100 g $ha^{-1}$ and clefoxydim at 50 and 75 g $ha^{-1}$. Pretilachlor at 1.0 kg $ha^{-1}$ gave significantly higher grain yield than pendimethalin at 1.0 Kg $ha^{-1}$ and weed free treatment. This was because of the toxicity of these treatments on the crop and is evident from the less number of panicles per square metre and less number of grains per panicle (Table 1). Quinclorac at 125 g $ha^{-1}$ gave significantly lower grain yield than quinclorac at 187.5 and 250 g $ha^{-1}$. This was due to the high degree of crop weed competition, less number of grains per panicle and 1000-grain weight in former than in latter treatments. Similar findings have also been reported by Bahar and Singh (2004).

The highest grain/straw ratio in weed free treatment was associated with the highest grain yield as well as highest straw yield. Pendimethalin at 1.0 kg $ha^{-1}$ and cyhalofopbutyl at 120 g $ha^{-1}$ give grain straw ratio at par with the weed free treatment though grain yield in these two treatments were significantly lower than that of weed free treatment. Thus it can be concluded that pendimethalin at 1.0 kg $ha^{-1}$ proved to be more promising than all other herbicides though none of the treatments could produce grain yield at par with weed free treatment.

**REFERENCES**

Bahar FA, Singh G (2004). Effect of herbicides on dry seeded rice (*Oryza sativa* L.) and associated weeds. Indian J. Weed Sci. 36(3,4):269-270.

Choubey NK, Kolhe SS, Tripathi RS (2001). Relative performance of cyhalofopbutyl for weed control in direct seeded rice. Indian J. Weed Sci. 2.33(3,4):132-135.

Kolhe SS, Tripathi RS (1998). Integrated Weed ManagementIn direct seeded rice. Indian J. Weed Sci. 30(1,2):51-53.

Malik RK, Singh S (1995). Littleseed canary grass (*phalaris minor*) resistance of isoproturon in India. Weed Technol. 9:419-425.

Singh GV, Singh P, Singh Y, Singh KM, Mortimer M, Johnson DE, White JL (2003). Efficacy of herbicides alone and in combinations in direct seeded rice. Biennial Conference ISWS, March 12-14, GBPUA & T, Pantnagar, P. 3.

# Effects of different maize (*Zea mays* L.) – soybean (*Glycine max* (L.) Merrill) intercropping patterns on yields and land equivalent ratio

**Matusso J. M. M., Mugwe J. N. and Mucheru-Muna M.**

Kenyatta University (KU), Agricultural Resources Management Department, P. O. Box 43844 -00100, Nairobi, Kenya.

**In central highlands of Kenya, the low soil fertility and inability to replenish it are amongst the major constraints affecting the productivity of maize and cash crops, leading to hunger and poverty. The adoption of ISFM technologies such as maize-soybean intercropping system is being promoted as one of the options to address low crop productivity among the farmers. This study intended to determine the effects of maize-soybean intercropping patterns on yields and to quantify the land equivalent ratio (LER) of different maize–soybean intercropping patterns in the two contrasting sites. The experiment was arranged in a randomized complete block design (RCBD) with four replications. The treatments were four maize (M) – soybean (S) intercropping patterns (conventional=1M:1S; MBILI-MBILI=2M:2S; 2M:4S; 2M:6S) and two sole crops of maize and soybean, respectively. The results showed that in both sites during the both seasons maize stover and grain yields were significantly affected by the intercropping pattern. During 2012 LR at Embu site the MBILI treatment produced significantly higher stover and grain yields (13.12 t ha$^{-1}$, $p$=0.0001 and 6.11 t ha$^{-1}$, $p$<0.0001, respectively) than all other treatments. During 2012 SR, still the MBILI treatment had recorded significantly the highest stover and grain yield (7.62 t ha$^{-1}$, $p$<0.0001 and 5.62 t ha$^{-1}$ $p$=0.0467, respectively) than all other treatments. During 2012 LR at Kamujine site the conventional treatment produced significantly the highest stover yield (3.87 t ha$^{-1}$, $p$=0.0461) than only the 2M:6S treatment. During 2012 SR at Kamujine site, the MBILI treatment had recorded significantly the highest stover and grain yield (6.55 t ha$^{-1}$, $p$=0.0005 and 3.55 t ha$^{-1}$ $p$=0.0006, respectively) than all other treatments. During both seasons in both sites, the soybean yield was significantly affected by the intercropping pattern. During the 2012 LR, the yields were reduced by 60 and 81% due to the intercropping with maize, at Embu and Kamujine, respectively; whereas, during the 2012 SR, the yields were reduced by 52 and 78% as effect of intercropping with maize, at Embu and Kamujine, respectively. During both seasons at both locations, the partial land equivalent ratio (LER) values were significantly affected by the intercropping patterns ($p$<0.0001).**

**Key words:** Maize-soybean yields, land equivalent ratio, intercropping patterns, central highlands, Kenya.

## INTRODUCTION

Smallholder farmers are the most important food security stakeholders in Sub-Saharan Africa (FAO, 2011), who mainly practice subsistence agriculture characterized by low crop productivity due to the soil nutrient depletion (Mugwe et al., 2007). The majority of these farmers lack financial resources to purchase sufficient amount of mineral fertilizers to replace soil nutrients removed through harvested crop products (Jama et al., 2000), crop residues, and through loss by runoff, leaching and as gases (Bekunda et al., 1997). Consequently, poor soil fertility has emerged as one of the greatest biophysical constraint to increasing agricultural productivity hence threatening food security in this region (Mugwe et al., 2009; Mugendi, 1997). Therefore, it is necessary to adopt improved and sustainable technologies in order to guarantee improvements in food productivity and thereby food security (Landers, 2007; Gruhn et al., 2000). Such technologies include the use of integrated soil fertility management practices (ISFM) such as intercropping cereals with grain legumes as one of its main components (Mucheru-Muna et al., 2010; Sanginga and Woomer, 2009). Cereal – grain legume intercropping has potential to address the soil nutrient depletion on smallholder farms (Sanginga and Woomer, 2009). The legumes play an important role in nitrogen fixation (Peoples and Craswell, 1992), and are important source of nutrition for both humans and livestock (Nandwa et al., 2011). In the central highlands of Kenya, cereal – legume intercropping is already being widely practiced by the smallholder famers. According to Sanginga and Woomer (2009) intercropping cereal and grain legume crops helps maintain and improve soil fertility, because crops such as cowpea, mung bean, soybean and groundnuts accumulate from 80 to 350 kg nitrogen (N) $ha^{-1}$ (Peoples and Craswell, 1992). For instance, soybean can positively contribute to soil health, human nutrition and health, livestock nutrition, household income, poverty reduction and overall improvements in livelihoods and ecosystem services, than many others leguminous grain crops (Rakasi, 2011; Raji, 2007). Improved intercropping systems are part of ISFM technologies (Mucheru-Muna et al., 2010; Sanginga and Woomer, 2009) and in central highlands of Kenya the information is scarce regarding to optimum cropping pattern of maize-soybean intercropping system.

## MATERIALS AND METHODS

### Study area

The experiment will be carried out in two sub counties of central highlands of Kenya, namely Embu West and Tigania East sub counties.

### Embu west sub county

Embu West District is located in Embu County, in the central highlands of Kenya, and occupies an area of 708 $Km^2$ and is bordered by Mbeere district to the East and South East, Kirinyaga to the West and Meru South to the North. The experimental site lies within N 0° 31′ 4.2′′ E 37° 27′ 20′′ and at the altitude of 1468 m above the sea level (ASL), at Embu Agricultural Staff Training College (Jaetzold et al., 2006). Diagnosis study carried out in the central highlands of Kenya have reported soil fertility constraints, particularly N and P deficiencies, low carbon content and low soil pH (Gachimbi et al., 2002). The major agro-ecological zone (AEZ) is Upper Midland 2 (UM 2), the soils are humic nitisols and the total arable land area is 478 $km^2$ with total available agricultural land area covering 371 $km^2$. Table 1 shows the soil characteristics of the soils in Kamujine. The average annual rainfall varies from 1230 to 909 mm with long rainy season between March and June and short rainy season between October and December, respectively. Rainfall for the two seasons in which the experiment was conducted is presented Figure 1.

### Tigania east sub county

Tigania East Sub County is located in Meru County, in the central highlands of Kenya and it occupies 108.6 $km^2$. The experimental site lies within N 0° 6′ 19.5′′ E 037° 64′ 39.6′′ and at the altitude of 935 m above the sea level (ASL), at Kamujine Dispensary in Mikinduri Division. The major agro-ecological zones are Lower Midlands 3 and Upper Midland 3 (LM 3 and UM 3), the soils are mainly eutric Nitisols and humic Cambisols. The annual average temperature varies from 19.2 to 22.9°C (Jaetzold et al., 2006). Table 1 shows the soil characteristics of the soils in Kamujine.

The average annual rainfall varies from 1000 to 2200 mm with long rainy season between March and June and short rainy season between October and December, respectively (Jaetzold et al., 2006). Rainfall for the two seasons in which the experiment was conducted is presented Figure 2.

### Management of the experiment

The fields were ploughed using hand hoe and left as such for two weeks. Plots measuring 7.0 by 4.5 m were marked just before planting. Pathways measuring 3.0 m and 2.0 m were left between the blocks and plots, respectively. At Embu-ATC, planting was done on the 23rd of March and 12th of October 2012 for the 1st and 2nd seasons, respectively. At Kamujine, planting was done on the 26th of March and 15th of October 2012 for the 1st and 2nd seasons, respectively. The sole maize (*Zea mays* L.) var. DK 8031 was planted at a spacing of 0.75 m 0.50 m inter and intra-row, respectively. The number of hills per row was 10 with three seeds per hill in order to ensure maximum plant population and to account for germination failure; and two weeks after germination the excess plants were thinned out to remain with two plants per hill. The sole soybean (*Glycine max* (L.) Merrill) var. Gazelle was hand drilled at a spacing of 0.45 m × 0.10 m in inter and intra – row spacing resulting to 62 plants per row to ensure maximum germination/population and the excess plants were thinned out to remain with the recommended population of 31 plants per row after 2 weeks of emergence. The following external nutrient replenishment inputs were applied per plot: 6 kg of manure equivalent to 30 kg N $ha^{-1}$, applied two weeks before planting; 94.5 g of CAN as source of N, equivalent to 30 kg N $ha^{-1}$, for soybean the Nitrogen (starter N) was applied at sowing while for maize it was applied when the crop has six leaves, as topdressing; 189 g of TSP as source of P, equivalent to 60 kg P $ha^{-1}$, which was applied at sowing. The fertilizers were applied accordingly to the recommendation from FURP (1987). Management practices were the same for both the monocrop and the maize – soybean intercrop.

### Experimental design and management

First, soil samples from the experimental sites were collected at 0 to 15 cm depth for analysis for organic carbon, total nitrogen using standard methods (Okalebo et al., 2002), extractable P, Ca, Mg, K, Na using Mehlich-1 (M1) extraction method, where P and $Mg^{2+}$ were determined colourimetrically in a spectrophotometer and $Ca^{2+}$, and $K^+$ were determined using flame photometer. The experiment was established in Embu-ATC (Embu West district) and in Kamujine (Tigania East district) and it was laid out as a randomized complete block design (RCBD) with four replicate blocks and plot sizes measuring 7 m × 4.5 m. The cropping system was of sole maize (*Zea mays* L.), sole soybean (*Glycine max* (L.) Merrill) and maize (M) – soybean (S) intercropping with cropping patterns (Table 2).

### Maize and soybean harvest and yields

Maize and soybean grain and stover was harvested at maturity from a net area of each treatment demarcated after leaving out two rows

**Table 1.** Soil characteristics at Embu – ATC and Kamujine sites, Kenya.

| Soil parameter | Embu – ATC site | Kamujine site |
|---|---|---|
| pH in water $_{(1:2.5)}$ | 5.30 | 5.50 |
| Total N (%) | 0.03 | 0.01 |
| Total soil organic carbon (%) | 2.64 | 1.88 |
| Extractable P (ppm) | 13.40 | 9.54 |
| Exchangeable Ca (C mol kg$^{-1}$) | 0.22 | 0.21 |
| Exchangeable Mg (C mol kg$^{-1}$) | 0.53 | 0.53 |
| Exchangeable K (C mol kg$^{-1}$) | 0.12 | 0.08 |
| Clay (%) | 65 | 45 |
| Sand (%) | 17 | 20 |
| Silt (%) | 18 | 35 |

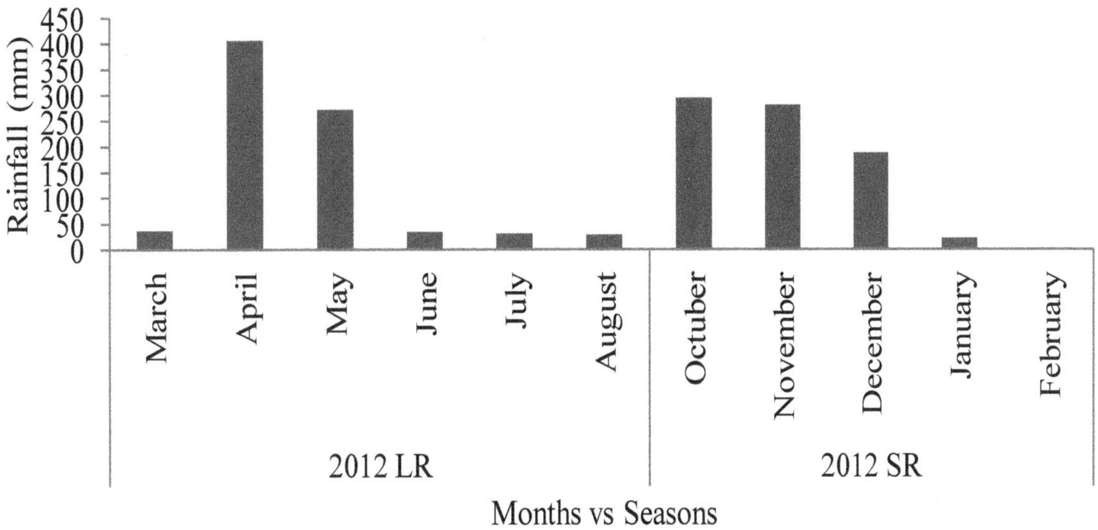


**Figure 1.** Rainfall amount during 2012 LR and 2012 SR at Embu-ATC, Embu west sub county, Kenya.

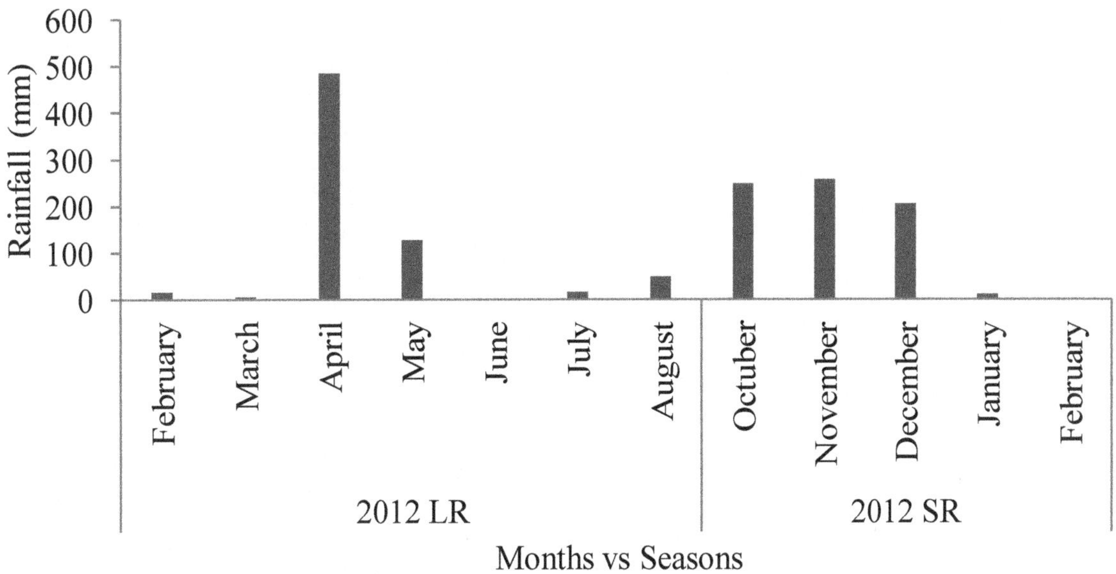


**Figure 2.** Rainfall amount during 2012 LR and 2012 SR at Kamujine site, Tigania East Sub county, Kenya.

**Table 2.** Treatments in the two sites (ATC-Embu and Kamujine).

| Treatment | Cropping system |
|---|---|
| T1 | Sole maize |
| T2 | Sole soybean |
| T3 | Maize-soybean (1:1) |
| T4 | Maize-soybean (2:2) |
| T5 | Maize-soybean (2:4) |
| T6 | Maize-soybean (2:6) |

on each side of the plot and the first two and the last two maize/soybean plants on each row to minimize the edge effect (Table 3). The entire plants on the plots was harvested by cutting at the ground level and weighted to represent the total fresh weight. Maize/soybean cobs/pods were manually separated from the stover, sun-dried, and packed in sacks before threshing. After threshing, moisture content of the grains was determined using a moisture meter and grain yield adjusted to 12% moisture content using the following formula. Similarly, the yields and harvest index were calculated using the following formulas.

$$Adjusted \ yield = measured \ yield * \frac{(100 - sample \ moisture \ content)}{(100 - s\tan dard \ moisture \ content)} \quad (1)$$

$$Yield \ (t/ha) = 10 * \frac{Dry \ weight \ (kg/m^2)}{Net \ area \ (m^2)} \quad (2)$$

$$Harvest \ Index \ (HI)(\%) = \frac{Grain \ yield \ (t/ha)}{Total \ biomass \ yield \ (t/ha)} * 100 \quad (3)$$

Performance of the intercropping maize – soybean systems

To evaluate the intercrop's performance, land equivalent ratio was calculated as shown as:

$$LER = \frac{X_{yi}}{X_{ys}} + \frac{Y_{yi}}{Y_{ys}} \quad (4)$$

Where, $X_y$ and $Y_y$ refer to X (maize) and Y (soybean) yields, respectively. The subscript "i", refers to the intercrop and the subscript "s" the sole crop.

#### Data analysis

Data of maize – soybean yields and on land equivalent ratio were subjected to analysis of variance using SAS version 8. To test the differences between different cropping pattern and conventional intercropping systems, the yields were subjected to *t-student* test at 95% of significance level ($p<0.05$).

## RESULTS AND DISCUSSION

### Maize yields

At Embu during the both seasons, maize stover and grain yields ware significantly affected by the intercropping pattern. For instance, during 2012 LR the MBILI treatment produced significantly higher stover and grain yields (13.12 t $ha^{-1}$, $p$=0.0001 and 6.11 t $ha^{-1}$, $p$<0.0001, respectively) than all other treatments, excluding the conventional (for the grain yield and stover yield) and also sole maize treatment (for the stover yield). During 2012 SR, still the MBILI treatment had recorded significantly the highest stover and grain yield (7.62 t $ha^{-1}$, $p$<0.0001 and 5.62 t $ha^{-1}$ $p$=0.0467, respectively) than all other treatments, except sole maize and conventional treatments. Only during 2012 LR that the harvest index was significantly affected by the intercropping patterns, where the 2M:6S treatment observed the highest HI of 37.30% than all other treatments. Slightly different situation was observed during 2012 SR, where although not significantly different ($p$=0.5398) from other treatments, the MBILI treatment recorded numerically the highest HI of 42.22%. These differences between two the seasons in term of HI may be related to variations in rainfall and/or differences in weed pressure.

During both season at Kamujine site, maize stover yield was significantly affected by the intercropping pattern. For instance, during 2012 LR the conventional treatment produced significantly the highest stover yield (3.87 t $ha^{-1}$, $p$=0.0461) than only the 2M:6S treatment. In this season, the grain yield was not significantly affected by the intercropping patterns ($p$=0.0704); however, the sole maize treatment had recorded numerically the highest value of 3.36 t $ha^{-1}$ and 2M:6S treatment with the lowest. The lower maize yields under 2M:6S treatment could be related to the lower plant density compared to other treatments. During the 2012 LR, the maize stover yields was strongly significantly ($p \leq 0.05$) positively correlated with the N uptake at all the sampling periods, excluding at 16 WAP and 20WAP. During 2012 SR, the MBILI treatment had recorded significantly the highest stover and grain yield (6.55 t $ha^{-1}$, $p$=0.0005 and 3.55 t $ha^{-1}$ $p$=0.0006, respectively) than all other treatments, except sole maize and conventional treatments. Differently from Embu site, the harvest index was not significantly affected by the intercropping patterns during both seasons ($p$=0.4243 and $p$=0.4908, respectively); however, the MBILI (during 2012 LR) and 2M:4S (during 2012 SR) treatments observed numerically the highest HI of 51.40 and 37.29%, respectively (Table 4).

Increased maize yield under MBILI treatment found at Embu site was also reported by Undie et al. (2012) in Nigeria under maize-soybean intercrop, Mucheru-Muna et al. (2010) in the central highlands of Kenya and Woomer et al. (2004) in western Kenya with maize-beans, and Solanki et al. (2011) in India with maize-blackgram. This, could be due to the fact that the MBILI intercrop arrangement offer better opportunities to the components to utilize available resources more effectively than the conventional (1M:1S) intercropping pattern. Rajat De and Singh (1979) stated that the modified 2M:2S system affords a better solar energy harvest than the 1M:1S crop arrangement, due to the fact that the former might have provided sufficient light to the

**Table 3.** Net plot area per treatment in both sites.

| Intercropping pattern | Total plot size ($m^2$) | Net plot ($m^2$) |
|---|---|---|
| Sole maize | 7.0m×4.5m=31.5 | 6.0m×3.5m=21.0 |
| Sole soybean | 7.0m×4.5m=31.5 | 4.95m×3.9m=19.3 |
| Maize-soybean (1M:1S) | 7.0m×4.5m=31.5 | 5.25m×3.5m=18.4 |
| Maize-soybean (2M:2S) | 7.0m×4.5m=31.5 | 4.8m×3.5m=16.9 |
| Maize-soybean (2M:4S) | 7.0m×4.5m=31.5 | 5.3m×3.5m=18.7 |
| Maize-soybean (2M:6S) | 7.0m×4.5m=31.5 | 4.5m×3.5m=15.8 |

**Table 4.** Effect of intercropping pattern on maize yields and harvest index during 2012 LR and 2012 SR at Embu and amujine sites.

| Location | Treatment | Stover yield (t/ha) | | Grain yield (t/ha) | | Harvest Index (%) | |
|---|---|---|---|---|---|---|---|
| | | 2012 LR | 2013 SR | 2012 LR | 2013 SR | 2012 LR | 2013 SR |
| | Sole maize | 11.73 | 8.14 | 4.95 | 4.58 | 29.95 | 36.02 |
| | Maize-soybean (1M:1S) | 12.64 | 7.74 | 5.49 | 5.16 | 30.45 | 39.27 |
| Embu | Maize-soybean (2M:2S) | 13.12 | 7.62 | 6.11 | 5.62 | 31.80 | 42.22 |
| | Maize-soybean (2M:4S) | 8.89 | 4.91 | 4.05 | 3.48 | 31.30 | 41.43 |
| | Maize-soybean (2M:6S) | 5.26 | 4.45 | 2.74 | 3.16 | 37.30 | 41.52 |
| *p-value* | | 0.0001*** | <0.0001*** | <0.0001*** | 0.0467* | 0.0168* | 0.5398 |
| $LSD_{(0.05)}$ | | 2.59 | 1.31 | 0.87 | 1.79 | 4.23 | 8.63 |
| | Sole maize | 3.81 | 6.00 | 3.36 | 2.98 | 47.05 | 32.90 |
| | Maize-soybean (1M:1S) | 3.87 | 6.34 | 3.09 | 3.44 | 44.59 | 35.03 |
| Kamujine | Maize-soybean (2M:2S) | 2.95 | 6.55 | 3.07 | 3.55 | 51.40 | 35.14 |
| | Maize-soybean (2M:4S) | 2.59 | 4.83 | 2.47 | 2.86 | 49.23 | 37.29 |
| | Maize-soybean (2M:6S) | 1.90 | 3.97 | 1.9 | 1.82 | 50.39 | 32.19 |
| *p-value* | | 0.0461* | 0.0005*** | $0.0704^{ns}$ | 0.0006*** | 0.4243 | 0.4908 |
| $LSD_{(0.05)}$ | | 1.43 | 1.00 | 1.07 | 0.64 | 8.24 | 6.55 |

$^{ns}$, Not significant; *significant at $p \leq 0.05$; **, significant at $p<0.01$; ***, significant at $p<0.001$.

lower leaves to continue photosynthesis. Furthermore, Brintha and Seran (2009) stated that productivity rates increases with LAI because of increased total light interception, but larger LAI values often cause no more increases and then decreases on a ground basis, probably due to respiratory $CO_2$ loss from heavily shaded leaves and stems. However, when the plant population is different the yield difference is determined by population density rather than crop arrangement (Rajat De and Singh, 1979), which was not the case in this particular situation. Similarly, Sikirou and Wydra (2008), Dapaah et al. (2008), Mpairwe at al. (2002) and Olufemi et al. (2001) reported higher cereal grain yield under intercropping system compared to its sole crop. Nzabi et al. (2000) stated that maize grain yield differs with different legume species and intercropping produces higher maize grain yield than in pure stand.

On the other hand, Thobatsi (2009) in RSA, Silwana and Lucas (2002) also observed reduction in maize yield under intercropping system compared to its monocrop, under similar environment of limited rainfall conditions, as it was observed at Kamujine site during 2012 LR. In this site it was necessary to supplement with irrigation water because the crop had started to suffer for water stress due to lack of rainfall during flowering period, as it can be seen in the Figure 2. Amede (1995) stated that one of the factors that reduces maize grain yield is dry conditions that occur specially during the flowering period. More total populations under intercrops compared to monocrop under stress conditions might result in intercrop yields than sole crop yields due to increased competition moisture (Natarajan and Willey, 1986). Yield reductions involving one or all intercropping components in intercropping could be associated to inter-specific competition for nutrients, moisture and/or space (Adaniyan et al., 2007). Moreover, Kamujine has a light texture soil, which has low moisture holding capacity (Jaetzold et al., 2006), resulting therefore in reduced yields under intercropping (Natarajan and Willey, 1986).

The efficiency of a crop variety to convert the dry matter into economic yield is determined by its harvest index. The higher the value, the higher will be dry matter conversion efficiency. The absence of significant differences in HI of maize observed during 2012 SR at

**Table 5.** Effect of intercropping pattern on soybean yields and harvest index during 2012 LR and 2012 SR at Embu and Kamujine sites.

| Location | Treatment | Stover yield (t ha$^{-1}$) | | Grain yield (t ha$^{-1}$) | | Harvest Index (%) | |
|---|---|---|---|---|---|---|---|
| | | 2012 LR | 2012 SR | 2012 LR | 2012 SR | 2012 LR | 2012 SR |
| | Sole soybean | 1.41 | 1.45 | 1.44 | 1.65 | 50.28 | 53.52 |
| | Maize-soybean (1M:1S) | 0.40 | 0.76 | 0.26 | 0.41 | 39.26 | 35.38 |
| Embu | Maize-soybean (2M:2S) | 0.53 | 0.55 | 0.35 | 0.42 | 39.98 | 43.17 |
| | Maize-soybean (2M:4S) | 0.60 | 0.83 | 0.46 | 0.76 | 42.95 | 47.98 |
| | Maize-soybean (2M:6S) | 1.03 | 1.17 | 0.88 | 1.01 | 45.99 | 46.60 |
| *p-value* | | 0.0015** | 0.0001*** | 0.0002*** | < 0.0001*** | 0.0004*** | 0.0003*** |
| $LSD_{(0.05)}$ | | 0.43 | 0.28 | 0.39 | 0.18 | 4.02 | 5.73 |
| | Sole soybean | 1.09 | 1.27 | 0.56 | 0.72 | 32.39 | 36.41 |
| | Maize-soybean (1M:1S) | 0.18 | 0.21 | 0.05 | 0.07 | 19.44 | 23.74 |
| Kamujine | Maize-soybean (2M:2S) | 0.19 | 0.24 | 0.04 | 0.08 | 19.29 | 24.31 |
| | Maize-soybean (2M:4S) | 0.52 | 0.58 | 0.17 | 0.30 | 22.42 | 36.56 |
| | Maize-soybean (2M:6S) | 0.78 | 0.83 | 0.24 | 0.37 | 22.55 | 33.65 |
| *p-value* | | < 0.0001*** | 0.0004*** | 0.0007*** | <0.0001*** | 0.0864 | 0.0988 |
| $LSD_{(0.05)}$ | | 0.13 | 0.39 | 0.20 | 0.08 | 10.17 | 12.52 |

$^{ns}$, not significant; *significant at $p \leq 0.05$; **significant at $p<0.01$; ***significant at $p<0.001$.

Embu and Kamujine sites was also reported by Haseeb-ur-Rehman et al. (2010) and Egbe et al. (2010) in maize-cowpea intercropping; Saleem et al. (2011) in maize-legume intercropping systems; and, Carruthers et al. (2000) in maize-soybean intercropping.

## Soybean yields and components

During both seasons in both sites, the soybean yield was significantly affected by the intercropping pattern. During the 2012 LR, the yields were reduced by 60 and 81% due to the intercropping with maize, at Embu and Kamujine, respectively; whereas, during the 2012 SR, the yields were reduced by 52 and 78% as effect of intercropping with maize, at Embu and Kamujine, respectively (Table 5). For instance, at Embu during 2012 LR, the sole soybean observed significantly the highest stover yield (1.41 t ha$^{-1}$, $p$=0.0015) than all the intercropping patterns, followed by the 2M:6S intercrop pattern with 1.03 t ha$^{-1}$ which was statistically different from MBILI (0.53 t ha$^{-1}$, $p$ = 0.0015) and conventional (0.40 t ha$^{-1}$, $p$ = 0.0015) intercropping patterns, but not significantly different from the 2M:4S intercropping pattern. Still at the same site and season, the sole soybean observed statistically the highest grain yield (1.44 t ha$^{-1}$, $p$=0.0002) than all the intercropping patterns, followed again by the 2M:6S treatment (0.88 t ha$^{-1}$, $p$=0.0002), which was significantly different from other intercropping patterns (conventional, MBILI and 2M:4S), and these were not statistically different among them. Similar situation was also observed during 2012 SR, where sole soybean treatment recorded statistically the highest stover yield (1.45 t ha$^{-1}$, $p$<0.0001) than all other treatments, followed by the 2M:6S treatment (1.17 t ha$^{-1}$, $p$<0.0001), which was statistically different from the rest of the intercropping patterns. Although the conventional treatment observed 28% higher stover yield than the MBILI treatment, they were not statistically different ($p$>0.05) among them. The monocropped soybean observed also the highest grain yield (1.65 t ha$^{-1}$, $p$<0.0001) than all other treatments, followed by the 2M:6S treatment with 1.01 t ha$^{-1}$ ($p$<0.0001), which significantly different from the rest of the treatments. During this season, there was significant positive correlation between the soybean grain yield and soil nitrates, and soil mineral N, with $r$=0.56 ($p$=0.0109) and $r$=0.45 ($p$=0.047), respectively. During both seasons, the harvest index of soybean was significantly ($p$=0.0004 and $p$=0.0003, respectively) reduced by the intercropping patterns. For instance, during 2012 LR the sole soybean treatment observed the highest HI of 50.28% which was significantly ($p$=0.0004) different from all other treatments. Similar situation was observed during 2012 SR, where still sole soybean recorded significantly ($p$=0.0003) the highest HI (53.52%) than all other treatments, excluding the 2M:4S treatment.

At Kamujine during 2012 LR, the sole soybean observed statistically the highest stover and grain yields (1.09 t ha$^{-1}$, $p$<0.0001 and 0.56 t ha$^{-1}$, $p$=0.0007, respectively) than all other treatments, followed by the 2M:6S treatment with 0.78 t ha$^{-1}$ for the stover yield, which was significantly different ($p$<0.0001) from the other intercropping patterns; but the MBILI intercrop pattern was not statistically different from the conventional intercropping pattern. For the grain yield, there were no significant differences among the intercropping patterns; however, the 2M:6S observed the highest yield (0.24 t ha$^{-1}$) than all the others. During this season, there was significant positive correlation between the soybean grain yield and soil nitrates, and soil mineral N, with *r*=0.73 (*p*=0.0002) and *r*=0.76 (*p*<0.0001), respectively. During 2012 SR, monocropped soybean treatment observed also significantly the highest stover yield (1.27 t ha$^{-1}$, *p*=0.0004) than all other treatments. Similarly, sole soybean treatment recorded the highest grain yield of 0.72 t ha$^{-1}$ (*p*<0.0001) than all other treatments. During this season, there was also significant positive correlation between the soybean grain yield and soil nitrates, and soil mineral N, with *r*=0.61 (*p*=0.0043) and *r*=0.60 (*p*<0.0053), respectively During both season, the harvest index of soybean was not significantly (*p*=0.0864; *p*=0.0988) affected by the intercropping patterns; however, it was numerically reduced by the intercropping patterns. In general, the yields were increased about 23 and 26% in the second season compared to the first season, at Embu and Kamujine, respectively. Probably due to the cumulative effects of the goat manure that was applied in the first and second seasons, and also due to root and nodule senescence.

The reduction on soybean yields under intercropping with maize or sorghum has also been reported by several researchers (Ijoyah and Fanen, 2012; Egbe, 2010; Muoneke et al., 2007; Muneer et al., 2004; Heibsch et al., 1995; Olufajo, 1992; Pal, Oseni, and Norman, 1992; Neupane, 1983). This reduction in soybean yields under intercropping could be due to interspecific competition between the intercrop components for water, light, air and nutrients, and also the aggressive effects maize ($C_4$ species) on soybean, a $C_3$ species (Egbe, 2010; Muoneke et al., 2007). According to Heibsch et al. (1995), crops with $C_4$ photosynthetic pathways have been known to be dominant when intercropped with $C_3$ species like soybean. The shading of soybean by the maize plants (taller) may also have contributed to reduction on the yields of intercropped soybean. Olufajo (1992) reported that shading by the taller plants in mixture could reduce the photosynthetic rate of the lower growing plants and thereby reduce their yields. It was confirmed in this study that the lower soybean yields in the rows is closer to the maize plant compared to the middle one, which was positively correlated with the photosynthetically active radiation (PAR) intercepted. In addition, Lesoing and Francis (1999) stated that water stress and shading are probably the two major factors, which contribute to reduced legume component yield under intercropping. Moreover, Natarajan and Willey (1980) and Sivakumar and Virmani (1980) reported that stover yield often shows a positive correlation with the amount of radiation intercepted by crops in intercropping systems. In this study it was observed positive correlation between grain yields with the amount of radiation intercepted by crops in intercropping treatments. The reduction of harvest index of the legume component observed in this study was also reported by other researchers (Alhassan et al., 2012; Zaefarian et al., 2007). The reduction was due to maize shading effects on soybean, which caused the legume component to allocate its photosynthates to vegetative growth and height increasing for competing with taller maize. On the other hand, the findings of this study at Kamujine site were in agreement with Carruthers et al. (2000) who reported that the HI for intercrop soybean was not significantly different from monocrop soybean.

**Effects of maize–soybean intercropping patterns on land equivalent ratio LER of maize and soybean**

During both seasons at both locations, the partial LER values were significantly affected by the intercropping patterns (*p*<0.0001). For instance, during 2012 LR at Embu site, the total LER values proved the greater yield advantage of intercropping maize and soybean over component sole crops, though yield advantage varying from 81 to 22%. During this season, the MBILI treatment observed significantly the highest LER value (1.81, *p*=0.0019) than all the other treatments. However, the partial LER for both maize and soybean were all less than 1.00, except for maize where the MBILI was 48 per cent significantly (*p*<0.0001) more productive than the conventional treatment which was 11% more productive than pure maize. Whereas, during 2012 SR the total LER values observed yield advantage varying from 43 to 20%; and, the partial LER for both maize and soybean were also less than 1.00, excluding for maize in MBILI and conventional treatments that was numerically 6 and 5%, respectively, more productive than pure maize. In this second season, conventional treatment proved to be significantly (1.43, *p*=0.0005) more productive than all other treatments, excluding the MBILI treatment. During 2012 LR at Kamujine site, the total LER values proved the greater yield advantage of intercropping maize and soybean over component sole crops, though yield advantage varying from 55 to 11%. During this period, the 2M:6S treatment observed statistically the highest LER value (1.55, *p*<0.0001) than all the other treatments. The partial LER for both maize and soybean were all less than 1.00. Whereas, during 2012 SR the total LER values observed yield advantage varying from 28 to 22%; and, the 2M:4S treatment showed significantly the highest LER value (1.29, *p*=0.0024) than the maize and soybean monocrops, followed by the MBILI treatment with LER of 1.28 (*p*=0.0024), but they were not statistically different from other intercropping patterns. The partial LER for both maize and soybean were also less than 1.00, excluding for maize in the MBILI treatment which significantly (1.12, *p*<0.0001) more productive than pure maize treatment (Table 6). Probably the greater LER of intercrops was mainly due to a greater resource use and resource complementarily than when the species were grown separated.

Intercropping maize with soybean showed advantages

**Table 6.** Effect of intercropping pattern on land equivalent ratio land equivalent ratio (LER) during 2012 LR and 2012 SR at Embu and Kamujine sites.

| Location | Treatment | **Partial** land equivalent ratio (LER) | | | | **Total** land equivalent ratio (LER) | |
|---|---|---|---|---|---|---|---|
| | | Maize | | Soybean | | | |
| | | 2012 LR | 2012 SR | 2012 LR | 2012 SR | 2012 LR | 2012 SR |
| Embu | Sole maize | 1.00 | 1.00 | - | - | 1.00 | 1.00 |
| | Sole soybean | - | - | 1.00 | 1.00 | 1.00 | 1.00 |
| | Maize-soybean (1M:1S) | 1.11 | 1.05 | 0.24 | 0.38 | 1.35 | 1.43 |
| | Maize-soybean (2M:2S) | 1.48 | 1.06 | 0.33 | 0.32 | 1.81 | 1.38 |
| | Maize-soybean (2M:4S) | 0.90 | 0.68 | 0.41 | 0.52 | 1.31 | 1.20 |
| | Maize-soybean (2M:6S) | 0.49 | 0.61 | 0.73 | 0.71 | 1.22 | 1.32 |
| *p-value* | | <0.0001*** | <0.0001*** | <0.0001*** | <0.0001*** | 0.0019** | 0.0005** |
| LSD $_{(0.05)}$ | | 0.27 | 0.18 | 0.25 | 0.09 | 0.35 | 0.19 |
| Kamujine | Sole maize | 1.00 | 1.00 | - | - | 1.00 | 1.00 |
| | Sole soybean | - | - | 1.00 | 1.00 | 1.00 | 1.00 |
| | Maize-soybean (1M:1S) | 0.97 | 1.08 | 0.14 | 0.14 | 1.11 | 1.22 |
| | Maize-soybean (2M:2S) | 0.99 | 1.12 | 0.14 | 0.16 | 1.13 | 1.28 |
| | Maize-soybean (2M:4S) | 0.95 | 0.86 | 0.41 | 0.43 | 1.36 | 1.29 |
| | Maize-soybean (2M:6S) | 0.93 | 0.65 | 0.63 | 0.60 | 1.55 | 1.25 |
| *p-value* | | <0.0001*** | <0.0001*** | <0.0001*** | <0.0001*** | <0.0001*** | 0.0024** |
| LSD $_{(0.05)}$ | | 0.07 | 0.12 | 0.09 | 0.17 | 0.12 | 0.16 |

$^{ns}$, Not significant; *significant at $p$=0.05; **significant at $p$=0.01; ***significant at $p$<0.001.

in land use efficiency expressed as LER when compared with their sole crops. A LER greater than 1.00 has also been reported with maize-soybean (Yusuf et al., 2012; Addo-Quaye et al., 2011; Pawar et al., 2011; Solanki et al., 2011; Ullah et al., 2007; Thole, 2007; Kipkemoi et al., 2001), maize-cowpea (Dahmardeh et al., 2010; Hugar and Palled, 2008), and maize-beans (Yilmaz et al., 2008; Odhiambo and Ariga, 2001). The lower LER for soybean observed in both sites can be explained by the findings of Ofori and Stern (1986) and Davis et al. (1984) who reported that light is the most important factor determining LER of maize and soybean intercropping and LER declines when legume becomes severely shaded. Furthermore, the higher productivity of the intercrop system compared to the sole crop may have resulted from complementary and efficient use of growth resources by the component crops (Li *et al.*, 2003, 2006). For instance, Sivakumar and Virmani (1980) observed that in maize-pigeon pea intercrop, dry matter production per unit of PAR intercepted was higher in the mixture than in sole crops, which was confirmed in this study though higher light interception in intercropping treatments compared to monocrops.

## Conclusions

The maize-soybean intercropping patterns had significant effect on maize stover and grain yields during both seasons at Embu site. The MBILI treatment recorded the highest stover and grain yields. During 2012 LR at Kamujine site only the maize stover yield was significantly affected by maize-soybean intercropping patterns. The conventional treatment observed the highest stover yield than 2M:6S treatment; but it was statistically a par with the one recorded by the MBILI treatment. And during the 2012 SR, the maize stover and grain yields maize-soybean intercropping patterns affected significantly the MBILI treatment recorded the highest stover and grain yields.

During both seasons at both localities, the soybean yields were significantly affected by the maize-soybean intercropping patterns. During the 2012 LR, the yields were reduced by 60 and 81% due to the intercropping with maize, at Embu and Kamujine, respectively; whereas during 2012 SR, the yields were reduced by 52 and 78% as effect of intercropping with maize at Embu and Kamujine sites, respectively.The total LER values greater than a unit proved the greater yield advantage of intercropping maize-soybean over component sole crops, at both localities and during the two seasons.

## RECOMMENDATIONS

From the results of this study can be recommended, to the farmers of central high lands of Kenya, the use of MBILI-MBILI maize-soybean intercropping pattern because it gave efficient resources.

## ACKNOWLEDGEMENT

The Agricultural Green Revolution for Africa, through the Grants Office at Kenyatta University under the Direction and management of Dr. Maina Mwangi, is highly appreciated for the financial support that enabled the authors to carry out their coursework and the research activities.

### REFERENCES

Adaniyan ON, Akande SR, Balogun M. O, Saka JO (2007). Evaluation of Crop Yield of African Yam Bean, Maize and Kenaf under Intercropping Systems. American – Eurasian J. Agric.Environ. Sci. 2(1):99-102.

Alhassan GA, Kalu BA, Egbe OM (2012). Influence of Planting Densities on the Performance of Intercropped Bamabra Groundnut with Cowpea in Makurdi, Benue State, Nigeria. Int. J. Develop. Sustain. 1(3):1-20.

Alvarenga AD, Rezende PM, Andrade MJ, de B, Andrade LA, de B (1998). Corn-Soybean Intercropping. V. Intercropping System Effects and Corn Planting. Ciência Rural, Santa Maria, v. 28(2):199-204.

Amede T (1995). Yield Gain and Risk Minimization in Maize Through Cultivar Mixtures in Semi-Arid Zones of the Rift Valley in Ethiopia. Exp. Agric. 31:161– 168.

Bekunda MA, Bationo A, Ssali H (1997). Soil Fertility Management in Africa: A Review of Selected Research Trials, p. 63-79. *In* R.J. Buresh *et al.* (ed.) Replenishing soil fertility in Africa. SSSA Spec. Publ. 51. SSSA, Madison, WI.

Brintha I, Seran TH (2009). Effect of Paired Row Planting of Radish (Raphanussativus L.) Intercropped with Vegetable Amaranthus (Amaranthus tricolor L.) on Yield Componets of Radish in Sandy Regosol. J. Agric. Sci. 4:19-28.

Carr PM, Gardner JC, Schatz BJ, Zullnger SW, Guldan J (1995). Grain Yield and Biomass of a Wheat – Lentil Intercrop. Agron. J. 87:547–579.

Carruthers K, Prithiviraj B, Fe Q, Cloutier D, Martin RC, Smith DL (2000). Intercropping Corn with Soybean, Lupin and Forages: Yield Component Responses. Europ J. Agron. 12:103-115.

Dapaah HK.m Asafu-agyei, JN, Ennin SA, Yamoah C (2008). Yield Stability of Cassava, Maize, Soybean and Cowpea Intercrops. J. Agric. Sci. 140:73-82.

Egbe OM, Alibo SE, Nwueze I (2010). Evaluation of Some Extra-Early- and Early-Maturing Cowpea Varieties for Intercropping with Maize in Southern Guinea Savanna of Nigeria. Agric. Biol. J. N. Am. 1(5):845-858.

FAO (2011). The State of The World's Land and Water Resources for Food and Agriculture.Managing systems at risk. Food and Agriculture Organization of the United Nations.Rome, 2011.

Gachimbi LN, de Jager A, van Kuelen H, Thuranira EG, Nandawa SM (2002). Participatory Diagnosis of Soil Nutrient Depletion in Semi-Arid Areas of Kenya. Managing Africa's Soils P. 26, IIED.

Ghanbari A, Dahmardeh M, Siahsar B, A, Ramroudi M (2010). Effect of Maize (*Zea mays* L.) – Cowpea (*Vigna unguiculata* L.) Intercropping on Light Distribution, Soil Temperature and Soil Moisture in Arid Environment. J. Food, Agric. Environ. 8(1):102-108.

Gruhn P, Goletti F, Yudelman M (2000). Integrated Nutrient Management, Soil Fertility, and Sustainable Agriculture: Current Issues and Future Challenges. International Food Policy Research Institute, 2033 K Street, N.W. Washington, D.C. 20006 U.S.A. September, 2000.

Haseeb-ur, R., Ali, A., Muhammad, W., Tanveer, A., Tahir, M., Nadeem, M. A., and Zamir, M. S. (2010). Impact of Nitrogen Application on Growth and Yield of Maize (*Zea mays* L.) Grown Alone and in Combination with Cowpea (*Vigna unguiculata* L.). *American-Eurasian J. Agric. Environ. Sci.* 7(1):43-47.

Hiebsch, C., Tetio-Kegho, F., Chirembo, F. P., (1995). Plant Density and Soybean Maturity in a Soybean-Maize Intercrop. *Agron. J.* 87: 965 – 970.

Jaetzold R, Schmidt H, Hornetz B, Shisanya CA (2006). Farm Management Handbook of Kenya. Natural Conditions and Farm Information, 2nd edition, vol. 11/C. Ministry of Agriculture/GTZ, Nairobi, Kenya (Eastern Province).

Jama B, Palm CA, Buresh RJ, Niang A, Gachengo C, Nziguheba G, Amadalo B (2000). *Tithonia* Diversifolia as a Green Manure for Soil Fertility Improvement in Western Kenya: a Review. Agrofor. Syst 49:201–221. doi:10.1023/A:1006339025728.

Landers JN (2007). Tropical Crop–livestock Systems in Conservation Agriculture: The Brazilian Experience. *Integrated Crop Management.* P. 5–2007.FAO.Rome, 2007.

Lesoing GW, Francis CA (1999). Strip Intercropping of Grain Sorghum/Soybean in Irrigated and Rainfed Environments. J. Prod. Agric. 12(4):601-606.

Mpairwe DR, Sabiiti EN, Ummana NN, Tegegne A, Osuju P (2002). Effects of Intercropping Cereal Crops with Forage Legumes and Source of Nutrients on Cereal Grain Yield and Fodder Dry Matter Yields. Afri. Crop Sci. J. 10:81– 97.

Mucheru-Muna M, Pypers P, Mugendi D, Kung'u J, Mugwe J, Merckx R, Vanlauwe B (2010). Staggered Maize–Legume Intercrop Arrangement Robustly Increases Crop Yields and Economic Returns in the Highlands of Central Kenya. Field Crops Research 115:132–139. 2009 Elsevier B.V. All rights reserved. Journal homepage: www.elsevier.com/locate/fcr.

Mugendi DN (1997). Tree Biomass decomposition, Nitrogen Dynamics and Maize growth Under Agroforestry conditions in the sub-humid highlands of Kenya. PhD Thesis, University of Florida, Gainesville, Florida, U.S.A.

Mugwe J, Mugendi D, Kung'u J Mucheru-Muna, M. (2007). Effect of Plant Biomass, Manure and Inorganic Fertilizer on Maize Yield in the Central Highlands of Kenya. Afri. Crop Sci. J. 15(3):111-126 ISSN 1021-9730/2007.Printed in Uganda. All rights reserved 2007, African Crop Science Society.

Mugwe J, Mugendi D, Mucheru-Muna M, Merckx R, Chianu J, andVanlauwe B (2009). Determinants of the Decision to Adopt Integrated Soil Fertility Management Practices by Smallholder Farmers in the Central Highlands of Kenya. Expl Agric. 45:61–75. Cambridge University Press doi:10.1017/S0014479708007072 Printed in the United Kingdom.

Muneer AP, Fida HM, Mumtaz AK, Muhammad IS (2004). Performance of Maize in Intercropping System with Soybean under Different Patterns and Nitrogen Levels. J. App. Sci. 4(2):201-205.

Muoneke CO, Ogwuche MAO, Kalu BA (2007). Effect of Maize Planting Density on the Performance of Maize/Soybean Intercropping System in a Guinea Savannah Agro-ecosystem. Afri. J. Agric. Res. 2(12):667-677, December 2007.

Natarajan M, Willey RW (1986). The Effects of Water Stress on Yield Advantages of Intercropping Systems. Field Crops Res. 13:117-131.

Neupane RK (1983). Effects of Planting Arrangement on Maize and Soybean Intercropping. J. Inst. Agric. Anim. Sci. 4:1-5.

Olufajo OO (1992). Response of soybean to intercropping with maize in a subhumid tropical environment. Trop. oil seeds J. 1:27-33.

Olufemi *et al.* (2001)

Pal UR, Oseni TO, Norman JC. (1992). Effect of Component Densities on Productivity of Soybean/Maize Soybean/Sorghum Intercrop. J. Agron. Crop Sci. 170:66-70.

Peoples MB, Craswell ET (1992). Biological nitrogen fixation: Investments, Expectations and Actual Contributions to Agriculture. *Plant and Soil* 141:13-39. Kluwer Academic Publishers.Printed in the Netherlands.PLSO SA02. In: Biological Nitrogen Fixation for Sustainable Agriculture. Developments in Plant and Soil Sciences. P. 49.

Rajat De Singh SP (1979). Management Practices for Intercropping Systems. P: 17-21. *In*: Proceedings of the International Workshop on Intercropping, 10 -13 Jan 1979, Hyderadab, India. Edited by Willey, R. W. and Garver, C.

Raji JA. (2007). Intercropping Soybean and Maize in a Derived Savanna Ecology. Afri J. Biotechnol. 6(16):1885-1887. http://www.academicjournals.org/AJB. ISSN 1684–5315 ©2007 Academic Journals.

Rakasi K.(2011). Nutritional and Health Benefits of Soybeans and Micro – Enterprise Opportunities. Quinto congress de la Soja de

Marcosur: 14 al 16 de Setembre/Rosario, Argentina.

Saleem R, Ahmed ZI, Ashraf M, Arif M, Malik MA, Munir M, Khan MA (2011). Response of Maize-Legume Intercropping System to Different Fertility Sources under Rainfed Conditions. Sarhad J. Agric. 27(4):503-511.

Sanginga N, Woomer PL (2009). Integrated Soil Fertility Management in Africa: Principles, Practices and Development Process. (eds.). Tropical Soil Biology and Fertility Institute of the International Centre for Tropical Agriculture. Nairobi. P. 263.

Sikirou S, Wydra K (2008). Effect of Intercropping Cowpea with Maize or Cassava on Cowpea Bacterial Blight and Yield . J. Plant Dis. Protect.114 (4):145-151.

Silwana TT, Lucas EO (2002). The Effect of Planting Combinations and Weeding and Yield of Component Crops of Maize/Bean and Maize/Pumpkin Intercrops. J. Agric. Sci. 138:193-200.

Sivakumar MVK, Virmani SM (1980). Growth and Resource Use of Maize, Pigeonpea and Maize/Pigeonpea Intercrop in an Operational Research Watershed. Exp. Agric. 16:377-386.

Solanki NS, Singh D, Sumeriya HK (2011). Resources Utilization in Maize (*Zea mays*)-based Intercropping System under Rainfed Condition. Indian J. Agric. Sci. 81(6):511-515.

Thobatsi T (2009). Growth and Yield Responses of Maize (*Zea mays* L.) and Cowpea (*Vigna Unguiculata* L.) in an Intercropping System. Msc. Thesis.University of Pretoria. P. 159.

Undie UL, Uwah DF, Attoe EE (2012). Growth and Development of Late Season Maize/Soybean Intercropping in Response to Nitrogen and Crop Arrangement in the Forest Agro-ecology of South Southern Nigeria. Inter. J. Agric. Res.7(1):1-16.

Woomer PL, Langa, M, Tungani JO (2004). Innovative maize–legume intercropping results in above- and below-ground competitive advantages for understory legumes. West Afri. J. Appl. Ecol. 6:85–94.

# A review on induced mutagenesis in soybean

**Mudasir Hafiz Khan and Sunil Dutt Tyagi**

Division of Plant Breeding and Genetics, Kisan (P.G) College, Simbhaoli, Ghaziabad (U.P), India.

**The history of mutations with early results in soybean has been well documented in the literature available. Mutation breeding has been used in recent years as a valuable supplement to other methods of plant breeding in generating new variability and development of crop varieties with new architecture, superior biochemical constitution and suitable growth and developmental rhythms. The utility of this method is evident from the fact that in several crops induced mutants have been released as new varieties. In this paper, efforts have been made to review the literature on induced mutations in soybean. The different mutagenic agents used for inducing mutations, effects of different mutagenic agents on yield, quality contributing characters and resistance to different diseases have been described.**

**Key words:** *Glycine max* (L.) Merrill, induced mutagenesis, crop varieties.

## INTRODUCTION

Soybean [*Glycine max* (L.) Merrill] has become the miracle crop of the 21st century. It is a triple beneficiary crop, which contains about 40% proteins, possessing high level of essential amino acids except methionine and cystine, 20% oil rich in poly unsaturated fatty acids specially omega-6 and omega-3 fatty acids, 6 to 7% total minerals, 5 to 6% crude fibre and 17 to 19% carbohydrates (Chauhan et al., 1988). Besides, it has good amount of iron, vitamin B-complex and isoflavones such as daidzein, genistein of glycitin. The presence of calcium and iron makes it highly suitable for women who suffer from osteoporosis and anemia. The isoflavones of soybean have been found to possess health benefits, as they exhibited properties like cancer preventing, combating menopausal problem and helping to recover from diabetics (Chauhan et al., 2002). In the present paper, efforts have been made to review up-to-date literature on induced mutations in soybean.

## EFFECT OF MUTAGENIC AGENTS

### Meiosis

Ahmad et al. (1977) reported paracentric inversions in the hybrids between the two species of soybean as the dose was increased. Qing et al. (1997) reported that after irradiation of soybean seeds for 3 days with 500 rad gamma rays, the number of mitochondria per cell decreased, while the number of vacuoles increased and cell structure changed dramatically with some organelles having disintegrated. Treatment with 5000 rad gamma rays caused significant cell damage and inhibited cell growth. Ping et al. (1998) studied cytomorphology of a male sterile mutant NJ89-1 in soybean and reported that the observations on anther and pollen development showed that NJ89-1 differed from msl-ms6 mutant in many aspects such as abortion stage, meiosis, tetrad formation, pollen wall and anther wall, etc. NJ89-1 displayed similar meiotic abnormalities of asynapsis or desynapsis to those of st2-st5 mutants by differed from st2-st5 mutants in female fertility with females of st2-st5 mutants being strongly impaired.

Bione et al. (2002) observed many univalent a few bivalents in diakinesis of a mutant line BR97-13774H. Telophase II exhibited a varied number of different sized nuclei; pollen sterility was estimated at 93.12%. Bione et al. (2002b) reported that many univalent, few or total absence of bivalents were found in diakinesis of

BR97-12986H, soybean mutant. Bivalents presented in or two terminal chiasmata, while univalents retained the sister chromated cohesion. Bivalents and most univalents congregated at the equatorial plats, although univalents frequently migrated to the poles prematurely. Laggards resulting from delay in chiasmata terminalization were also recorded. Pollen sterility was estimated at 91.2% segregation ratio for sterility in this line and its progenies reached 3:1.

### Seed germination

Hassan et al. (1985) irradiated seeds of Bragg, Hodgson and Lee-74 containing 1113% moisture content with 100 to 500 Gy gamma rays and 5 to 30 Gy fast neutrons and reported that growth inhibition increased with increasing doses and germination was inhibited only at the higher doses. Lee-74 was the most sensitive variety to gamma radiation and Bragg the most sensitive to fast neutron doses above 20 Gy, as revealed by differences in epicotyl length. Bhatnagar et al. (1989) treated seeds of soybean cv. Bragg with EMS (ethyl methanesulfonate) and gamma rays, with or without additional exposure to UV light for2 h at 260 rim and reported that mutation frequency in the M2 ranged from 2.24 to 22.85%. Among the mutants obtained T-214 was from the 25 kR gamma radiation + UV treatment. It exceeded the parent in germinability by 15% and was 5 days earlier in maturity.

### Seedling survival

$M_2$ seedling survival decreased (in comparison with untreated control) following radiation and chemical treatment. Zakri and Jalani (1986) treated two cultivars (Palmetto and Acadian) of soybean with ethyl methanesulfonate and gamma rays and reported that the cultivar Palmetto showed higher survival percentages following either treatment. Li (1988) reported that when dry seeds of several soybean varieties were treated with electrons at various doses, marked effects were observed in the M1, on seedling height and survival percentage. Wang et al. (1989) treated seeds of the soybean line LF81-837 with 0.001, 0.002 and 0.005 M $NaNO_3$ at pH 3 and reported that plant survival was reduced by 49.9% in the M2, generation when seeds were treated with 0.005 M sodium azide solution. Li et al. (1994) irradiated dried seeds of soybean cv. Changnong 5 with gamma rays (120 to 200 Gy, 2.28 Gy/min) or an electron beam (180 to 450 Gy) and reported that dressing treated seeds with benzamide resulted in a lower survival rate and fertility, but a higher chromosomal aberration rate compared with those treated with gamma rays or the electron beam alone.

Satpute and Fultambkar (2012) treated the dormant seeds of soybean cultivars MAUS-71 and JS-335 with varied concentration of chemical mutagen (EMS) and physical mutagen (Gamma rays). They reported a dose dependant decrease in most of the characters in M1 generation. They further reported that the reduction in germination percent over control was noticed in all mutagenic treatments in both the cultivars, while increased pollen sterility was associated with corresponding increases in dose/ concentration of mutagens

### Chlorophyll mutations

Constantin et al. (1974) reported that fast neutrons and ethyl methane sulphonate (EMS) were the most effective inducers of chlorophyll deficiencies and morphological mutants. Fujii and Tano (1987) reported that the somatic M, mutations induced by EMS (ethyl methanesulfonate) and segregation of chlorophyll deficient mutants in the $M_2$ were studied in the strain T-219. The results indicated that in the $M_2$ generation, chlorophyll mutation frequencies were estimated as 0.14, 0.61 and 0.41% per µg EMS at treatments with 0.5, 0.1 and 0.2% EMS, respectively. Harb (1990) treated the two cultivars of soybean with gamma rays and reported that there was large reduction in the chlorophyll and carotenoid contents of mutants.

### Leaf mutants

Badaya and Mehrotra (1974) reported that treatment of presoaked Clark-63 seed with ethyl methanesulphonate, ethyleneimine, and gamma rays resulted in a mutation spectrum for leaf shape and size, leaflet number, and testa colour. Kiang and Halloran (1975) reported that the frequency of mutations induced by ethyl methane-sulphonate in the $M_1$, and $M_2$ generations for leaflet number were in the range 1 to 5% over all characters. Fujii et al. (1983).

Reported that when soybean seeds were treated with caffeine solutions, the number of mutant spots per leaf on the resulting plants ranged from 3.7 at 0.01% to 24.8 at 0.05% caffeine. However, when the seeds were treated with 12-O-tetradecanoylphorbol-l3-acetate (TPA) plus caffeine the number of leaf spots decreased, significantly so in the case of 0.03% caffeine. TPA alone at concentrations of 1 to 20 µg/ml did not induce any mutations. Fu (1986) treated soybean varieties (("SF" 7919-61 and "SF" 7910-3) with 15 krad of 10Co gamma rays and reported two special mutation types that is two opposite trifoliate leaves per node ("SF"7919-61) and 4-7 leaflets instead of 3 per leaf ("SF"7910-3)). He further reported that these lines showed a higher growth rate, increase in leaf area, stronger stem and lodging resistance than tiefeng No. 18.

Culbertson et al. (1991) reported that seeds of soybean genotype T31 were treated with 2.5 mM nitrosomethyl urea [N-methyl-N-nitrosourea] for 3 h. In the $M_4$ generation, 32 plants exhibited puberulence and smooth seed coats. Li et al. (1995) reported that no changes were found in photosynthesis or chlorophyll content with 0, 25 or 50% (v/v) methanol during 1, 2 or 3 weekly applications. Xue-Bai et al. (2000) while studying the mutagenic effect of 60 Co gamma irradiation on soybean observed that the M2 seedling were with one or three primary leaves and joined or wrinkled cotyledons.

## Altered stem structure

### Plant height

Nicolae and Nicolae (1977) treated the seeds of the soybean line B-89/J I with gamma rays and thermal neutrons and observed that most of the mutant lines in M2-M4 were taller than the control line of B-89/11. Fahmy et al. (1997) reported that the increasing doses of gamma-rays were negatively associated with plant height. Kundi et al. (1997) reported that after irradiation of three varieties of soybean viz PK 416, SL96 and PB Soybean No.1 with three doses of gamma rays viz 10, 20 and 30 kR, there was marginal increase in mean values for plant height.

### Growth type

Wang and Yu (1988) reported that three genotypes were exposed to 8 doses of gamma radiation at 2 developmental stages. Following radiation at the VE-V1 stage, growth of the $M_1$, generation was reduced more when the same dose was administered at a lower rate. Reductions in growth rate and number of morphological abnormalities were greater at doses of 50 Gy and above. Khvostova (1967) reported gamma rays induced short stemmed and lodging mutants. Mutations for internodal length in XZ generation were recorded by Humphrey (1951).

### Stem thickness

Mutations for stem size were recorded in $M_2$ generations of X-ray irradiated soybean varieties (Humphery, 1951).

### Branching

More densely branched mutants were recorded by Zacharias (1956) in X-ray irradiated material.

### Lodging

Weber and Fehr (1967) studied maturity period in both irradiated and segregating populations and donot observed any difference for lodging.

## Alteration of ripening time

### Maturity

Szyrmer and Boros (1986) studied the maturation period of soybean lines and reported that the maturation period of the mutant was not significantly shorter than in the control varieties. Bhatnagar et al. (1989) treated seeds of soybean cv. Bragg with ethyl methane-sulfonate (EMS) and reported that the mutants were 5 days earlier in maturity. Tulmann and Peixoto (1990) observed some very early mutants with the same level of productivity as in the parent cultivar (Parana) when seeds were treated with 22 kR gamma rays. Mehta et al. (1994) treated a local cultivar of soybean (Kalitur) with gamma rays and EMS. They reported that two mutants namely 7 and 13 were isolated in $M_2$ and the maturity period of these mutants were superior as compared to the control treatment of Kalitur. Khudhair et al. (2002) irradiated the seeds of soybean with 0, 100, 200, and 300 kR of gamma rays and selected two mutants which were superior in their yield components and earliness in maturity compared with the variant H226 and other control cultivars.

### Number of pods per plant

Zakri and Jalani (1986) treated two cultivars (Palmetto and Acadian) of soybean with ethyl methane sulfonate and gamma rays and reported that the mutant P630-2 had 86 pods/plant compared with 22 for the control.

### Pod setting

Birnberg et al. (1987) reported that when leaves of soybean cultivars Evans and Lincoln were treated with gibberellic acid ($GA_3$) about 3 days before anthesis, the fraction of flowers on the associated node that set pods was reduced by 28 to 76%.

### Yield

Skorupska (1984) treated seeds of 14 soybean varieties with 10 kR of gamma rays administered alone or in conjunction with $NaNO_3$ solution, and reported that morphological abnormalities were frequent in all the treatments in $M_2$ while, in $M_3$, the traits showed the greatest variation for seed yield/plant. Krausse (1989) developed 6 mutant lines, out of which Dorado was the best mutant line having higher yields (1320 to 1560 kg/ha as compared to control line 1205 kg/ha).

Bhatnagar et al. (1990) reported that seeds of black-seeded soybean cv. Bhat were treated with gamma rays (15 to 25 kR), with or without additional exposure to UV radiation (2 h at 260 nm). The results indicated that among the useful mutants identified, T154 from the 20 kR + UV treatment surpassed the parent and local standards for yield. Bhatnagar et al. (1992) irradiated seeds of the soybean cultivars Punjab-1, Gaurav and NRC-I with 15 and 20 kR gamma-rays and observed that in the $M_3$ generation, Gaurav showed higher and more significant variability for seed yield and 100-seed weight than Punjab I and NRC-1. Pavadai and Dhanavel (2005) reported that 100-seed weight and yield per plant decreased with increasing dose of gamma rays.

### *Seeds/plant*

Increase in average seeds/plant in $x_2$ was reported by Sebok et al. (1963). According to Georgiev and Topcieva (1970), the average number of seeds/plant failed to increase with increase in dose but in the variety Adams, dose 4 and 8 kR increased the seeds/plant over the un-irradiated control. Rajput (1987) reported depressive effect of gamma rays on the mean value for seeds/plant which could have been due to poly-genornic mutations. These observations were based on $M_2$ of 10, 15, 20 and 25 kR doses kR gamma treated varieties viz., Loppa, T-I S and Columbus. Prakash et al. (1984) observed changes in seed coat colour from black to yellow, dull brown or brown whey they treated Birsa soybean-1, a spontaneous mutant of sepaya black with gamma rays. Also a yellow seeded M2 plant gave 16 promising M3 plants in the same experiment.

### *Plant vigour*

Seven plants showing marked increase in vigour in $X_2$ generation over normal plants were observed by Humphrey (1951). Further in subsequent generation, Humphrey (1954) confirmed the same.

## Alteration of seed storage substances

### *Protein content and oil content*

Dahiya (1973) reported that with radiation treatment (gamma rays) the quality of proteins changed. Some of the M2 progenies showed a greater range of variation in the content of methionine and tryptophan. Hiraiwa et al. (1975) reported that treatment of the varieties Mutsumejiro, Raiden and Miyagishirome with 8 and 16 kR gamma rays, the mean protein contents of the selected $M_4$ populations of first two varieties, treated with8 and 16 kR gamma rays, and of the last variety, treated with 8 kR gamma rays, were significantly greater than those of the untreated controls. In the $M_5$ generation, mean protein contents of selected populations of Mutsumejiro and Raiden were significantly higher than those of the controls. Qiu and Gao (1988) reported higher frequencies of mutants with high protein and oil contents using EMS as compared with fast neutrons and these were more highly heritable in $M_2$ and $M_3$ lines. Nilegaonkar and Agte (1989) reported that when Kalitur, a black seeded soybean cultivar was exposed to physical and chemical mutagens and subsequently studied for changes in electrophoretic and solubility behaviour of proteins, trypsin inhibitor activity and proximate analysis. Mutation induced changes in protein structure and lowered fat percentage. The mutant cv. MACS 107 had a trypsin inhibitor activity significantly lower, by 15%, than Kalitur.

Wang et al. (1989) treated seeds of the soybean line LF81-837 with 0.2 or 0.4% solutions of ethyl methane-sulfonate (EMS) at pH 7 and reported that treatment of the seeds with 0.4% EMS followed by selection in the early generations is recommended for improving protein content. Eskins et al. (1991) grow normal green (Clark LI) and mutant yellow (Clark y9y9) mutants in full-spectrum solar irradiation and reported that response of the mutant to light quality indicated that blue light slightly enhanced expression of the mutation at higher irradiances. Rubisco (ribulose-bisphosphate carboxylase) proteins and rubisco activity (leaf area basis) were directly related to irradiance level but were enhanced in blue light over equal irradiance red light. This enhancement was not shown in the presence of far-red light. Qing et al. (1996) reported that gamma rays increased the plant peroxides activity at higher doses. Protein content also increased with increasing doses of irradiation.

## Resistance to disease

Tsai et al. (1974) reported that seed of six varieties of soybean were treated with 1530 kR gamma rays or 1% ethyl methane-sulphonate solution, and about 5000 $M_3$ to $M_6$ lines, derived from apparently resistant plants, tested for *Phakopsora pachyrhizi* resistance under natural conditions, followed by repeated selection, five lines showed appreciable resistance, two of which, CHI-41 and CH3-77, were finally selected for production. Smutkupt et al. (1974) reported that seeds of Sansai and SJ2 were treated with 5 to 30 kR gamma rays and seven lines originating from the $M_2$ of both varieties were selected in the $M_4$ on the basis of lodging resistance. Nicolae (1979) reported that Cerag I was selected in Algeria from a collection of induced mutants from Romania. Compared with its parent B107/I0 (T) is more resistant to cold and drought. Oh (1983) reported that from 430 $M_3$ soybean lines, five were selected as highly resistant and 20 as moderately resistant. Kwon and Oh (1983) reported that seeds of the soybean varieties Kwangkyo and Kangrim were irradiated with 15 and 20 kR gamma rays. In the M3, 18 mutants from Kwangkyo were selected which showed moderately resistance against soybean mosaic virus. Wang et al. (1986) reported that a cultivar Heinong 26 derived from Dongnong- 4 following gamma irradiation was tolerant to drought and of low temperatures during the seedling stage.

Tulmann et al. (1988) treated three soybean cultivars with gamma irradiation or treated with ethyl methanesulfonate in order to induce resistance to tobacco ring spot *nepovirus* (causal agent of bud blight) and *Phakopsora pachyrhizi* and reported that resistant mutants were selected in the $M_2$ and $M_3$ generation. Oh et al. (1988) reported that seeds of soybean were gamma-irradiated with 15 and 25 kR and mutants were screened for resistance to soybean mosaic *potyvirus*. Five lines were subsequently selected which showed highly resistance.

Smutkupt et al. (1988) treated seeds of I1 cultivars with gamma-irradiation at 15 and 30 kR and reported that $M_3$ bulk and single populations and $M_2$ bulk populations were screened for resistance to *Phakopsora pachyrhizi* at 2 locations. After further selection, 16 lines were selected as *P. pachyrhizi* tolerant mutants. Nazim et al. (1988) reported that seeds of the cultivars Calland, Columbus and Williams were gamma-irradiated with 20 kR. $M_2$ plants were screened against *Drechslera* [Cochliobolus] *australiensis* and *Alternaria alternata*. Their reactions suggested that the induced resistances were polygenically inherited. In the $M_3$, 30 and 25 mutants were true breeding for resistance to C. *australiensis* and *A. alternata,* respectively.

## Resistance to chemicals

Hendratno (1988) reported seeds of cv. Orba treated with fast neutrons, gamma-rays, ethyl methanesulfonate and sodium azide and observed one mutant which showed high tolerance of Al toxicity and exceeding Orba in yield by 17%. Sebastian (1989) reported that seed mutagenesis (using N-nitroso-N-methylurea [N-methyl-N-nitrosourea] and ethyl methanesulfonate) followed by selection for resistance to chlorsulfuron yielded a soybean mutant with a high degree of resistance to both post- and pre-emergence applications of a variety of sulfonylurea (SU) herbicides.

## Inheritance of quantitative characters

### *Heritability*

Singh et al. (1980) reported that when ten varieties were irradiated with 10 to 20 kR gamma rays, the estimates of heritability (broad sense) ranged from 0 to 51% for days to flowering, 0 to 58% for days to maturity, 0 to 80% for plant height, 0 to 49% for primary branches, 0 to 81% for pods per plant, 0 to 92% for seed per pod and 0 to 80% for yield per plant. Geetha and Vaidynathan (1998) reported an increase was noticed in heritability and genetic advance for some economic traits like seed yield per plant and 100-seed weight in M2 generations. Kumar and Lal (2001) reported that the phenotypic and genetic coefficient of variation and the estimates of heritability in the broad sense significantly increased in the mutagenic populations. Pavadai et al. (2010) reported that variability, heritability and genetic advance as per cent of mean was recorded high for mutagen treated plants than the untreated plants for all the generation. 50 KR of gamma rays treatment was effective than the other mutagenic treatments compared to control.

### *Genetic divergence*

Mehetre et al. (1996) studied the gamma induced genetic divergence in $M_2$ generation of soybean and reported that the genetic diversity was independent of varieties and doses of gamma rays. Mehetre et al. (1996) opined that the genetic diversity was independent of varieties and doses of gamma rays. They further observed the sufficient amount of variability due to induced mutations for different polygenic characters over the parent variety in $M_2$ families.

## ACKNOWLEDGMENT

The financial support of this research from the Kisan P.G. College, Simbhaoli, Ghaziabad, U.P, India and Ch. Charan Singh, University, Meerut, U.P, India is appreciated.

#### REFERENCES

Ahmad QN, Britten EJ, Byth DE (1977). Inversion bridges and meiotic behaviour in species hybrids of soybean. J. Hered. 68:360-364.

Badaya SN, Mehrotra HN (1974). Morphological variations induced by physical and chemical mutagens in soybean *(Glycine max L.)*. Symposium on use of radiations and radioisotopes in studies of plant productivity. Agron. Abstr. p. 31.

Bhatnagar PS, Prabhakar Tiwari SP, Sandhu JS (1989). Improvement of soybean variety 'Bragg' through mutagenesis. Mutat. Breed Newsletter 33:15-16.

Bhatnagar PS, Tiwari S, Prabhakar P (1990). Application of mutagenesis for the improvement of an indigenous black seeded soybean variety of India. Mutat. Breed. Newsletter 36:8.

Bhatnagar PS, Tiwari SP, Singh C (1992). Differential dose-response of soybean genotypes to mutagenesis with gamma rays. Biovigyanam 18(2):111-113.

Bione NCP, Pagliarini MS, Almeida LA.-de (2002). An asynaptic mutation in soyabean *[Glycine max. (L.)* Merrill] associated with total absence of sister chromatid cohesiveness. *Cytologia.* 67(2):177-183.

Bione NCP, Pagliarini MS, Almeida LA.de (2002b). A new and

distinctive malesterile, female-fertile desynaptic mutant in soybean *(Glycine max.).* Heredita 136(2): 97-103.
Birnberg PR, Cordero RF, Brenner ML (1987). Characterization of vegetative growth of dwarf soybean genotypes including a gibberellin-insensitive genotype with impaired cell division. Am J. Bot. 74(6):868-876.
Chauhan GS, Verma NS, Bains GS (1988). Effect of extrusion processing on the nutritional quality of protein in rice-legume blends. Nahrung 32(1):43.
Chauhan OP, Chauhan GS, Singh G, Kumbhar BK, Mishra, DP (2002). Varietal variability in the contents of nutrients and anti-nutrients in different parts of soybean seeds. J. Rural Agric Res. 2(2):42-50.
Constantin MJ, Klobe WD, Skold LN (1974). Mutation induction in soybean. Mutation Breeding Newsletter, 3: 9-10.
Culbertson RDR, Harper JE, Hymowitz T (1991). Chemical mutagenesis of T31 soybean. Soybean Genetics Committee Report. 18:229-233.
Dahiya BS (1973). Improvement of Mungbean through induced mutations. Indian J. Genet. Plant Breed. 33(3):460-468.
Eskins K, Jiang CZ, Shibles R (1991). Light-quality and irradiance effects on pigments, light-harvesting proteins and Rubisco activity in a chlorophyll- and lightharvesting-deficient soybean mutant. Physiol. Plantarum. 83:(1):47-53.
Fahmy EM, Rashed MA, Sharabash MTM, Hammad AHA, El-Demerdash HM (1997). Effect of gamma rays on yield and its components for some soybean cultivars *(Glycine max.* L.). Arab Univ. J. Agric. Sci. 5(1):57-68.
Fu LQ (1986). Study of mutants with opposite trifoliate leaves and multi-leaflet leaves in soybean. Mutat. Breed. Newsletter 27:11.
Fujii T, Shizaki M, Fujiki H, Sugimura T (1983). Effect of TPA on the mutagenicity of caffeine in the soybean mutation test. Mutat. Res.110(2):263-269
Fujii T, Tano S (1987). Mutagenic activities of EMS on somatic and recessive mutations in the soybean test system: Annual Rep. Natl. Inst. Genet. Japan. 36:66-68.
Geetha K, Vaidyanathan P (1998). Studies on induction of mutations in soybean *(Glycine max.* L. Merill) through physical and chemical mutagens. Agric. Sci. Digest. Karnal. 18(1):27-30
Georgiev Z, Topcieva A (1970). *Genetica i Selekcija,* 3: 403-409.
Harb RKH (1990). Frequency, spectrum, anatomical and chemical studies on soybean chlorophyll mutations. Bull. Faculty Agric. Univ. Cario. 41(3): 595-608
Hassan S, Mohammad T, Khan S, Brunner H (1985). The effect of gamma and fast neutron irradiations on M, seedling growth in soybean. Nucleus Pakistan. 22(1-2):19-22.
Hendratno K 1988. Induced mutations in soybean for earliness and tolerance to high concentrations of aluminium in high acid *soil. Improvement of grain legume production using* induced mutations. Proceedings of a workshop, Pullman, Washington, USA, 1-5 July, p. 463.
Hiraiwa S, Nkamura S, Tanaka S (1975). Induction of mutants with high protein content in soybean. Mutat. Breed. Newsletter 6: 8.
Humphrey LM (1951). Effects of neutron irradiation on soybeans. Soybean Digest. 12:11-12.
Humphrey LM (1954). Induced mutations in soybean. Soybean Digest. 14:18-19.
Khudhair HA, Jasim AM, Nassralla AY, Suhail AK, Abass HA, Hassan AX, Hussain AJ, Mahmood JN (2002). Induction new suitable soybean genotypes for Iraqi environmental conditions. Dirasat. Agric. Sci. 29(1):65-73.
Khvostova VV (1967). Erwin baur. Ged. Vorl. 4:21-27.
Kiang LC, Halloran GM (1975). Chemical mutagenesis in soybean *(Glycine max (L.)* Merill) using ethyl methane-sulphonate and hydroxylamine hydrochloride. Mutat. Res. 2(3):373-382.
Krausse GW (1989). Early ripening, productive soybean mutant variety suitable for combine harvesting. Mutat. Breed. Newsletter 34:3-4.
Kumar S, Lal JP (2001). Response of selection for grain yield in mutagenic population in lentil *(Lens culinaris* Medik). Progress. Agric. 1(1):59-62.
Kundi RS, Gill MS, Singh, TP, Phaul PS (1997). Radiation induced variability for quantitative traits in soybean *(Glycine max.* (L) Merrill). Crop Improv. 24(2):*23*-1234.
Kwon SH, Oh JH (1983). Induced mutation for soybean mosaic virus disease resistance in soybean. Induc. Mutat. Dis. Resist. crop Plants-II. 5:183-191.
Li GQ (1988). Study of the effect of electron beam on soybean radiation mutation. Acta Agric. Univ. Jilinesis. 10(3):57-62.
Li Y, Gupta G, Joshi, JM, Siyumbano AK (1995). Effect of methanol on soybean photosynthesis and chlorophyll. J. Plant Nutr. 18(9):1875-1880.
Li-GuoQuan,Yu-ShaoHua Li-Meng, Piao-TieFu, Chen-Guang, Yuan-YaPing, Bao-HePing, Xu-YaoKui, Li GQ, Yu SH, Li M, Piao TF, Chen G., Yuan YP, Bao HP, Xu YK (1994). Effects of benzamide on radiative mutation in soybean. Acta Agric. Nucleatae.Sinica. 8(3):141-148.
Mehetre SS, Mahajan CR, Hajare DN, Desai NS (1996). Assessment of gamma induced genetic divergence in M3 generation of soybean. Adv. Plant Sci. 9(1):39-42.
Mehetre SS, Mahajan CR, Shinde RB, Ghatge RD (1996). Assessment of Gamma induced genetic divergence in $_{M2}$ generation of Soybean. Indian J. Genet. 56(2):186-190.
Mehta AK, Mehta SK, Tiwari AS (1994). Improvement of genetic architecture of a local cultivar Kalitur of soybean by mutation. Indian J. Genet. Plant Breed. 54(4): 357-359:
Nazim M, Hosary AA. Rady MS (1988). Induced mutations for disease resistance to leaf spot fungi in soybean. Improvement of grain legume production using induced mutations. Proceedings of a workshop, Pullman, Washington, USA. 9:383-397.
Nicolae I (1979). A soybean mutant variety in Algeria. Mutat. Breed. Newsletter. 14:5.
Nicolae I, Nicolae F (1977). Inducing favourable mutations by irradiating soybean. Annales de P Institute. National Agronomigua (EL-Harrach), 7(3):51-57.
Nilegaonkar S, Agte V (1989). Induced variation in chemical composition of black seeded soybean variety - Kalitur. J. Food Sci. Technol. 26(5):256-258.
Oh JH (1983). Induced mutations for soybean mosaic virus resistance. Mutat. Breed. Newsletter 21:6.
Oh JH, Kwon SH, Song HS, Kim JR (1988). Induced mutations for varietal improvement in soybean. Improvement of grain legume production using induced mutations. Proceedings of a workshop, Pullman, Washington, USA. 7:355-370.
Pavadai P, Dhanavel D (2005). Effect of gamma rays on yield and its components in soybean *[Glycine max (L.)* Merrill. Var. Co-1]. Crop Res. 30(3):459-461.
Pavadai P, Girija M, Dhanavel D (2010). Effect of Gamma Rays on some Yield Parameters and Protein Content of Soybean in M2, M3 and M4 Generation. J. Exper. Sci. 1(6):08-11.
Ping YS, Yi GJ, Qing XH (1998). A genetic cytomorphological study on the male sterile mutant NJ 89-1 in soybean. Soybean.Sci. 17(1):32-38.
Prakash R, Trivedi HBP, Kerketta V, Haque-MF, Prakash R (1984). Mutation breeding research in soybean in India. Soybean Genet. Newsletter 11:43-44.
Qing-Yu-Zhang (1997). Studies on and application of the variations in characters in soyabean progeny produced by irradiation with 60Co gamma rays. J. Henan Agric. Sci. 1:68.
Qing-Y-Z, Zhang FH, Fan-Zhan B, Huang-XL (1996). Studies on the radiation effect of 60Co gamma -rays on the soyabean plant. J. Henan Agric. Sci. 6:15-17.
Qiu GJ, Gao S (1988). Studies on physically and chemically induced soybean mutations of high protein and oil content and their genetic pattern. Improvement of grain legume production using induced mutations. Proceedings of a workshop, Pullman, Washington, USA, 1-5 July, pp. 421-433.
Rajput MA (1987). Induction of polygenic variance in soybean. Soybean Newsletter, 14: 117-119.
Satpute RA, Fultambkar RV (2012). Effect of mutagenesis on germination, survival and pollen sterility in M1 generation of soybean [*Glycine max* (L.) Merill]. Int. J. Recent Trends in Sci. Technol. 2(3):30-32.
Sebastian SA, Fader GM, Ulrich JF, Forney DR, Chaleff RS (1989). Semi-dominant soybean mutation for resistance to sulfonylurea herbicides. Crop Sci, 29(6):1403-1408.

Singh BB, Upadhyayam HD, Corbin FT (1980). Induced genetic variability for qualitative and quantitative characters in soybean. *World* Soybean Research Conference II, Abstracts pp. 42-43.
Skorupska H (1984). Identification and evaluation for mutation of agricultural characters in soybean. Soybean Genet. Newsletter 11:53-59.
Smutkupt S, Phrek G, Ramanujan S, Iyer. RD (1974). Improvement of soybean protein by mutation breeding in Chiang Mai, Thailand. Proceedings of the Second General Congress of the Society for the Advancement of Breeding Researches in Asia and Oceania. Session X. Breeding for protein nutritive quality. 7:662-668.
Smutkupt S, Wongpiyasatid A,Lamseejan S (1988). Improvement of rust resistance in soybean by mutation breeding. Improvement of grain legume production using induced mutations. Proceedings of a workshop, Pullman, Washington, USA, 7:371381
Szyrmer J, Boros L (1986). Performance of early maturing soybean lines obtained from a mutation breeding programme. Improvement of grain legume production using induced mutations. Proceedings of a workshop, Pullman, Washington, USA, 1-5 July, pp. 411-419.
Tsai KH, Lu YC, Oka HL, Oka-H (1974). Mutation breeding of soybean for the resistance to rust disease. SABRAOJ.6(2):181-191.
Tulmann, NA, Menten JOM, Ando A, Alberini J, Peixoto, TC (1988). Induced mutations for disease resistance and other agronomic characteristics in bean *(Phaseolus vulgaris* L.) and soybean *(Glycine max (L.)* Merrill). Improvement of grain legume production using induced mutations. Proceedings of a workshop, Pullman, Washington, USA, 26:167-187.
Tulmann NA, Peixoto TC. (1990). Early maturing and good yield mutants in soybean (*Glycine max*. (L.) Merr.) in Brazil. Mutat. Breed. Newsletter 36:9.
Wang BR, Wang LZ, Weng XY, Chen Y, Wu HL, Xu XC, Wang PY (1986). Soybean variety Heinong No. *26.* Mutat. Breed. Newsletter 27:10-11.
Wang PY, Yu BS (1988). Preliminary study on gamma-rays chronic radiation for growing plants in soybean. Soybean Genet. Newsletter. 18:82-85.
Wang PY, Wang LZ, Zhang JZ (1989). Induced protein content mutation in soyabean. Soybean Genet. Newsletter 16:38-40.
Weber CR, Fehr WR (1967).Effect of hybridization and thermal neutron irradiation on quantitative characters of soybeans. Crop Sci. 7:78-81.
Xue-Bai, Meng-Li Fen, Zhao-XiaoNan, Guo-YuHong, Liu-BinHong (2000). Mutagenic effect of 60Co gamma irradiation on soybean plants. Soybean Sci. 19(2):150-153.
Zacharias M (1956). Mutationsversuche an Kulturpflanzen VI Rontgenbestrahlungen der Sojabohne (Glycine soja (L. ) Sieb. et Zucc ) Zuchter. 26:321-338.
Zakri AH, Jalani BS (1986). Improvement of soybean through mutation breeding. Improvement of grain legume production using induced mutations Proceedings of a workshop, Pullman, Washington, USA, 1-5 July, pp. 451-461.

# The importance of cereals (Poaceae: Gramineae) nutrition in human health: A review

**Muhammad Haroon Sarwar[1], Muhammad Farhan Sarwar[2], Muhammad Sarwar[3], Niaz Ahmad Qadri[4] and Safia Moghal[4]**

[1]Faculty of Medicine and Allied Medical Sciences, Isra University, Hyderabad, Sindh, Pakistan.
[2]Government College Samanabad, Faisalabad, Punjab, Pakistan.
[3]Pakistan Atomic Energy Commission, Nuclear Institute of Agriculture, Tandojam-70060, Sindh, Pakistan.
[4]Hameeda Medicare, Ghulam Muhammadabad, Faisalabad, Punjab, Pakistan.

**This article presents some of the latest research endeavors that aim to improve our understanding of how the various grain components can be manipulated to improve contributions of cereals to human health. The health benefits of wholegrain cereal products are now widely recognized and considered due to the presence of a wide range of bio-active components. Cereals in their natural form (as whole grain) are rich source of vitamins, minerals, carbohydrates, fats, oils, and protein. Also, such cereals are chosen for consumption and they have the higher fiber which is an important nutrient that helps to prevent weight gain and heart disease. For best results, fruits, vegetables and whole grains may be eaten every day to provide extra nutrition and fiber. The meal can be made healthier by eating high-fiber whole-grain cereals that has low sugar. However, eating a variety of cereals rather than just feeding on one item is more helpful to health.**

**Key words:** Cereal, grain, human, health, diet, Pakistan.

## INTRODUCTION

The term cereal is a derivative from Latin word 'cerealis' meaning 'grain' which is botanically, a type of fruit called a caryopsis, composed of the endosperm, germ, and bran. The cereals are annual common grass members of the grass family (a monocot family Poaceae, also known as Gramineae), which usually have long, thin stalks, such as wheat, rice, maize, sorghum, millet, barley and rye, whose starchy grains are used as food. The term cereal is not limited to these grains, but, also refers to foodstuff prepared from the starchy grains of cereal like flours, breads and pasta. Cereal science is a study concerned with all technical aspects of cereal. It is to study the nature of the cereals and the changes that occur naturally, and as a result of handling and processing.

Amazingly, the foods human beings eat most are grasses all around the world. Cereal grains were the first agricultural attempts by early man, and people still enjoy them today depending on where they live and what grows there well. Cereal grains are grown in greater quantities and provide more food energy worldwide than any other type of crops; they are therefore staple food crops. In their natural form (as in whole grain), they are a rich source of vitamins, minerals, carbohydrates, fats, oils and protein. However, when refined by the removal of the bran and germ, the remaining endosperm is mostly carbohydrate and lacks the majority of the other nutrients. As human food, cereals are usually marketed in raw grain form or as ingredients of food products. As animal feed, they are consumed mainly by livestock and poultry, which are eventually rendered as meat, dairy and poultry products for human consumption. They are also used industrially in the production of a wide range of substances, such as glucose, adhesives, oils and alcohols. Cereals provide more food for human consumption than any other crops. Generally, four miscellaneous groups of foods prepared from the cereals grains are:

(1) Baked products, made from flour or meal, include breads, pastries, pancakes, cookies and cakes.
(2) Milled grain products, made by removing the bran and usually the germ (or embryo of the seed), include polished rice, farina, wheat flour, cornmeal, hominy, corn grits, pearled barley, semolina (for macaroni products), prepared breakfast cereals, soup, gravy and other thickenings.
(3) Beverages made from fermented grain products and from boiled, roasted grains.
(4) Whole-grain products include rolled oat, brown rice, popcorn, shredded and puffed grains, and breakfast foods (Sarwar, 2008).

## NUTRITIONAL FACTS OF CEREALS

All cereal grains have high energy values, mainly from the starch fraction, but, also from the fat and protein portions. Apart from moisture content and inedible substances such as cellulose, cereal grains contain carbohydrates- mainly starches (comprising 65 to 75% of their total weight), as well as proteins (6 to 12%) and fat (1 to 5%) along with traces of minerals and vitamins.

## HOW TO PICK OUT A HEALTHY CEREAL

Some cereals are made from whole grains that have been ground into flours and combined with thickeners, flavorings, vitamins and other ingredients during the cereal-making process. These cereals still may be very healthy, but, when heavily processed, these lose some of the nutritional benefits. Studies have shown that "intact" grains- grain kernels that are eaten in their pure form are better as lowering cholesterol and maintaining heart health than highly processed whole grains. There are many delicious whole grain cereal options to choose from a variety of cereals, so, a wide variety of nutrients from these tasty grains could be taken. Here is what to look for when taking a cereal:

(1) At least 3 gm of fiber per 100 calories (so if a cereal has 200 calories per serving, it should have at least 6 of fiber) - this is a sign that the cereal is truly whole grain.
(2) Six or less gm of sugar per serving (cereals with raisins or other dried fruit will have more), that is alright.
(3) Organic or natural brands that are made with responsibly grown ingredients (natural foods sector offer healthier cereals made with healthier ingredients). Raisins or other dried fruits will add a lot of grams of sugar to the listing on the nutrition panel; they are not distinguished from added sugars, so the amounts can only be estimated.

## WHOLE GRAIN CEREALS

The most healthful cereals are made with whole grains. If a person is trying to lose weight, controls cholesterol or diabetes, or just needs a lot of energy, the best bet is a hot cooked cereal of whole grains. It can be flavored with raisins or other dried fruits, or may be a handful of nuts. If someone prefers cold cereal, it is needed to check the list of ingredients carefully. And the first ingredient should be a whole grain. It is recommend that foods with partially hydrogenated oils (or trans fats), and they still show up in a few cereals should be avoided. Cereals and wholegrain foods can reduce the risk of developing certain diseases including coronary heart disease, colon cancer, diabetes and diverticular disease.

### Nutritional contents of whole grain cereals

The grains consist of three major parts which are:

(1) Bran: The outer layer of the grain (fiber omega-3 fatty acids, vitamins and dietary minerals).
(2) Endosperm: The main part of the grain (mainly starch), and
(3) Germ: The smallest part of the grain (vitamin E, folate, thiamine, phosphorus, magnesium).

The whole grains contain all three layers of the grain. The wholegrain cereals are a rich source of many essential vitamins, minerals and phyto-chemicals. The typical cereal food is low in saturated fat, but, is a source of polyunsaturated fats, including omega-3 linolenic acid, cholesterol free, high in both soluble and insoluble fiber and resistant starch, an excellent source of carbohydrates, a significant source of protein, a good source of B-complex vitamins, including folate, a good source of many minerals such as iron, magnesium, copper, phosphorus; zinc, and a good source of antioxidants and phytochemicals that can help to lower blood cholesterol levels. The wholegrain cereals contain many different phytochemicals that have been linked to significant health benefits. These phytochemicals include:

(1) Lignans: Can lower the risk of coronary heart disease, and regress or slow cancers in animals.
(2) Phytic acid: Reduces the glycaemic index of food, which is important for people with diabetes, and helps to protect against the development of cancer cells in the colon.

(3) Saponins, phytosterols, squalene, oryzanol and tocotrienols have been found to lower blood cholesterol.
(4) Phenolic compounds: Have antioxidant effects (Sarwar, 2008, 2009; Sarwar et al., 2004).

Cereal grains have been the principal component of human diet and have played a major role in shaping human civilization for thousands of years. Around the world; rice, wheat, and maize, and to a lesser extent, sorghum and millet are important staples and critical to daily survival of billions of people. More than 50% of world's daily caloric intake is derived directly from cereal grains consumption. Most of the grains used for human food are milled to remove the bran (pericarp) and germ, primarily to meet sensory expectations of consumers. The milling process strips the grains of important nutrients including dietary fiber, phenolics, vitamins and minerals which are beneficial to health (Awika, 2011). The studies have shown that whole grains consumption are associated with diminished risk of serious diet-related diseases. These conditions include coronary heart disease, certain cancers (especially of the large bowel), inflammatory bowel disease and disordered laxation. Carbohydrates are important contributors to the health benefits of whole grains. Insoluble non-starch polysaccharides (major components of dietary fiber) are effective laxatives. While soluble non-starch polysaccharides (especially mixed-link *β*-glucans) has lower plasma cholesterol and so can reduce heart disease risk, but, the effect is inconsistent (Topping, 2007).

Non-starch polysaccharides occur naturally in many foods. The physiochemical and biological properties of these compounds correspond to dietary fiber. Non-starch polysaccharides show various physiological effects in the small and large intestine and therefore have important health implications for humans. The remarkable properties of dietary non-starch polysaccharides are water dispersibility, viscosity effect, bulk and fermentibility into short chain fatty acids. These features may lead to diminish risk of serious diet related diseases which are major problems in Western countries and are emerging in developing countries with greater affluence. These conditions include coronary heart disease, colo-rectal cancer, inflammatory bowel disease, breast cancer, tumor formation, mineral related abnormalities, and disordered laxation. Insoluble non-starch polysaccharides (cellulose and hemicellulose) are effective laxatives, whereas, soluble non-starch polysaccharides (especially mixed-link β-glucans) lower plasma cholesterol levels and help to normalize blood glucose and insulin levels, making these kinds of polysaccharides a part of dietary plans to treat cardiovascular diseases and diabetes.

Moreover, a major proportion of dietary non-starch polysaccharides escapes the small intestine nearly intact, and is fermented into short chain fatty acids by commensal microflora present in the colon and cecum, and promotes normal laxation. Short chain fatty acids have a number of health promoting effects and are particularly effective in promoting large bowel function. Certain non-starch polysaccharides through their fermented products may promote the growth of specific beneficial colonic bacteria which offer a prebiotic effect (Kumar et al., 2012).

Wheat is considered a good source of protein, minerals, B-group vitamins and dietary fiber that is an excellent health-building food. Thus, it has become the principal cereal, being more widely used for the making of bread than any other cereal because of the quality and quantity of its characteristic protein called gluten. Gluten makes bread dough, sticks together and gives it the ability to retain gas. Wheat has several medicinal virtues; starch and gluten in wheat provide heat and energy, the inner bran coats, phosphates and other mineral salts, the outer bran, the much-needed roughage, indigestible portion that helps easy movement of bowels, the germ, vitamins B and E, and protein of wheat helps to build and repair muscular tissue. The wheat germ, which is removed in the process of refining, is also rich in essential vitamin E, and its lack can lead to heart disease. The loss of vitamins and minerals in the refined wheat flour has led to widespread prevalence of constipation and other digestive disturbances and nutritional disorders. The whole wheat, which includes bran and wheat germ, therefore, provides protection against diseases such as constipation, ischaemic, heart disease, disease of the colon called diverticulitis, appendicitis, obesity and diabetes (Kumar et al., 2011).

Wheat is considered as utmost among the cereals largely due to the fact, that its grain contains protein with unique chemical and physical properties. Besides being a rich source of carbohydrates, wheat contains other valuable components such as protein, minerals (P, Mg, Fe, Cu and Zn), and vitamins like thiamine, riboflavin, niacin and vitamin E (Siddiqui and Sarwar, 2002; Sarwar and Sattar, 2007). Certain wheat varieties of Pakistan are grown over a wide agro-climatic range and as such are anticipated to exhibit good yield and quality production. It is therefore, necessary to investigate the nutritional status of wheat varieties in terms of biochemical and physiochemical characteristics available for food and nutritional purposes in Pakistan or elsewhere. Such study is significant in providing an opportunity to explore the available wheat varieties and to further improve their nutritional excellence and also essential for setting nutritional regulations for domestic and export purposes. Physiochemical characteristics like ash, total grain mass and moisture are important parameters in the study of nutritional and agricultural aspects (Ali et al., 2009, 2011).

## CONCLUSION

The most important consideration for a good health is to incorporate at least 30 min of daily physical exercise in conjunction with eating cereal that is a good way to burn more calories than to consume. Eating the same foods all the time could result in boredom, so, a person should always try different kinds of cereal to keep things exciting. Eating only cereal for two meals a day could result in a lack of consumption of foods from other food groups. Diabetics can eat all kinds of food, including many commercial cereals. According to experts the diabetics should eat cereals with 3 g or more of dietary fiber and 5 g or less of total sugar. When choosing a healthy breakfast cereal, usually limit intake of added sugar by skipping granolas and cereals with marshmallows, frosting, dried fruits and chocolate flavoring. Also, choose cereals that are higher in fiber. Fiber is an important nutrient which is helpful to prevent weight gain and heart disease. For best results, eat fruit, vegetables and whole grains every day to provide extra nutrition and fiber. The meal can be made healthier by eating high-fiber wholegrain cereals that has low sugar. Also, eating a variety of cereals is helpful to health rather than just feeding on one item.

## REFERENCES


Ali A, Sarwar M, Khanzada S, Abro GH (2009). Reaction of Certain Wheat Varieties to the Action of Red Flour Beetle, *Tribolium castaneum* (Herbst) (Coleoptera) Under Insectary Conditions. Pak. J. Zool. 41(1):51-56.

Ali A, Sarwar M, Khanzada S, Abro GH (2011). Evaluating Resistance of Wheat Germplasms to Attack by Red Flour Beetle, *Tribolium castaneum* (Herbst) (Coleoptera). Pak. J. Zool. 43(4):793-797.

Awika JM (2011). Major Cereal Grains Production and Use around the World. In: Advances in Cereal Science: Implications to Food Processing and Health Promotion (Editors: J.M. Awika, V. Piironen, S. Bean. 1089:1-13.

Kumar P, Yadava RK, Gollen B, Kumar S, Verma RK, Yadav S (2011). Nutritional Contents and Medicinal Properties of Wheat: A Review. Life Sci. Med. Res. 22:1-10.

Kumar V, Amit KS, Gudrun B, Klaus B (2012). Dietary Roles of Non-Starch Polysachharides in Human Nutrition: A Review. Crit. Rev. Food Sci. Nutr. 52(10):899-935.

Sarwar M (2008). Laboratory studies on different wheat genotypes for their resistance against Khapra Beetle *Trogoderma granarium* Everts (Coleoptera: Dermestidae). Pak. J. Seed Technol. 2(11&12):46-53.

Sarwar M (2009). Evaluating wheat varieties and genotypes for tolerance to feeding damage caused by *Tribolium castaneum* (Herbst) (Coleoptera: Tenebrionidae). Pak. J. Seed Technol. 2(13&14):94-100.

Sarwar M, Ahmad N, Rajput AA, Tofique M (2004). Search for Varietal Resistance within Stored Wheat Genotypes against the Infestation of Red Flour Beetle, *Tribolium castaneum* Herbst (Coleoptera: Tenebrionidae). Proce. 4th Inter. Congress of Entomological Sciences, University of Agric., Faisalabad, September, 22-23. pp. 23-27.

Sarwar M, Sattar M (2007). Varietals assessment of different wheat varieties for their resistance response to Khapra beetle *Trogoderma granarium.* Pak. J. Seed Technol. 1(10):1-7.

Siddiqui QH, Sarwar M (2002). Pre and Post harvest losses in wheat. Pak. Gulf Econ. XXI(6):30-32.

Topping D (2007). Cereal complex carbohydrates and their contribution to human health. J. Cereal Sci. 46:220-29.

# Yield response of three groundnut (*Arachis hypogaea L.*) varieties intercropped with maize (*Zea mays*) in the guinea savanna zone of Ghana

**Konlan S.[1] , Sarkodie-Addo, J.[2], Kombiok . M. J.[3], Asare, E.[2] and Bawah, I.[3]**

[1]Cocoa Research Institute of Ghana (CRIG), New Tafo-Akim, Ghana.
[2]Kwame Nkrumah University of Science and Technology (KNUST), Kumasi, Ghana.
[3]Savanna Agricultural Research Institute (SARI), Nyankpala, Ghana.

**A field experiment was conducted under rain-fed conditions on the research farm of Savanna Agricultural Research Institute, Nyankpala, during the 2007 and 2008 seasons. The objective was to study the response of three new groundnut varieties (Jenkaar, Kpanieli and Nkosuor) to row intercropping with maize. The experiment was laid in randomized complete block design with four replicates. Treatments comprised sole groundnut, sole maize, single row groundnut alternated with single row maize (G1M1), single row groundnut alternated with double row maize (G1M2), double row groundnut alternated with single row maize (G2M1) and double row groundnut alternated with double row maize (G2M2). The population densities of groundnut and maize in the intercrops affected their yield performance. Row intercropping arrangement that encouraged large leaf area in groundnut supported the formation of more pods per plant and subsequently larger dry pod yield. The highest groundnut and maize yields in the intercrop in both years were obtained respectively from G2M1 and G1M2. The G1M1 and G1M2 row intercropping arrangement was therefore the most advantageous in both years, achieving a land equivalent ratio values greater than 1. Combinations of the Kpanieli variety and maize which achieved land equivalent ratio greater than 1 was also more advantageous than intercropping maize with the Jenkaar and Nkosuor varieties in both years.**

**Key words:** Guinea savanna, rain-fed, row intercropping, yield performance, land equivalent ratio.

## INTRODUCTION

The groundnut (*Arachis hypogaea* L.) is an important oilseed crop of the semi-arid tropics (Tarimo, 1997; ICRISAT, 2008) that ranks thirteenth (13$^{th}$) in importance among world crops (Hatam and Abbasi, 1994). Groundnut is a staple food in a number of developing countries much valued for its protein content and as source of income for small holder farmers (Peanut CRSP, 1990). It is also a good source of edible oil for humans, as well as a nutritive feed supplement for livestock (Abulu, 1978; Goldsworthy and Fisher, 1987). In Ghana, groundnut is grown by farm families on small scale, both in pure stands and in crop mixtures, especially with cereals (Tsigbey et al., 2003). Yields obtained from the crop are traditionally low due to a combination of factors including unreliable rains, little technology available to small scale farmers, pest and disease occurrences, poor seed technology and agronomic practices, as well as increased cultivation on marginal lands (Konlan et al.,

2013a). Despite the numerous problems facing groundnut cultivation, it ranks as the number one grain legume grown in the Guinea savanna zone of Ghana (Tsigbey et al., 2003; Naab et al., 2005).

Incidentally, maize (*Zea mays*) is the most important and widely cultivated cereal in Ghana. However, competition between the two crops for limited land holdings by farm families who need to produce both the maize to feed the family and groundnut for income to cater for the health, education and other needs of the family is intense. To enable the farm family meet its household food needs and cash requirements, many subsistence farmers therefore practice intercropping in which groundnut frequently forms an important part of the system (Tsigbey et al., 2003; Naab et al., 2005). Groundnut-maize intercropping, as a common practice among farmers in dry land areas is well documented in Ghana (Atuahene-Amankwah et al., 1990; Tsigbey et al., 2003; Naab et al., 2005) and elsewhere (Molatudi and Mariga, 2012; Siddig et al., 2013; Mehdi, 2013; Reddy et al., 1987). In most of these reports, groundnut-maize intercropping achieved land equivalent ratios (LER) greater than 1 and gave higher economic returns. The yields obtained from the intercrops were found to relate directly to their population densities (Langat et al., 2006), giving an indication that the overall plant population can be skewed to favour one crop over the other in the intercrop depending on the farmer's priority or individual crop profitability. After successfully establishing the advantages of groundnut-maize intercropping systems through decades of scientific research, new promising maize and groundnut varieties should as a matter of principle, be evaluated and only released to farmers based on their ability to meet the demands of current intercropping systems. This will help meet the two-pronged need of the farm family to produce food and obtain cash income from the same piece of land. The yield response and nitrogen fixing capacities of these groundnut varieties were earlier evaluated under high density planting in sole systems (Konlan et al., 2013a; 2013b). This study was therefore set up to evaluate their growth and yield response to row intercropping arrangement with maize in the Guinea savanna zone under rain-fed conditions.

## MATERIALS AND METHODS

### Experimental design and treatments

The experiment was laid out in randomized complete block design with four replicates and six treatments. The three new groundnut varieties (Jenkaar, Kpanieli and Nkosuor) were intercropped with maize (*Obatanpa* variety) under different row intercropping arrangements. The treatments evaluated were:

1. Sole maize planted at 60 x 40 cm giving plant population density of 41, 667 plants per hectare.
2. Sole groundnut planted at 30 x 15 cm (Konlan et al., 2013a) giving plant population density of 222,222 plants per hectare.
3. G1M1: 1 row of groundnut (90 x 15 cm) alternated with 1 row of maize (90 x 40 cm) giving a final plant population of 74,074 groundnut plants / ha and 27, 778 maize plants / ha.
4. G2M1: 2 rows of groundnut (67.5 x 15 cm) alternated with 1 row of maize (135 x 40 cm) giving final plant population of 98, 765 groundnut plants / ha and 18, 518 maize plants / ha.
5. G1M2: 1 row of groundnut (165 x 15 cm) alternated with 2 rows of maize (82.5 x 40 cm) giving a final plant population of 40, 404 groundnut plants / ha and 30, 487 maize plants / ha.
6. G2M2: 2 rows of groundnut (105 x 10 cm) alternating with 2 rows of maize with maize (105 x 40 cm) giving a plant population of 95, 238 groundnut plants / ha and 23, 809 maize plants / ha.

A single plough operation, followed by a single harrowing was carried out using a tractor prior to lining and pegging. Two seeds and one seed per hole respectively of maize and groundnut were planted on flats and the first weeding done with a hand hoe 4 weeks after sowing (WAS). 60 kg N / ha of NPK (23:10:5) was applied to the maize plants 2 WAS. The fertilizer was placed in holes drilled closed to the maize plants and covered with soil. A top dressing of 50 kg Sulphate of Ammonia per hectare was applied to the maize at 6 WAS just after the second weed management operation using the same localized placement method.

### Growth parameters and yield measurements

#### *Plant stand and height*

Plant stand ($m^2$) data was taken 2 WAS. All plants within the net plots were counted and subjected to square root transformations to obtain plant stand per every 4 $m^2$. Five plants of each net plot were randomly selected and identified with a tag. Heights of these selected plants were monitored at two weeks interval from 4 to 8 WAS in both years. The height of each plant was measured using a measuring tape. Measurement was done from the ground level to the tip of the longest leaf blade or tassel (maize) or the tip of the apical meristem for groundnut.

#### *Leaf area index*

Five groundnut plants from each border rows were cut at the ground level and all the leaves stripped. The fresh leaves were weighed ($W_f$) and the weights recorded. Fifty leaf discs of the fresh leaves were made using a 1.0 cm diameter cork borer. These were also weighed ($W_b$). Since the diameter of the cork borer was known, the area of each leaf disc was estimated as:

$$A = \pi r^2$$

The total area of 50 leaf discs was then determined to be the product of the area of one disc and the number of discs in $cm^2$. By relating the area of 50 discs to the weight, it was possible to calculate the leaf area as $W_f$ / $W_b$ x area of leaf discs. The LAI of groundnut was then calculated as the ratio of the total leaf area and the area of ground space covered by each plant as described by Watson (1952).

The lengths and breadths at the widest portions of the leaves of five maize plants randomly selected and tagged from each maize plot were also measured. The product of the length and breadth gave the leaf area, which was then multiplied by a factor of 0.75 as described by Kamprath and Moll (1977) to get the total leaf area. The ratio of the total leaf area and the area of ground space covered by each plant gave the LAI (Watson, 1952).

**Table 1.** Plant stands of groundnut and maize as affected by intercropping and row arrangement two weeks after sowing.

| Int. partner | Groundnut stand ($m^2$) | | Maize stand ($m^2$) | |
|---|---|---|---|---|
| | 2007 | 2008 | 2007 | 2008 |
| Sole crop | 4.3 | 4.4 | 2.1 | 2.0 |
| Maize/Jenkaar | 2.2 | 2.3 | 2.0 | 2.0 |
| Maize/Kpanieli | 2.0 | 2.2 | 2.0 | 2.0 |
| Maize/Nkosuor | 2.1 | 2.0 | 2.0 | 2.0 |
| Lsd $_{0.05}$ | 1.4 | 0.2 | ns | ns |
| Row pattern | | | | |
| G1M1 | 2.9 | 3.1 | 4.2 | 4.1 |
| G2M1 | 4.3 | 4.4 | 4.0 | 4.0 |
| G1M2 | 2.4 | 3.2 | 3.8 | 3.7 |
| G2M2 | 4.0 | 4.2 | 3.0 | 3.1 |
| Lsd $_{0.05}$ | ns | ns | 0.3 | 0.3 |

Plant stand means are square root transformations. G1M1, 1 row groundnut, 1 row maize; G2M1, 2 rows groundnut, 1 row maize; G1M2, 1 row groundnut, 2 rows maize; G2M2, 2 rows groundnut, 2 rows maize, Int partner, intercrop partner.

### *Yield and yield components*

Yield data collected included number of seeds per pod, number of pods per plant, mean (100) seed weight, shelling outturn (%) and dry pod yield (t $ha^{-1}$). To determine shelling outturn, pods and cobs from the five randomly tagged groundnut and maize plants respectively were each put in open bags and air dried thoroughly to a moisture level of 13 % before shelling. These were then weighed before shelling (Wp and Wc respectively). After shelling, the shelled seeds were weighed and recorded. The shelling percentage was determined as the weight of groundnut seed (Ws) divided by weight of pods (Wp).

Shelling outturn (%) = $\frac{Ws}{Wp}$ x 100 for groundnut

Where, Ws = weight of groundnut seed, and Wp = weight of groundnut pods

In the case of the maize, shelling outturn was determined as the weight of maize seed (Ws) divided by weight of cobs (Wc), thus;

Shelling outturn (%) = $\frac{Ws}{Wc}$ x 100

Where, Ws = weight of maize seed, and Wc = weight of maize cobs

### *Land equivalent ratio*

The Land Equivalent Ratios (LER) of the groundnut and maize intercropping systems were calculated by expressing the intercrop grain yield as a ratio of the sole crop grain yield. This was done following the procedure described by Wiley and Osiru (1972) as;

$$LER = L_a + L_b = \frac{Ya}{Sa} + \frac{Yb}{Sb}$$

Where, $L_a$ and $L_b$ are LERs of crop variety a and b; $Y_a$ and $Y_b$ are the individual crop yields in the intercrops, and $S_a$ and $S_b$ are the sole crop yields.

By this procedure, a partial LER for each crop component was obtained. To determine the total LER, the partial LERs of the component crops were added.

## RESULTS

### Plant stand, height and leaf area index

Intercropping groundnut with maize significantly (α = 0.05) reduced the population densities of both crops compared to their respective sole systems (Table 1). The overall maize densities were not significantly affected by intercropping compared to groundnut whose population density was significantly reduced. Row arrangement was also not found to significantly (α = 0.05) influence groundnut population density (Table 1) although plant arrangement that favoured more rows of groundnut resulted in relatively higher plant densities for groundnut. Densities of maize was however significantly (α = 0.05) influenced by row arrangement with higher densities resulting from the G1M1 and G2M1 arrangement which were significantly (α = 0.05) different from the densities recorded by the G2M2 arrangement in both years (Table 1). The height of groundnut varieties were not significantly (α = 0.05) influenced by intercropping them with maize as well as by the row arrangement at 4 WAS (Table 1). At 6 WAS, intercropping significantly (α = 0.05) reduced the height of the Jenkaar and Nkosuor varieties. Intercropped Kpanieli which recorded similar height as the sole groundnut crop was also found to be significantly taller (α = 0.05) than plants of intercropped Jenkaar and Nkosuor (Table 2). At 8 WAS, height of intercropped Jenkaar and Nkosuor were similar to the sole crop all of which were significantly (α=0.05) shorter than the intercropped

**Table 2.** Groundnut and maize height at 4, 6 and 8 weeks after sowing as affected by intercropping and row arrangement.

| Int. partner | Groundnut height (cm) | | | | | | Maize height (cm) | | | | | |
|---|---|---|---|---|---|---|---|---|---|---|---|---|
| | 2007 | | | 2008 | | | 2007 | | | 2008 | | |
| | 4 WAS | 6 WAS | 8 WAS | 4 WAS | 6 WAS | 8 WAS | 4 WAS | 6 WAS | 8 WAS | 4 WAS | 6 WAS | 8 WAS |
| Sole crop | 9.0 | 19.2 | 33.2 | 8.5 | 28.3 | 47.5 | 45.4 | 88.8 | 194.5 | 39.5 | 139.1 | 192.1 |
| Maize/Jenkaar | 8.5 | 17.9 | 33.4 | 7.1 | 20.5 | 35.8 | 37.6 | 97.8 | 182.1 | 38.8 | 125.5 | 169.8 |
| Maize/Kpanieli | 9.0 | 19.7 | 39.4 | 7.9 | 29.2 | 48.9 | 43.6 | 96.7 | 182.5 | 38.2 | 134.2 | 175.9 |
| Maize/Nkosuor | 8.6 | 18.4 | 34.1 | 7.4 | 20.6 | 36.5 | 44.7 | 98.0 | 180.4 | 38.8 | 121.1 | 167.8 |
| Lsd $_{0.05}$ | ns | 1.5 | 3.1 | ns | 5.3 | 9.7 | ns | ns | ns | ns | ns | ns |
| | | | | | | | | | | | | |
| Row pattern | | | | | | | | | | | | |
| G1M1 | 9.6 | 21.8 | 39.3 | 7.1 | 23.8 | 39.2 | 44.4 | 97.8 | 191.7 | 38.1 | 128.3 | 165.1 |
| G2M1 | 8.2 | 18.5 | 34.8 | 6.9 | 17.8 | 30.3 | 42.1 | 95.8 | 179.2 | 35.3 | 126.9 | 164.4 |
| G1M2 | 8.8 | 20.7 | 37.8 | 7.2 | 21.3 | 36.5 | 45.7 | 94.0 | 185.4 | 35.2 | 136.3 | 169.5 |
| G2M2 | 8.0 | 18.5 | 35.1 | 7.8 | 18.3 | 35.6 | 45.8 | 98.6 | 188.5 | 35.8 | 126.3 | 165.7 |
| Lsd $_{0.05}$ | ns | 1.7 | 0.7 | ns | 2.7 | 2.3 | ns | ns | ns | ns | ns | ns |

WAS, weeks after sowing; G1M1, 1 row groundnut, 1 row maize; G2M1, 2 rows groundnut, 1 row maize;, G1M2, 1 row groundnut, 2 rows maize; G2M2, 2 rows groundnut, 2 rows maize; cm, centimeters; Int partner, intercrop partner.

Kpanieli (Table 2). Row arrangement did not significantly (α = 0.05) influence groundnut height at 4 WAS in both years (Table 2). At 6 and 8 WAS in both years, a single row of groundnut alternated with single row maize (G1M1), and single row groundnut alternated with double row maize (G1M2) recorded similar height measurements both of which were significantly (α = 0.05) different from the double row groundnut alternated with single (G2M1) and double (G2M2) row maize. The height of maize plants in both years was not significantly (α = 0.05) influenced by either intercropping with the groundnut varieties or row arrangement (Table 2).

Leaf area indices of groundnut varieties were significantly (α = 0.05) reduced by intercropping with maize (Table 3). All intercropped groundnut varieties recorded significantly (α = 0.05) smaller LAI compared to the sole groundnut crop at 4, 6 and 8 WAS in both years (Table 3). Also, the LAI of intercropped Kpanieli was significantly (a = 0.05) larger than those of intercropped Jenkaar and Nkosuor during the three data collection occasions in both years. Leaf area indices of the groundnut varieties were significantly (α = 0.05) improved by double groundnut row intercropping (G2M1 and G2M2) which were found to be significantly larger at 4, 6 and 8 WAS in both years than the single groundnut row (G1M1 and G1M2) intercropping (Table 3). Maize leaf area index was not influenced by intercropping or row arrangement in both years (Table 3).

## Yield, yield components and correlations

The number of seeds per pod of groundnut varieties was not significantly influenced by either intercropping or row arrangement (Table 4). The number of pods per plant of the intercropped Kpanieli was similar to that of the sole groundnut, both of which were significantly (α = 0.05) higher than that recorded for Jenkaar and Nkosuor (Table 4).

The two double row groundnut arrangement improved the number of pods per plant significantly (α = 0.05) when compared to the single row arrangements. The mean seed weights of the intercropped groundnut varieties were all similar and significantly lower than that of the sole groundnut in both years. The two double row groundnut arrangement also significantly (α = 0.05) improved mean seed weight compared to the single row arrangements in both years (Table 4). Also, shelling outturn of all groundnut varieties were significantly reduced by intercropping with maize in both years. The shelling outturn of the Kpanieli variety was however, found to be significantly (α = 0.05) higher than that of the Nkosuor variety in both years. Like mean seed weight, shelling outturn was favoured by the double groundnut row arrangement which recorded significantly (α = 0.05) higher values compared to the single groundnut row arrangement (Table 4). Intercropping groundnut varieties with maize significantly reduced the pod yield of the varieties in both years. Pod yield of the Kpanieli variety was also found to be significantly (α = 0.05) higher than those of the Jenkaar and Nkosuor varieties. The yield differences between the Jenkaar and Nkosuor varieties were also significant (α = 0.05). Following the pattern of mean seed weight and shelling outturn, double row (G2M1 and G2M2) groundnut arrangement resulted in significantly higher (α = 0.05) pod yields when compared to the single (G1M1 and G1M2) row arrangements which recorded similar pod yield values (Table 4). Maize mean

**Table 3.** Groundnut and maize leaf area indices at 4, 6 and 8 weeks after sowing as affected by intercropping and row arrangement.

| Int. partner | Groundnut LAI | | | | | | Maize LAI | | | | | |
|---|---|---|---|---|---|---|---|---|---|---|---|---|
| | 2007 | | | 2008 | | | 2007 | | | 2008 | | |
| | 4 WAS | 6 WAS | 8 WAS | 4 WAS | 6 WAS | 8 WAS | 4 WAS | 6 WAS | 8 WAS | 4 WAS | 6 WAS | 8 WAS |
| Sole crop | 1.64 | 3.86 | 3.97 | 1.93 | 4.18 | 4.40 | 1.34 | 3.17 | 4.31 | 1.25 | 1.60 | 2.39 |
| Maize/Jenkaar | 1.22 | 2.04 | 2.85 | 1.43 | 2.45 | 3.10 | 1.37 | 2.86 | 3.32 | 1.32 | 1.64 | 2.44 |
| Maize/Kpanieli | 1.31 | 2.66 | 3.41 | 1.64 | 2.78 | 3.64 | 1.28 | 2.90 | 3.30 | 1.48 | 1.89 | 2.98 |
| Maize/Nkosuor | 1.21 | 2.03 | 2.84 | 1.46 | 2.45 | 3.10 | 1.23 | 3.10 | 3.55 | 1.39 | 1.75 | 2.35 |
| Lsd $_{0.05}$ | 0.02 | 0.02 | 0.02 | 0.04 | 0.02 | 0.03 | ns | ns | ns | ns | ns | ns |
| Row pattern | | | | | | | | | | | | |
| G1M1 | 1.24 | 1.43 | 2.87 | 1.50 | 1.91 | 3.32 | 1.55 | 3.83 | 3.81 | 1.67 | 1.77 | 2.68 |
| G2M1 | 1.30 | 2.70 | 3.71 | 1.62 | 3.22 | 4.03 | 1.47 | 3.80 | 3.69 | 1.71 | 1.81 | 2.48 |
| G1M2 | 0.98 | 1.40 | 1.88 | 1.04 | 1.45 | 1.95 | 1.85 | 3.12 | 3.73 | 1.58 | 1.73 | 2.36 |
| G2M2 | 1.33 | 2.81 | 3.71 | 1.55 | 3.21 | 3.72 | 1.53 | 3.26 | 3.53 | 1.66 | 1.82 | 2.59 |
| Lsd $_{0.05}$ | 0.15 | 0.02 | 0.02 | 0.01 | 0.02 | 0.04 | ns | ns | ns | ns | ns | Ns |

LAI, leaf area index; WAS, weeks after sowing; G1M1, 1 row groundnut, 1 row maize; G2M1, 2 rows groundnut, 1 row maize;, G1M2, 1 row groundnut, 2 rows maize; G2M2, 2 rows groundnut, 2 rows maize; Int partner, intercrop partner.

**Table 4.** Groundnut yield and yield components as affected by intercropping with maize and row arrangement.

| Int. Partner | Seeds pod$^{-1}$ | | Pods plant$^{-1}$ | | Pod yield (tha$^{-1}$) | | MSW (g) | | Shelling outturn (%) | |
|---|---|---|---|---|---|---|---|---|---|---|
| | 2007 | 2008 | 2007 | 2008 | 2007 | 2008 | 2007 | 2008 | 2007 | 2008 |
| Sole groundnut | 1.8 | 1.8 | 11.1 | 10.3 | 2.145 | 2.127 | 45.1 | 47.4 | 53.8 | 49.0 |
| Maize/Jenkaar | 1.8 | 1.8 | 8.2 | 9.1 | 0.834 | 0.824 | 43.5 | 44.7 | 45.3 | 44.3 |
| Maize/Kpanieli | 1.8 | 1.8 | 10.4 | 10.7 | 1.163 | 1.101 | 41.5 | 45.7 | 47.8 | 47.1 |
| Maize/Nkosuor | 1.8 | 1.8 | 8.7 | 9.3 | 0.732 | 0.703 | 41.2 | 44.2 | 43.8 | 43.9 |
| Lsd $_{0.05}$ | ns | ns | 2.7 | 1.7 | 0.35 | 0.12 | 1.5 | 2.9 | 3.3 | 1.8 |
| Row pattern | | | | | | | | | | |
| G1M1 | 1.7 | 1.8 | 9.6 | 9.1 | 0.665 | 0.618 | 42.1 | 42.4 | 42.8 | 41.3 |
| G2M1 | 1.7 | 1.8 | 9.7 | 9.8 | 0.926 | 0.928 | 43.7 | 45.4 | 47.4 | 47.2 |
| G1M2 | 1.7 | 1.8 | 6.5 | 7.3 | 0.381 | 0.306 | 41.9 | 43.8 | 37.6 | 35.6 |
| G2M2 | 1.7 | 1.8 | 9.6 | 9.1 | 0.431 | 0.400 | 43.6 | 45.7 | 48.3 | 47.3 |
| Lsd $_{0.05}$ | ns | ns | 1.3 | 1.8 | 0.47 | 0.13 | 1.7 | 1.7 | 3.9 | 2.1 |

Seeds pod$^{-1}$, Number of seeds per pod); Pods plant$^{-1}$, number of pods per plant; tha$^{-1}$ , tons per hectare; MSW, mean seed weight in grams; G1M1, 1 row groundnut, 1 row maize; G2M1, 2 rows groundnut, 1 row maize;, G1M2, 1 row groundnut, 2 rows maize; G2M2, 2 rows groundnut, 2 rows maize; Int partner, intercrop partner.

seed weight was not significantly ($\alpha$ = 0.05) influenced by either intercropping or row arrangement in both years (Table 5). Grain yield of maize was significantly ($\alpha$ = 0.05) reduced by intercropping with the groundnut varieties (Table 5). The G1M2 and G1M2 row arrangements produced similar yields both of which were significantly ($\alpha$ = 0.05) higher than the yields recorded by the G2M1 and G2M2 row arrangements (Table 5).

The number of pods per plant and ultimately dry pod yield were positive and strongly correlated to groundnut leaf area index in 2007 (Figures 1a and 1b) and 2008 (Figures 2a and 2b). Maize grain yield also showed strong and positive correlations with plant population densities in both years (Figure 3a and 3b) although yield indices were unaffected by either intercropping or row arrangement.

## Land equivalent ratio

Land equivalent ratios (LER) ranged between 1.01 and 1.11 in 2007 and 0.96 and 1.11 in 2008 as a result of

**Table 5.** Maize yield and yield components, and land equivalent ratio as affected by intercropping and row arrangement.

| Int. partner | MSW (g) | | Shelling outturn (%) | | Grain yield (tha$^{-1}$) | | LER | |
|---|---|---|---|---|---|---|---|---|
| | 2007 | 2008 | 2007 | 2008 | 2007 | 2008 | 2007 | 2008 |
| Sole maize | 32.3 | 32.1 | 79.0 | 77.3 | 3.220 | 3.920 | - | - |
| Maize/Jenkaar | 32.7 | 33.1 | 79.2 | 78.0 | 2.305 | 2.117 | 1.10 | 0.98 |
| Maize/Kpanieli | 33.5 | 32.6 | 81.4 | 77.9 | 2.476 | 2.115 | 1.11 | 1.11 |
| Maize/Nkosuor | 33.3 | 33.3 | 79.5 | 78.1 | 2.375 | 2.190 | 1.01 | 0.96 |
| Lsd $_{0.05}$ | ns | ns | ns | ns | 0.31 | 0.47 | | |
| Crop pattern | | | | | | | | |
| G1M1 | 33.8 | 34.2 | 79.7 | 76.7 | 2.876 | 3.056 | 1.10 | 1.11 |
| G2M1 | 32.5 | 32.2 | 80.4 | 78.1 | 1.476 | 1.720 | 0.97 | 0.84 |
| G1M2 | 33.0 | 32.5 | 80.4 | 80.6 | 2.718 | 2.867 | 1.14 | 1.12 |
| G2M2 | 33.1 | 33.1 | 79.7 | 76.7 | 1.182 | 1.453 | 1.08 | 0.99 |
| Lsd $_{0.05}$ | ns | ns | ns | ns | 0.39 | 0.44 | | |

MSW, mean seed weight; tha$^{-1}$, tons per hectare; LER, land equivalent ratio; G1M1, 1 row groundnut, 1 row maize; G2M1, 2 rows groundnut, 1 row maize;, G1M2, 1 row groundnut, 2 rows maize; G2M2, 2 rows groundnut, 2 rows maize; Int partner, intercrop partner.

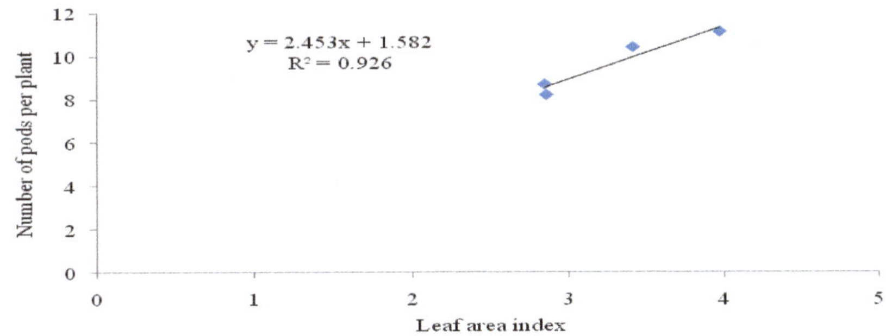


a

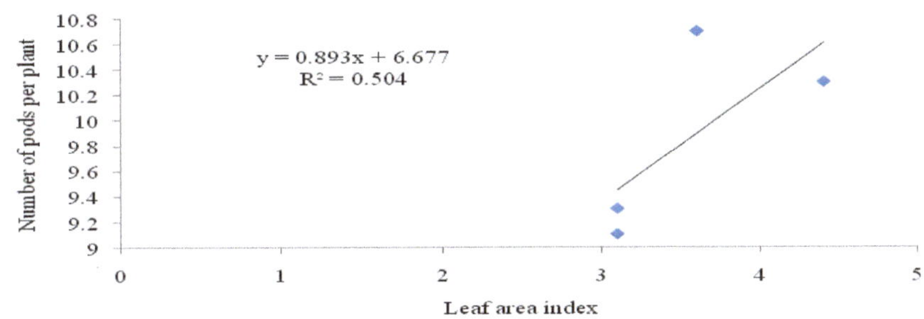


b

**Figure 1.** The relationship between groundnut leaf area index and number of pods per plant in 2007 (a) and 2008 (b) was strong and positive in both years.

intercropping groundnut and maize (Table 5). In both years, the highest LER was recorded by the Kpanieli-maize intercrop. Row arrangement with its consequences ranging from 0.97 to 1.14 in 2007 and 0.99 to 1.12 in 2008. Land equivalent ratios greater than 1 was achieved in both years by the G1M1 and G1M2 row intercropping (Table 5), highlighting the contribution of maize yield to LER in the intercrop.

## DISCUSSION

Intercropping generally resulted in reductions in both groundnut and maize population densities with a simple

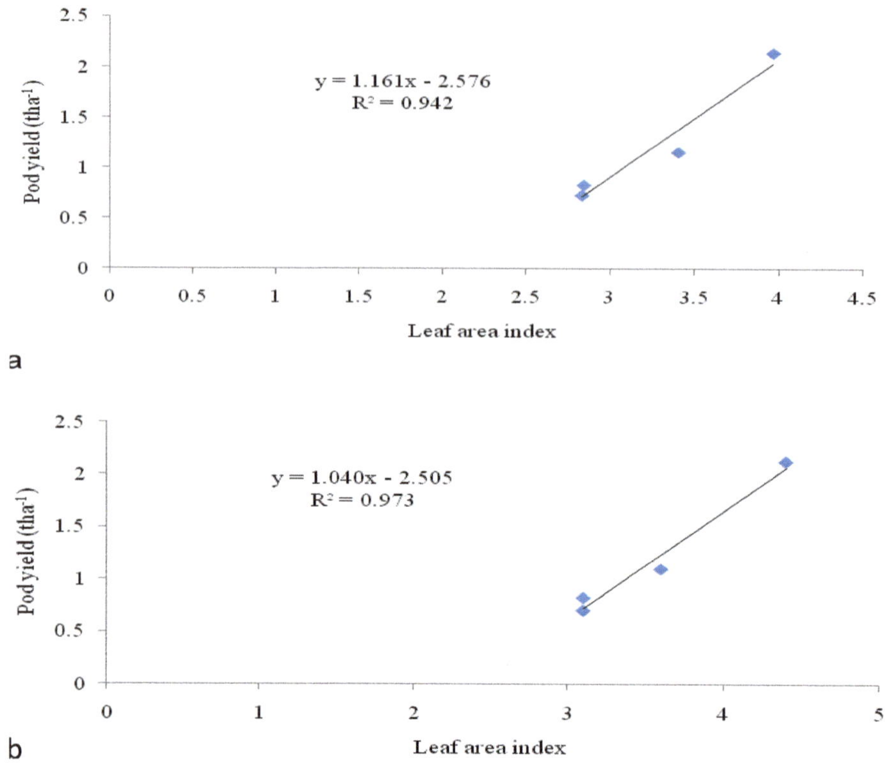


**Figure 2.** The relationship between groundnut leaf area index and dry pod yield in 2007 (a) and 2008 (b) shows that dry pod yield responded positively and strongly to increases in groundnut leaf area index. ($tha^{-1}$, tons per hectare).

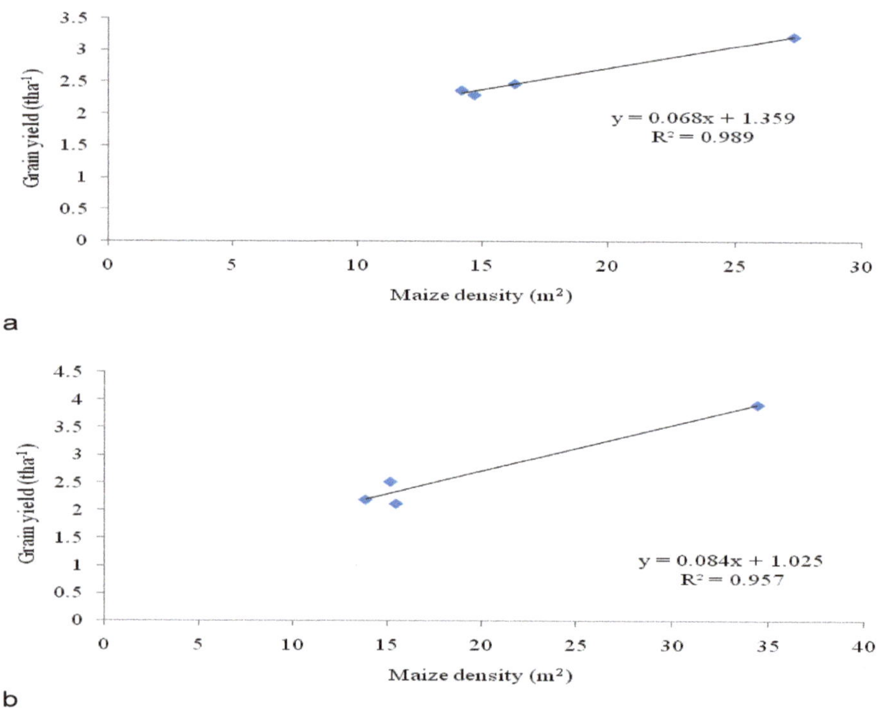


**Figure 3.** The relationship between maize population density and grain yield ($tha^{-1}$) in 2007 (a) and 2008 (b) shows that grain yield responded positively and strongly to increases in maize population density in the intercrop in both years.($tha^{-1}$, tons per hectare; $m^2$, square metres).

inverse relationship between the population densities of the two crops. This was so because the population densities of both crops could not be increased at the same time in the planting pattern evaluated (Molatudi and Mariga, 2012; Siddig et al., 2013; Mehdi, 2013). The lack of differences in the height of intercropped and sole groundnut crop were probably because of a strong influence on height by the genotypes which was not significantly influenced by the modified environment due to the presence of the taller and potentially shading maize.

Also maize shade was not so dense as to elicit vigorous vertical growth by the groundnut varieties. Consequently, row arrangement in the groundnut-maize intercrop that allowed more sunlight to penetrate encouraged relatively shorter plants. On the other hand, a single row of groundnut embedded between single and double rows of maize with relatively reduced aerial space and light penetration produced relatively taller plants in response to a relatively more dense shade by the maize plants (Konlan et al., 2013a).

The reduction in leaf area indices of intercropped groundnut was probably as a result of reduced photosynthesis due to the shading effect of the maize plants. Less dry matter was therefore available to support new leaf production and development, leading to reduced LAI in the intercropped groundnut compared to its sole counterpart in both years (Dalley et al., 2004). The behaviour of the intercropped Kpanieli which supported larger LAI and taller plants compared to the other two varieties in both years was probably a varietal characteristic (Ahmad and Mohammad, 1997). In spite of the fact that groundnut-maize intercropping led to significant reductions in groundnut LAI, row arrangements that allowed in more sunlight supported groundnut LAI that were similar to the sole groundnut crop in both years.

This suggests that sunlight availability was critical for photosynthate production and played a major role in the formation of leaves and development of larger LAI which then supported more photosynthesis (Dalley et al., 2004).The lack of differences in the number of seeds recorded per pod was probably due to the fact that it is controlled by the plant genotype (Ahmad and Mohammad, 1997) and was not influenced by intercropping with maize.

The number of pods per plant was however easily influenced by modification of the crop environment due to intercropping, suffering significantly from limited light reception and nutrient availability. This probably led to the Jenkaar and Nkosuor varieties suffering reduced number of pods per plant in both years. The generally reduced LAI of intercropped groundnut which might have affected dry matter production and hence the availability of assimilates for pod production and filling probably led to the lower mean seed weights and shelling outturn recorded by the intercropped groundnut varieties compared to their sole crop. However, row arrangement in the intercrop that allowed more ground space and light penetration led to formation of larger LAI and supported higher number of pods per plant, mean seed weight and shelling outturn in both years. This could be attributed to reduced competition for nutrients and potentially higher photosynthetic rates resulting in the availability of dry matter for pod formation and filling (Ahmad et al., 2007). The intense competition experienced by the groundnut varieties in the groundnut-maize intercrop which affected the yield components translated into reduced dry pod yields in both years. On the other hand row arrangement that made more space available to the groundnut, potentially increased groundnut population density in the intercrop while at the same time reducing competition for nutrients and allowing the interception of more Photosynthetically active radiation (PAR), leading to higher dry pod yields.

The lack of differences in height of maize under the different row intercropping arrangements was probably due to the absence of significant competition as a result of intercropping with the groundnut. This could be attributed to the fact that the groundnut varieties probably supplied their own nitrogen requirement through biological nitrogen fixation (Giller, 2001) and did not need to compete with the maize for the available nitrogen (Francis, 1986; Rwamugira and Massawe, 1990). Also, the relatively shorter groundnut plants in the intercrop meant that they were growing below the maize plants, effectively removing competition for light. Maize LAI was therefore unaffected by either intercropping or row arrangement in the intercrop.

Maize yield was however, directly proportional to the maize population density, with row arrangement favouring higher maize population densities producing higher seed yield in both years. In the groundnut, row arrangement giving larger LAI resulted in higher number of pods per plant and larger dry pod yield in both years. This was probably because such large leaf areas in groundnut is known to intercept more incident solar radiation for photosynthesis, which then makes assimilates available for pod formation and pod filling (Ali and Malik, 1992; Wells et al., 1993; Jaaffar and Gardner, 1988). Even though increase in the population of maize plants per unit area has been shown to reduce individual plant yield, the additional plants per square metre probably more than compensated for the reduction in the yield of individual plants, leading to higher grain yield when maize population density was increased. The higher LER obtained from the Kpanieli-maize intercrop was basically due to the higher pod yield of the Kpanieli groundnut variety in the intercrop since maize yield when intercropped with the other groundnut varieties were similar. The relatively higher maize yield obtained from the G1M1 and G1M2 row intercropping arrangement was the major determining factor leading to the achievement of land equivalent ratios greater than 1 by the two treatments in both years.

## Conclusion

In cropping systems where groundnut-maize intercropping is considered in the Guinea savanna zone of Ghana, it would be more beneficial to choose the Kpanieli groundnut variety and establish the intercrop in either a single row groundnut (90 x 15 cm) alternated with single row maize (90 x 40 cm) or single row groundnut (165 × 15 cm) alternated with double row (82.5 x 40 cm) maize.

## ACKNOWLEDGEMENT

The authors are grateful to the Director of Savanna Agricultural Research Institute (SARI), Nyankpala, who hosted the study on the institute's research farm in addition to providing human resource support and other critical facilities for two full seasons.

### REFERENCES

Abulu GO (1978). An economic analysis of groundnut production in Northern Nigeria.In: proceedings of the national seminar on groundnut production. Pp. 83-90. Kano, Nigeria.

Atuahene-Amankwah G, Hossain MA, Assibi MA (1990). Groundnut Productionand Improvement in Ghana. First ICRISAT Regional Groundnut Meeting for West Africa, 13-16 Sept. 1988. Niamey, Niger, pp. 45-47.

Ahmad N, Mohammad R (1997). Evaluation of promising groundnut (*Arachis hypogaea*) varieties for yield and other characters. Crop and Soil Science, P. 251.

Ahmad N, Mohammad R, Ulas K (2007). Evaluation of different varieties, seed rates and row spacing of groundnut, planted under agro-ecological conditions of Malakand Division. J. Interacademia. 9(4):178-183.

Ali N, Malik NS (1992). Performance of short duration ground-nut lines suitable to increase cropping intensity in rainf-ed areas of Pakistan. In: Nigam S.N. (ed.), Groundnut - A Global Perspective. International Crop Research Institute for the Semi-Arid Tropics.

Dalley CB, Kells JJ, Renner KA (2004). Effect of glyphosate application timing change row spacing on weed growth in corn (*Zea mays*) and soybean (*Glycine max*). Weed Technol. 18:177-182. Patancheru, India, P. 433.

Francis CA (1986). Multiple cropping systems. Macmillan, New York. 378 pp.

Giller KE (2001). Nitrogen fixation in tropical cropping systems. 2nd Edition. CABI, Wallingford, UK. P. 405.

Goldsworthy MG, Fisher MN (1987). Physiology of tropical field crops. John Wiley and Sons. pp. 474-476.

Hatam M, Abbasi GQ (1994). History and economic importance of groundnuts(Arachis hypogaea L.). In: Crop production. Bashir, E. and Bantel, R. (Eds). Pub NBF., pp. 350-351

ICRISAT (2008). ICRISAT West Africa programs' Annual report. 2007. Niger: ICRISAT

Jaaffar Z, Gardner FP (1988). Canopy development, yield and market quality in peanut as affected by genotype and planting pattern. Crop Sci. 28:299-305.

Kamprath EJ, Moll RH (1977). Effects of population density upon agronomic traitsassociated with genetic increases in yield of maize. (*Zea mays* L). Agron. J. 69:81-84.

Konlan S, Sarkodie-Addo J, Asare E, Kombiok MJ (2013a). Groundnut (*Arachis hypogaea* L.) varietal response to spacing in the Guinea savanna agro-ecological zone of Ghana: Growth and yield. Afr. J. Agric. Res. 8(22):2769-2777.

Konlan S, Sarkodie-Addo J, Asare E, Kombiok MJ (2013b). Groundnut (*Arachis hypogaea* L.) varietal response to spacing in the Guinea savanna agro-ecological zone of Ghana: Nodulation and nitrogen fixation. Agric. Biol. J. North America 4(3):324-335.

Langat MC, Okiror MA, Ouma JP, Gesimba RM (2006). The effect of intercropping groundnut (*Arachis hypogea* L.) with sorghum (*Sorghum bicolor* L. Moench) on yield and cash income. Agricultura Tropica Et Subtropica 39(2):87-90.

Molatudi RL, Mariga IK (2012). Grain yield and biomass response of a maize/dry bean intercrop to maize density and dry bean variety. Afr. J. Agric. Res. 7(20):3139-3146.

Mehdi D (2013). Intercropping Two Varieties of Maize (*Zea mays* L.) and Peanut (*Arachis hypogaea* L.): Biomass Yield and Intercropping Advantages. Int. J. Agric. Forest. 3(1):7-11.

Naab JB, Tsigbey FK, Prasad PVV, Boote KJ, Bailey JE, Bradenberg RL (2005).Effects of sowing date and fungicide application on yield of early and late maturing peanut cultivars grown under rainf-ed conditions in Ghana. Crop Protect. 24(1):107-110.

Peanut CRSP (1990). The Peanut CRSP: Improving global production and use of peanut. Research Summary, 1982-1990. University of Georgia, Griffin. P. 16.

Rwamugira WP, Massawe RD (1990). Groundnut/maize intercrop: effect of spatial arrangement on yield and its components. Pages 149-153. In: proceedings of the fourth Regional Groundnut Workshop for Southern Africa, 19-23 March 1990, Arusha, Tanzania. Patancheru, A. 502:324 India: ICRISAT.

Reddy MS, Kelly G, Musanya JC (1987). Recent Agronomic developments in groundnut investigations in Zambia. in proceedings of the second regional Groundnut Workshop for Southern Africa, 10-14 Feb. 1986, Harare, Zimbabwe. Patancheru, A. P. 502:324 India: ICRISAT. pp. 57-64

Siddig A, Mohamed A, Adam A, Mohamed A, Bahar H, Abdulmohsin RK (2013). Effects of Sorghum (Sorghum Bicolor (l) Moench) and Groundnut (Arachis Hypogaea l) Intercropping on Some Soil Chemical Properties and Crop Yield Under Rain-Fed Conditions. ARPN J. Sci. Technol. 3(1):69-74.

Tarimo AJP (1997). Physiological response of groundnut to plant population density. Afr. Crop Sci. J. 5(3):267-272.

Tsigbey FK, Brandenburg RL, Clottey VA (2003). Peanut production methods in northern Ghana and some disease perspectives. Online J. Agron. 34(2):36-47

Watson DJ (1952). The physiological basis of variation in yield. Adv. Agron. 4:101-145.

Wells RJ, Burton JW, Kilen TC (1993). Soybean growth and light interception: Response to differing leaf and stem morphology. Crop Sci. 33:520-524.

Wiley RW, Osiru DSO (1972). Studies on mixture of maize and beans (*Phaseolus vulgaris*) with particular reference to plant population. J. Agric. Sci. (Cambridge) 79:519-529.

# Effect of processing on some biochemical characteristics and storability of locust bean condiment

**Daramola B. and Osanyinlusi S. A.**

Department of Food Technology Federal Polytechnic, P. M. B. 5351, Ado-Ekiti, Ekiti State, Nigeria.

**Storage life of 'green' locust beans condiment is less than 72 h due to inherent high moisture content (>55%) and vulnerability to peroxidation due to high fat content (> 37%). Therefore, elimination or significant reduction of moisture and fat should lead to increase in storage life hence studied. 'Green' locust beans condiment was dried and delipidated using three solvents. Assessment of some physiochemical, biochemical, sensory characteristics and shelf-life preliminary study on the preparation were conducted. Moisture content (<10%) of samples was at the level that samples are stable to spoilage microorganism. Lipids were extracted using three solvents in the order of diethylether > hexane > methanol with concomitant increase in proportion of protein in the delipidated sample in comparison to non-delipidated sample. Analysis of sensory scores (p = 0.05) showed that diethylether extracted sample was the most preferred and methanol extracted sample the least preferred. Using off-odour perception as accelerated storage instrument, shelf-life model insight (scale factor = 28 days, shape factor 3.8) was obtained using Weilbull distribution.**

**Key words:** Locust beans condiment, poor storability, processing, biochemical characteristics, shelf-life model insight.

## INTRODUCTION

The choicest condiment of natural origin for enhancing flavour of food products by the people of Western Nigeria is fermented locust beans (*Parkia. biglobosa*) indigenously called 'Iru'. Being the most important food condiment in West and Central Africa, its frequent usage suggests it could play a contributory role to the intake of proteins, peptides, amino acids and B-vitamins (Fetuga et al., 1973). A lot of research works notably by Odunfa (1981), Odunfa and Adewuyi (1985b) had been carried out on the process of production of locust bean condiment largely from microbiological or fermentation perspective.

At harvest, fermented locust bean is at its 'green' form. The condiment is characterised, first by high moisture content, hence vulnerable to spoilage by microorganism. Second, by high fat content, therefore, prone to fat deterioration and its undesirable consequences. Third, the condiment is marked by offensive odour probably of phenolic and ammoniacal origin. These are thought to be the barrier responsible for poor storability and unfriendly odour of locust bean condiment. Thus, constitute a challenge and should be addressed. Therefore, it is envisioned that if fat and moisture could be eliminated or reduced significantly objectionable odour could be reduced and shelf life could be extended from its present 2 to 3 days and the envisioned spurred this study.

The objectives of the present study were to dry and delipidate locust bean condiment, evaluate effects on some physiochemical, biochemical compositions and determine shelf-life preliminarily.

## MATERIALS AND METHODS

Locust beans condiment was obtained from a local producer at the point of harvest. Hexane, diethylether and methanol were of analytical grade.

### Preparation of sample and solvent extraction

Locust bean condiment was dried using air oven at a temperature of 50°C for 12 h, and subsequently divided into 3 portions: 1st portion was milled, 2nd portion was grated into average particle size of 4 mm while the third portion was not reduced in size. Each (40 g) of the different particle size group was later extracted with 400 ml of three solvents namely: hexane, diethylether and methanol for 8h using soxhlet extractor. After solvent extraction process, the samples were dried in air oven to expel remnant solvents from samples. These were used for subsequent analysis.

### Methods of analysis

#### *Proximate composition*

Proximate composition was determined using the method of the Association of Official Analytical Chemists (AOAC, 1990) method. Carbohydrate was obtained by difference.

#### *Determination of free amino acids*

Free amino acids in the partially defatted locust bean condiment and non-defatted samples were extracted with 80% ethanol (v/v) in accordance with the method of Odibo et al. (1990). The free amino acids in the ethanolic extract were estimated using the ninhydrin colorimetric method (Rosen, 1957) using glycine as standard.

#### *Determination of reducing power*

Reducing power of each sample was determined in accordance with the method of Oyaizu (1986). Simply, each sample (1 mg/ml) in methanol (2.5 ml) was mixed with sodium phosphate buffer (pH 6.6). the buffered sample was mixed with conditioning reagents (1%$K_3$-Fe-$CN_6$, 10% TCA, 0.1% $FeCl_3$) centrifuged, diluted using distilled water and absorbance was measured at 700 nm. Higher absorbance indicates a higher reducing power.

#### *Measurement of pH*

The pH of samples was measured using Omega H. $HP_x$ digital pH meter. Standardization of the meter was done using buffer solutions of pH 4 and 9.

#### *Determination of titratable acidity*

Partially defatted locust bean condiment and non-defatted samples were prepared for and titrated with standard alkaline. Results were expressed as lactic acid equivalent (AOAC, 1990).

#### *Shelf-life study*

Off- odour was choosing as accelerated storage instrument for shelf-life model study, because off-odour release is a critical characteristic of locust bean condiment even when not spoilt hence could serve as accelerated study instrument for the condiment. Ten gram of each sample was tested over time for off-flavour development. Three panellists that are familiar with locust bean condiment were used to evaluate each sample for off-flavour development on a seven-point rate scale with category 1 indicating no-off flavour to category 7 indicating very strong off-flavour. On this scale an average of score of 2.5 or more was considered a high-enough off-flavour score for the product to fail. Using this procedure, failure times were obtained for the products. The data obtained were subsequently used to compute shelf-life parameters of importance using Weilbull hazard plots (Maxima and Jagar, 1984).

#### *Sensory evaluation*

Using ranking method, partially defatted locust bean condiment and non-defatted sample (control) was carried out by 10 panelists comprised of students of the Department of Food Technology, Federal Polytechnic Ado-Ekiti. The panellists were selected based on the history of their familiarity with processing of locust beans condiment. Sensory attributes evaluated were: Colour, odour, taste, flow characteristics and overall acceptability. Ranks were converted to scores according to the method of Fisher and Yates (1942) as reported by Larmond (1979)

### Statistical analysis

The converted scores were subjected to analysis of variance (ANOVA). When significant ($P = 0.05$) difference was observed, means were separated using Tukey's test (Snedecor, 1956).

## RESULTS AND DISCUSSION

The results of some physiochemical and biochemical assessment of the partially defatted locust beans condiment is shown in Table 1. The moisture content (%) ranged from 6.0 to 7.4 which suggest that the product is in water condition stable to food spoilage microorganisms. The value is also similar to each other within the efficiency of the air oven used. Similarly, the pH (6.2-6.9) of the products was similar to each other. The value is lower to the pH (8.9) of fresh fermented locust bean condiment earlier reported in literature by Kok et al. (1986) and Omafuvbe et al. (2004). The difference could be explained in terms of elimination of ammoniacal compounds responsible for the high pH of fresh condiment which was evaporated by the drying process carried out on the product.

Titrable acidity (µg lactic acid equivalent /g sample) of the defatted samples was high (12.07 to 18.89) in comparison to the low (6.30 to 12.60) values of the partially defatted samples. This suggests that the solvent and process used in lipid removal facilitate the removal of some acids most probability fatty acid present in the condiment. The importance of this is that it is important to carry out sensory test of the products since such fatty acids may have contributory role in condiment taste. A rough estimate of extractible lipid (%) by the examined solvents using weight difference suggests that all the

**Table 1.** Some physiochemical and biochemical characteristics of partially and non-delipidated locust bean condiment.

| Sample | Treatment description | Moisture content | Amino acids | Reducing power | pH | Extractible lipid | Titratable acidity |
|---|---|---|---|---|---|---|---|
| 1 | Whole beans hexane | 7.90 | 541.80 | 208.33 | 6.5 | 4.20 | 12.60 |
| 2 | Milled beans hexane | 6.00 | 680.30 | 223.96 | 6.2 | 27.95 | 10.24 |
| 3 | Coarse beans hexane | 6.70 | 320.22 | 109.38 | 6.5 | 23.48 | 10.79 |
| 4 | Whole bean diethylether | 6.50 | 380.50 | 177.08 | 6.6 | 19.65 | 10.79 |
| 5 | Milled beans diethyether | 6.70 | 671.10 | 119.79 | 6.2 | 35.00 | 10.18 |
| 6 | Coarse bean diethylether | 6.30 | 458.70 | 62.50 | 6.4 | 35.20 | 6.30 |
| 7 | Milled $CHCl_3$ – methanol | 6.0 | 588.00 | 93.75 | 6.4 | 35.13 | 6.57 |
| 8 | Whole $CHCl_3$ – methanol | 6.20 | 301.75 | 109.38 | 6.9 | 16.00 | 8.87 |
| 9 | Coarse $CHCl_3$ - methanol | 6.0 | 292.50 | 93.75 | 6.7 | 21.78 | 9.44 |
| 10 | Milled control | 7.0 | 347.90 | 244.79 | 6.7 | NA | 18.89 |
| 11 | Coarse control | 7.4 | 255.59 | 130.20 | 6.7 | NA | 15.11 |
| 12 | Whole control | 7.2 | 283.30 | 151.00 | 6.6 | NA | 12.07 |

Amino acid =µg glycine/g sample; reducing power = µg ferulic acid equivalent /g sample; extractible lipid (%) titrated acidity = µg lactic acid equivalent /g sample; moisture content (%), NA = not applicable.

**Table 2.** Some aspects of proximate composition (%) of partially and non-delipidated locust beans condiment (milled-samples).

| Sample | Ash | Ether extract | Moisture | Protein | *CHO |
|---|---|---|---|---|---|
| Diethyl ether | 6.14 | 3.25 | 8.00 | 70.59 | 12.02 |
| Methanol | 5.68 | 9.06 | 7.50 | 63.68 | 14.08 |
| Hexane | 4.54 | 13.75 | 8.00 | 59.50 | 14.21 |
| Control | 3.69 | 34.77 | 8.50 | 34.50 | 13.54 |

*Obtained by difference.

solvent remove lipids from samples in which diethylether solvates the highest amount of lipids. Because of highest surface area, the extracted lipid (%) (Table 1) in mild samples was higher in comparison to samples with smaller surface area (coarse and whole been samples). Removal of lipid should exert resistance of product to lipid susceptible deterioration such as peroxidation.

In addition, the effect of partial removal of lipids on amino acid concentration, one of the process effects of fermentation in sample was assessed. It was found that delipidation resulted to higher amino acid concentration in the partially defatted samples. The values of amino acid for milled sample were higher than the value of the other samples. This is by proportion and not quantity because of the samples had the highest amount of fat/lipid extracted from them by the used solvents. Amino acids assessment is important in the sense that it gave the clue to the fact that solvent and process did not lead to leaching of total amino acids into the extraction solvents. Amino acids play nutritional and taste (sensory) roles in locust beans condiment.

The samples exhibited reducing power, and index of antioxidative activity. This suggests that drying and lipid removal of locust bean do not abrogates exhibition of antioxidative activity of the condiment. This argument is important from nutrition stand-point. However, a cursory look on the result on Table 1 showed that methanol extracted sample was characterised by the least reducing power. Considering the values obtained for reducing power, generally, removal of lipid did not erode the reducing power of the samples. With the exception of methanol extracted samples. This suggests that some of the contributors to reducing activity of locust bean condiment are polar in nature, and probably have been extracted by the polar solvent, methanol.

The proximate composition of the milled samples is shown in Table 2. The proximate composition of the controlled sample was similar to an earlier report by Omafuvbe et al. (2004) except for the moisture content. This is because dry sample was used in this study. A comparison of the proximate composition of de-lipidated sample and control sample revealed the efficacy of the lipid extraction solvents on the locust bean condiment sample with similar particle size and extraction conditions (time and temperature). All the solvents extracted fat from the samples when compared with the lipid content of the control sample (Table 1). Diethylether showed the highest propensity for removing lipid with hexane removing the least amount of lipid. This result suggests that some of the lipid in locust bean is likely to be of higher polarity than that could be extracted by hexane and with less polarity to than that could be extracted by methanol. Therefore, leaving more lipids to be extracted by diethylether.

The variation in values obtained for ash and protein contents was not due to quantitative changes in the samples but due to concentration changes. This is because removal of fat does not increase the quantity of protein but increases its proportion in these samples.

## Shelf-life studies

The length of time during which an item or a food product performs satisfactorily is called its life (shelf). Shelf life of a product is of paramount importance for to the consumer and the producer. The changes that occur in physical or pure chemical substances can be described by Arrhenius equation. However, this model could not adequately describe changes that occur in biological materials such as food hence, the use of Weilbull distribution a versatile life distribution that has been used for food product. The details of mathematical principles and illustration of application of Weilbull distribution are well documented in Maxima and Jagar (1984) as follows.

## Mathematical theory of shelf- life model

This was accomplished using facile experimental procedure. Treaty on statistical principles on Weilbull distribution is not aimed but to allude to transformation of the distribution to linear plot described as follows.

If the life of a product is governed by Weilbull distribution its probability density function (pdf) and cumulative density function (cdf) is given by

$$F(x) = (\beta/\alpha\beta)\, x^{\beta-1} \exp(-(x/\alpha)\beta) \quad (1)$$

$x > 0$ since life of a product cannot be negative; where $\alpha$ and $\beta$ are scale and shape parameter of Weilbull curve. And

$$F(x) = 1 - \exp(-(x/\alpha)\beta) \quad (2)$$

When the Weilbull distribution is skewed the hazard function is given by

$$h(x) = (\beta/\alpha^{\beta})x^{\beta-1} \quad x>0 \quad (3)$$

And cumulative hazard function is:

$$H(x) = (x/\alpha)\beta \quad x > 0 \quad (4)$$

Equation (4) could be re-written to express time x as a function of the cumulative hazard function

$$H: X = \alpha\, (H(x)^{1/\beta} \quad (5)$$

Or in log units of time x

$$\text{Log } x = \log \alpha + (1/\beta) \log H(x) \quad (6)$$

Equation (6) describes a straight line with log $\alpha$ as the intercept and1/$\beta$ as the slope. Simply, these are the mathematical bases (linear equation) for hazard plot in which near linearity of plot suggest appropriateness or adequacy of the Weilbull model. Table 3 shows the age at failure, and hazard of the samples used in this study. The hazard plot cumulative hazard against time of failure of this data was accomplished using Weilbull hazard plot graph papers and the nearly obtained straight line was used to estimate scale ($\alpha$) of the x-axis and shape of the curve ($\beta$). No study in which shelf-life of locust beans condiment had been modelled using Weilbull probability distribution, hence unable to compare results. The scale value ($\alpha$ = 35 days) gives insight to the span of shelf-life of tested product.

The scale value ($\alpha$) represents the life of the product as governed by the Weilbull probability distribution. The probability density function can be used to calculate the probability that the life of a product does not exceed chosen amount of time which imply the probability of a products failure at or before time x. This could be used for predicting the shelf-life of the product.

Whereas, the shape value ($\beta$ = 3.8) suggests that Weilbull pdf tends to be symmetrical around the median which approximates to 35 days. The data plot could be used to estimate the probability of product failure with view to take action if necessary. For instance, estimation from the plot revealed that chances exist that 50% of product will fail after the 28$^{th}$ day. Bearing in mind that this is a preliminary study there are chances that the shelf life can be increased from the value determined herein by total delipidation and application of packaging and more importantly, the assessment was based on accelerated storage test using off-flavour that does not necessarily imply microbiological spoilage.

## Sensorial assessment

The result of sensory scores of the milled samples and control samples is presented in Table 4. Regarding colour, methanol extracted sample is statistically different from other samples. However the score obtained for diethylether sample was the highest which implied the most preferred. Regarding flow characteristics or dryness or resistance to cake-up. There is a difference between the control sample and other samples. Diethylether extract had highest sensory score and significantly different from other samples and the control sample had

**Table 3.** Storage-time failure data for partially and non-delipidated locust bean condiment.

| Sample No. | Number of termination | Age at termination | Hazard 100/k | Cumulative hazard |
|---|---|---|---|---|
| 1 | 12 | 14 | 8.33 | 8.33 |
| 2 | 11 | 16 | 9.09 | 17.42 |
| 3 | 10 | 17 | 10.00 | 27.42 |
| 4 | 9 | 23 | 11.11 | 38.53 |
| 5 | 8 | 29 | 12.50 | 51.03 |
| 6 | 7 | 32 | 14.29 | 65.32 |
| 7 | 6 | 34 | 16.67 | 81.99 |
| 8 | 5 | 36 | 20.00 | 101.99 |
| 9 | 4 | 40 | 25.00 | 126.99 |
| 10 | 3 | 45 | 33.33 | 160.32 |
| 11 | 2 | 53 | 50.00 | 210.32 |
| 12 | 1 | 53 | 100.00 | 310.32 |

**Table 4.** Sensory scores and visual description of partially and non-delipidated locust beans condiment (milled – samples).

| Sensory characteristics | Samples | | | |
|---|---|---|---|---|
| | Hexane | Diethylether | Methanol | Control |
| Colour | $0.32^{a}$ | $0.57^{a}$ | $-1.03^{b}$ | $0.14^{a}$ |
| Flow property/fluidity | $0.10^{b}$ | $0.91^{a}$ | $0.02^{b}$ | $-1.03^{c}$ |
| Odour | $1.13^{a}$ | $0.44^{ab}$ | $-0.22^{bc}$ | $-0.79^{c}$ |
| Taste | $0.72^{a}$ | $1.03^{a}$ | $0.50^{a}$ | $-0.54^{b}$ |
| Overall acceptability | $-0.1^{b}$ | $0.91^{a}$ | $-1.03^{c}$ | $0.22^{b}$ |
| Visual description | Light yellow | Bleached white | Dark | Light dark |

*Values not followed by the same letter in a row are significantly (p=0.05) different.

the least score. The cake characteristics of the control sample could be accounted for by the presence of lipid, fat and oil in the sample which bound sample particles together. The odour of samples extracted using hexane and diethylether were preferred to the methanol and control samples. This suggests that lipids responsible for odouriferous compounds in locust beans condiment are probably non-polar. Because these are the group of lipids extracted using hexane and diethlether solvents.

All the delipidated samples significantly taste better than the non- delipidated sample. This is reasonable since removal of lipid lend increase in concentration of protein, amino acid and peptide which are responsible for flavour of the tastant. In overall, sample extracted using diethylether, was most preferred while methanol extracted sample was least preferred, ranked behind control sample. This human judgment was informed by the appearance of methanol extracted sample which appeared black and as a result not appealing to the panels as shown by their comments in the questionnaire. It is known that when locust bean condiment is dried it turns dark in colour due to reaction between amino acid and carbonyl group of carbohydrate or fat as observed in this study, but the sample extracted using methanol was much darker. This suggests that methanol has promoted the carbonyl-amino reaction to produce the black pigments. The veracity of this speculation needed to be established. Visual colour of samples extracted with diethylether and hexane is also presented in Table 4.

## Conclusion

Drying and partial delipidation resulted to product with minimal amount of moisture and fat with increased proportion of protein and ash content and consequently prolonged the shelf life of fermented locust beans condiment as revealed by shelf-life preliminary study. The process significantly improved sensory characteristics notably colour and odour of the condiment.

## REFERENCES

AOAC (1990). Official Methods of Analysis 14th Edition. Association of Official Analytical Chemists, Washington DC.

Fetuga BL, Babatunde GM, Oyenuga VA (1973). Protein quality of some Nigerian food stuffs-chemical assay of nutrients and amino acids composition. J. Sci. Food. Agric. 24:1505-1510.

Kok R, Ikenebomeh MJ, Ingram JM (1986). Processing and fermentation of the African locust bean. (*Parkia filicoidea* Welw.) to produce dawadawa. J. Sci. Food. Agric. 37:272-282.

Larmond E (1979). Laboratory Method of Sensory Evaluation of Food () department of Agriculture, Ottawa, Canada. P. 1637

Maxima CG, Jagar S (1984). Statistical methods in food and consumer research. Academic Press Inc Orlando Florida. pp. 274-311.

Odibo FC Nwabunnia F Osuigwe DI (1990). Biochemical changes during fermentation of *Telfaria* seeds for ogiri production. World J. Microbiol. Biotechnol. 6:425-427.

Odunfa SA (1981). Microorganisms associated with fermentation of African locust bean during in preparation. J. Plant Foods 3:245-250.

Odunfa SA, Adewuyi EY (1985b). Optimization of process conditions for the fermentation of African locust bean (*Parkia biglobosa*) H. Effect starter cultures. Chem, Mikrobiol, Technol Lebensm 9:118-122.

Omafuvbe BO, Falade OS, Osuntogun BA, Adewusi SRA (2004). Chemical and Biochemical changes in African locust bean (*Parkia biglobosa*) and melon (*Citrullus vulgaris*) seeds during fermentation to condiments. Pak. J. Nutr. 3:140-145.

Oyaizu M (1986). Studies on extracts of browning reaction: Antioxidative activities of extracts of browning reactions prepared from glucosamine. Japan J. Nutr. 44:307-315.

Rosen H (1957). A modified ninhydrin Colorimetric Analysis for amino acids. Arch. Biochem. Biophys. 67:10-15.

Snedecor GW (1956). Statistical Methods Applied to Experiments in Agriculture and Biology, 5th ed. Ames, IA: State College Press.

# Effect of processing germinated shea kernels on the quality parameters of shea (*Vitellaria paradoxa*) butter

**Jude U. Obibuzor , Roland D. Abigor, Vivien Omoriyekemwen, Emmanuel A. Okogbenin and Tonbra Okunwaye**

Biochemistry Division, Nigerian Institute For Oil Palm Research (NIFOR), PMB 1030, Benin City , 300001, Edo State, Nigeria.

**The processing methods, kernel storage and phytosanitory conditions are some of the major challenges leading to production of low quality shea butter. The effect of processing germinated shea kernels on the overall quality of processed butter was examined. Shea fruits were picked under different shea trees, pooled, and divided into two portions: one portion was processed immediately; the other was germinated for 9 days. The butter expressed from the respective portions was analyzed using standard methods. The yield of the extracted butter from the germinated and ungerminated kernel was 52 and 49%, respectively. The physicochemical properties of butter from the ungerminated and germinated nuts were free fatty acid (FFA) (0.22 and 1.66%), saponification value (SV), (223 and 239 mg KOH/g), unsaponifiable matter (7.77 and 9.39%), iodine value (Wijs) (IV) (33 and 30), and peroxide value (PV) (0.10 and 8.14 meq/kg), respectively. Germination process depleted the lipid content of shea kernel, and elevated the FFA by 7-folds, the PV by 81, while IV decreased. Again, germination mobilized linoleic acid more than the other constituent fatty acids, and the elemental composition remained generally unaffected. Prompt sterilization and sorting out the germinated kernels are necessary steps to improving the quality of shea butter.**

**Key words:** Germinated shea kernel, butter quality, desaturation, fatty acids, elemental analysis.

## INTRODUCTION

Shea tree (*Vitellaria paradoxa*) is one of the most important sources of vegetable oil in rural areas of the savanna zone of West Africa (Nikiema and Umali, 2007). Nigeria is the leading producer of seeds: 355,000 t in 1999, 58% of the African production, but 10,000 t lower than in 1996. Mali and Burkina Faso are other leading producers; at the end of the 1990s they produced 85,000 and 70,000 t/year, respectively, followed by Ghana Ghana (55,000 t), Côte d'Ivoire (20,000 t), Benin (15,000 t) and Togo (6500 t) (Nikiema and Umali, 2007). Contrary, Nigeria is correspondingly not the leading exporter of shea butter. Shea tree stores fat in the kernel of its seed. The approximate chemical composition of the kernel per 100 g dry matter, which varies considerably across the shea distribution region of West Africa (Okullo et al., 2010) is fat 31 to 62 g, protein 7 to 9 g, carbohydrate 31 to 38 g, unsaponifiable matter 2.5 to 12 g (Nikiema and Umali, 2007). The kernels are processed into shea butter and pressed cake. Food processing in general involves synergism between different physical processes to transform raw animal/plant materials into consumer-ready products. In addition, consumers judge food quality based on the sensory and nutritional characteristics such as texture, flavour, aroma, shape, colour calorie content, vitamins as well as shelf life (Norton and Sun, 2008). Currently, the processors use three main methods namely, artisanal, mechanical and solvent, which have different extraction efficiencies, to extract shea butter from the kernel. Carette et al, (2009) had reported that the manual method is preferred by the processors although less efficient. This is because manual method was found to be more profitable compared to mechanized shea butter extraction due to the fuel cost, maintenance cost, and cost of the shea kernels that are involved with the mechanized process. In artisanal, for example, the first step is to break the kernels into small pieces using mortar and pestle. Then, these nut pieces are roasted, and when roasted, the pieces are pounded again with a mortar and pestle to obtain a brown-black paste. The roasted black paste is grinded with a grinding stone on a flat stone surface and beaten into paste with the addition of warm water to a desired consistency. This mass is then washed once or twice before boiling the creamy mass in a cooking pot to yield the butter. Shea butter is reputed to have wide range of applications in the pharmaceutical, cosmeceutical (AarhusKarlshamn, 2012), nutraceuticals, etc industries as well as in the folk medicine. The health benefits of the shea butter are attributed to the influence of the constituents of the unsaponifiable.

The quality of fats and oils is dictated by several physical and chemical parameters that are dependent on the source of oil; geographic, climatic, and agronomic variables of growth in the case of plant oils as well as processing and storage conditions (Shahidi, 2005). In the case of shea butter production, the fruits are collected from the ground but must not be allowed to stand for long on the ground as they germinate quickly. This is because the shea fruit has no dormancy period, it germinates within few days of dropping from the tree (Jøker, 2000). This means that is for high quality butter, prompt processing is an important requirement but currently neglected by the local processors due to the difficulties associated with fruit collection. The process of germination is capable of modifying the lipid composition by the endogenous enzymes leading to poor understanding of the state of the nut, prior to processing. The state of the plant materials prior to processing could affect or modify all or some of the food qualities parameters listed above besides other factors such as processing methods and conditions (Olaniyan and Oje, 2007). It has been reported that changes in the expression of various fatty acid metabolizing enzymes can result in changes in seed oil compositions (Napier, 2007; Knutzon, 1992). For example, the predominant off-flavour perceived in some products made from soybeans comes from the activity of endogenous lipoxygenase enzymes found in soybeans (Iassonova et al., 2009).

Germination process in seed sometimes may be a nuisance due to the depletion of the target reserves or a strategy to enhance some nutritive constituents of the seed. According to Munshi et al. (2007), the quantity of phospholipids, glycolipids and sterols in cotyledons and embryonic axes in fast germinating seeds increased progressively between the 1 to 6 days after sowing (DAS) compared with the slow growing seeds. The fatty acid composition in cotyledons of fast growing seeds showed increased levels of palmitic and oleic acids 6 and 8 DAS, while a decline in palmitic and stearic acids as well as accumulation of oleic and linoleic acids were observed in slow growing seeds. On the other hand, Urbano et al. (2004) studied the effects of germination of *Pisum sativum* L, for 2, 4 or 6 days, with and without light, on the proteolytic activity, the contents of soluble protein and non-protein nitrogen, and the amount of available starch of *P. sativum* L. as well as their nutritive utilization by growing rats, and concluded that germination of peas for 2 days would be sufficient to significantly improve the palatability and nutritive utilization of protein and carbohydrates from *P. sativum* L.

The timely picking of shea fruits from the bush is one of the most important steps in obtaining high quality butter ultimately. The picking activity requires contingent arrangement; it is tedious, and clumsy in addition to being a dangerous undertaking because the picking crews are constantly faced with some challenges such as snake bite, poor visibility, covering long distances, harassment by monkeys and other reptiles (Nahm, 2010). Because of these challenges associated with shea fruit collection, the processors insist on processing every kernel picked irrespective of its state (germinated) or integrity.

This present work seeks to evaluate the effect of processing germinated shea kernels on the quality of butter via physicochemical analysis.

## MATERIALS AND METHODS

### Shea kernel

Shea kernels were obtained from Bida, Niger State, the city is located at latitude 9.07459600 (9428.55N), longitude 6.00850500 (6 0 30. 62 E) and elevation of 590 ft above sea level, and is a semi-arid town. Shea fruits (20 kg) were picked under about 10 trees in Bida, Niger State, Nigeria in July, 2012.

### Chemicals and reagents

All the chemicals and reagents used in this study were of analytical grade (AnalaR). Shea fruits of about 13 kg, were picked under 10 different shea trees within 500 m radius, and were pooled. The picked shea fruits were immediately transported in black polythene bags to Nigerian Institute for Oil Palm Research (NIFOR) Shea Tree Substation, Bida and divided into 2 portions. One portion was parboiled at 100°C for 90 min to stop the activities of endogenous enzymes during transportation to Benin City (about 9 h trip), while the second was not parboiled but allowed to germinate. The boiled fruits were allowed to cool and then manually de-pulped. The de-pulped fruits were spread on a wooden floor overnight before transporting to Biochemistry Division Laboratory where the extraction and characterization experiments were performed. The

**Figure 1a.** Germinated shelled shea kernels.

**Figure 1b.** Ungerminated shelled shea kernels.

**Figure 1c.** Freshly germinated unshelled shea nuts.

**Figure 1d.** Germinated shelled shea kernel showing dried radicle.

unparboiled portion was heaped for 9 days behind Room 76 of Biochemistry Division building, NIFOR, Benin City at ambient temperature 29 ± 3°C to induce germination(Figure 1a). The freshly germinated fruits (Figure 1c) (5.85 kg) were selected and, for the butter extraction treated exactly as the fresh ungerminated sample (Figure 1b).

Both the germinated and ungerminated fruits were respectively de-pulped and dried at 105°C in an oven (Gallenkamp Plus II, England) until it could be easily cracked (14 h). The dried nuts were cracked (shelled) manually (Figure 1d), and were further oven dried at 105°C for a total of 18 h. Then, weighed portions of the dried nuts were processed into shea butter by respectively expressing the oil with a hand operated stainless expeller. The expressed oil was centrifuged at 3500 revolution per minute (rpm) to remove the artifact and debris to yield very clear golden yellow oil. The emphasis here was not on the yield of the oil but purity.

### Proximate analysis of the kernels

Samples of kernels from each category were selected at random, pooled and examined for moisture, lipid, crude protein, nitrogen free extract (NFE) and ash contents. The moisture content was determined by AOCS method (AOCS, 1997). The shea kernels were milled with a blender (lexus, China) and oil extracted for 18 h with 96% pure n-hexane (Scharlau, Spain) in a soxhlet apparatus and the lipid percentage was computed by weight difference. Ash and nitrogen content were measured according to AOAC methods 40.1.08 and 40.1.06, respectively (AOAC, 2010). The ash content was determined by heating the samples in a muffle furnace at 525°C, cooled in a desiccator and weighed with an analytical balance. The protein content on the other hand was obtained according to the Kjeldahl method and converted to protein percentage by multiplying with the Atwater factor (conversion). NFE was quantified by difference: hundred minus the sum of percentage moisture, lipids, ash and crude protein. The elemental analysis of the pressed cake was carried out using the standard atomic absorption spectroscopy (AAS) and titrimetric methods.

The extracted butter samples (solidified on cooling) were respectively analyzed using the standard methods of analysis (AOCS,1997) as follows: free fatty acid (FFA) (Ca-5a), moisture (Ca 2c-25), peroxide value (PV) (Cd 8 -90), saponification value (SV) (Cd-3-25), iodine value(wijs) (IV) (Cd 1-25), density(Cc10b -25), Lovibond colour (Cc 13e) unsaponifiable (UM) (Ca 6a 40) and p-anisidine Value (AnV) (Cd 18-90) and melting range (Cc 3-25). We computed total oxidation, Totox using (2PV+AnV).

### Statistical analysis

We performed all chemical analyses in triplicate. The data generated were analyzed using the SPSS software version 17. We computed the means and standard deviation and, performed the analysis of variance (ANOVA) $P < 0.05$ to find significant differences between means.

## RESULT

Tables 1 and 2 respectively show the proximate composition of the kernels and the physicochemical characteristics of the expressed butter from the ungerminated and germinated kernels. Table 3 shows the fatty acid composition of the butter, while the result of the elemental analysis was presented in Table 4.

**Table 1.** Proximate composition of different shea kernel (dry weight basis).

| Parameter (%) | Germinated | Ungerminated |
|---|---|---|
| NFE | 38.34 ± 0.1[a] | 36.88 ± 0.003[b] |
| Lipid | 48.87 ± 0.03[a] | 51.52 ± 0.05[b] |
| Protein | 10.34 ± 0.04[a] | 8.58 ± 0.02[b] |
| Ash | 2.95 ± 0.02[a] | 3.10 ± 0.01[a] |

Rows with different alphabets are significantly different (P < 0.05).

**Table 2.** Physicochemical characteristics of the expressed shea butter.

| Parameter/Units | Ungerminated | Germinated |
|---|---|---|
| FFA (%) | 0.22 ± 0.02[a] | 1.66± 0.06[b] |
| SV (mg/KOH/g) | 223.33 ± 0.58 [a] | 239.0 ± 2.0[b] |
| PV (meq oxygen/kg oil) | 0.10 ± 0.02 [a] | 8.14 ± 0.59[b] |
| p-Anisidine value (meq oxygen/kg oil) | 0.03 ± 0.11[a] | 3.17 ± 0.21[b] |
| Totox | 0.23 | 19.45 |
| IV (wijs) | 33.06 ± 0.57 [a] | 30.29 ± 0.4 [b] |
| Density (g per ml) at 40°C | 0.880 ± 0.00 [a] | 0.870 ± 0.00[b] |
| Unsaponifiable matter (%) | 7.77 ± 0.07 [a] | 9.39 ± 0.03 [b] |
| Moisture (%) | 0.004 ± 0.002[a] | 0.004 ± 0.001[a] |
| *Melting range (°C) | 32 - 33 | 33 - 35 |
| Colour (Lovibond 5¼"cell) | 1.2Red[a] 13Yellow[a] | 1.3Red[a] 12.4Yellow[a] |

Rows with different alphabets are significantly different (P < 0.05). *, Was not subjected to ANOVA analysis.

**Table 3.** Fatty acid composition of the butter extracted from the germinated and ungerminated shea kernel.

| Fatty acid (wt %) | Ungerminated | Germinated |
|---|---|---|
| Palmitic acid (C16: 0) | 5.89 ± 0.02[a] | 5.88 ± 0.04[b] |
| Stearic (C18:0) | 41.87 ± 0.01[a] | 42.81 ± 0.03[b] |
| Oleic (C18:1) | 45.12 ± 0.03[a] | 44.20 ± 0.02[b] |
| Linoleic (C18:2) | 5.34 ± 0.04[a] | 3.65 ± 0.03[b] |
| Linolenic (C18:3) | 0.94 ± 0.04[a] | 0.40 ± 0.07[b] |
| Arachidonic (C20:0) | 0.84 ± 0.03[a] | 0.49 ± 0.05[b] |
| Total | 99.47 | 98.57 |

Rows with different alphabets are significantly different while rows with the same alphabets are not significantly different.

## DISCUSSION

Shea tree is exploited because of its high content of solid fat (shea butter) stored in the kernel reserve. The reserves are mainly in the form of oil, protein, and starch in embryo or endosperm, and their relative amounts vary in different species (Ming-Feng et al., 2009). The seed germination and the subsequent seedling growth need large amounts of energy and nutrition, which can be provided only by seed reserves, because the germinating seed lacks a mineral uptake system and photosynthetic apparatus (Bewley, 1997). Therefore, any biochemical action or otherwise that tends to negatively affect the butter depletion or yield must be properly managed. Table 1 shows that germination process in shea kernel for a period of 9 days significantly modified the constituents of the kernel. The lipid content with respect to the ungerminated decreased from 51.52 to 48.87%, the protein content increased from 8.58 to 10.34%, while the NFE also increased in the germinated kernel. However,

**Table 4.** Elemental composition of the pressed cake shea kernels.

| Parameter | Germinated | Ungerminated |
|---|---|---|
| Calcium (%) | $0.29 \pm 0.04^a$ | $0.24 \pm 0.7^a$ |
| Magnesium (%) | $0.04 \pm 0.04^a$ | $0.03 \pm 0.03^a$ |
| Sodium (%) | $1.2 \pm 0.1^a$ | $1.68 \pm 0.00^b$ |
| Potassium (%) | $23.32 \pm 0.00^a$ | $23.52 \pm 0.02^b$ |
| Chromium (ppm) | $0.04 \pm 0.00^a$ | $0.05 \pm 0.01^a$ |
| Manganese (ppm) | $0.11 \pm 0.02^a$ | $0.17 \pm 0.03^a$ |
| Iron (ppm) | $3.37 \pm 0.82^a$ | $2.20 \pm 0.4^a$ |
| Cobalt (ppm) | $0.03 \pm 0.01^a$ | $0.03 \pm 0.01^a$ |
| Nickel (ppm) | $0.06 \pm 0.03^a$ | $0.07 \pm 0.04^a$ |
| Copper (ppm) | $2.13 \pm 031^a$ | $2.87 \pm 0.12^b$ |
| Zinc (ppm) | $2.2 \pm 0.10^a$ | $2.07 \pm 0.11^a$ |
| Phosphorous (ppm) | $9.4 \pm 0.92^a$ | $8.93 \pm 0.65^a$ |
| Sulphur (ppm) | $2.07 \pm 0.31^a$ | $1.83 \pm 0.12^a$ |
| Selenium (ppm) | $0.03 \pm 0.01^a$ | $0.02 \pm 0.00^b$ |

Rows with different alphabets are significantly different while rows with the same alphabet are not significantly different

there was no significant difference between the ash content of the germinated and ungerminated shea kernels.The finding in this study, in relation to the depletion of butter in the germinated kernel, suggests that butter is mobilized, and consumed during the germination process. This implies that for maximum butter yield, all the conditions necessary for shea kernel germination, must be eliminated or reduced to the barest minimum. This is because from this study, a shea kernel processor that processed predominantly germinated feedstock will incur a butter loss of 26.5 Kg per metric tonne of kernel. It was observed in the course of this study that germination of shea kernel occurs when the shea fruit are heaped prior to processing. This may not be noticed without a cursory look at the kernels because when the germination process is no longer sustainable in the heap, the plumule or radicle dies back, dries up and ultimately detached in the process of packing or sort (where it is applicable). The kernel from the external will look unperturbed and intact (Figure 1).

Fatty acid formation due to the hydrolysis of the triacylglycerol in shea butter is one of the major constraints the butter processors strive to minimize in order to produce high quality butter. Table 2 shows that germination increased the FFA by 7-folds from 0.22% in the ungerminated to 1.66%, however, FFA values for both the germinated and ungerminated kernel were within grade A according to UEMOA standard. In addition to the fact that UEMOA standard is due for review, the 7-fold increase in ffa means much to a butter refiner that will lose 14.4 kg butter as soap stock during neutralization stage. Again, very low ffa improves shelf life of butter therefore attracts a premium price. The elevated free fatty value could be attributed to the germination process even though that it is a known fact that hydrolysis could proceed via microbial, enzymatic and autocatalytic pathways. This is because the prompt sterilization of the kernel used in this study must have eliminated the possibility of microbial activities, and the time lag prior to analysis may not be enough for auto hydrolysis. This finding is in line with those reported by other researchers. For example, Ming-Feng et al. (2009) reported that during germination, triacylglycerol stored in seeds are converted into glucose and a variety of essential metabolites. The SV, unsaponifiable matter content, density, PV and anisidine value and total oxidation (Totox) were elevated in the germinated kernel with respect to the ungerminated, while IV decreased. These values generally indicated that germination lowers the quality parameters of shea butter, although in this study, the values expectedly appeared to be small given the fact that the kernels were processed 9 days after picking. This is particularly true for p-anisidine and PVs which respectively measure the secondary and primary oxidative states of the butter. These values will definitely increase in the actual current processing conditions in the field. The melting range was also wider. The moisture content was not affected. The rise in SV suggests the presence of lower molecular weight fatty acids, while the decreased IV on the other hand suggests the action of desaturase enzymes.

The influence of germination on shea butter oxidation was evaluated by measuring the PV of the extracted butter. The data show that the PV was significantly ($p < 0.05$) higher in the germinated kernel (8.14 ± 0.59) than the ungerminated kernel (0.10 ± 0.02). This may indicate that germination process encourages lipid oxidation albeit both values fell within acceptable range by the international market. It is therefore envisaged that prolonged germination period may have more deteriorating effect with respect to lipid oxidation on the quality of butter extracted thereof.

The percentage unsaponifiable matter in both the ungerminated (7.77%) and germinated (9.39%) was significantly ($p < 0.05$) different. This difference in the unsaponifiable matter content could be attributed to the exudation of extractable unsaponifiable metabolite produced during the germination process. It may also be one of the reasons for the wide variation in the range of unsaponifiable matter reported for shea butter. This is because reports in the literature seldom state that condition of the shea kernel employed the work. The composition of the unsaponifiable matter in germinating shea kernel will be investigated to ascertain the source of the increment.

The fatty acid composition of the extracted butter from germinated and ungerminated shea kernels was determined and presented in Table 3. In this study, there was a significant ($P < 0.05$) decrease in the percentage fatty acid composition of the usual six major fatty acids found in shea butter namely [palmitic, (C16:0) stearic (C20:0), oleic (C18:1), linoleic (C18:2) linolenic (C18:3) and arachidonic acids (C18:0)] in the germinated kernel compared to the ungerminated. The fatty acids from both samples of butter contained oleic acid (C18:1) and linoleic acid (C18:2) in the highest amount, followed by palmitic acid (C16:0) and stearic acid (C18:0), respectively. After 9 days germination period, the content of linoleic acid (C18:2) decreased from 5.34 to 3.65% of total fatty acid. These results are similar to the works of Ming-Feng et al. (2009) and Kristina et al. (2007) which suggest the utilization of unsaturated fatty acid in the germination activities of the shea kernel. This altered fatty acid composition due to germination process observed in this study will definitely have impact on other intrinsic properties, applications and health benefits of the butter. Besides the tree-to-tree variation inherent in shea tree and its fruits, processing of germinated shea kernels may be a source variation in the butter quality. This however will depend largely on the ratio of the germinated to the ungerminated kernel in a process batch to be obvious.

### Elemental analysis

In Table 4, save for potassium copper, sodium and selenium which showed significant ($p < 0.05$) variation in other elements analyzed for in the press cake showed no significant difference. The observed differences may be due to the variation in the soil composition within the sample collection area. This indicates that germination had minimal effect on the elemental constituents of the shea kernel.

### Conclusion

Germination contributes significantly to quality depreciation of shea butter especially with respect to FFAs, PV, and the fatty acid composition. The study showed that linoleic acid was mobilized during germination more than the other constituent fatty acids and the elemental composition was not generally affected. The overriding implication of this finding remained that processing shea kernel batch that has high ratio of germinated compared to ungerminated will result in low quality butter with significantly altered fatty acid composition. It is clear therefore that to improve the quality of shea butter prompt sterilization and sorting out, the germinated kernels is needed.

## ACKNOWLEDGMENTS

The authors express their gratitude to the Acting Executive Director, NIFOR, for his support and permission to publish this work, and Mr. George Mbaeyi for the statistical analysis.

## REFERENCES

Bewley JD (1997) Seed germination and dormancy Plant Cell, 9:1055-1066.

Carette C, Malotaux M, van Leeuwen M, Tolkamp M. (2009). Shea nut and butter in Ghana: Opportunities and constraints for local processing Wageningen University, Wageningen.

Fereidoon Shahidi (2005) Quality Assurance of Fats and Oils Bailey's Industrial Oil and Fat Products, Sixth Edition, Six Volume Set (ed) Fereidoon Shahidi, John Wiley & Sons, Inc.

Iassonova DR, Johnso LA, Hammond EG, Beattie SE (2009). Evidence of an enzymatic source of off flavors in lipoxygenase-null soybeans J. Am. Oil. Chem. Soc. 86:59 -64.

Jøker D (2000) *Vitellaria paradoxa* Gaertn.f. SEED LEAFLET No. 50, www.dfsc.dk.

Knutzon DS, Thompson GA, Radke SE, Johnson WB, Knauf VC, Kridl JC (1992) Modification of *Brassica* seed oil by antisense expression of stearoyl-acyl carrier protein desaturase gene. Proc. Natl. Acad. Sci. 89:2624-2628.

Kristina C, Gretchens S, Jillian D, Margaret D, Emile G, Tracy H, Dan W, Nathan S, Paul H, Ming-Feng Y, Yu-Jun L, Yun L, Hui C, Fan C, Shi-Hua S. (2009) Proteomic Analysis of Oil Mobilization in Seed Germination and Post-germination Development of *Jatropha curcas* J. Proteome. Res. 8(3):1441-1451. DOI: 10.1021/pr800799s.

Munshi SK, Sandhu S, Sharma S (2007). Lipid composition in fast and slow germinating sunflower (*Helianthus annuusl*) Seeds Gen. Appl. Plant Physiol. 33(3-4):235-246.

Nahm HS (2011). Quality Characteristics of West African Shea Butter (*Vitellaria paradoxa*) and Approaches to Extend Shelf-Life Graduate School-New Brunswick Rutgers, The State University of New Jersey.

Napier JA (2007). The production of unusual fatty acids in transgenic Plants, Annu. Rev. Plant Biol. 58:295-319.

Nikiema A, Umali BE (2007) *Vitellaria paradoxa* C.F.Gaertn. In: van der Vossen, H.A.M. and Mkamilo,G.S.(Editors) PROTA14:Vegetable oils/Oléagineux [CD-Rom].PROTA, Wageningen, Netherlands.

Norton T, Sun DW (2008) Recent Advances in the use of high pressure as an effective processing technique in the food industry, Food Bioproc. Technol. 1 (1):2-34.

Official methods and Recommended Practices of The American oil Chemists' Society (1997). 5th edn. AOCS Press, Champaign Illinois.

Official methods of analysis of AOAC International (2010) 18th edn AOAC International, suite 500, 481 North Frederick Avenue, Gaithersburg, Maryland 20877-2417, USA.

Okullo JBL, Omujal F, Agea, JG, Vuzi PC, Namutebi A, Okello JBA, Nyaazi SA (2010). Characterization of Shea butter *Vitellaria paradoxa* C.F. Gaertn) oil from the shea Districts of Uganda, African J. Food Agric. Nut. Develop. 10(1):2070-2084.

Olaniyan AM, Oje K (2007) Quality Character of Shea butter recovered from shea kernel through dry extraction process, J. Food Sci. Technol. 44 (4):404-407.

PORIM Test Method (2004) Determination of fatty acid composition, MPOB p3.4 Palm Oil Research Institute of Malaysia.

Ted L, (2007) Development, fatty acid composition, and storage of drupes and seeds from the endangered pond berry (L *inderamelissifolia)*, Biological Conservation, 137:489-496.

Union Economique Monetaire Ouest Africaine (UEMOA) (accessed on 7 March, 2012) Standards for Unrefined shea Butter. http://www.prokarite.org.

Urbano G, Aranda P, Vı´lchez A, Aranda C, Cabrera L, Porres JM, Lo´pez-Jurado M (2004) Effects of germination on the composition and nutritive value of proteins in Pisumsativum, L. Food Chem. 93:671–679.

# Affiliation of seed size with germination aspects and morphological traits in safflower (*Carthamus tinctorius* L.)

**Mirshekarnezhad B.[1] , G. Abbas Akbari[1], G. Ali Akbari[1] and H. Sadeghi[2]**

[1]Department of Agronomy and Plant Breeding, Aboureihan College, University of Tehran, Iran.
[2]Seed and Plant Certification and Registration Institute, Karaj, Iran.

**Affiliation of seed size with germination rate and percentage, seedling emergence rate and percentage, 50% mean emergence time, cumulative emergence rate, through 4 Safflower cultivars (Isfahan, Goldasht, Padideh, and Sina), in three different sizes of seed, were investigated under two separate experimental conditions in the field and laboratory of Seed and Plant Certification and Registration Institute Karaj-Iran, during 2008 to 2010. Both experiments were conducted in a factorial arrangement based on randomized complete block design, with three replications. The field results showed that cultivars Goldasht and Padideh were superior compare to other cultivars and large seeds with the highest seedling emergence rate, 1000-SW, and seed yield were superior to control and small. In the laboratory, morphological traits consisted of germination percentage, fresh and dry weight of root and shoot were significantly different, so that cultivars Goldasht and padideh with large seeds displayed better reaction. Thus, due to display better performance by cultivars Goldasht and Padideh, also creating vigorous seedlings by large seeds, using cultivars with adequate food storage as well as selecting large seeds with more germination potential to obtain higher yield and good plant establishment should take into consideration.**

**Key words:** Cultivar, Goldasht, mean germination time, morphological traits, vigorous seedling.

## INTRODUCTION

Safflower is a versatile crop that can add considerable value to cropping systems by playing various strategic roles. Safflower was particularly important as an oil and pigment in southern Asia (Iran, Afghanistan, and India), As for all crops, planting seed should be genetically pure, free from seed borne diseases, have a high germination percentage (>80%) and be free from weed and other crop seeds. Many suppliers can provide a copy of equality certificate on request, otherwise samples can be sent to an accredited laboratory for testing. The longevity of oilseeds under normal silo conditions is limited, so fresh seed should be used where possible (http://www. Australianoilseeds.com/-data/assets/pdf, 2013). In general, seed size has effects on many characters both in the field and laboratory tests. Seed size is an important seed quality characteristic affected by variety, environment and management practices. Seed may be viable and germinate but not emerge. A crust may form and inhibit emergence particularly on certain soils or if clear and windy days follow planting.

Crust formation may also be a problem when planting into cold soils, where seed germination and emergence are slowed by low soil temperature (Mike, 2004). Percent of seed emergence and seedling emergence rate are the characteristics that could be observed in the field and could be used as an indicator of seed vigor. These characters usually differ under field stresses such as low temperature, wet or crusted soil in which small seeds of soybean and common bean perform better than large seeds due to seedling from small seeds were less damaged than large seeds (Hoy and Gamble, 1987; Sexton et al., 1994), while the large seeds of winter wheat (Chastain et al., 1995) and mung bean (Amin, 1999) under the same conditions tended to do better in germination. Variation in seed size of chickpea is associated with geographical effects on genetic factors, irregular flowering, number of seeds per pod, and number of pods per plant (Pedersen et al., 1961; Smithson et al., 1985; Miller et al., 2002).

Also, several studies reported that, different sizes of sorghum seeds (Suh et al., 1974), and soybean seeds (Sexton et al., 1994) did not affect yield of these crops. (Singh et al., 1972) reported that, large seeds of soybean had greater supply of stored energy to support early seedling growth and subsequently affected plant growth and development. Similar result obtained by Roy et al. (2008) in rice, implied that, germination rate and seedling vigor index increased with the increase of seed size, while Kaya and Day (2008) reported that, the use of large seeds produced more vigorous seedlings under non saline conditions but the superiority was used up under saline conditions.

However, seed size has been considered to be a significant factor only during the early stage of plant growth, nevertheless, Amin (1999) reported that, 50% of large-seeded mung bean matured earlier than that of small-seeded type. Despite large seed has an advantage of having higher stored energy supply but not all reports demonstrate the effects of seed size on the yield. According to previous studies, planting seed from crops which experience an extend period of warm, wet weather prior to harvest should be avoided. These conditions favor the development of Alternaria which can infect seed and transmit the disease to the next crop. Alternaria can cause newly planted seed to rot in the soil or the damping off of seedlings. The germination percentage of seed can also be markedly reduced by rain on mature crops which causes sprouting in the head (http://www.Australianoilseeds.com/-data/assets/pdf, 2013).

Furthermore, undesirable seed and seeds with low viability which create feeble seedlings lead to minimum germination percentage while, seeds with proper size cause to stability in field emergence during next stages of plant growth. Therefore, this study was conducted to evaluate the affiliation of seed size with germination and its effect on seedling emergence rate, yield and yield components of safflower.

## MATERIALS AND METHODS

This research was carried out via two separate experiments, in the laboratory and research farm of Seed and Plant Certification and Registration Institute, Karaj – Iran, during (2008 to 2010 season). The farm is located at 35°59' N latitude, 50° 75' E longitude, and 1313.00 m of altitude. Soil texture in research farm was loamy clay (Zahedi et al., 2012).

### Laboratory trial

In the laboratory, experimental treatments involved 4 safflower cultivars (Isfahan, Goldasht, Padideh and Sina), in 3 sizes of seed. The seed samples were obtained from Seed and Plant Breeding Institute Karaj-Iran. The seeds were prepared according to their diameters using proper sieves:

(1) Small: 2.0 to 3.0 mm
(2) Large: >3.0 mm
(3) Control: no sorting

After treating the seeds with fungicide, 16 replicates of 25 seeds for each cultivar of 3 different sizes were prepared in 3 replicates, the samples were placed between 2 rolled filter papers (Whatman #2) and dampened with water then the whole rolled papers were put into a sealed plastic bag and placed in growth chamber at 25°C. A seed was considered to be germinated when the emerging radical elongated to 2.0 mm. Germination percentage (GP) recorded every 24 h for 10 days and germination rate was computed by using the following formula as stated by Maguire (1962). After fixing the number of sprouts at the end of 10$^{th}$ day, randomly 10 sprouts were selected, after that sprout length, root and shoot lengths were measured in centimeter, also fresh and dry weights of root and shoot were separately determined in grams.

$$\text{Germination rate} = \frac{\text{No. of normal seedlings}}{\text{Days to first count}} + \ldots + \frac{\text{No. of normal seedlings}}{\text{Days to final count}}$$

### Field trial

By using a randomized complete block design in factorial arrangement with 3 replications, randomly the treatments were distributed to experimental plots formed by 3.00 m long with rows at 5.0 m spacing. Then in each plot the second row was separated into two distances (0.5 and 2.5 m lengths) so that per 2.5 m length involved 40 seeds at a space of 5.0 cm between them. The sprouts in this line were counted daily to determine seedling emergence rate and percentage (SER), (SEP), 50% mean emergence time (MET), and cumulative emergence rate (CER). Moreover to determine yield and yield components, after discarding border rows about 16 plants from each plot were harvested (1.0 m$^2$). Three plants of each plot were selected randomly, then the characters comprised of plant number per plot, head number per plant, seed number per head, seed number per plant, 1000-seeds mass (in grams), head diameter (in millimeter) and plant height (in centimeter) were evaluated.

### Statistical analysis

Analysis of variance (ANOVA) for all investigated parameters and correlation analysis were performed by using MSTAT-C computer software, version 2.10 (Michigan State University), and SAS

**Table 1.** Analysis of variance for measured characters in the laboratory.

| S.O.V | df | Mean squares (ms) Germination percentage (%) | Germination rate (ng/d)* | Fresh weight of shoot | Dry weight of Shoot | Fresh weight of root | Dry weight of root | Sprout length | Shoot length | Root length |
|---|---|---|---|---|---|---|---|---|---|---|
| R(Rep) | 2 | 25.86 | 7.244$^{ns}$ | 0.5045$^{ns}$ | 0.000004 | 0.000784$^{ns}$ | 0.0000485ns | 4.4955** | 5.5151** | 0.2773$^{ns}$ |
| Cultivar (C) | 3 | 896.07** | 53.317** | 2.2588** | 0.003155** | 0.050143** | 0.0000419ns | 1.8474* | 2.7507** | 3.2101** |
| Seed size (S) | 2 | 146.19** | 17.780** | 0.812** | 0.009691** | 0.037417* | 0.0013908** | 0.5139$^{ns}$ | 0.0043$^{ns}$ | 0.2338$^{ns}$ |
| C×S | 6 | 16.27$^{ns}$ | 0.810$^{ns}$ | 0.0347$^{ns}$ | 0.000221$^{ns}$ | 0.00071$^{ns}$ | 0.0000017$^{ns}$ | 0.1869$^{ns}$ | 0.2574$^{ns}$ | 0.2539$^{ns}$ |
| Error | 22 | 12.53 | 2.843 | 0.09165 | 0.0001458 | 0.006765 | 0.0000303 | 0.5437 | 0.3966 | 0.4959 |
| C.V.(%) | | 4.66 | 9.29 | 11.59 | 9.14 | 22.92 | 18.91 | 4.32 | 11.61 | 6.13 |

** = Significant at 1% level, * = significant at 5 % level, ns = non-significant, *= number of germination per day.

**Table 2.** Means comparison of laboratory characters using Duncan multiple range test (DMRT).

| Variable | S.O.V | Germination percentage (%) | Germination rate (ng/day)* | Fresh weight of shoot (g) | Dry weight of shoot (g) | Fresh weight of root (g) | Dry weight of root (g) | Sprout length (cm) | Shoot length (cm) | Root length (cm) |
|---|---|---|---|---|---|---|---|---|---|---|
| Cultivar | (Esfahan) | 80.56$^{b}$ | 20.41$^{a}$ | 2.16$^{c}$ | 0.114$^{c}$ | 0.31$^{b}$ | 0.02944$^{a}$ | 16.57$^{b}$ | 5.59$^{a}$ | 10.79$^{c}$ |
| | (Goldasht) | 74.22$^{c}$ | 17.95$^{b}$ | 2.88$^{b}$ | 0.157$^{a}$ | 0.28$^{b}$ | 0.04933$^{a}$ | 17.52$^{a}$ | 5.88$^{a}$ | 11.26$^{bc}$ |
| | (Padideh) | 86.00$^{a}$ | 19.42$^{ab}$ | 3.18$^{a}$ | 0.135$^{b}$ | 0.44$^{a}$ | 0.04433$^{a}$ | 17.36$^{a}$ | 5.60$^{a}$ | 11.77$^{ab}$ |
| | (Sina) | 62.78$^{d}$ | 14.83$^{c}$ | 2.23$^{c}$ | 0.122b$^{c}$ | 0.40$^{a}$ | 0.03089$^{a}$ | 16.79$^{ab}$ | 4.62$^{b}$ | 12.16$^{a}$ |
| Seed size | (Control) | 73.75$^{b}$ | 19.54$^{a}$ | 2.34$^{b}$ | 0.105$^{c}$ | 0.304$^{b}$ | 0.0197$^{c}$ | 16.07$^{a}$ | 5.43$^{a}$ | 11.4$^{a}$ |
| | (Large) | 79.92$^{a}$ | 17.24$^{b}$ | 2.86$^{a}$ | 0.162$^{a}$ | 0.415$^{a}$ | 0.0408$^{a}$ | 16.22$^{a}$ | 5.12$^{a}$ | 11.13$^{a}$ |
| | (Small) | 74.00$^{b}$ | 17.68$^{b}$ | 2.64$^{a}$ | 0.129$^{b}$ | 0.358$^{ab}$ | 0.0268$^{b}$ | 17.07$^{a}$ | 5.4$^{a}$ | 11.55$^{a}$ |

Means within the same column, followed by the same letters are not significantly different at 0.05 level by DMRT. *= Number of germination per day.

procedures (SAS Institute., 1988). Significant differences among means were compared by Duncan's multiple range test ($P < 0.05$).

## RESULTS AND DISCUSSION

### Laboratory

The analysis of resulting data for germination of safflower cultivars in 3 sizes indicated that, there were significant differences among cultivars for all measured traits except dry weight of root (Table 1). According to means comparison, the cultivar Padideh showed the highest amount of GP (86%), and greater fresh weight of shoot (3.18 g), and the cultivar Goldasht stood out with greatest dry weight of shoot (0.157) (Table 2). Also among seed sizes the large seeds showed the highest amount of GP (79.92%), fresh and dry weight of root presented better development compare to control and small. The results are in agreement with the observations of Kaya and Day (2008), Saranga et al. (1998) on sunflower, Roy et al. (2008) on rice and Chiangmai et al. (2006) on mung bean, they found better behavior of large seeds for most of the investigated parameters.

### Field

The results of investigated parameters showed significant differences ($P < 0.01$) among cultivars for all traits except number of plant per plot (Table

**Table 3**. Analysis of variance for measured characters in the field.

| S.O.V | df | Mean squares (MS) Seedling emergence percentage (%) | 50% Mean germination time | Seedling emergence rate (nse/d)* | Cumulative emergence rate (nse/d)* | Plant height (cm) | 1000-seed weight (g) | Plant number (n) | Head per plant (n) | Seed per head (n) | Head diameter (mm) | Seed yield (g/m²) |
|---|---|---|---|---|---|---|---|---|---|---|---|---|
| R(Rep) | 2 | 261.78** | 0.043** | 0.72** | 72.181** | 5.58$^{ns}$ | 4.565$^{ns}$ | 91.194* | 37.75$^{ns}$ | 24.78$^{ns}$ | 1.08$^{ns}$ | 1406.86** |
| Cultivar(C) | 3 | 1100.8** | 0.071** | 2.31** | 209.465** | 814.92** | 1132.9** | 42.59$^{ns}$ | 1681.0** | 1306.296** | 57.124** | 14285.36** |
| Seed size (S) | 2 | 435.53** | 0.085** | 2.48** | 84.866** | 64.33$^{ns}$ | 340.57** | 17.528$^{ns}$ | 112.58$^{ns}$ | 36.11$^{ns}$ | 3.961$^{ns}$ | 6161.86** |
| C×S | 6 | 27.194$^{ns}$ | 0.005$^{ns}$ | 0.048$^{ns}$ | 2.040$^{ns}$ | 20.22$^{ns}$ | 17.43$^{ns}$ | 20.23 $^{ns}$ | 449.81* | 42.520$^{ns}$ | 1.148$^{ns}$ | 224.08$^{ns}$ |
| Error | 22 | 33.051 | 0.004 | 0.108 | 3.964 | 62.49 | 13.84 | 20.982 | 124.93 | 68.475 | 4.164 | 196.89 |
| C.V.(%) | | 7.74 | 8.20 | 8.87 | 9.39 | 7.01 | 8.03 | 24.39 | 29.54 | 18.08 | 7.11 | 6.80 |

**= Significant at 1% level; * = significant at 5% level, ns = non-significant, * = number of seedling per day.

**Table 4.** Means comparison of field characters using Duncan multiple range test (DMRT).

| Variable | S.O.V | Seedling emergence percentage (%) | Mean germination time ng/d)* | Seedling emergence rate (g) | Cumulative emergence rate (g) | Plant height | 1000-SW | Head per plant | Seed per head | Head diameter (cm) | Seed yield (g/m$^2$) |
|---|---|---|---|---|---|---|---|---|---|---|---|
| Cultivar | (Esfahan) | 75.889$^b$ | 0.802$^b$ | 22.189$^b$ | 22.189$^b$ | 125.333$^a$ | 37.463$^c$ | 38.000$^b$ | 53.222$^a$ | 28.200$^b$ | 204.778$^b$ |
| | (Goldasht) | 75.556$^b$ | 0.803$^b$ | 22.167$^b$ | 22.167$^b$ | 114.111$^b$ | 62.771$^a$ | 23.556$^c$ | 43.444$^b$ | 30.956$^a$ | 255.444$^a$ |
| | (Padideh) | 86.222$^a$ | 0.902$^a$ | 25.967$^a$ | 25.967$^a$ | 103.111$^c$ | 42.358$^b$ | 33.556$^{bc}$ | 56.667$^a$ | 30.333$^a$ | 207.111$^b$ |
| | (Sina) | 59.444$^c$ | 0.684$^c$ | 14.478$^c$ | 14.478$^c$ | 108.444$^{bc}$ | 42.730$^b$ | 56.222$^a$ | 29.778$^c$ | 25.378$^c$ | 157.889 |
| Seed size | (Control) | 68.833$^b$ | 0.879$^a$ | 4.125$^a$ | 23.950$^a$ | 110.08$^a$ | 41.379$^c$ | 35.0$^a$ | 47.17$^a$ | 29.38$^a$ | 186.0$^c$ |
| | (Large) | 80.750$^a$ | 0.711$^c$ | 3.225$^c$ | 18.642$^c$ | 114.25$^a$ | 51.967$^a$ | 37.42$^a$ | 43.83$^a$ | 28.46$^a$ | 230.750$^a$ |
| | (Small) | 73.250$^b$ | 0.804$^b$ | 3.783$^b$ | 21.008$^b$ | 113.92$^a$ | 45.645$^b$ | 41.08$^a$ | 46.33$^a$ | 28.32$^a$ | 202.167$^b$ |

Means within the same column, followed by the same letters are not significantly different at 0.05 level by DMRT. *= Number of germination per day.

3. Also significant differences observed among different sizes of seed for characters SER, SEP, MET, CER, 1000-SW and seed yield, for which cultivar Padideh with higher amount of SEP (80.75%), MET (0.902%), SER (25.96), CER (25.96), then cultivar Goldasht with greater 1000-SW (62.77 g) and seed yield (255.4 g/m$^2$) were superior to other cultivars. Among seed sizes large seeds with the highest amount of SEP (80.7%), 1000-SW (51.96 g), seed yield (230.7 g/m$^2$), plant height (114.25 cm) were superior, after that control seeds performed better reaction compare to small in each cultivar as shown in Table 4. Similar results obtained by Elliott and Rakow (1999) on oilseed rape and Matrinelli- Seneme et al. (2000) on corn seed. Apart from seed size variation among varieties, each variety may have unique features in which it can adapt to a specific condition (Amin, 1999). Large seeds could produce larger bean sprouts including sprout weight, head diameter, and stem diameter as well. This can be explained that, larger cotyledons had produced larger hypocotyls Black (1956), and Burris et al. (1971). Seed size × cultivar interaction

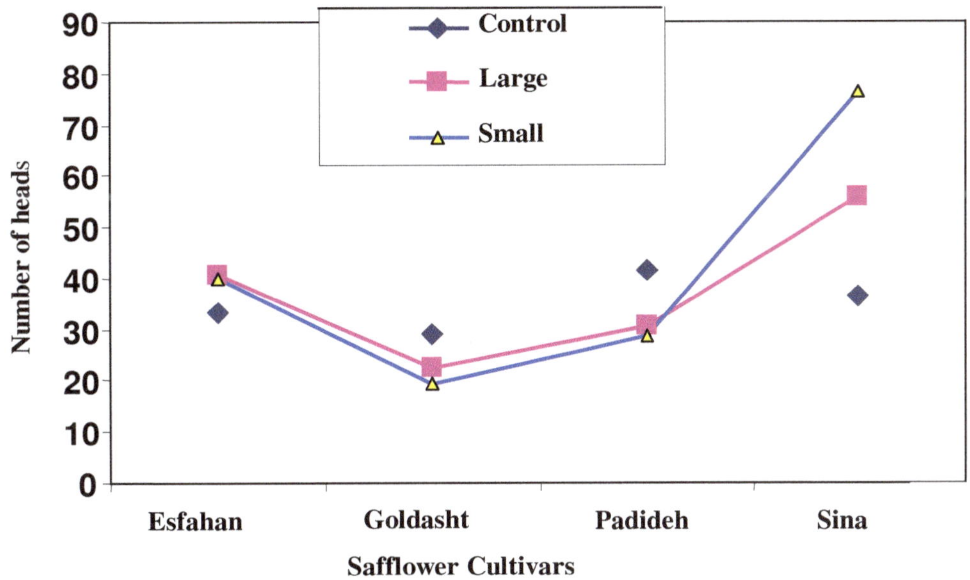


**Figure 1**. Cultivar × Seed size interaction on number of head per plant.

only for the number of heads per plant was significant ($P < 0.05$), and its performance is shown on Figure 1. The coefficient of variation were from 6.8% (seed yield) to 29.5% (head per plant) which is an indicator of accurate performance of experiments and records of agronomic characters.

## Correlation coefficient between laboratory and field traits

The evaluation of correlation coefficients between investigated parameters in the field and laboratory disclosed that, SEP in the field was positively correlated with GP and fresh weight of shoot in the laboratory as shown in Table 5. Positive correlation observed between SEP in the field with shoot length in the laboratory, on the other hand MET had positive correlation with GR but revealed negative correlation with dry weight of root. SER in the field with GR and shoot length had positive correlation, but indicated negative correlation with dry weight of root in the laboratory. There was positive correlation between CER in the field with GR, GP and shoot length under laboratory condition. Number of seeds per head was positively correlated with GR and GP. In addition positive correlation was found between head diameter with the characters including GR and GP, sprout length and shoot length. It figured out that, large seed with more germination vigor was the character which could be able to produce sprouts with better germination behavior and had compensated seed yield and other yield components. Seed yield disclosed positive correlation with fresh and dry weight of root and shoot length in the laboratory that originated from producing sprouts with higher weight by large seeds due to their ability for having more efficient photosynthesis under non-stress condition in the laboratory.

One observation on the effect of seed size and seed number was that, both characteristics affected yield. However, the characters such as number and seed size may both affect and compensate each other in determining yield (Spaeth and Sinclair, 1984; Board et al., 1999) and the main factor which directly contributed to yield was dependent upon plant species and varieties. The correlation between yield and yield components with seed size has been estimated in different varieties and crop plants but not all reports presented the same conclusions. Length and width of soybean seeds were positively correlated with final seed size (Fraser et al., 1982) but 1000-kernel weight of seeds harvested and the seeds per head of spring wheat were not significantly correlated with initial seed size Austenson and Walton (1970).

## Conclusion

In conclusion, regarding background of each cultivar that influences seed vigor and plant establishment in the field, cultivars Goldasht and Padideh were superior to other cultivars for most of the investigated characters associated with viability. Subsequently, germination characteristics were significantly affected with seed size and among 3 sizes of seed, large seeds displayed better performance than other sizes of seed. Considering grain as the final sink during loading process with assimilate at the final step of ripening; imply that larger seeds with bigger area of cotyledon could be able to store much sufficient food in this period rather than small.Due to use of excessive seed at planting time by farmers that arisen

**Table 5**. Correlation coefficients among measured characters in the field and laboratory.

| Field \ Lab | Germination percentage (%) | Germination rate (ng/d)* | Fresh weight of shoot (g) | Dry weight of shoot (g) | Fresh weight of root (g) | Dry weight of root (g) | Sprout length (cm) | Shoot length (cm) | Root length (cm) |
|---|---|---|---|---|---|---|---|---|---|
| Seedling emergence percentage (%) | 0.936** | 0.477 | 0.762** | 0.501 | 0.260 | 0.377 | 0.166 | 0.658* | -0.462 |
| Mean germination time (%) | 0.454 | 0.777** | 0.209 | -0.360 | -0.297 | -0.672* | 0.526 | 0.489 | -0.093 |
| Seedling emergence rate (nse/day)* | 0.491 | 0.836** | 0.163 | -0.401 | -0.453 | -0.696* | 0.533 | 0.622* | -0.263 |
| Cumulative emergence rate (nse/day) | 0.675* | 0.856** | 0.365 | -0.203 | -0.300 | -0.503 | 0.498 | 0.685* | -0.317 |
| Plant height (cm) | 0.108 | 0.275 | -0.473 | -0.007 | -0.498 | 0.016 | -0.503 | 0.327 | -0.788** |
| 1000-Seed weight (g) | -0.015 | -0.303 | 0.508 | 0.811** | -0.163 | 0.485 | 0.471 | 0.401 | -0.130 |
| Plant number | 0.376 | 0.373 | 0.021 | 0.355 | -0.358 | 0.151 | 0.028 | 0.662* | -0.575 |
| Head per plant (n) | -0.442 | -0.499 | -0.465 | -0.342 | 0.346 | 0.062 | -0.443 | -0.665* | 0.358 |
| Seed per head (n) | 0.836** | 0.866** | 0.359 | -0.127 | -0.090 | -0.259 | 0.135 | 0.574 | -0.499 |
| Head diameter (mm) | 0.617* | 0.631* | 0.616* | 0.261 | -0.323 | -0.149 | 0.730** | 0.804** | -0.322 |
| Seed yield (g/m$^2$) | 0.527 | 0.226 | 0.616* | 0.778** | -0.205 | 0.437 | 0.348 | 0.793** | -0.563 |

** = Significant at 1 % level, * = Significant at 5 % level, * = Number of germination per day, * = Number of seedling per day.

from uncertainty of field emergence and undesirable density, the selection of seeds based on proper size with more germination potential that would decrease the utilization of seed per hectare is recommended.


## ACKNOWLEDGEMENTS

The author appreciates the faculty of Abourihan College, University of Tehran for practical guidance and the staff of Seed and Plant Certification and Registration Institute of Karaj-Iran, for providing accommodations.


### REFERENCES


Amin AD (1999). Influence of seed size on the performance of mung bean varieties under postrice and upland cropping systems. Asia regional center-AVRDC [online URL:www.arc.avrdc-org/pdf-files/Amin (17-N). pdf] accessed on July 13, 2006.

Austenson HM, Walton PD (1970). Relationships between initial seed weight and mature plant characters in spring wheat. Can. J. Plant Sci. 50:53-58.

Black JN (1956). The influence of seed size and depth of sowing on pre-emergence and early vegetative growth in subterranean clover (*Trifolium subterraneum* L.). Aust. J. Agric. Res. 7:93-109.

Board JE, Kang MS, Harville BG (1999). Path analyses of the yield formation process for late-planted soybean. Agron. J. 91:128-135.

Burris JS, Edje OT, Wahab AH (1971). Effects of seed size on seedling performance in soybeans. II seedling growth and photosynthesis and field performance. Crop sci. 13:207-210.

Chastain T, Ward GKJ, Wysocki DJ (1995). Stand establishment responses of soft white wheat to seedbed residue and seed size. Crop sci. 35:213-218.

Chiangmai PN, Laosuwan P, Anyuwat AW (2006). The effect of mung bean seed size on germination ability, bean sprout production and agronomic characters. Silpakorn Univer. Int. J..6:1-2.

Elliott RH, Rakow GFW (1999). Influence of seed size on the agronomic performance of oilseed rape. AAFC Research center, 107 science place, Saskatoon, saskach ewan, Canada S7N. X2.

Fraser J, Egli DB, Leggett JE (1982). Pod and seed development in soybean cultivars with differences in seed size. Agron. J. 74:81-85.

Hoy DJ, Gamble EE (1987). Field performance in soybean with seeds of different size and density. Crop sci. 27:121-126.

http://www. Australianoilseeds.com/-data/assets/pdf-file; visited on June 26, 2013.

Kaya MD, Day S (2008). Relationship between seed size and Nacl on germination, seed vigor and early seeding growth of sunflower. Afr. J. Agric. Res. 3(11):787-791.

Maguire JD (1962). Seed of germination-aid in selection and evaluation for seedling emergence and vigor. Crop Sci. 2:176-177.

Matrinelli-Seneme A, Zenotto MD, Nakagawa J (2000). Seed size and shap effect on corn seed quality, cultivar AL-34. Revista Brasilevia de Sementes 22(1):232-238.

Mike O (2004). Seeding Rates for Small Grains in Arizona,Cooperative Extension, College of Agriculture and Life Sciences. The University of Arizona.

Miller P, McKay K, Jenks B (2002). Montguide: Montan State University, Extension Service; 2002. Growing Chickpea in the Northern Great Plains. MT200204 AG 3/2002, (Available from: http://www.montana.edu/wwwpb/pubs/mt200204.pdf)

Pedersen MW, Jones LG, Rogers TH (1961). Producing Seeds of the Legumes. In: Stefferud A, editor. Seeds. US Department of Agriculture. pp. 171–181. Yearbook.

Roy SKS, Hamid A, Miah MG, Hashem A (2008). Seed size variation and its effects on germination and seedling vigour in Rice. J. Agron. Crop Sci. 176(2):79-82.

Saranga Y, Levi A, Horcicka P, Wolf S (1998). Large sunflower seeds are characterized by low embryo vigor. J. Am. Soc. Hort. Sci. 123(3):470-474.

Sexton PJ, White JW, Boote KJ (1994). Yield-determining processes in relation to cultivar seed size of common bean. Crop Sci. 34:84-91.

Singh JN, Tripathi SK, Negi PS (1972). Note on the effect of seed size on germination, growth and yield of soybean (*Glycine max* L.). Indian J. Agric. Sci. 42(1):83-86.

Smithson JB, Thompson JA, Summerfield RJ (1985). Chickpea (*Cicer arietinum* L.) In: Summerfield RJ, Roberts EH, editors. Grain Legume. Crops.London: Collins Publications, pp. 312–390.

Spaeth SC, Sinclair TR (1984). Soybean seed growth. II. Individual seed mass and component compensation. Agron. J. 76:128-133.

Suh H, Casady AJ, Vanderlip RL (1974). Influence of sorghum seed weight on the performance of the resulting crop. Crop Sci. 14:835-836.

Zahedi H, Shirani AH, Tohidi Moghadam HR (2012). Effect of zeolite and selenium foliar application on growth, production and some physiological attributes of three canola (*Brassica Napus* L.) cultivars subjected to drought stress. Revista Cientifica UDO Agricola 12(1):135-142.

# Resistant starch content, molecular structure and physicochemical properties of starches in Virginia-grown corn, potato and mungbean

**Yixiang Xu[1]*, Cory Grizzard[1], Edward N. Sismour[1], Harbans L. Bhardwaj[1] and Zhenxing Li[2]**

[1]Agricultural Research Station, Virginia State University, Petersburg, VA 23806, USA.
[2]Food Safety Lab, Ocean University of China, Qingdao, Shangdong Province, P. R. 266003 China.

**Starches were isolated from Virginia-grown corn, potato, and mungbean, and their resistant starch content, molecular structure and physicochemical properties were investigated for potential applications. All starches, extracted with combination of chemical and physically method exhibited high purity with low protein, fat and ash, and high carbohydrate. Potato starches had the highest resistant starch content, while mungbean starches showed the highest amylose content. Amylose content as well as the starch granule size and structure were responsible for resistance to digestibility. Compared to their mungbean and corn counterparts, potato starches had the highest amylopectin molecular weights and largest granular size. A typical A-and B-type crystalline structure was assigned to corn and potato starches, respectively, while mungbean starches had a $C_A$-type crystalline pattern. Both potato and mungbean starch granules were smooth, oval and irregular ellipsoids, while corn starches had polyhedral granules. The gelatinization transition temperatures ($T_o$, $T_p$, and $T_c$) of the starches were significantly different, with the order of corn> mungbean> potato. Water absorption capacity of corn starches was lower than that of potato and mungbean starches. The results would assist food scientists in determining the potential end-uses of starches.**

**Key words:** Resistant starch, molecular structure, physicochemical properties, starch, Virginia.

## INTRODUCTION

Starch is the most abundant storage reserve carbohydrate, and is mainly found in the seed, root, tubers, and fruits of plants. It is the most important carbohydrate in the human diet, and accounts for 60 to 70% of total dietary energy intake (National Nutrition Monitoring Bureau, 1991). Historically, starch has been disparaged as an underlying cause of weight gain and obesity, because it was considered to be fully digestible and absorbable in the small intestine. However, recent human studies have indicated that a starch fraction, termed 'resistant starch', is indigestible in the small intestine and enters the large intestine where it is fermented by colonic bacteria to short-chain fatty acids (Sajilata et al., 2006), which has the potential to prevent colon cancer, diabetes, atherosclerosis, and obesity-related complications (Sharma et al., 2008; Topping and Clifton, 2001).

Corn and potato are starch-rich traditional crops grown in Virginia with starch accounting for up to 70% of total solids in kernels and tubers (Shapours and Salassi, 2006; Stark and Love, 2003). The productions of these two crops in 2010 were approximately 21 million bushels and 952,000 cwt, and ranked 2nd and 7th in production value, respectively (USDA-NASS, 2011). Beside the traditional crops, non-traditional food legume crops in Virginia are being researched extensively as alternative crops for Virginia farmers to promote income diversification due to changes in tobacco production (Bhardwaj et al., 1996, 1999). Tobacco had been the historical mainstay of many local farm-based economies in Virginia until production fell significantly from the early 1990's to 2009 (USDA-NASS, 2011). Parallel initiatives should be taken during

*Corresponding author. E-mail: yixu@vsu.edu

adoption or establishment of new crops to develop or enhance their utilization to increase cash value. Mungbean, chickpea and pigeon pea are non-traditional legume crops in Virginia that show great potential for further development. Generally, protein and carbohydrate are the two main constituents in legume seed. Legume seeds contain about 60% carbohydrate with starch as the major carbohydrate component. In addition, compared to other sources such as cereals, legume starches have reduced digestibility since they are characterized by high resistant starch contents (Hoover and Zhou, 2003).

Composition and properties of starch vary among crop cultivars due to genetics, environmental factors and agronomic practices (Hoover and Ratnayake, 2002). Currently, there is no information regarding the compositions and properties of starches from Virginia-grown corn, potato and mungbean. Therefore, the objectives of this study were to evaluate the compositions and properties of starches extracted from selected Virginia-grown corn, potato and mungbean varieties, with the intent to provide the baseline information for: (1) plant breeders to develop or select cultivars with their starches having certain functional properties, and (2) food scientists to determine potential end-uses of these starches.

## MATERIALS AND METHODS

Two corn varieties (Southern States 731 and Pioneer 35F37) were obtained from a local farmer. Two mungbean cultivars (Berken and TexSprout) were cultivated at Randolph Farm of Virginia State University. The corn and mungbean were harvested in the fall of 2011. The kernels and seeds were cleaned and dried prior to starch extraction. Two potato cultivars (Atlantic and Superior), were harvested from Virginia Polytechnic Institute and State University's Eastern Shore Agricultural Research and Extension Center in the summer of 2012. All chemicals used are reagent-grade.

### Starch isolation

Starches were extracted from the corn kernels, mungbean seeds, and potato tubers following the method described by Vasanthan (2001) with some modifications. Briefly, corn kernels and mungbean seeds were softened by steeping in water containing 0.2% (w/v) $SO_2$ at 50°C for 24 h with a steep water-to-seed ratio of 2:1 (w/w). Seed coats were removed manually from the mungbean seeds. The samples were macerated in water with ice in a blender until smooth slurry formed. Starch slurries were filtering through nylon mesh. Potato tubers were peeled and macerated in water with ice as described above. No steeping was necessary for potato tubers. Starch was recovered by washing, filtering, and centrifuging. The recovered starch was oven dried at 40°C prior to analysis.

### Proximate composition

Analyses of crude protein, crude lipid, and ash were performed using the methods as described by Association of Official Analytical Chemists (AOAC, 2000). Crude protein content was measured using the combustion method (AOAC, 2000) with a Vario MAX CN (Elementar Americas, Inc., Mt. Laurel, NJ, USA). Organic N in the samples was converted into nitric oxide by combustion, which was reduced further to molecular nitrogen. Crude protein content was calculated by multiplying nitrogen concentration by a conversion factor of 6.25. Oil was extracted from dried and ground samples using hexane in a Soxhlet apparatus for 8 hrs (AOAC method 948.22, 2000). Hexane was separated from the extracted oil by evaporation at ambient temperature in a fume hood. The cups were cooled and the crude lipid content weighed. Ash was determined using dried samples following AOAC method 950.49 (2000). The samples were weighed into separate porcelain crucibles and placed in a preheated muffle furnace (600°C) for 2 h. The crucibles were transferred into desiccators, cooled and weighed. Carbohydrate content was determined by subtracting the total percentage of other components from 100.

### Determination of resistant starch

Resistant starch contents of all samples were determined using Megazyme® Resistant Starch Assay Kits (Megazyme International Ireland Ltd. Wicklow, Ireland) based on AOAC Method 2002.02 (AOAC International, 2002). In brief, a sample (100 mg) was mixed with 4 ml α-amylase containing amyloglucosidase followed by incubating in a shaking water bath for 16 h to hydrolyze digestible starch. The resistant portion was washed using ethanol (95%) and centrifuged at 3600 rpm for 20 min. The solid was twice washed with 50% ethanol followed by centrifuging at 3600 rpm for 20 min. The supernatant from each wash was collected to measure hydrolysable starch. Potassium hydroxide solution (2 M) was added to solubilize the resistant starch, and the pH of the solution was adjusted to 4.75 using 8 ml of 1.2 M sodium acetate buffer (pH 3.8). After incubation with amyloglucosidase (0.1 ml, 3300 U/ml) at 50°C for 30 min, the samples were centrifuged at 3600 rpm for 10 min. Three milliliters of glucose-oxidase-peroxidase-aminoantipyrine was added to aliquots (0.1 ml) of the supernatant, and the mixture was incubated at 50°C for 20 min. Sodium acetate buffer (100 ml, 100 mM, and pH 4.5) was added to the hydrolyzed samples collected from the ethanol washings, and aliquots (0.1 ml) were incubated with 10 µl of dilute amyloglucosidase (3300 U/ml) solution at 50°C for 20 min. Glucose-oxidase-peroxidase-aminoantipyrine (3 ml) was then added, and tubes were incubated at 50°C for a further 20 min. Absorbance for both resistant and hydrolysable starch was measured using an evolution 60 s spectrophotometer (Thermo Scie) at 510 nm for glucose concentration).

### Amylose content

Amylose content of the starches was determined following the method of Hoover and Ratnayake (2001) with some modifications. A starch sample was first de-fatted using hexane in a Soxhlet apparatus for 6 h, followed by oven drying for 12 h. Lipid-free starch (20 mg) was added into a round-bottom tube containing 8 ml 90% DMSO and mixed vigorously for 5 min using a vortex mixer. The starch dispersion was heated in a water bath at 85°C for 15 min, followed by cooling to the room temperature. The sample subsequently was diluted to 25 ml with water in a volumetric flask. An aliquot of diluted sample (1.0 ml) was transferred to a 50 ml volumetric flask, and 5 ml iodine solution was added. The volume was adjusted to 50 ml and the absorbance was measured at 600 nm. Amylose concentration was determined from a standard curve developed using amylose and amylopectin blends.

### Molecular weight distribution

Molecular weight ($M_W$) distribution of the starches was determined using high-performance size-exclusion chromatography (HPSEC) as described by Ratnayake and Jackson (2007). For each assay, sample of 50 mg (dry basis) was dispersed in 10 ml of 90% (v/v) DMSO/water solution and maintained at room temperature for 5 days on a multi-tube rotator (Model: 4632Q, Thermo Scientific, Essex, UK) with a fixed shaker speed of 30 rpm. Dispersed samples were filtered through a 1.2 µm Magna nylon supported membrane (GE Osmonics, Minnetonka, MN), and 100 µL of filtrate was injected into the HPSEC system (equipped with Shimadzu LC-20AD pump, Shimadzu CTO-20A column oven, Shodex RI-101 detector). Size exclusion columns (Shodex OHPack SB-807G, SB-807 HQ, SB-806 M HQ, SB-804 HQ and SB-802.5 HQ) used in the system were connected in series and maintained at 50°C. Degassed distilled water was used as the mobile phase at 1 ml/min flow rate. Pullulan standards (Showa Denko K.K., Tokyo, Japan) P-5, P-10, P-20, P-50, P-100, P-200, P-400, and P-800 representing molecular weights of $0.53\times10^4$, $1.2\times10^4$, $2.08\times10^4$, $4.67\times10^4$, $9.54\times10^4$, $19.4\times10^4$, $33.8\times10^4$, and $75.8\times10^4$, respectively, were used to create the standard curve. The molecular weights of samples were calculated using the following equation ($R^2 = 0.987$):

$$\text{Molecular weight} = 10^{-0.2905RT + 14.759}$$

where RT = retention time (min)

### Crystalline structure

Crystalline structure of the starches was analyzed as described by Xu et al. (2004) using a Panalytical X'pert Pro MPD X-Ray Diffractometer (XRD, Panalytical B.V., Almelo, the Netherlands). Samples were scanned with a CuKa target at 40 kV and 30 mA from $2\theta = 2$ to 40° with a scanning speed of 2° $min^{-1}$. The degree of crystallinity of samples was quantitatively estimated by following the method of Nara and Komiy (1983).

### Morphology structure

Morphological structure of the starch granules was determined using a SU-70 scanning electron microscope (Hitachi, Japan) (Li et al., 2011). The samples were mounted on an aluminum specimen holder using double-sided tape and were coated with gold palladium to a thickness of 30 nm. The coated samples were examined at an accelerating potential of 2 to 5 kV.

### Thermal properties

Thermal properties of the starches were measured using a differential scanning calorimeter (DSC-2000) (TA Instruments, New Castle, DE) as described by Singh et al. (2004). Starch was weighed into high pressure DSC pans, and distilled water was added to make suspensions with 70% moisture content. Pans were hermetically sealed and equilibrated for 1 h at ambient temperature before heating in the DSC. The heating range and rate were 30 to 100°C and 10°C/min, respectively. Indium and zinc were used for calibration and an empty pan was used for reference. Onset temperature ($T_o$), peak temperature ($T_p$), conclusion temperature ($T_c$), gelatinization temperature range ($\Delta T$), and enthalpy ($\Delta H$) for gelatinization were determined.

### Water absorption capacity

Water absorption capacity (WAC) was determined following AACC method 56-20 (1983) with some modifications. Samples (about 2 g) were dispersed in 30 ml of distilled water in pre-weighed centrifuge tubes. Starch was agitated using stir bars in the centrifuge tubes at room temperature for 1 h and were then centrifuged at 5000 × g 40 min. Free water was removed, and the wet starch was drained for 5 min and weighed. WAC was calculated as:

$$\text{WAC}=\frac{weight\ of\ wet\ starch - weight\ of\ drystarch}{weight\ of\ drystarch\,(d.b.)}x100\%$$

### Statistical analyses

Three replications were used to obtain average values and standard deviations for all tests. All results were analyzed with SAS version 9.2 statistical software (SAS Institute Inc., Cary, NC). ANOVA with Duncan's Multiple Range test was used to evaluate the differences in composition and properties. Probability ($P$) ≤ 0.05 indicates significance.

## RESULTS AND DISCUSSION

### Proximate composition and starch content

Proximate compositions, specifically crude protein, crude lipid, ash, and total carbohydrate, of starches isolated from Virginia-grown corn, potato, and mungbean cultivars are presented in Table 1. All starches were characterized by low protein, fat and ash, and high carbohydrate. This indicated that starches extracted from each crop had high purity and were suitable for further evaluation. Potato starch had significantly ($p<0.05$) higher resistant starch content (49.4% for Atlantic and 46.1% for Superior) compared to starches extracted from mungbean (31.5% for Berken and 35.0% for TexSprout) and corn (28.3% for Pioneer and 20.0% for Southern State). These results are in agreement with previous reports that tuber and legume starches contain higher amount of resistant starch than cereal starches (Chen et al., 2010; Liu et al., 2006). There are four types of resistant starch (RS), termed $RS_1$, $RS_2$, $RS_3$, and $RS_4$ based on differences in composition, structure, and response to processing methods. The type of resistant starch in the native starch granules is $RS_2$, due to intact and ungelatinized granular structure. The differences in resistant starch contents between different varieties for both potato and mungbean were not significant ($p>0.05$) while a significant difference was observed between the two corn varieties.

Virginia-grown corn, potato and mungbean starches had amylose contents of 19.9 to 26.9%, 23.0 to 26.7%, and 33.6 to 37.9%, respectively, which were close to those grown at other locations as reported in other studies, e.g. 16.9 to 21.3% for corn starch (Sandhu and Singh, 2007), 15.0 to 23.1% for potato starch (Kaur et al., 2007), and 29.9 to 33.6% for mungbean starch (Kaur et al., 2011). Mungbean starches had significantly ($p<0.05$) higher amylose content than corn and potato starches. The differences in amylose content between different cultivars for all crops were significant ($p<0.05$). Within the same crop, there was positive correlation between amylose content and resistant starch, for example corn starches ($r = 0.97$, $p <0\ 0.01$), potato starches ($r=0.91$,

**Table 1.** Proximate composition, resistant starch composition, and amylose content (g · 100 $g^{-1}$ dry weight) of starches isolated from Virginia-grown corn, potato, and mungbean cultivars

| Cultivar | Crude protein (%) | Crude oil (%) | Ash (%) | Total carbohydrate (%) | Resistant starch (%) | Amylose content (%) |
|---|---|---|---|---|---|---|
| **Corn** | | | | | | |
| Southern States 731 | 0.37±0.01$^{a}$ | 0.11±0.03$^{a}$ | 0.03±0.00$^{bc}$ | 99.5±0.03$^{a}$ | 20.0±0.58$^{d}$ | 19.9±0.66$^{e}$ |
| Pioneer 35F37 | 0.39±0.02$^{a}$ | 0.12±0.03$^{a}$ | 0.01±0.00$^{c}$ | 99.5±0.03$^{a}$ | 28.3±1.21$^{c}$ | 26.9±0.44$^{c}$ |
| **Potato** | | | | | | |
| Atlantic | 0.25±0.03$^{b}$ | 0.03±0.01$^{b}$ | 0.18±0.01$^{a}$ | 99.5±0.03$^{a}$ | 49.4±1.14$^{a}$ | 26.7±1.19$^{c}$ |
| Superior | 0.24±0.03$^{b}$ | 0.04±0.01$^{b}$ | 0.26±0.01$^{a}$ | 99.5±0.04$^{a}$ | 46.1±1.79$^{a}$ | 23.0±1.47$^{d}$ |
| **Mungbean** | | | | | | |
| Berken | 0.28±0.01$^{b}$ | 0.09±0.04$^{a}$ | 0.05±0.01$^{b}$ | 99.6±0.04$^{a}$ | 31.5±0.57$^{bc}$ | 33.6±1.50$^{b}$ |
| TexSprout | 0.28±0.02$^{b}$ | 0.12±0.05$^{a}$ | 0.04±0.01$^{b}$ | 99.6±0.03$^{a}$ | 35.0±0.79$^{b}$ | 37.9±2.44$^{a}$ |

Data are expressed as mean ± standard deviation (n=3), Means followed by the same letter within a column indicate no significant (p>0.05) difference among cultivars.

**Table 2.** Structural and morphological properties of starches isolated from Virginia-grown corn, potato and mungbean cultivars.

| Cultivar | Molecular weight (Da) | | Granular size (µm) | Percentage of crystallinity (%) |
|---|---|---|---|---|
| | Amylopectin (×$10^{8}$) | Amylose (×$10^{5}$) | | |
| **Corn** | | | | |
| Southern States 731 | 3.06±0.34$^{c}$ | 2.03±0.65$^{b}$ | 16.8±1.65$^{d}$ | 31.0±0.36$^{d}$ |
| Pioneer 35F37 | 2.56±0.20$^{c}$ | 1.79±0.34$^{b}$ | 15.6±0.89$^{d}$ | 29.3±0.30$^{e}$ |
| **Potato** | | | | |
| Atlantic | 57.3±13.7$^{a}$ | 8.50±0.78$^{a}$ | 24.0±1.37$^{b}$ | 38.2±0.34$^{a}$ |
| Superior | 37.5±8.22$^{b}$ | 7.45±0.29$^{a}$ | 28.2±4.83$^{a}$ | 35.7±0.46$^{b}$ |
| **Mungbean** | | | | |
| Berken | 3.58±0.03$^{c}$ | 6.19±0.16$^{a}$ | 19.3±1.36$^{c}$ | 33.9±0.16$^{c}$ |
| TexSprout | 3.39±0.05$^{c}$ | 7.01±0.51$^{a}$ | 22.7±1.18b$^{c}$ | 33.4±0.30$^{c}$ |

Data are expressed as mean ± standard deviation (n=3), Means followed by the same letter within a column indicate no significant (p >0.05) difference among cultivars.

p<0.01) and mungbean starches (r = 0.886, p < 0.05). This is consistent with the results obtained by previous studies (Polesi et al., 2011; Themeier et al., 2005). However, compared to their corn and mungbean counterparts, potato starches had intermediate amylose content but the highest resistant starch levels. Large size of granules might be responsible for the high resistance of raw potato starch (Leszczyñski, 2004). Granular size of potatoes starches was found to be significantly (p<0.05) larger than those of corn and mungbean starches (Table 2). It is well known that starch hydrolysis requires enzyme adsorption on the surface of starch granules (Leloup et al., 1992). The large potato starch granular size results in smaller surface area, therefore, lowing degree of enzyme adsorption on the surface.

## Structural and morphological properties

Molecular weight (Mw) distributions of Virginia-grown corn, potato, and mungbean starches are summarized in Table 2. Two peaks, identified as amylopectin and amylose, were present in the all samples. Potato starches had significantly (p<0.05) higher amylopectin Mw (10 to 22 times) than corn and mungbean starches. The difference in amylopectin $M_W$ between corn and mungbean starches was not significant. The starches from Virginia-grown mungbean had similar amylopectin

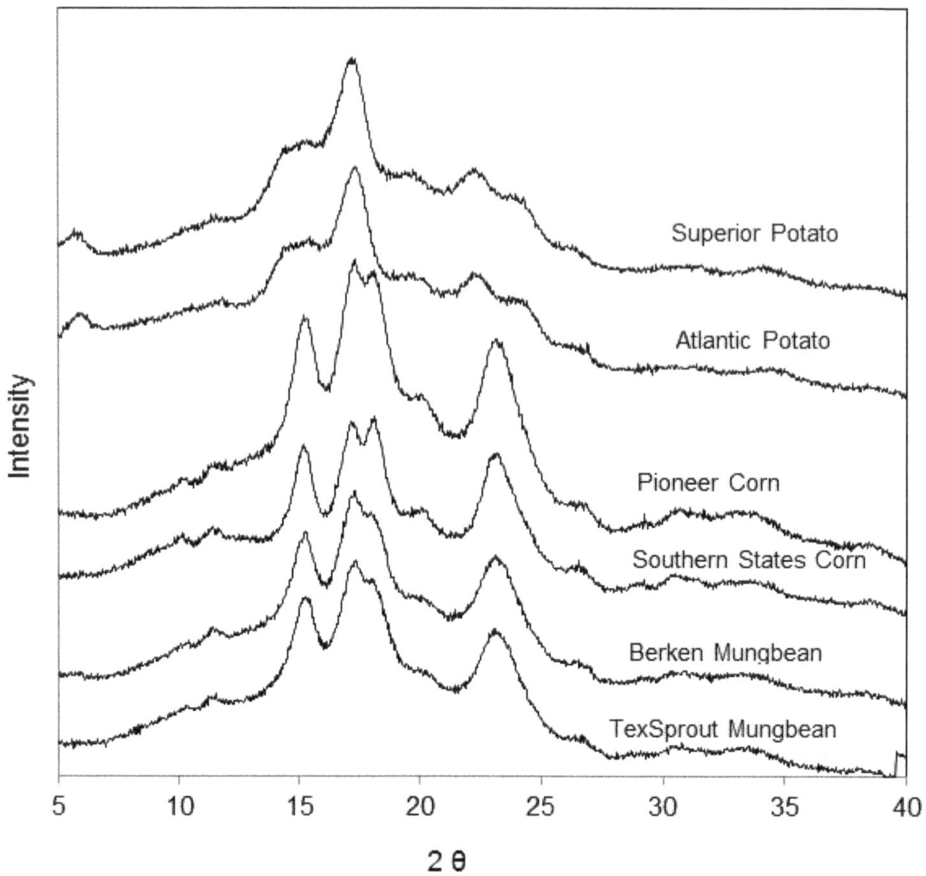


**Figure 1.** X-ray diffraction patterns of starches extracted from Virginia-grown corn, potato, and mungbean cultivars.

Mw ($3.8 \times 10^8$ Da) as reported for starches isolated from mungbean grown at other locations. Compared to corn and potato starches isolated from crops grown at other locations, Virginia-grown corn starches had lower amylopectin Mw while a significantly higher Mw was observed for starches from Virginia-grown potatoes (Yoo and Jane, 2002). Amylopectin $M_W$ was found to be positively correlated to relative crystallinity of the starch ($r = 0.861$, $p < 0.05$). This is consistent with the result obtained in our previous study of chickpea starch (Xu et al., 2013). No difference was found for amylose $M_W$ among potato and mungbean starches, while amylose Mw of corn starches was significantly lower.

X-ray diffraction was used to study the crystalline structure of the starch granules. The diffraction patterns of corn, potato and mungbean starches are shown in Figure 1. Corn starches showed a typical A-type crystalline structure with the diffraction peaks at 2θ around 15 and 22.8°, an unresolved doublet at 17.0 and 18.2°, and a small peak at 19.5°, whereas a B-type pattern was observed for potato starches with a strong reflection peak at 2θ of 17° and several medium peaks at 2θ about 5.4, 14.4 and 21.7° (Jiranuntakul et al., 2011). Although mungbean starches exhibited a C-type crystalline structure (a mixture of A- and B-type structure), the A-type crystalline predominated. A unique peak at 2θ of about 5°, which is typically found in the B-type crystal, was not detected. Furthermore, although split peaks at 2θ about 17.0 and 18° were observed in our mungbean starches, they were not obvious as supposed to be for a typical A-type crystal. This implied that mungbean starches had a $C_A$-type crystalline pattern. These results are in agreement with previous reports regarding the crystalline structure of mungbean starches (Hoover et al., 1997; Kim et al., 2007). The percent crystallinity of the starches was significantly different (Table 3). Crystal size, amount of crystalline regions, orientation of the double helices within the crystalline domains, and extent of interaction between double helices might be responsible for the differences in degree of crystallinity between starches (Hoover and Ratnayake, 2002), all of which could be influenced by the location of origin of the crop and environmental growth conditions (Huang et al., 2007).

The morphological structures of starch granules from Virginia-grown corn, potato and mungbean examined by SEM show significant variations in shape and size (Figure 2 and Table 3). Corn starches granules exhibited polyhedral appearance whereas potato and mungbean

**Table 3.** Thermal and water absorption properties of starches extracted from Virginia-grown corn, potato, and mungbean cultivars.

| Cultivar | Thermal properties | | | | | Water absorption (%) |
|---|---|---|---|---|---|---|
| | $T_o$ (°C) | $T_p$ (°C) | $T_c$ (°C) | $\Delta T$ ($T_c$-$T_o$, °C) | $\Delta H$ (J/g) | |
| **Corn** | | | | | | |
| Southern States 731 | 69.6±0.36$^{ab}$ | 74.3±0.33$^{a}$ | 80.4 ±0.57$^{a}$ | 10.8 ±0.57$^{ab}$ | 2.03±0.09$^{c}$ | 77.9±0.37$^{b}$ |
| Pioneer 35F37 | 70.1±0.64$^{a}$ | 74.1±0.16$^{a}$ | 81.5±0.47$^{a}$ | 11.4 ±0.11$^{a}$ | 2.27±0.06$^{b}$ | 76.1±2.80$^{b}$ |
| **Potato** | | | | | | |
| Atlantic | 66.3±0.69$^{d}$ | 69.0 ±0.54$^{c}$ | 75.6±0.52$^{c}$ | 9.28 ±0.12$^{ab}$ | 2.75±0.54$^{a}$ | 86.1±1.62$^{a}$ |
| Superior | 65.4±0.15$^{d}$ | 68.2±0.18$^{c}$ | 74.5±0.15$^{c}$ | 8.97 ±0.35$^{ab}$ | 2.57±0.28$^{ab}$ | 84.8±1.96$^{a}$ |
| **Mungbean** | | | | | | |
| Berken | 67.8±0.16$^{c}$ | 72.2±0.47$^{b}$ | 77.1±0.57$^{b}$ | 9.37±0.40$^{ab}$ | 1.46±0.08$^{d}$ | 83.6±2.59$^{a}$ |
| Texas Sprout | 69.1±0.30$^{bc}$ | 72.7±0.39$^{b}$ | 77.3±0.24$^{b}$ | 8.25±0.06$^{b}$ | 1.14±0.11$^{e}$ | 85.8±1.98$^{a}$ |

Data are expressed as mean as mean ± standard deviation (n=3), Means followed by the same letter within a column indicate no significant (p>0.05) difference among cultivars.

granules had smooth, oval or irregular ellipsoidal appearance. Corn starch granules were the smallest with mean size of 15.6 µm for Pioneer and 16.8 µm for Southern States. On the other hand, potato starches had the largest granules with mean granular sizes of 24.0 µm for Atlantic and 28.2 for Superior, followed by mungbean starches with mean granular sizes of 19.3 µm for Berken and 22.7 µm for TexSprout. The granular size of starches from Virginia-grown corn, potato and mungbean *fell* within a *range* similar to those already reported (Kim et al., 2007; Swinkels, 1985).

## Thermal and water absorption properties

The gelatinization transition temperatures [$T_o$ (onset), $T_p$(peak), and $T_c$ (conclusion)], and enthalpy of gelatinization ($\Delta H$) for starches from corn, potato and mungbean are presented in Table 3. DSC thermographs of the starches are shown in Figure 3. Significant differences were observed in $T_o$, $T_p$, and $T_c$ among the starches with the order of corn> mungbean> potato. However, the differences between cultivars of each crop for $T_o$, $T_p$ and $T_c$ were not significant. $T_o$, $T_p$ and $T_c$ of corn starches were in the ranges of 69.6-70.1, 74.1-74.3 and 80.4-81.5°C, respectively. Mungbean starches had an $T_o$ of 67.8-69.1°C, $T_p$ of 72.2-72.7°C, and $T_c$ of 77.1-77.3°C, while the ranges of $T_o$, $T_p$ and $T_c$ of potato starches were 65.4-66.3, 68.2-69.0°C and 74.5-75.6°C, respectively. These values were in agreement with the previous reports for the starches from corn, mungbean, and potato (Kaur et al., 2007, 2011; Li and Yeh, 2001; Sandhu and Singh, 2007).

A variety of factors, including form and distribution of starch granules, internal arrangement of starch fractions within the granule, and amylose content and size, could contribute to the differences in gelatinization temperature (Sandhu and Lim, 2008). The differences in the gelatinization temperature range ($\Delta T$) among different starches were not significant. Enthalpy of gelatinization ($\Delta H$) for the starches ranged from 1.14 to 2.75 J/g, which was much lower than those values reported by others for corn, potato, and mungbean starches (Kaur et al., 2007, 2011; Li and Yeh, 2001; Sandhu and Singh, 2007). Potato starches had the highest enthalpy followed by corn and mungbean. The lower enthalpy of gelatinization is suggested to be mainly due to the disruption of the double helices A high $\Delta H$ suggests that more energy must be applied to unravel and melt strongly associated double helices (formed by the outer branches of adjacent amylopectin chains) within the native granule during gelatinization (Hoover et al., 1997). Our results suggest a lower degree of association between double helices in starches from these Virginia-grown crops compared to similar crops from other locations.

Water absorption capacity (WAC) ranged from 76.1% for Pioneer corn starch to 86.1% for Atlantic potato starch (Table 3). The difference in WAC of the starches among the different cultivars may be attributed to the starch structure and extent of interactions between starch chains and water. A-type crystalline structure is more dense and binds less water than B- and C type structures, which results in lower WAC of corn starches compared to potato and mungbean starches (Wang et al., 1998).

## Conclusions

All extracted starches had high purity and were characterized by low protein, fat and ash, and high carbohydrate. Compared to corn and mungbean starches, potato starches had the highest resistant starch content but intermediate amylose content. The resistance of starch

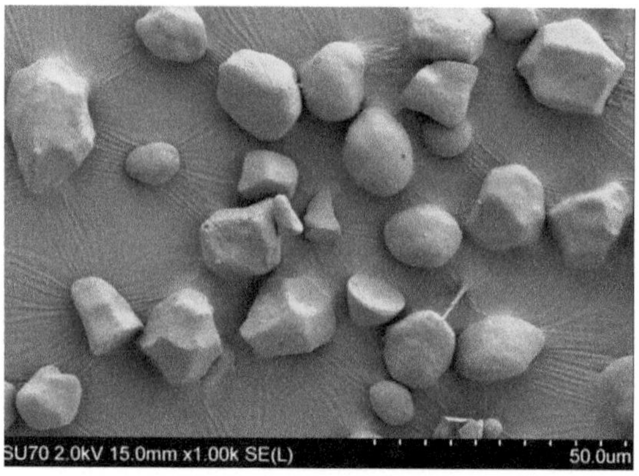


Pioneer 35F37 Corn

Southern States 731 Corn

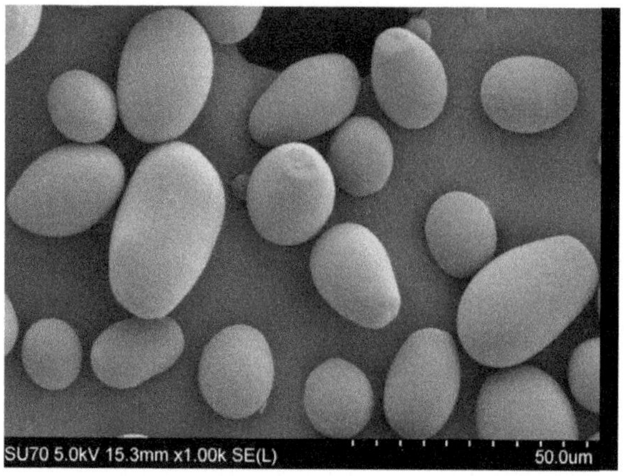


Atlantic Potato

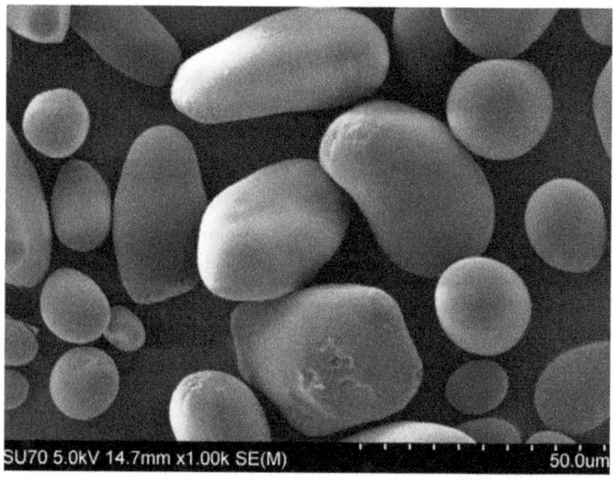


Superior Potato

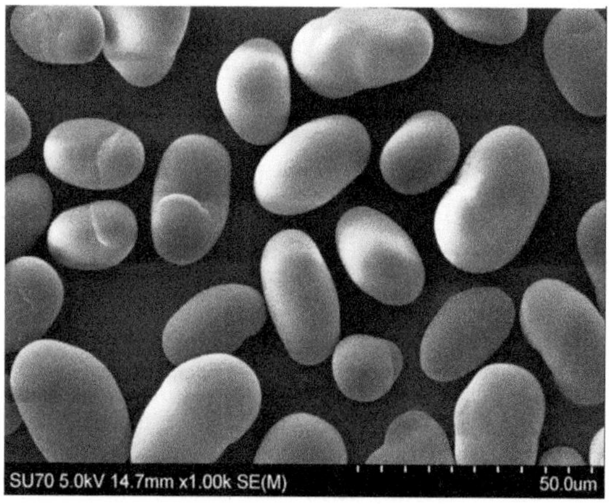


TexSprout Mungbean

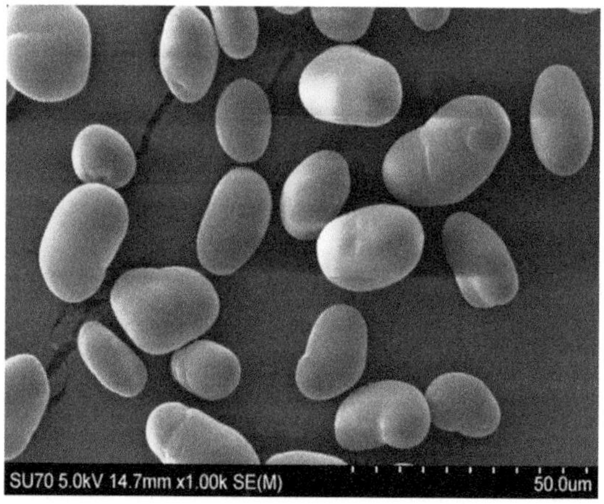


Berken Mungbean

**Figure 2.** Scanning electron micrographs of starches extracted from Virginia-grown corn, potato, and mungbean cultivars.

to enzymatic hydrolysis was dependent on amylose content and the size and structure of starch granules. Higher amylopectin molecular weight and large particle size also contributed to high resistance of potato starches.

Crystallinity, granular morphology, and thermal properties of the starches from Virginia-grown corn, potato and mungbean were in agreement with those already reported for same crop. The results not only provide

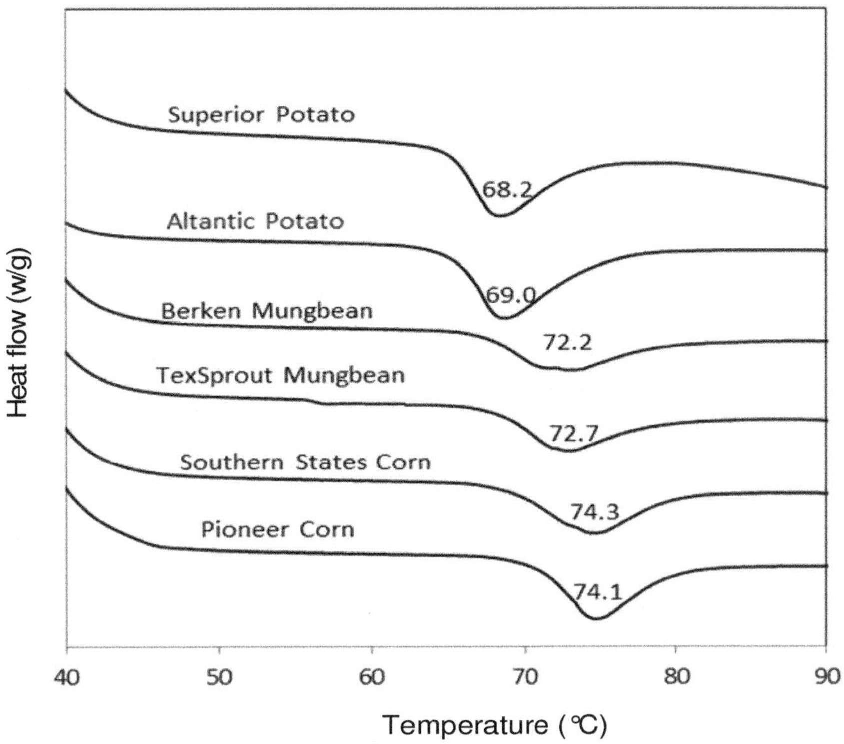


**Figure 3.** DSC thermographs of starches extracted from Virginia-grown corn, potato, and mungbean cultivars.

baseline information for plant breeders in developing new cultivars with improved properties, but also assist food scientists in determining the potential end-uses of the starches.

## ACKNOWLEDGMENTS

The authors thank Dr. Anwar Hamama, Mrs. Zaid Abraha-Eyob, and Mrs. Naomi Pearson of the VSU Agricultural Research Station, Wajira Ratnayake of Food Processing Center at the University of Nebraska-Lincoln, and Drs. Everett E Carpenter and Dimitri Pestov of the Virginia Commonwealth University Nano materials Core Characterization Facility for technical support. We thank Drs. Glenn Chappell and Josh Freeman for providing corn kernels, potato tubers, and mungbean seeds, respectively. Funding was provided through the USDA Evans-Allen program and the research was conducted at Virginia State University Agricultural Research Station (Journal Series number 301).

### REFERENCES

AACC International (1983). Approved Methods of Analysis, 8th Ed.Method 56-20, Hydration capacity of pregelatinized cereal products. AACCI, St. Paul, MN.

AOAC International (2000). Approved Methods of Analysis, 17th Ed. Method 948.22, Lipid and oil; Method 950.49, Ash; Method 990.03, Crude protein; Association of Official Analytic Chemists International: Gaithersburg, MD, USA.

AOAC International (2002). Method 2002.02, Resistant starch in starch and plant materials. J. AOAC Int. 85:1103.

Bhardwaj HL, Hankins A, Mebrahtu T, Mullins J, Rangappa M, Abaye O, Welbaum GE (1996). Alternative crops research in Virginia. Prog. in New Crops, J. Janick, ed. ASHS Press: Alexandria, VA. pp. 87-96.

Bhardwaj HL, Rangappa M, Hamama AA (1999). Chickpea, faba bean, lupin, mungbean, and pigeonpea: potential new crops for the mid-Atlantic Region of the United States. Pers. on New Crops and New Uses, J. Janick, ed. ASHS Press: Alexandria, VA. pp. 202-205.

Chen L, Liu R, Qin C, Meng Y, Zhang J, Wang Y, Xu G (2010). Sources and intake of resistant starch in the Chinese diet. Asia Pac. J. Clin. Nutr. 19(2):274-282.

Hoover R, Li YX, Hynes G, Senanayake N (1997) Physicochemical characterization of mungbean starch. Food Hydrocolloid. 11(4):401-408.

Hoover R, Ratnayake WS (2001). Determination of total amylose content of starch. Curr. Prot. Food Anal. Chem. E 2.3.

Hoover R, Ratnayake WS (2002). Starch characteristics of black bean, chick pea, lentil, navy bean and pinto bean cultivars grown in Canada. Food Chem. 78:489-498.

Hoover R, Zhou Y (2003). *In vitro* and *in vivo* hydrolysis of legume starches by a-amylase and resistant starch formation in legumes - A review. Carbohyd. Polym. 54:401-417.

Huang J, Schols HA, van Soest JJG, Jin Z, Sulmann E, Voragen A G J (2007). Physicochemical properties and amylopectin chain profiles of cowpea, chickpea and yellow pea starches. Food Chem. 101:1338-1345.

Jiranuntakul W, Puttanlek C, Rungsardthong V, Puncha-arnon S,Uttapap D (2011). Microstructural and physicochemical properties of heat-moisture treated waxy and normal starches. J. Food. Eng. 104:246-258.

Kaur A, Singh N, Ezekiel R, Guraya HS (2007). Physicochemical, thermal and pasting properties of starches separated from different potato cultivars grown at different locations. Food Chem. 101:643-

651.
Kaur M, Sandhu KS, Singh N, Lim ST (2011). Amylose content, molecular structure, physicochemical properties and in vitro digestibility of starches from different mungbean (*Vigna radiate* L.) cultivars. Starch/Stärke. 63:709-716.
Kim SH, Lee BH, Baik MY, Joo MH, Yoo SH (2007). Chemical structure and physical properties of mungbean starches isolated from 5 domestic cultivars. J. Food Sci. 72(9): C471-477.
Leloup VM, Colonna P, Ring SG (1992). Physico-chemical aspects of resistant starch. J. Cereal Sci. 16: 253-266.
Leszczyñski W (2004). Resistant starch - classification, structure, production. Pol. J. Food Nutr. Sci. 13: 37-50.
Li JY, Yeh AI (2001). Relationship between thermal, rheological characteristics and swelling power for various starches. J. Food Eng. 50:141-148.
Li S, Gao Q, Ward R (2011). Physicochemical properties and in vitro digestibility of resistant starch from mungbean (*Phaseolus radiatus*) starch. Starch/Stärke 63:171-178.
Liu Q, Donner E, Yin Y, Huang, RL, Fan MZ (2006). The physicochemical properties and *in vitro* digestibility of selected cereals, tubers and legumes grown in China. Food Chem. 99(3):470-477.
Nara S, Komiy T (1983). Studied on the relationship between water saturated state and crystallinity by the diffraction method for moistened potato starch. Starch 35:407-410.
National Nutrition Monitoring Bureau (1991). Trends in nutrition. Nutr. News 12(3):1-4.
Polesi LF, Sarmento SBS, Anjos, CBP (2011). Composition and characterization of pea and chickpea starches. Braz. J. Food Tech. 14(1):74-81.
Ratnayake WS, Jackson DS (2007). A new insight into the gelatinization process of native starches. Carbohydr. Polym. 67:511-529.
Sajilata MG, Singhal RS, Kulkarni PR (2006). Resistant starch-a review. Compr. Rev. Food Sci. Food Safety 59(1):1-17.
Sandhu KS, Singh N (2007). Some properties of corn starches II: physicochemical, gelatinization, retrogradation, pasting and gel textural properties. Food Chem. 101:1499-1507.
Sandhu KS, Lim ST (2008). Digestibility of legume starches as influenced by their physical and structural properties. Carbohydr. Polym. 71:245-252.
Sharma A, Yadav BS, Ritika (2008). Resistant starch: physiological roles and food applications. Food Rev. Int. 24:193-234.
Shapours H, Salassi M (2006). The economics feasibility of ethanol production from sugar in the United States. A report through a cooperative agreement between OEPNU, OCE, USDA and LSU. Available online: (November 27, 2012). <http://www.usda.gov/oce/reports/energy/EthanolSugarFeasibilityReport3.pdf>
Singh N, Sandhu KS, Kaur M (2004). Characterization of starches separated from Indian chickpea (*Cicer arietinum* L.) cultivars. J. Food Eng. 63:441-449.
Stark JC, Love SL (2003). Tuber quality. Potato Production Systems. JC Stark, SL Love, eds. Univ. Idaho Extension: Moscow. pp. 329-343.
Swinkels JJM (1985). Composition and properties of commercial native starches. Starch/Starke 37:S.1-5.
Themeier H, Hollmann J, Neese U, Lindhauer MG (2005). Structural and morphological factors influencing the quantification of resistant starch II in starches of different botanical origin. Carbohydr. Polym. 61:72-79.
Topping DL, Clifton PM (2001). Short-chain fatty acids and human colonic function: roles of resistant starch and nonstarch polysaccharides. Physiol. Rev. 81(3):1031-1064.
USDA-NASS (2011). Quick Stats database - Virginia. United States Department of Agriculture, National Agricultural Statistics Service, http://www.nass.usda.gov (November 27, 2012).
Vasanthan T (2001). Overview of laboratory isolation of starch from plant materials. Curr. Prot. Food Anal. Chem. E2.1.1-E2.1.6.
Wang TL, Bogracheva TY, Hedley CL (1998). Starch: as simple as A, B, C? J. Exp. Bot. 49:481-502.
Xu, YX, Miladinov V, Hanna MA (2004). Synthesis and characterization of starch acetates with high degree of substitution. Cereal Chem. 81(6):735-740.
Xu Y, Sismour EN, Narina SS, Dean D, Bhardwaj HL, Li Z (2013). Composition and properties of starches from Virginia-grown kabuli chickpea (*Cicer arietinum* L.) cultivars. Int. J. Food Sci. Tech. doi:10.1111/j.1365-2621.2012.03217.x. 48(3):539-547.
Yoo SH, Jane JL (2002). Molecular weight and gyration radii of amylopecitn determined by high-performance size exclusion chromatography equipped with multi-angle laser light scattering refractive index detectors. Carbohydr. Polym. 49:307-314.

# Growth performance, carcass characteristics and economics of production of broilers fed diets with two sources of protein and two levels of wheat offal

**Dieumou F. E. , Adegbola T. A. and Doma U. D.**

Animal Production Programme, School of Agriculture, Abubakar Tafawa Balewa University,
P. M. B. 0248 Bauchi, Nigeria.

**A feeding trial was conducted to investigate the effects of diets containing two sources (S) of protein, soybean meal (SBM) and groundnut cake (GNC) combined with 7.5 and 15% levels (L) of inclusion of wheat offal on growth performance, carcass characteristics and economics of production of broilers. A total of 204 chicks of Anak line were assigned to four dietary treatments each replicated three times in a factorial experiment. Birds were provided with feed and water *ad libitum* and the experiment lasted for seven weeks. Protein sources affected the relative organ weights as well as the carcass characteristics except for the caeca, gut and liver. Broilers fed SBM based diets had significantly higher ($P < 0.05$) daily feed intake (94.64 vs 110.71 g), ($P < 0.001$) daily weight gain (33.90 vs 52.2 g) and better ($P < 0.001$) feed conversion ratio (2.12 vs 2.82) than those on GNC based diets. The cost of feed intake per bird (N/kg feed) was significantly higher ($P < 0.001$) on SBM based diets than those containing GNC, but the cost of feed per kilogram of bird sold (N/kg gain) was significantly lower ($P < 0.001$) on SBM based diets because birds on these diets achieved significantly greater ($P < 0.001$) weight gain than those on diets containing GNC. Wheat offal levels had no effects on any of the parameters studied. In conclusion, soybean meal is a better protein source than groundnut cake and broiler feeds could contain up to 15% level of wheat offal at the starter phase without reduction of growth performance.**

**Key words:** Broilers, economics of production, groundnut cake, growth performance, soybean meal.

## INTRODUCTION

Inadequate feeding and poor management practices are some of the major problems affecting animal production in developing countries. Animal nutritionists are challenged on a daily basis to assess the nutritive values of feedstuffs available in their communities and to propose the possible combinations of these feedstuffs into complete feeds that meet the nutrient requirements of livestock at a reasonable cost. Feed alone accounts for up to 70 to 80% of the total cost of broiler production (Akdeniz et al., 2006) and adequate nutrition is one of the major inputs necessary for the full expression of the genetic potentials of poultry and the prevention of stress (Adene, 1989). Poultry accounts for 19.4% of the world's meat production compared to beef (38.3%), pork (34.9%), mutton (4.6%), Chevon (1.4%), buffalo meat (1.0%) and horse meat (0.4%) (Considine, 1982). Moreover, poultry industry is an area where scientific innovations have served to make production one of the most technologically advanced protein production systems in the world (Dibner et al., 1998). Consequently, poultry production is the most economic route for producing high quality animal protein within the shortest possible time for the rapidly increasing human population across the world, and especially in the sub-Saharan African countries. This is because of their short generation interval and the intensive method of production, which allows high stocking density. Broiler chickens are fast growing animals, which are able to achieve market weight within the short period of 6 to 7 weeks nowadays. Poultry meat has many desirable nutritional benefits such as low lipid content with a relatively high concentration of polyunsaturated fatty acids and higher protein content (Kinsella et al., 1990; Nettleton, 1991). An incentive to greater broilers production will therefore go a long way in solving the problem of low animal protein intake (3.25 - 6.8 g per person per day) of the average Nigerian (Shaib et al., 1997; Abubakar, 1998). Poor productivity and high mortality of livestock, which characterize this industry is largely explained by the inadequacy of feeding the right quantity and quality of feeds to the various livestock species. Feed insufficiency is due to stiff competition with need for human food, particularly for the fast growing and prolific monogastric species poultry and pigs, and for concentrate mixes for ruminants. The food insufficiency is further compounded by the huge foreign debt and currency devaluation, which militate against feed importation (Olumide and Mpoko, 2001). This study was therefore designed to ascertain the effects of two sources of plant protein and two wheat offal levels based diets on the growth performance attributes, carcass characteristiccs, and economics of production of broiler chickens.

## MATERIALS AND METHODS

### Experimental birds and management

A total of 204 day old chicks of *Anak* 2000 line purchased from ECWA farms in Jos were received at the brooding room of the University Poultry Research Farm, brooded for one week before being transferred into the experimental pens. Four experimental diets and water were offered *ad libitum* every day. Brooding continued for the next three weeks using 200 watts electric bulbs with adjustable height above the floor. The feeding trial lasted for seven weeks during which feed intake, weight gain and feed conversion ratio were monitored on weekly basis. Vaccinations were given when due. Seven cases of mortality were recorded through out the feeding trial all occurring in the starter phase due to physical and mechanical injuries, giving an overall percentage mortality of 3.43%.

### Feed preparation

Four iso-nitrogenous diets with varying energy densities were used both at the starter phase (23% crude protein) and the finisher phase (20% crude protein) using soybean meal and groundnut cake as test protein sources and two levels of wheat offal of 7.5 and 15%. The starter phase began from the second week of age and lasted for three weeks and the finisher phase lasted from 5 to 8 weeks of age (4 weeks).

From the various treatment combinations, the following diets were obtained:

Diet 1: Groundnut cake and wheat offal at 7.5% level of inclusion,
Diet 2: Groundnut cake and wheat offal at 15% level of inclusion,
Diet 3: Soybean meal and wheat offal at 7.5% level of inclusion,
Diet 4: Soybean meal and wheat offal at 15% level of inclusion.

The composition of diets are shown in Table 1 for the starter and finisher phases.

### Experimental design, data collection and statistical analysis

The 204 chicks of Anak 2000 line were assigned to four treatments using 51 chicks per treatment, each at the start of the experiment. Each experimental diet was replicated three times with 17 birds per replicate in deep litter pens. The birds were arranged in a 2 x 2 factorial experiment of a completely randomised design. At eight day of age, the chicks were weighed and randomly allotted to the dietary treatments. Data collected were subject to the analysis of variance (ANOVA) option of SPSS statistical software version 11.0 as described by Nie et al. (1970). Significantly different means among treatments were separated as per the standard method of Duncan (1955) at 5% level of probability ($P < 0.05$).

## RESULTS

The main effects (Table 2) shows that birds fed SBM based diets during the experimental period consumed significantly ($P < 0.001$) more feed (94.64 vs 110.71 g) than those fed GNC based diets. The better performance of birds on SBM based diets observed in this work could be due to better nutrient digestibility coupled with a more correctly balanced amino acid profile of soybean meal. The wheat offal levels in diets did not influence feed consumption in birds throughout the feeding trial neither did interaction between protein sources (S) and wheat offal levels (L) had any effects on the pooled performance of broilers.

Table 3 presents the main effect of protein sources and wheat offal levels on relative organs weight (expressed in percentage body weight) and carcass characteristics of broilers. There were significant ($P < 0.001$) effects of the protein sources on live weight, eviscerated weight, plucked weight, carcass weight and dressing percentage of broilers. The protein sources also had a very highly significant ($P < 0.001$) effect on the heart, abdominal fat and the neck. The protein sources and wheat offal levels had no significant effect on the liver, gut and caeca indicating that the diets were devoid of any toxic materials. The wheat offal levels in the diets had no significant effects on either the carcass characteristics or on the internal organs and offal. Birds fed SBM based diets had very highly ($P < 0.001$) significant weights of all the carcass parameters, the weight of the heart, the abdominal fat pad and the legs; a significantly greater ($P < 0.01$) weight of the lungs, the gizzard, the kidneys and the head compared to those fed GNM based diets.

The effects of protein sources and wheat offal levels interaction on the carcass characteristics (kg) and relative organs weight (expressed in percentage body weight) of

**Table 1.** Percentage composition of experimental diets in the starter and finisher phases.

| Ingredient | Starter phase (Diet) | | | | Finisher phase (Diet) | | | |
|---|---|---|---|---|---|---|---|---|
| | 1 | 2 | 3 | 4 | 1 | 2 | 3 | 4 |
| Yellow maize | 52.00 | 46.00 | 53.00 | 53.00 | 59.70 | 53.80 | 60.60 | 54.60 |
| Soybean meal (SBM) | - | - | 32.70 | 32.70 | - | - | 28.10 | 26.60 |
| Groundnut cake (GNC) | 33.70 | 32.20 | - | 59.70 | 29.00 | 27.40 | - | - |
| Fish meal (FM) | 03.00 | 03.00 | 03.00 | 03.00 | - | - | - | - |
| Wheat offal (WO) | 07.50 | 15.00 | 07.50 | 15.00 | 07.50 | 15.00 | 07.50 | 15.00 |
| Bone meal | 02.00 | 02.00 | 02.00 | 02.00 | 02.00 | 02.00 | 02.00 | 02.00 |
| Limestone | 01.00 | 01.00 | 01.00 | 01.00 | 01.00 | 01.00 | 01.00 | 01.00 |
| Sodium chloride(NaCl) | 00.25 | 00.25 | 00.25 | 00.25 | 00.25 | 00.25 | 00.25 | 00.25 |
| Vit/Min premix[k] (0.25) | 00.25 | 00.25 | 00.25 | 00.25 | 00.25 | 00.25 | 00.25 | 00.25 |
| Lysine | 00.10 | 00.10 | 00.10 | 00.10 | 00.10 | 00.10 | 00.10 | 00.10 |
| Methionine | 00.20 | 00.20 | 00.20 | 00.20 | 00.15 | 00.15 | 00.15 | 00.15 |
| Totals | 100.00 | 100.00 | 100.00 | 100.00 | 100.00 | 100.00 | 100.00 | 100.00 |
| **Calculated nutrients content** | | | | | | | | |
| Metabolizable energy (Kcal/kg) | 2866.67 | 2722.90 | 2927.92 | 2781.74 | 2925.75 | 2822.87 | 2978.74 | 2872.57 |
| Crude protein (%) | 23.00 | 23.00 | 23.00 | 23.00 | 20.00 | 20.00 | 20.00 | 20.00 |
| Crude fibre (%) | 03.47 | 04.01 | 04.35 | 04.84 | 03.46 | 04.00 | 04.21 | 04.70 |
| Calcium (%) | 01.35 | 01.36 | 01.35 | 01.36 | 01.18 | 01.19 | 01.18 | 01.18 |
| Phosphorus (%) | 00.86 | 00.88 | 00.86 | 00.92 | 00.77 | 00.84 | 00.77 | 00.83 |
| Lysine (%) | 01.01 | 01.03 | 01.31 | 01.32 | 00.82 | 00.83 | 01.08 | 01.08 |
| Methionine (%) | 00.54 | 00.53 | 00.59 | 00.58 | 00.42 | 00.41 | 00.46 | 00.45 |

[K]Each 2.5 kg premix contained the followings: Vit. A 10,000,000 IU; Vit. $D_3$ 3,000,000 IU; Vit. E 30,000 IU; Vitamin $K_2$ 3 g; Vit. $B_1$ 1.7 g; Vit $B_2$ 5.0 g; Vit. $B_6$ 3.1 g; Vit. $B_{12}$ 16 mg; Biotin 60 mg; Niacin 31.0 g; Pantothenic Acid 8 g; Folic Acid 0.8 g; Manganese 85 g; Zinc 50 g; Iron 25 g; Copper 6 g; Iodine 1.1 g; Selenium 120 mg; Cobalt 220 mg; B.H.T 60 g; Ethoxyquin 65 g; Choline Chloride 200 g.

**Table 2.** Main effects of protein sources and wheat offal levels on the growth performance of broiler chickens (2 to 8 weeks of age).

| Parameter | Protein sources (PS) | | Wheat offal levels (WOL) | | Levels of significance | |
|---|---|---|---|---|---|---|
| | GNM | SBM | 7.5% | 15% | PS | WOL |
| Daily feed intake (g/bird) | 94.64 | 110.7 | 101.86 | 103.49 | * | ns |
| Daily weight gain (g/ bird) | 33.90 | 52.20 | 42.40 | 43.70 | *** | ns |
| Feed conversion ratio | 2.82 | 2.12 | 2.52 | 2.42 | *** | ns |
| Mortality | 0 | 3 | | | | |

Mean values within a row under protein sources (PS) differ significantly. Mean values in the same row under wheat offal levels (WOL) are all statistically similar.

**Table 3.** Main effects of protein sources and wheat offal levels on carcass characteristics (kg) of broiler chickens and relative organs weight expressed as (% body weight).

| Parameter | Protein sources (PS) | | Wheat offal levels (WOL) | | Levels of significance | |
|---|---|---|---|---|---|---|
| | GNM | SBM | 7.5% | 15% | PS | WOL |
| Live weight | 2.16 | 2.88 | 2.54 | 2.50 | *** | ns |
| Plucked weight | 1.90 | 2.58 | 2.26 | 2.22 | *** | ns |
| Eviscerated weight | 1.69 | 2.32 | 2.02 | 1.99 | *** | ns |
| Carcass weight | 1.47 | 2.05 | 1.77 | 1.75 | *** | ns |
| Dressing percentage | 67.8 | 71.05 | 69.38 | 69.48 | *** | ns |
| Lungs | 0.50 | 0.40 | 0.44 | 0.44 | ** | ns |
| Liver | 1.53 | 1.47 | 1.46 | 1.54 | ns | ns |
| Heart | 0.48 | 0.33 | 0.40 | 0.41 | *** | ns |
| Gizzard | 2.21 | 1.70 | 1.94 | 1.96 | ** | ns |
| Kidneys | 0.23 | 1.18 | 0.19 | 0.20 | ** | ns |
| Abdominal fat pad | 1.68 | 0.59 | 1.17 | 1.09 | *** | ns |
| Gut | 2.63 | 2.42 | 2.55 | 2.49 | ns | ns |
| Caeca | 0.51 | 0.48 | 0.51 | 0.48 | ns | ns |
| Head | 2.53 | 2.23 | 2.37 | 2.39 | ** | ns |
| Neck | 3.71 | 2.99 | 3.30 | 3.40 | *** | ns |
| Legs | 3.66 | 3.42 | 3.60 | 3.48 | ns | ns |

Mean values within a row under protein sources (PS) differ significantly ($P < 0.05$) except for the liver, gut, caeca and legs. Mean values in the same row under wheat offal levels (WOL) are all not significant ($P > 0.05$).

**Table 4.** Effects of protein sources and wheat offal levels interaction on the carcass characteristics (kg) and relative organs weight (expressed in % body weight) of broiler chickens.

| Parameter | Protein sources | Groundnut meal (%) | | Soyabean meal (%) | | SEM |
|---|---|---|---|---|---|---|
| | Wheat offal levels | 7.5 | 15 | 7.5 | 15 | |
| | Diets | 1 | 2 | 3 | 4 | |
| Live weight | | 2.22 | 2.11 | 2.87 | 2.90 | $0.15^{ns}$ |
| Plucked weight | | 1.96 | 1.84 | 2.57 | 2.60 | $0.15^{ns}$ |
| Eviscerated weight | | 1.74 | 1.64 | 2.30 | 2.35 | $0.13^{ns}$ |
| Carcass weight | | 1.51 | 1.42 | 2.02 | 2.07 | $0.11^{ns}$ |
| Dressing percentage | | 68.23 | 67.40 | 70.54 | 71.57 | $1.21^{ns}$ |
| Lungs | | 0.48 | 0.51 | 0.41 | 0.38 | $0.05^{ns}$ |
| Liver | | 1.54 | 1.53 | 1.39 | 1.56 | $0.11^{ns}$ |
| Heart | | 0.46 | 0.51 | 0.34 | 0.33 | $0.03^{ns}$ |
| Gizzard | | 2.11 | 2.31 | 1.77 | 1.62 | $0.26^{ns}$ |
| Kidneys | | 0.21 | 0.24 | 0.18 | 0.17 | $0.02^{ns}$ |
| Abdominal fat | | 1.67 | 1.70 | 0.68 | 0.49 | $0.34^{ns}$ |
| Gut | | 2.58 | 2.68 | 2.53 | 2.32 | $0.23^{ns}$ |
| Caeca | | $0.49^{ab}$ | $0.53^{ab}$ | $0.54^{a}$ | $0.42^{b}$ | 0.06* |
| Head | | 2.45 | 2.60 | 2.29 | 2.17 | $0.17^{ns}$ |
| Neck | | 3.56 | 3.86 | 3.03 | 2.95 | $0.29^{ns}$ |
| Legs | | 3.75 | 3.57 | 3.44 | 3.39 | $0.25^{ns}$ |

[ab]Means bearing no superscripts in the same row are all statistically similar ($P > 0.05$).

broiler chickens had no effects on any of the parameters studied except for the caeca ($P < 0.05$) of broilers Table 4. This variation in the relative caeca weight indicates that the effect of wheat offal levels in the diets depended on the protein sources.

The economics of production of birds fed experimental

**Table 5.** Economics of production of broilers fed diets with two sources of protein and two levels of wheat offal (2 to 8 weeks of age).

| Parameter | Diet 1 | Diet 2 | Diet 3 | Diet 4 | SEM |
|---|---|---|---|---|---|
| Daily feed intake (g/ bird) | 93.76 | 95.51 | 109.96 | 111.46 | $1.33^{ns}$ |
| Daily weight gain (g/ bird) | 32.36 | 35.45 | 52.44 | 51.96 | $1.54^{ns}$ |
| Feed conversion ratio | 2.94 | 2.70 | 2.09 | 2.15 | $0.15^{ns}$ |
| Total feed cost (₦) | $159.62^b$ | $156.60^b$ | $207.52^a$ | $207.2^a$ | 2.25*** |
| Total weight gain (kg/bird) | $1.75^b$ | $1.74^b$ | $2.57^a$ | $2.50^a$ | 0.04*** |
| Feed cost (₦/kg gain) | $91.24^a$ | $90.25^a$ | $80.66^b$ | $82.77^{ab}$ | 1.22*** |
| Mortality | 0 | 0 | 2 | 1 | - |

[ab]Means bearing common superscripts in the same row are not significantly different (P > 0.05).

diets is shown in Table 5. The cost per kilogram feed and total cost of feed consumed per bird were significantly higher (P < 0.001) on the SBM based diets. In like manner, the total weight gain was significantly higher (P < 0.001) in birds fed SBM based diets than those fed GNC based diets. The feed cost per kilogram gain on diets 1, 2 and 4 were statistically similar and were significantly higher (P < 0.001) than for that on diet 3. However, diets 3 and 4 were statistically similar indicating that SBM based diets were cheaper and more economical for broilers production than GNC based diets.

## DISCUSSION

Generally, the growth performance attributes of birds on SBM were significantly better (P < 0.05) than those on GNM. These results agree with the findings of Mustafa et al. (2012) who fed broilers with diets including different percentage of animal protein and plant protein sources and obtained significant (P < 0.05) effects on their growth and economic parameters. Three cases of mortality were recorded on SBM based diets due to physical and mechanical injuries. This mortality rate falls within the accepted range of 0-5% for good management practices and safety of diets as advocated by ISA (1996).

There was a significant (P < 0.05) protein source and wheat offal level interaction effect on the caeca relative weight alone meaning that the effect of wheat offal levels in the diets depended on the protein sources. This is a reflexion of the fibre digesting ability attributed to the caecum in non ruminant animals and it agrees with the report of Steenfeldt et al. (2007) who found fibre digestibility to vary from 24 to 29% in layer hens fed large amounts of roughage, thus indicating that caecal fibre digestion can be significant in layer hens.

The cost per kilogram feed and total cost of feed consumed per bird were significantly higher (P < 0.001) on the SBM based diets. In like manner, the total weight gain was significantly higher (P < 0.001) in birds fed SBM based diets than those fed GNM based diets. The feed cost per kilogram gain on diets 1, 2 and 4 were statistically similar and were significantly higher (P < 0.001) than for that on diet 3. However, diets 3 and 4 were statistically similar indicating that SBM based diets were cheaper and more economical for broilers production than GNM based diets. These results are in line with the report of Ojewola et al. (2005) who compared the utilization of three animal protein sources by broiler chickens and found significant treatment effects (P < 0.05) in gross margins.

## Conclusion

It was concluded that soybean meal based diets gave higher carcass yield at eight weeks of age with concomitant reduction in feed cost regardless of dietary contents of wheat offal and produced better returns compared to groundnut cake based diets. Therefore, iso-nitrogenous as well as iso-caloric diets containing soybean meal with up to 15% level of inclusion of wheat offal could bring about efficient growth performance in broilers.

## REFERENCES

Abubakar MM (1998). Utilization of unconventional feedstuffs for sustainable livestock production. Inaugural Lecture Series Number 9. Abubakar Tafawa Balewa University, Bauchi, Nigeria.

Adene DF (1989). Feed-related deficiencies, diseases and anomalies in livestock poultry. Feed Mill Management in Nigeria. Federal Livestock Department. Ibadan University Press, Ibadan, pp. 174-197.

Akdeniz RC, Ak I, Boyar S (2006). Feed Industry and Problems in Turkey, Turkish Agricultural Engineering VI. Technical Congress Proceedings Ankara 2:935-960.

Considine DM (1982). Foods and Food Production. Encyclopaedia, VanNorstrand, New York.

Dibner JJ, Knight CD, Ivey FJ (1998). The feeding of neonatal poultry. World Poult. 14(5):36-40.

Duncan DB (1955). Multiple range and F-test. Biometrics 11:1-42.

ISA (Institut de Selection Animale) (1996). Guide d'Elevage des Poulets de Chair. Service Production et d'Assistance Technique, Quintin, France. pp. 3-10.

Kinsella JE, Lokesh B, Stone RA (1990). Dietary n-3 polyunsaturated fatty acids and amelioration of cardiovascular disease: Possible mechanism. Am. J. Clin. Nutr. 52:1-28.

Mustafa R A, Hakim AS, Azizullah M, Rab NS, Hidayatullah SN, Ahmed

U (2012). Effect of various protein source feed ingredients on the growth performance of broiler. Int. J. Med. Plant Res. 1(4):038-044.
Nettleton JA (1991). n-3 Fatty acids composition of plant and seafood sources in human nutrition. J. Am. Diet. Ass. 91:331-337.
Nie N, Bent DH, Hull CH (1970). Statistical Package for Social Sciences survival manual: a step by step guide to data analysis using SPSS. New York, Mc GrawHill, 1st edition, P. 343.
Ojewola GS, Okoye FC, Ukoha OA (2005). Comparative Utilization of Three Animal Protein Sources by Broiler Chickens. Int. J. Poult. Sci. 4(7):462-467.
Olumide OT, Mpoko B (2001). Post-harvest technologies in nigeria's livestock industry: status, challenges and capacities. A Presentation at the GFAR – GIPhT Workshop, 17–21 September 2001, Entebbe Uganda.
Shaib B, Adedipe NO, Aliyu A, Jir MA (1997). Integrated Agricultural Production in Nigeria. In: Strategies and Mechanisms for Food Security. Proceedings of the National Workshop on Nigeria's Position at the World Food Summit, Abuja, July 31$^{st}$-August 2nd, 1996. pp. 11-74.
Steenfeldt, S, Kjaer JB, Engberg RM (2007). Effect of feeding silages or carrots as supplements to laying hens on production performance, nutrient digestibility, gut structure, gut microflora and feather pecking behaviour. Br. Poult. Sci. 48(4):454-468.

# Effects of plant density of intercropped soybean with tall sorghum on competitive ability of soybean and economic yield at Otobi, Benue State, Nigeria

**O. M. Egbe**

Department of Plant Breeding and Seed Science, University of Agriculture, P. M. B. 2373, Makurdi, Benue State, Nigeria. E-mail: onyiloegbe@yahoo.co.uk.

**Field experiment was conducted for three years to evaluate the influence of plant population density of intercropped soybean with sorghum on its competitive ability and economic yield at Otobi, Benue State, Nigeria. The experiment was laid out in split-split plot with three replications. The main plot treatment comprised of cropping systems with two levels (sole cropping, intercropping), while the sub-plot treatment was soybean variety at three levels (TGX 536-O2D, Samsoy 2, TGX 923-2E) and the sub-sub-plot treatment was plant population density of soybean at three levels (200,000 plants/ha; 333,000 plants/ha and 400,000 plants/ha). Planting intercropped soybean at 333,000 plants/ha gave significantly higher seed yield than planting at 400,000 plants/ha, which in turn had greater seed yield than planting at 200,000 plants/ha during the three years of experimentation. Increased density of soybean beyond 333,000 plants/ha in the intercrop reduced sorghum yield. The results showed that all the intercrop combinations had land equivalent ratio (LER) and area- time equivalency ratio( ATER) above unity (1.63 - 1.97) and (1.41 - 1.47),respectively, under all the densities of soybean tested, suggesting a considerable benefit for intercropping soybean with sorghum. LER figures decreased with increase in soybean density. ATER was highest at 333.000 plants/ha, suggesting higher productivity at this density. Aggressivity values were consistently negative at 200,000 plants/ha with a mean value of -0.25 and it was inconsistent at 333,000plants/ha and 400,000 plants/ha. The competitive ratio of soybean increased (0.76 - 1.15) with increasing density of the soybean in the intercrop combinations, indicating higher competitiveness at higher densities than the sorghum component. The competitive ratio of sorghum had the opposite response (1.23 - 0.76). Relative crowding coefficient (K) was inconsistent at all densities, while land equivalent coefficient (LEC) was above 0.25 indicating intercrop advantages for all combinations. 'Soybean yield equivalent' was also highest at 333,000 plants/ha (1.70) implying general suitability of soybean to intercropping at this density. Dominance analysis revealed that sole crop treatments were dominated. Intercropped soybean planted at 333,000 plants/ha and 400,000 plants/ha gave significantly higher net benefits than those planted at 200,000 plants/ha. Marginal analysis showed that planting soybean at 333,000 plants/ha gave the highest marginal rate of return and this was 2825%.These results suggest that growing soybean under intercropping at Otobi is biologically efficient at 333,000 plants/ha and more profitable.**

**Key words:** Intercropped soybean, density, competition, Otobi.

## INTRODUCTION

Intercropping might positively impact on the future food problems in developing countries (Egbe, 2005).This may be through efficient use of solar energy and other growth resources. Also, optimization of land resource use could be achieved when crops are grown under intercropping and plant population density increased. Growing legumes and cereals together for food is not only popular among subsistence farmers in the tropics, who produce the bulk of food in developing countries, but it is also expanding to the warmer regions in the tropics (Fujita and Ofosu-Budu,

1996). Intercropping is receiving attention because it offers potential advantages for resource utilization, decreased inputs and increased sustainability in crop production, but our understanding of interactions among intercropped species is still very limited. The basic physiological and morphological differences between non-legume and legume benefit their mutual association (Akunda, 2001).The differences in the depth of rooting, lateral root spread and root densities are some of the factors of competition between the component crops in an intercropping system for water and nutrients, and hence input use efficiency. Soybean is considered a crop of enormous potential for improving human diet as well as animal feed and features prominently as raw material base for agro-industries. In Benue State, located in Southern Guinea Savanna of Nigeria, soybean is the most important oilseed crop grown during the wet season and it is currently grown on over 88,000 ha of land with an annual production of over 173,000 tons of grain and a mean yield of 1.95 t/ha (BNARDA, 2007). It is traditionally grown under crop mixtures with varying densities of maize, sorghum, cassava and citrus (Egbe, 1995; BNARDA, 2003).Yield advantages accruing from such crop mixtures had been reported (Egbe, 1995, Muoneke et al., 2007). But several studies on plant population densities in cereal/legume intercropping had reported more yield depression of the legumes than the cereal component (Muoneke et al., 2007; Fujita et al., 1992). Such studies had often used sub-optimal densities of the legume and optimal densities of the cereal, often attributed to farmers' preference for the cereal component.

Akunda (2001) demonstrated that higher soybean populations provided a way to optimizing growth and yields in soybean/millet intercropping systems. Thus, plant population can be used as a tool to manage crop growth, maximize biomass, the time required for canopy closure and yield (Akunda, 2001). In Otobi, located in the Southern Guinea Savanna of Nigeria, growing soybean by peasant farmers for home consumption and the market began in the mid 1980s, and intercropping soybean with the tall traditional sorghum is very popular (BNARDA, 2003). Information on the effects of component densities of maize and sorghum on the yield of soybean are available (Pal et al., 1992; Muoneke et al., 2007), but documented information on the optimum plant population density of the various recommended soybean varieties for intercropping with tall traditional sorghum is scanty. Also information on the profitability of soybean/ sorghum intercropping cropping systems in this region is lacking. It was hypothesized that increased density of soybean intercropped with sorghum could increase its yield, competitive ability and profitability. The present study was undertaken to provide documented information on the effects of densities of intercropped soybean on its economic yield with a view to improve the productivity of soybean/sorghum intercropping system and enhance food security in Southern Guinea Savanna.

## MATERIALS AND METHODS

A field experiment were conducted for three years (2001, 2004 and 2005) at the National Root Crops Research Institute Sub-station, Otobi {07°10′ N, 08°39′ E, elevation 105.1 m} in Benue State, located in Southern Guinea Savanna of Nigeria (Kowal and Knabe, 1972). The experiments were carried out to determine the effects of increased population densities of intercropped soybean on its competitive ability and economic yield at Otobi. The precipitations at the experimental site within the period of June - November were: 1605.1, 1712.0 and 1570.2 mm for 2001, 2004 and 2005, respecttively. The texture of the top soil (0 - 30 cm) of the experimental site was sandy loam. The soil at the experimental site was classified as Typic Paleustalf (USDA). Eight core samples of soil were collected from different parts of the experimental field and bulked into a composite sample and used for the determination of the chemical and physical properties of the soil before planting. The level of total nitrogen (N) was 520 mg $kg^{-1}$ soil, phosphorus and potassium averaged 68.20 and 43.50 mg $kg^{-1}$ soil, respectively. The experiment was laid out in split-split plot with three replications. The main plot treatment comprised of cropping systems with two levels (sole cropping, intercropping), while the sub-plot treatment was soybean variety at three levels (TGX 536-O2D, Samsoy 2, TGX 923-2E) and the sub-sub-plot treatment was plant population density of soybean at three levels (200,000; 300,000 and 400,000 plants/ha designated as P1-0.5 × 0.1, P2-0.5 × 0.06 and P3-0.5 × 0.05, respectively). Gross plot size was 6 × 4 m. Soybean varieties (sole as well as intercrop) were sown on the same day with sorghum (traditional, photoperiodic-sensitive, red-colored grain) during the last week of June of each experimental year. In both sole- and inter-cropped plots, sorghum population density was maintained at 40,000 plants per hectare (0.5 × 0.5 m × 1 plant/hill). Intercropping had a 1:1 (soybean: sorghum) row proportion. All plots received a basal application of 100 kg of NPK: 15:15:15 at the rate of 15 kg N, 6.55 kg P and 12.45 kg K per hectare by broadcasting. The sole- and inter –cropped sorghum were top-dressed four weeks after planting (w.a.p) with 46 kg N per hectare by opening the soil around each plant and banding at 5 - 8 cm depth and covering with the dug-out soil. Two manual weeding were done at 3 w.a.p and 6 w.a.p., respectively. Harvesting of both crops was done from the inner 4 × 3 m at physiological maturity and this represented yield per plot.

Intercrop advantage was calculated by the determination of land equivalent ratio (LER) (Ofori and Stern, 1987). The LER, an accurate assessment of the biological efficiency of the intercropping situation, was calculated as:

$$LER = (Yab/Yaa) + (Yba/Ybb)$$

Where $Yaa$ and $Ybb$ are yields as sole crops of $a$ and $b$ and $Yab$ and $Yba$ are yields as intercrops of $a$ and $b$. Values of LER greater than 1 are considered advantageous.

The relative dominance of one species over the other in this intercropping study was estimated by the use of relative crowding coefficient (K) (Banik et al., 2006). K was calculated as:

$$K = (K_{soybean} \times K_{sorghum})$$

Where, $K_{soybean} = Yab \times Zba / (Yaa-Yab) \times Zab$
$K_{sorghyum} = Yba \times Zab / (Ybb-Yba) \times Zba$,

where, $Yab$ and $Yba$ were the yields of soybean and sorghum in the intercrop, respectively, $Yaa$ and $Ybb$ ,were the yields of soybean and sorghum in sole crop, respectively and $Zab$ and $Zba$ were the respective proportions of soybean and sorghum in the in the intercropping systems. When the value of K is greater than 1.00,

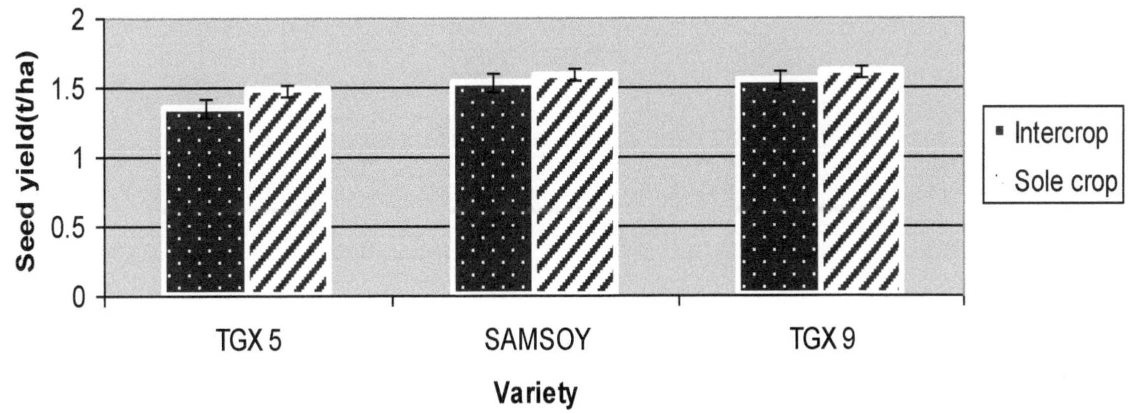


**Figure 1.** Effect of cropping system with variety on the seed yield of soybean intercropped with sorghum at Otobi. TGX 5=TGX 536-02D, TGX 9=TGX 923-2E, SAMSOY=SAMSOY 2.

there is intercrop advantage; when K is equal to 1.00, there is no yield advantage; and when K is less than 1.00, there is a disadvantage. Land equivalent coefficient (LEC), a measure of interaction concerned with the strength of relationship was calculated thus,

$$LEC = La \times Lb$$

Where, $La$ = LER of main crop and $Lb$ = LER of intercrop (Adetiloye et al., 1983). For a two-crop mixture the minimum expected productivity coefficient (PC) is 25%, that is, a yield advantage is obtained if LEC value exceeds 0.25. Area-time equivalency ratio (ATER), the ratio of number of hectare-days required in monoculture to the number of hectare-days used in the intercrop to produce identical quantities of each of the components, was computed as:

$$ATER = \frac{(Rya \times ta) + (Ryb \times tb)}{T}$$

Where, Ry = Relative yield of species 'a' or 'b' i.e., yield of intercrop/yield of main crop, t= duration (days) for species 'a' or 'b' and T = duration (days) of the intercropping system (Hiebisch and Mc Collum, 1987).

Competitive ratio (CR) indicates the number of times by which one component crop is more competitive than the other. Relative species competition is often evaluated using competitive ratios (Putnam et al., 1984).This was calculated as:

$$Ra = La/Lb \times zba/zab$$

Where $Ra$ is the competitive ratio of crop *a* and $La$ and $Lb$ are the LERs of crops *a* and *b* respectively, $zba$ is the proportion of crop *a* in the *ab* intercrop and $zab$ is the proportion of crop *b* in the *ab* intercrop. If $Ra < 1$, there is a positive benefit and the crop can be grown in association; if $Ra > 1$, there a negative benefit. The reverse is true for $Rb$.

Aggressivity is another index that represents a simple measure of how much the relative yield increase in crop *a* is greater than that of crop *b* in an intercropping system. It was calculated as:

$$Aab = (Yab/YaaZab) - (Yba/YbbZba)$$

Where $Yaa$ and $Ybb$ are yields as sole crops of *a* and *b* and $Yab$ and $Yba$ are yields as intercrops of *a* and *b*. $Zab$ and $Zba$ are the sown proportions of *a* and *b*, respectively.

If $Aab = 0$, both crops are equally competitive; if $Aab$ is positive, *a* is dominant; if $Aab$ is negative, *a* is the dominated crop (Ghosh et al., 2006).

Soybean yield equivalent was calculated as described by Prasad and Srivastava (1991).

$$\text{Soybean equivalent (t/ha)} = \frac{\text{Yield of intercrop}}{\text{Market price of soybean}} \times \text{Market price of intercrop}$$

The economic performance of the intercropping were evaluated to decide if soybean yield at intercropped with sorghum in 2001,2004 and 2005 at Otobi at different densities and additional sorghum yield justified adoption of this intercropping system by farmers in Southern Guinea Savanna of Nigeria. Therefore, returns per naira invested (RI) was computed as RI = gross margin (GM) / total variable cost (TVC) The higher the value of RI the more profitable is the cropping system. Furthermore, the economic analysis was carried out as described by CIMMYT (1998) to estimate the benefit-cost ratio. A dominance analysis was done and the nondominated treatments were further subjected to marginal rate analysis.

Standard procedures were followed to collect data and analyzed using GENSTAT Release 7.23 (2007), following standard analysis of variance procedures (Gomez and Gomez, 1984) and least significant difference (LSD) test at 5% probability level was used to compare the treatment means. It must be noted that data for each trait measured for the three years were pooled and analyzed to determine the year effect before being analyzed for the respective years of 2001, 2004 and 2005.

## RESULTS

The rainfalls during each of the three years of experimentation were adequate for crop growth and development.

Intercropping significantly depressed yield of TGX 536-02D, but the yield reductions in the other two varieties were not significant (Figure 1). Intercropped soybean produced lower seed yield than their sole crop counterparts at the various densities tested. Samsoy 2 and TGX

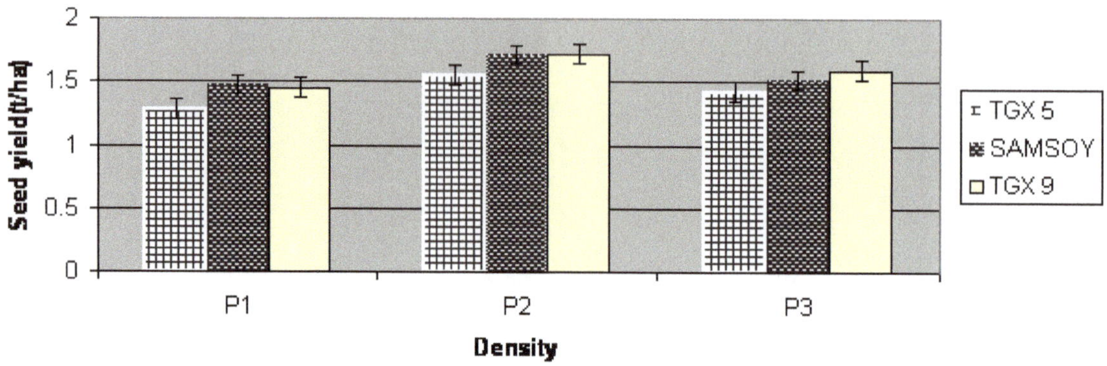


**Figure 2.** Influence of variety with density on the seed yield of soybean intercropped with sorghum at Otobi.

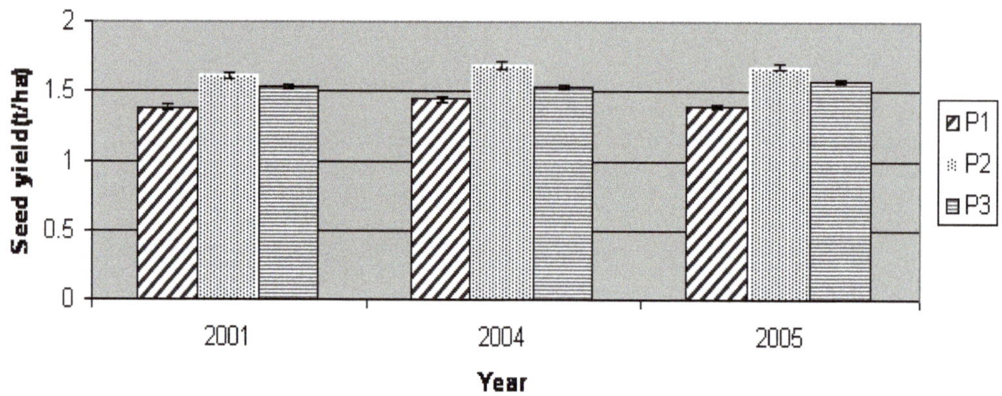


**Figure 3.** Effects of density on the seed yield of soybean intercropped with sorghum at Otobi.

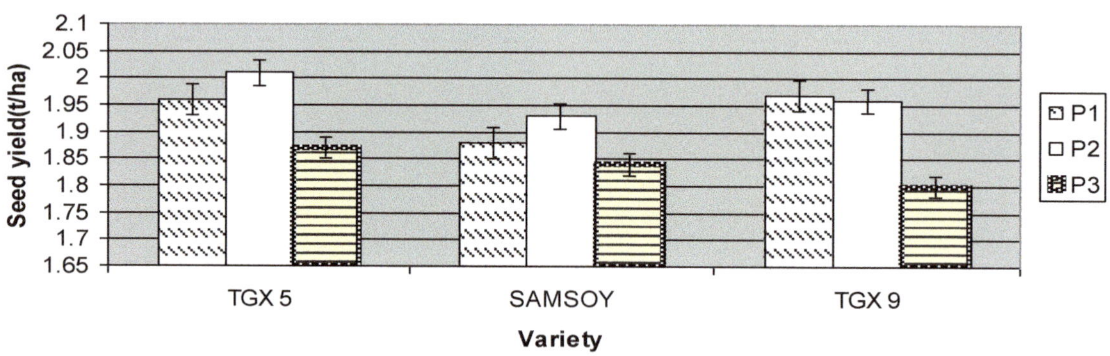


**Figure 4.** Influence of variety and density on the seed yield of sorghum intercropped with soybean at Otobi.

923-2E gave significantly higher seed yield than TGX 536-O2D at P1, P2 and P3 (Figure 2). The main effect of density on the seed yield of soybean intercropped with sorghum was consistently significant ($p \leq 0.05$) in 2001, 2004 and 2005. Planting soybean at P2 gave significantly higher seed yield than planting at P3, which in turn had greater seed yield than planting at P1 during the three years of experimentation (Figure 3). Sorghum seed yield was significantly higher in P1 and P2 than in P3 under the three soybean varieties tried (Figure 4). Sorghum seed yield exceeded 2.0 t/ha only when grown with TGX 536-02D at P2. The results showed that all intercrop combinations had LER above unity at the densities of soybean tested in this study (Figure 5). The LER figures

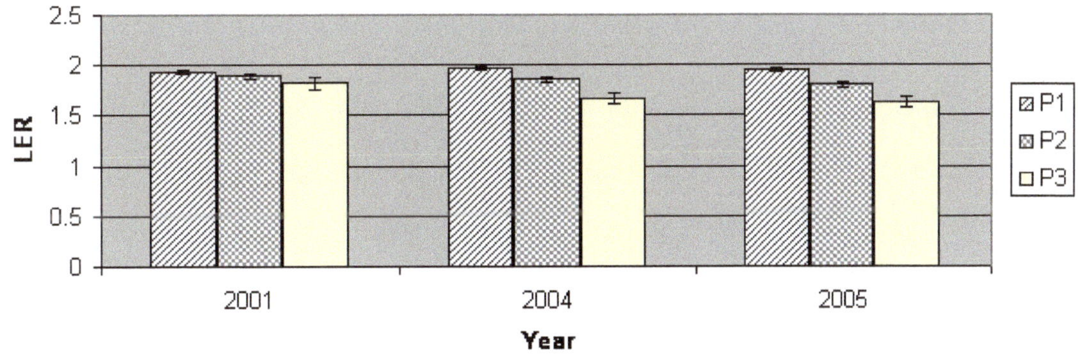


**Figure 5.** Influence of density on LER of soybean intercropped with sorghum at Otobi.

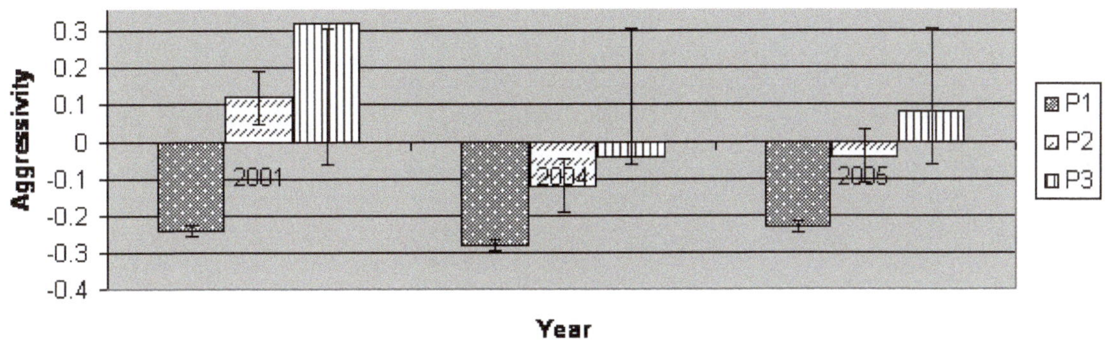


**Figure 6.** Influence of density on aggressivity of soybean intercropped with sorghum at Otobi.

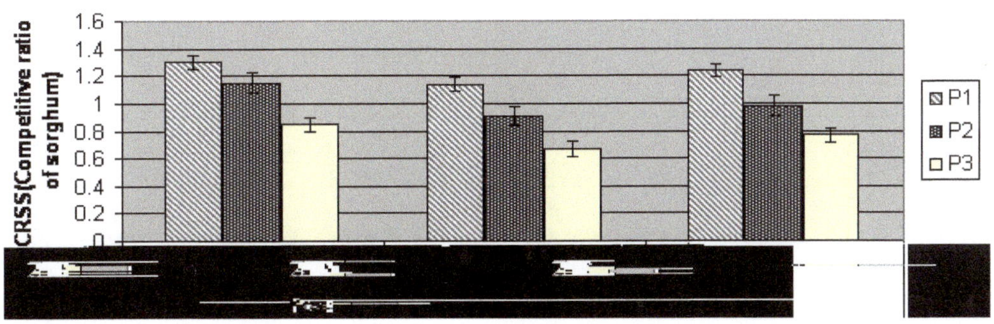


**Figure 7.** Effects of density on competitive ratio (CRSS) of sorghum intercropped with soybean at Otobi.

decreased with increase in soybean density from P1 to P3 during the three years of the work. LER figures were highest at P1 and lowest at P3 (Figure5). Aggressivity values of soybean were consistently negative at P1 with a mean value of -0.25 and it was inconsistent at P2 and P3 during the course of the work for three years (Figure 6). Figure 7 presents the influence of density on the competitive ratio of sorghum (CRSS) at Otobi in 2001, 2004 and 2005. The competitive ratio of sorghum decreased with increasing density of soybean in the intercrop combinations used in this study. P1 consistently gave higher values of CRSS than P2, which in turn had higher values than P3.The competitive ratio of soybean (CRSSO) increased with increasing density of the soybean in the intercrop combinations (Figure 8). CRSSO was highest at P3 an d least at P1. The main effect of density significantly influenced the total relative crowding coefficient (K) of soybean varieties intercropped with sorghum. The effect of density on the total relative crowding coefficient of soybean intercropped with sorghum at Otobi was inconsistent during the three years of the study (Figure 9). The interaction effects of variety with density,

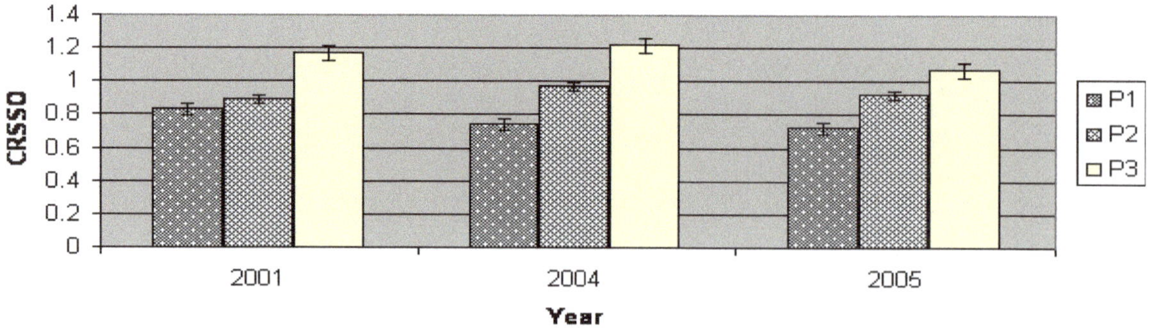


**Figure 8.** Influence of density on the competitive ratio (CRSS) of soybean intercropped with sorghum at Otobi.

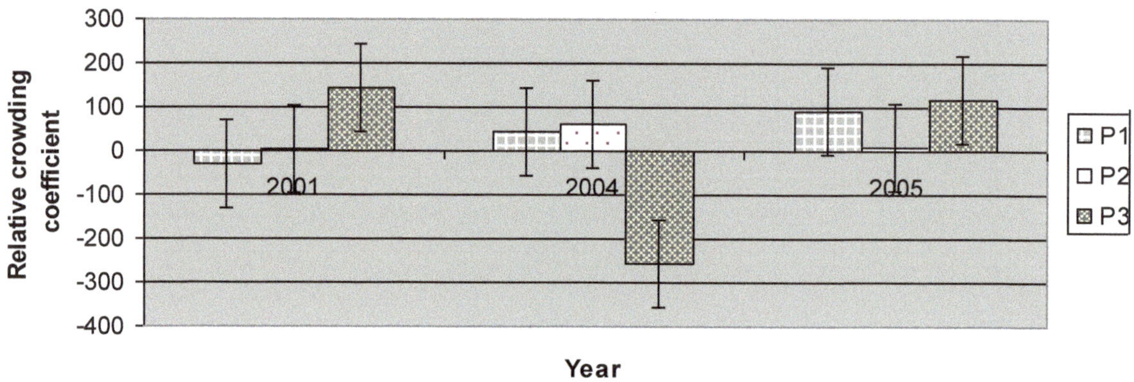


**Figure 9.** Effect of density on the relative crowding coefficient of soybean intercropped with sorghum at Otobi

**Table 1.** Effect of density on the land equivalent ratio (LEC), area-time equivalency ratio (ATER) and the soybean equivalent yield (SOYEQUIV.) on soybean intercropped with sorghum at Otobi.

| Density | LEC | | | | ATER | | | | SOYEQUIV. | | | |
|---|---|---|---|---|---|---|---|---|---|---|---|---|
| | 2001 | 2004 | 2005 | Mean | 2001 | 2004 | 2005 | Mean | 2001 | 2004 | 2005 | Mean |
| P1 | 0.98 | 0.79 | 0.84 | 0.87 | 1.45 | 1.42 | 1.41 | 1.43 | 1.87 | 1.60 | 1.42 | 1.63 |
| P2 | 0.91 | 0.86 | 0.83 | 0.87 | 1.48 | 1.47 | 1.46 | 1.47 | 1.89 | 1.60 | 1.60 | 1.70 |
| P3 | 0.98 | 0.80 | 0.78 | 0.85 | 1.43 | 1.41 | 1.40 | 1.41 | 1.72 | 1.47 | 1.54 | 1.58 |
| Mean | 0.96 | 0.82 | 0.82 | 0.86 | 1.45 | 1.43 | 1.42 | 1.44 | 1.83 | 1.53 | 1.52 | 1.63 |
| LSD(0.05) | 0.13 | 0.09 | 0.18 | | 0.13 | 0.03 | 0.05 | | 0.05 | 0.03 | 0.07 | |

main effects of variety and density on the LEC of soybean intercropped with sorghum at Otobi were not significant in any of the years of the study. Mean LEC values varied from 0.85 at P1 to 0.87 at P2 and P3. The main effect of density significantly influenced ATER and 'soybean yield equivalent' (SOYEQUIV.) (Table1). ATER values were above 1.00 in all treatment combinations, and varied from 1.41 - 1.47 with a mean of 1.44. Although, ATER values did not vary significantly in 2001, soybean intercropped at P2 produced the highest significant ATER figures in 2004 and 2005. 'Soybean equivalent yield' (SOYEQUIV) figures were significantly higher at P1 and P2 than at P3 in 2001 and 2004. However, in 2005, P2 and P3 gave significantly higher 'soybean yield equivalent' than P1 (Table 1).The highest mean value of 'soybean equivalent yield' was obtained at P2. TVC (total variable cost) indicated that sole cropping produced higher values (₦52,480.00) than intercropping (₦46,730.00) (Table 2). Table 3 showed that inter-cropping consistently gave higher values of gross margin (GM) and RI. The values of GM ranged from ₦16, 310.00/ha to ₦17, 230.00/ha under sole cropping, and from ₦33,960.00/ha to ₦43,080.00/ha under inter-cropping. RI values varied from ₦ 0.31 to ₦ 0.91 under

**Table 2.** TVC (₦'000/ha) of soybean intercropped with sorghum at Otobi in 2001, 2004 and 2005.

| Cropping system | TVC | | | |
|---|---|---|---|---|
| | 2001 | 2004 | 2005 | Mean |
| Sole cropping | 50.89 | 52.90 | 53.66 | 52.48 |
| Intercropping | 45.59 | 47.61 | 46.98 | 46.73 |
| Mean | 48.24 | 50.26 | 50.32 | 49.61 |
| FLSD(0.05) | 0.16 | 0.17 | 0.16 | |

**Table 3.** Effect of cropping system on the GM (₦'000/ha) and RI of soybean intercropped with sorghum at Otobi in 2001, 2004 and 2005.

| Cropping system | GM | | | | RI | | | |
|---|---|---|---|---|---|---|---|---|
| | 2001 | 2004 | 2005 | Mean | 2001 | 2004 | 2005 | Mean |
| Sole cropping | 17.23 | 16.31 | 17.15 | 16.90 | 0.34 | 0.31 | 0.91 | 0.52 |
| Intercropping | 33.96 | 42.07 | 43.08 | 39.70 | 0.75 | 0.89 | 1.37 | 1.00 |
| Mean | 25.60 | 29.19 | 30.12 | 28.30 | 0.55 | 0.60 | 1.14 | 0.76 |
| LSD(0.05) | 0.74 | 0.81 | 0.86 | | 0.02 | 0.02 | 0.12 | |

GM: gross margin.
RI: returns per investment.

**Table 4.** Effect of variety on the GM (₦'000/ ha) and RI of soybean intercropped with sorghum at Otobi in 2001, 2004 and 2005.

| Variety | GM | | | | RI | | | |
|---|---|---|---|---|---|---|---|---|
| | 2001 | 2004 | 2005 | Mean | 2001 | 2004 | 2005 | Mean |
| TGX536-O2D | 25.11 | 28.56 | 29.51 | 27.73 | 0.53 | 0.58 | 0.48 | 0.53 |
| SAMSOY2 | 26.24 | 31.93 | 32.37 | 30.18 | 0.56 | 0.64 | 0.66 | 0.62 |
| TGX923-2E | 25.43 | 27.66 | 28.47 | 27.19 | 0.54 | 0.57 | 0.58 | 0.56 |
| Mean | 25.59 | 29.38 | 30.12 | 28.37 | 0.54 | 0.59 | 0.57 | 0.57 |
| LSD (0.05) | 0.90 | 0.99 | 1.06 | | 0.02 | 0.01 | 0.02 | |

**Table 5.** Effect of density on the GM (₦'000/ ha) and RI of soybean intercropped with sorghum at Otobi in 2001, 2004 and 2005.

| Variety | GM | | | | RI | | | |
|---|---|---|---|---|---|---|---|---|
| | 2001 | 2004 | 2005 | Mean | 2001 | 2004 | 2005 | Mean |
| P1 | 22.80 | 25.89 | 27.51 | 25.40 | 0.48 | 0.53 | 0.49 | 0.50 |
| P2 | 26.64 | 30.93 | 31.56 | 29.71 | 0.57 | 0.63 | 0.65 | 0.62 |
| P3 | 27.35 | 30.75 | 31.29 | 27.09 | 0.58 | 0.63 | 0.58 | 0.60 |
| Mean | 25.97 | 29.19 | 30.12 | 27.40 | 0.54 | 0.59 | 0.57 | 0.57 |
| LSD (0.05) | 0.90 | 0.99 | 1.06 | | 0.02 | 0.02 | 0.02 | |

sole system and ₦0.75 to ₦1.37 under intercropping. GM and RI also differed with the varieties of soybean (Table 4). SAMSOY 2 gave higher GM and RI than TGX 923-2E and TGX 536-O2D. Soybean intercropped with sorghum at P1 gave consistently lower values of GM and RI than those planted at P2 and P3 (Table 5). Table 6 presents the results of dominance analysis of soybean varieties intercropped with sorghum at Otobi. The results showed that the sole crop systems were dominated and there were no significant differences between the intercropped soybean varieties in both TVC and the net benefits that accrued. Table 7 indicates the dominance analysis of soybean intercropped at different densities with sorghum. Intercropped soybean planted at P2 and P3 gave

**Table 6.** Dominance analysis, soybean varieties intercropped with sorghum at Otobi.

| Treatment | TVC(₦'000/ha) | Net benefits(₦'000/ha) |
|---|---|---|
| Intercropped TGX923-2E | 46.73 | 37.92 |
| Intercropped TGX 536-02D | 46.73 | 38.85 |
| Intercropped Samsoy 2 | 46.73 | 42.34 |
| Sole TGX 536-02D | 52.48 | 16.60 D |
| Sole TGX 923-2E | 52.49 | 16.46 D |
| Sole crop Samsoy2 | 52.49 | 17.40 D |
| LSD(0.05) | 0.79 | 5.63 |

**Table 7.** Dominance analysis, soybean intercropped at different densities with sorghum at Otobi.

| Treatment | TVC(₦'000/ha) | Net benefits(₦'000/ha) |
|---|---|---|
| Intercropped soybean planted at P1 | 46.71 | 35.83 |
| Intercropped soybean planted at P2 | 46.73 | 41.48 |
| Intercropped soybean planted at P3 | 46.75 | 41.80 |
| Sole crop soybean planted at P1 | 52.48 | 14.96 D |
| Sole crop soybean planted at P3 | 52.50 | 17.79 D |
| Sole crop soybean planted at P2 | 52.50 | 17.94 D |
| LSD (0.05) | 0.79 | 5.32 |

**Table 8.** Marginal analysis, soybean intercropped at different densities with sorghum at Otobi.

| Treatment | Total variable cost(₦/ha) | Marginal costs(₦/ha) | Net benefits(₦/ha) | Marginal benefits(₦/ha) | Marginal rate of return (%) |
|---|---|---|---|---|---|
| Intercropped soybean at P1 | 46,710.00 | | 35,830.00 | | |
| | | 20.00 | | 5,650.00 | 28250 |
| Intercropped soybean at P2 | 46,730.00 | | 41,480.00 | | |
| | | 20.00 | | 320.00 | 1600 |
| Intercropped soybean at P3 | 46,750.00 | | 41,800.00 | | |

significantly higher net benefits than those planted at P1. Marginal analysis showed that planting soybean at P2 gave the highest marginal rate of return (Table 8) and this was 2825%.

## DISCUSSION

The yield reduction of the intercropped soybean might be associated with interspecific competition between the intercrop components for growth resources (light, water, nutrients, air, etc.) and the depressive effects of sorghum. Pal et al. (1992) and Muoneke et al. (2007) had observed similar yield reductions in soybean intercropped with maize and sorghum in Benue State, Nigeria and associated the yield depression to interspecific competition and the depressive effect of the cereals. Ghosh (2004) further explained that because of the differences in canopy height of soybean and sorghum, the two species not only competed for nutrient water but also for sunlight. Samsoy 2 and TGX 923-2E recorded higher seed yield than TGX 536-O2D at the various densities. Such differential responses might be due to inherent genotypic capabilities of these varieties to withstand shade. Egbe and Kalu (2009) had reported similar observations when different pigeon pea varieties were evaluated under intercropping with sorghum in Benue State, Nigeria. Plant densities of soybean seemed to have had the dominant effect on the seed yield of soybean intercropped with sorghum at Otobi. Pal et al. (1992) had indicated that yields of component crops in the intercrop varied significantly with the components population density. Intercropped soybean planted at P2 (333,000 plants $ha^{-1}$) produced significantly higher seed yield than those planted at the other two densities. These results contradicted earlier findings (BNARDA, 1993; Egbe, 1995) which had recommended a lower density (200,000 plants $ha^{-1}$) of soybean for intercropping with sorghum. The better performance of soybean at P2 than P3 might be ascribed to reduced competitiveness for growth resources at P2 as compared to P3. This was further elucidated by the competitive ratio values for soy-bean (CRSSO), which was less than 1.00 at P1 and P2 but greater than 1.00 at P3, suggesting that competition became more severe at P3 (400,000 plants/ha). Ghosh et al. (2006) had indicated that if competitive ratio was less than 1, there is a positive benefit and the crop can be grown in association, but if greater than 1, there was negative benefit. According to Willey and Rao (1980), competitive ratio (CR) gives a better measure of competitive ability of crops and can prove a better index as compared to aggressivity. Seed yield of sorghum was significantly higher at P1 and P2 than at P3, further indicating adverse influence of interspecific competition at P3. Generally, the cereal component is considered a suppressing crop in legume/cereal associations like soybean/maize (Quainoo et al., 2000; Muoneke et al., 2007). This was shown to be completely true only when soybean was intercropped at sub-optimal densities (e.g. P1) as indicated by aggressivity figures in this study. Aggressivity values were inconsistent beyond P1. The consistent negative aggressivity values at P1 suggested that soybean was dominated at that population density. Although LER figures were above unity under all intercrop combinations in this study, it decreased with increasing density of soybean, suggesting decreased efficiency in land resource utilization with increasing density of soybean. The K values were generally inconsistent; however, the negative K figures obtained at P1 in 2001 and at P3 in 2004 suggested potential yield reductions at these densities as compared to P2. For a two-crop mixture the minimum expected productivity coefficient (PC) is 25%s, that is, a yield advantage is obtained if LEC value exceeds 0.25 (Adetiloye et al., 1983). All intercrop combinations in this study had LEC values above 0.25, suggesting yield advantages. Similar to LER, ATER values were above 1.00 indicating intercrop benefits. In particular, soybean intercropped at P2 gave the highest ATER of 1.47, further indicating that intercropping soybean at this density might be more productive. 'Soybean yield equivalent' was also highest at P2, implying general suitability of soybean to intercropping at this density. Intercropped soybean gave higher values of GM and RI and net benefits than the sole systems, probably because of the additional yield and values of the sorghum component. Egbe (2005) had made similar observations in pigeon pea/sorghum and pigeon pea-/maize intercropping studies at Otobi. Farmers are mostly concerned with the profitability of their farm enterprises. Njoroge et al. (1993) estimated the net benefit of intercropping coffee with food crops by subtracting the total variable costs from the gross profits. Similarly, Egbe (2005) had estimated the total profit and the marginal benefit; cost ratio from investment on different farm inputs used in pigeon pea/sorghum intercropping system by computing returns per naira invested (RI). Intercropped samsoy 2 was more profitable than the other soybean varieties probably because it attracted a higher price than the others and it also gave higher seed yield. The greater profit obtained at P2 than at the other densities might probably have arisen from the higher yield of soybean at this population density. These results suggest that growing soybean under intercropping is more profitable and biologically efficient at 333,000 plants/ha. Samsoy 2 is the preferred variety because of its grain yield and higher price in the market.

## Conclusions

Planting intercropped soybean at 333,000 plants/ha gave significantly higher seed yield than planting at 400,000 plants/ha, which in turn had greater seed yield than planting at 200,000 plants/ha. The competitive ratio of soybean increased with increasing density of the soybean in the intercrop combinations, but the competitive ratio of sorghum had the opposite response. Land equivalent ratio (LER),area – time equivalency ratio(ATER), land equivalent coefficient(LEC ) and 'soybean yield equivalent' indicated yield advantages for all intercrop combinations, especially at P333,000 plants/ha. Relative crowding coefficient (K) was inconsistent at all densities, while aggressivity was inconsistent at 333,000 and 400,000 plants/ha. Soybean intercropped with sorghum at 333,000 plants/ha gave higher net benefits and marginal rate of return than those planted at 200,000 and 400,000 plants/ha. These results suggest that growing soybean under intercropping is biologically efficient at 333,000 plants/ha and more profitable.

### REFERENCES

Adetiloye PO, Ezedinma FOC, Okigbo BN (1983). Aland equivalent coefficient concept for the evaluation of competitive and productive interactions on simple complex mixtures. Ecol. Modelling., 19: 27-39.

Akunda EM (2001). Intercropping and population density effects on yield component, seed quality and photosynthesis of sorghum and soybean. J. Food Tech. (Africa) 6: 170-172

Banik P, Midya BK, Sarkar, Ghose SS (2006). Wheat and chickpea intercropping systems in an additive series experiment: advantages and weed smothering. Eur. J. Agron. 24: 325-332.

BNARDA (1993). Benue State Agricultural and Rural Development Authority.Annual Report, 1993, BNARDA, Makurdi, Nigeria, p. 32

BNARDA (2003). Benue State Agricultural and Rural Development Authority. Annual Report, 2003, BNARDA, Makurdi, Nigeria, p. 25.

BNARDA (2007). Benue State Agricultural and Rural Development

Authority. Annual Report, 2007, BNARDA, Makurdi, Nigeria, p.28

CIMMYT (1998). From Agronomic Data to Farmer Recommendations: An Economic Training Manual. Completely revised edition. Mexico. D.F.

Egbe OM (1995). Effects of plant densities of intercropped local sorghum with soybean varieties and productivity of soybean/sorghum intercropping in Southern Guinea Savanna. M.Sc. thesis, University of Agriculture, Makurdi, Nigeria.

Egbe OM (2005). Evaluation of some agronomic potentials of pigeonpea genotypes for intercropping with maize and sorghum in Southern Guinea Savanna. Ph.D. thesis, University of Agriculture, Makurdi, Nigeria.

Egbe OM, Adeyemo MO (2006). Estimation of the effects of intercropped pigeonpea on the yield and yield components of maize in Southern Guinea Savanna of Nigeria. J. Sustain. Dev. Agric. Environ., 2: 107-119.

Egbe OM, Kalu BA (2009).Evaluation of pigeonpea(*Cajanus cajan* (L.) Millsp.) genotypes for intercropping with tall sorghum (*Sorghum bicolor* (L.) Moench.) in Southern Guinea Savanna of Nigeria. ARPN. J. Agric. Biol. Sci., 4: 54-65.

Fujita K, Ofosu-Budu KG, Ogata S (1992). Biological nitrogen fixation in mixed legume-cereal cropping systems. Plant soil, 141: 155-175.

Fujita K, Ofosu-Budu KG (1996). Significance of legumes in intercropping systems. *In*: Ito O, Katayama K, Johansen C, Kumar Rao JVDK, Adu-Gyamfi JJ, Rego TJ (eds). Roots and nitrogen in cropping systems of the semi-arid tropics. JIRCAS Int. Agric. Ser. No.3 Japan., pp. 19-40.

GENSTAT (2007). GENSTAT Release 7.23 (Copyright 2007) Lowes Agricultural Trust Rothamsted Experimental Station.

Ghosh PK (2004). Growth, yield, competition and economics of groundnut/cereal fodder intercropping in the semi-arid tropics of India. Field Crops Res., 88: 227-237.

Ghosh PK, Manna MC, Bandyopadhyay KK, Ajay, Tripathi AK, Wanjari RH, Hati KM, Misra AK, Acharya Subba Rao CL (2006). Interspecific interaction and nutrient use in soybean/sorghum intercropping system. Agron. J., 98: 1097-1108.

Gomez KA, Gome AA (1984). Statistical procedures for agricultural research. 2nd ed. John Willey & Sons, Toronto, ON, Canada.

Hiebisch CK, McCollum RE (1987). Area x time equivalency ratio: a method of evaluating the productivity of intercrops. Aggro. J., 79: 15-22.

Kowal JM, Knabe DT (1972). An agroclimatological atlas of the Northern States of Nigeria with explanatory notes. Ahmadu Bello University Press, Zaria.

Muoneke CO, Ogwuche MAO, Kalu BA (2007). Effect of maize planting density on the performance of maize/soybean intercropping system in a guinea savanna agroecosystem. Afr. J. Agric. Res., 2: 667-677.

Njoroge JM, Waithaka K, Chweya JA (1993).Effects of intercropping young plants of the compact Arabica coffee hybrid cultivar. Ruiru II with potatoes, tomatoes, beans and maize on coffee yields and economic returns in Kenya. Expl. Agric., 29: 373-377.

Ofori F, Stern WR (1987). Cereal –legume intercropping systems. Adv. Agron., 41: 41-90.

Pal UR, Oseni TO, Norman JC (1992).Effect of component densities on productivity of soybean/maize soybean/sorghum intercrop. J. Agron. Crop Sci., 170: 66-70.

Prasad K, Srivastava RC 91991). Pigeonpea (*Cajanus cajan*) and soybean (Glycine max) intercropping system under rainfed situation. Indian J. Agric. Sci.. 61: 243-246.

Putnam DH, Herbert SJ, Vargas A (1984). Intercropped corn soybean density studies. 1. Yield complementarity. Expl. Agric. 21: 41-51.

Quainoo AK, Lawson IYD, Yawson A (2000). Intercrop performance of maize, sorghum and soybean in response to planting pattern. J. Ghana Sci. Ass., 2: 31-35.

Willey RW, Rao MR (1980). A competitive ratio for quantifying competition between intercrops. Expl. Agric., 17: 257-264.

# Permissions

All chapters in this book were first published in JCO, by Academic Journals; hereby published with permission under the Creative Commons Attribution License or equivalent. Every chapter published in this book has been scrutinized by our experts. Their significance has been extensively debated. The topics covered herein carry significant findings which will fuel the growth of the discipline. They may even be implemented as practical applications or may be referred to as a beginning point for another development.

The contributors of this book come from diverse backgrounds, making this book a truly international effort. This book will bring forth new frontiers with its revolutionizing research information and detailed analysis of the nascent developments around the world.

We would like to thank all the contributing authors for lending their expertise to make the book truly unique. They have played a crucial role in the development of this book. Without their invaluable contributions this book wouldn't have been possible. They have made vital efforts to compile up to date information on the varied aspects of this subject to make this book a valuable addition to the collection of many professionals and students.

This book was conceptualized with the vision of imparting up-to-date information and advanced data in this field. To ensure the same, a matchless editorial board was set up. Every individual on the board went through rigorous rounds of assessment to prove their worth. After which they invested a large part of their time researching and compiling the most relevant data for our readers.

The editorial board has been involved in producing this book since its inception. They have spent rigorous hours researching and exploring the diverse topics which have resulted in the successful publishing of this book. They have passed on their knowledge of decades through this book. To expedite this challenging task, the publisher supported the team at every step. A small team of assistant editors was also appointed to further simplify the editing procedure and attain best results for the readers.

Apart from the editorial board, the designing team has also invested a significant amount of their time in understanding the subject and creating the most relevant covers. They scrutinized every image to scout for the most suitable representation of the subject and create an appropriate cover for the book.

The publishing team has been an ardent support to the editorial, designing and production team. Their endless efforts to recruit the best for this project, has resulted in the accomplishment of this book. They are a veteran in the field of academics and their pool of knowledge is as vast as their experience in printing. Their expertise and guidance has proved useful at every step. Their uncompromising quality standards have made this book an exceptional effort. Their encouragement from time to time has been an inspiration for everyone.

The publisher and the editorial board hope that this book will prove to be a valuable piece of knowledge for researchers, students, practitioners and scholars across the globe.

# List of Contributors


**Payam Moaveni**
Islamic Azad University, Shahr-e-Qods Branch, Iran

**Ahmad Ebrahimi**
Islamic Azad University, Iranshahr Branch, Iran

**Hossein Aliabadi Farahani**
Islamic Azad University, Takestan Branch, Iran

**Hailegiorgis Biramo Allaro**
Haramaya University, College of Agricuture and Environmental Science, Department of Agricultural Economics P. O. Box: 138, Dire Dawa, Ethiopia

**Muhammad Sarwar**
Nuclear Institute of Agriculture, Tandojam-70060, Sindh, Pakistan

**Mohammad Nasri**
Agronomy and Plant Breeding Department, Islamic Azad University, Varamin Branch, Iran

**Mansoureh Khalatbari**
Agricultural Researcher at Varamin, Iran

**Hossein Aliabadi Farahani**
Member of Young Researchers Club, Islamic Azad University, Shahr-e-Qods Branch, Iran

**Muhammad Sarwar**
Pakistan Atomic Energy Commission, Nuclear Institute of Agriculture, Tando Jam-70060,Pakistan

**Muhammad Ashfaq**
University of Agriculture, Faisalabad-38040, Pakistan

**Latief Ahmad**
Allahabad Agricultural Institute-Deemed University Allahabad,-U.P

**M. Kaleem**
Allahabad Agricultural Institute-Deemed University Allahabad,-U.P

**R. A. Bhat**
Allahabad Agricultural Institute-Deemed University Allahabad,-U.P

**Maryam Ashofteh Beiragi**
Department of Agronomy and Plant Breeding, College of Agriculture, Zabol University, Zabol, Iran

**Mohsen Ebrahimi**
Department of Agronomy & Plant Breeding- College of Abouraihan - University of Tehran, Iran

**Khodadad Mostafavi**
Islamic Azad University- Karaj Branch Iran

**MohammadGolbashy**
NanoBioTechnology- University of Tehran Iran

**Saied Khavari Khorasani**
Scientific membrane of khorasan Razavi Agricultural Research and Natural Resources Center Mashhad, Iran

**D. O. Oke**
Department of Agricultural and Bio-Environmental Engineering, School of Engineering, Federal Polytechnic, P. M. B 5351, Ado - Ekiti, Ekiti State, Nigeria

**K. F. Omotayo**
Department of Agricultural and Bio-Environmental Engineering, School of Engineering, Federal Polytechnic, P. M. B 5351, Ado - Ekiti, Ekiti State, Nigeria

**V. Rameeh**
Agriculture and Natural Resources Research Center of Mazandran, Sari, Iran

**Muhammad Sarwar**
Nuclear Institute of Agriculture, Tandojam-70060, Sindh, Pakistan

**A. M. Olaniyan**
Department of Agricultural and Biosystems Engineering, Faculty of Engineering and Technology, University of Ilorin, P. M. B.1515, Ilorin 240003, Kwara State, Nigeria

**K. Oje**
Department of Agricultural and Biosystems Engineering, Faculty of Engineering and Technology, University of Ilorin, P. M. B.1515, Ilorin 240003, Kwara State, Nigeria

**Ritu Mahajan**
School of Biotechnology, University of Jammu, Jammu, India

**Syed Tabia**
School of Biotechnology, University of Jammu, Jammu, India

**Gourav Raina**
School of Biotechnology, University of Jammu, Jammu, India

**Neha Mangotra**
School of Biotechnology, University of Jammu, Jammu, India

**Muhammad Sarwar**
Pakistan Atomic Energy Commission, Nuclear Institute of Agriculture, Tando Jam-70060, Pakistan

**Oluwatosin K. Yusuf**
Trypanosomosis Research Unit, Department of Biochemistry, Federal University of Technology, P. M. B. 65, Minna, Nigeria

**Clement O. Bewaji**
Department of Biochemistry, University of Ilorin, Ilorin, Nigeria

**Dimberu G. Atinafu**
Department of Chemistry, College of Science, Bahir Dar University, Bahir Dar, Ethiopia

**Belete Bedemo**
Department of Chemistry, College of Science, Bahir Dar University, Bahir Dar, Ethiopia

**T. Radhamani**
Agricultural College and Research Institute, Madurai-625104, India

**R. Ushakumari**
Agricultural College and Research Institute, Madurai-625104, India

**R. Amudha**
Agricultural College and Research Institute, Madurai-625104, India

**K. Anjani**
Directorate of Oilseeds Research, Hyderabad, India

**Ijarotimi Oluwole Steve**
Department of Food Science and Technology (Human Nutrition Division), Federal University of Technology, Akure, Ondo State, Nigeria

**Aung Zaw Oo**
Tokyo University of Agriculture and Technology, Saiwaicho 3-5-8, Fuchu, Tokyo, 183-8509, Japan

**Khin Thuzar Win**
Tokyo University of Agriculture and Technology, Saiwaicho 3-5-8, Fuchu, Tokyo, 183-8509, Japan

**Lam Nguyen**
Hanoi University of Agriculture, Trau Quy, Gia Lam, Hanoi, Vietnam

**Sonoko Dorothea Bellingrath Kimura**
Tokyo University of Agriculture and Technology, Saiwaicho 3-5-8, Fuchu, Tokyo, 183-8509, Japan

**Joseph Adubofuor**
Department of Food Science and Technology, Kwame Nkrumah University of Science and Technology, Kumasi, Ghana

**Wilfred Sefah**
Department of Hotel, Catering and Institutional Management, Kumasi Polytechnic, Kumasi, Ghana

**James Henry Oldham**
Department of Biochemistry and Biotechnology, Kwame Nkrumah University of Science and Technology, Kumasi, Ghana

**B. Aliyu**
Department of Biological Sciences, Gombe State University, Gombe, Nigeria

**S. S. Hati**
Department of Chemistry, Gombe State University, Gombe, Nigeria

**G. A. Dimari**
Department of Chemistry, University of Maiduguri, Maiduguri, Nigeria

**P. O. Donli**
Department of Biological Sciences, University of Maiduguri, Maiduguri, Nigeria

**G. A. Abu**
Department of Agricultural Economics, College of Agricultural Economics, Extension and Management Technology, University of Agriculture, Makurdi, Benue, Nigeria

**S. Porpavai**
Agricultural Engineering College and Research Institute, Kumulur, Trichy 621 712, India

**P. Devasenapathy**
Tamil Nadu Rice Research Institute, Aduthurai, 612 101, India

**K. Siddeswaran**
Agricultural College and Research Institute, Coimbatore, 641003, India

**T. Jayaraj**
Tamil Nadu Rice Research Institute, Aduthurai, 612 101, India

**Muhammad Hamayoon Khan**
Nuclear Institute of Agriculture, Tandojam-70060, Pakistan

**Nazir Ahmad**
Nuclear Institute of Agriculture, Tandojam-70060, Pakistan

**S. M. Masoom Shah Rashdi**
Nuclear Institute of Agriculture, Tandojam-70060, Pakistan

**Muhammad Ismail**
Nuclear Institute of Agriculture, Tandojam-70060, Pakistan

**Imran. Rauf**
Nuclear Institute of Agriculture, Tandojam-70060, Pakistan

**Muhammad Tofique**
Nuclear Institute of Agriculture, Tandojam-70060, Pakistan

**M. Y. Moubarak**
Wheat Research Department, Field Crops Research Institute, A.R.C. Giza, Egypt

**M. F. Abdel-Monaim**
Plant Pathology Research Institute, A.R.C. Giza, Egypt

**M. Sudha Rani**
Plant Breeding, Seed Research and Technology Centre, Rajendranagar, Hyderabad-30, India

**P. S. Rao**
Plant Physiology, Institute of Biotechnology, Rajendranagar, Hyderabad-30, India

**K. Kanaka Durga**
Plant Breeding, Seed Research and Technology Centre, Rajendranagar, Hyderabad-30, India

**V. Bharathi**
Plant Pathology, Seed Research and Technology Centre, Rajendranagar, Hyderabad-30, India

**K. Bharathi**
Seed Research and Technology Centre, Rajendranagar, Hyderabad -30, India

**A.Vishnuvardhan Reddy**
Seed Research and Technology Centre, Rajendranagar, Hyderabad -30, India

**M. Govindappa**
Department of Biotechnology, Shridevi Institute of Engineering and Technology, Sira Road, Tumkur-572 106, India

**V. Ravishankar Rai**
Department of Studies in Microbiology, University of Mysore, Manasa Gangothri, Mysore-570 006, India

**S. Lokesh**
Department of Studies in Applied Botany, Seed Pathology and Biotechnology, University of Mysore, Manasa Gangothri, Mysore-570 006, India

**Fayaz Ahmed Bahar**
Department of Agronomy, G. B. Pant University of Agriculture and Technology, Pantnager-263145 (Uttaranchal), India

**Zahida Rashid**
Department of Agronomy, G. B. Pant University of Agriculture and Technology, Pantnager-263145 (Uttaranchal), India

**J. M. M. Matusso**
Kenyatta University (KU), Agricultural Resources Management Department, P. O. Box 43844 -00100, Nairobi, Kenya

**J. N. Mugwe**
Kenyatta University (KU), Agricultural Resources Management Department, P. O. Box 43844 -00100, Nairobi, Kenya

**M. Mucheru-Muna**
Kenyatta University (KU), Agricultural Resources Management Department, P. O. Box 43844 -00100, Nairobi, Kenya

**Mudasir Hafiz Khan**
Division of Plant Breeding and Genetics, Kisan (P.G) College, Simbhaoli, Ghaziabad (U.P), India

**Sunil Dutt Tyagi**
Division of Plant Breeding and Genetics, Kisan (P.G) College, Simbhaoli, Ghaziabad (U.P), India

**Muhammad Haroon Sarwar**
Faculty of Medicine and Allied Medical Sciences, Isra University, Hyderabad, Sindh, Pakistan

**Muhammad Farhan Sarwar**
Government College Samanabad, Faisalabad, Punjab, Pakistan

**Muhammad Sarwar**
Pakistan Atomic Energy Commission, Nuclear Institute of Agriculture, Tandojam-70060, Sindh, Pakistan

**Niaz Ahmad Qadri**
Hameeda Medicare, Ghulam Muhammadabad, Faisalabad, Punjab, Pakistan

**Safia Moghal**
Hameeda Medicare, Ghulam Muhammadabad, Faisalabad, Punjab, Pakistan

**S. Konlan**
Cocoa Research Institute of Ghana (CRIG), New Tafo-Akim, Ghana

**J. Sarkodie-Addo**
Kwame Nkrumah University of Science and Technology (KNUST), Kumasi, Ghana

**M. J Kombiok**
Savanna Agricultural Research Institute (SARI), Nyankpala, Ghana

**E. Asare**
Kwame Nkrumah University of Science and Technology (KNUST), Kumasi, Ghana

**I. Bawah**
Cocoa Research Institute of Ghana (CRIG), New Tafo-Akim, Ghana

**B. Daramola**
Department of Food Technology Federal Polytechnic, P. M. B. 5351, Ado-Ekiti, Ekiti State, Nigeria

**S. A. Osanyinlusi**
Department of Food Technology Federal Polytechnic, P. M. B. 5351, Ado-Ekiti, Ekiti State Nigeria

**Jude U. Obibuzor**
Biochemistry Division, Nigerian Institute For Oil Palm Research (NIFOR), PMB 1030, Benin City,300001, Edo State, Nigeria

**Roland D. Abigor**
Biochemistry Division, Nigerian Institute For Oil Palm Research (NIFOR), PMB 1030, Benin City,300001, Edo State, Nigeria

**Vivien Omoriyekemwen**
Biochemistry Division, Nigerian Institute For Oil Palm Research (NIFOR), PMB 1030, Benin City,300001, Edo State, Nigeria

**Emmanuel A. Okogbenin**
Biochemistry Division, Nigerian Institute For Oil Palm Research (NIFOR), PMB 1030, Benin City,300001, Edo State, Nigeria

**Tonbra Okunwaye**
Biochemistry Division, Nigerian Institute For Oil Palm Research (NIFOR), PMB 1030, Benin City,300001, Edo State, Nigeria

**B Mirshekarnezhad**
Department of Agronomy and Plant Breeding, Aboureihan College, University of Tehran, Iran

**G. Abbas Akbari**
Department of Agronomy and Plant Breeding, Aboureihan College, University of Tehran, Iran

**G. Ali Akbari**
Department of Agronomy and Plant Breeding, Aboureihan College, University of Tehran, Iran

**H. Sadeghi**
Seed and Plant Certification and Registration Institute, Karaj, Iran

**Yixiang Xu**
Agricultural Research Station, Virginia State University, Petersburg, VA 23806, USA

**Cory Grizzard**
Agricultural Research Station, Virginia State University, Petersburg, VA 23806, USA

**Edward N. Sismour**
Agricultural Research Station, Virginia State University, Petersburg, VA 23806, USA

**Harbans L. Bhardwaj**
Agricultural Research Station, Virginia State University, Petersburg, VA 23806, USA

**Zhenxing Li**
Food Safety Lab, Ocean University of China, Qingdao, Shangdong Province, P. R. 266003 China

**F. E. Dieumou**
Animal Production Programme, School of Agriculture, Abubakar Tafawa Balewa University,
P. M. B. 0248 Bauchi, Nigeria

**T. A Adegbola**
Animal Production Programme, School of Agriculture, Abubakar Tafawa Balewa University,
P. M. B. 0248 Bauchi, Nigeria

**U. D. Doma**
Animal Production Programme, School of Agriculture, Abubakar Tafawa Balewa University,
P. M. B. 0248 Bauchi, Nigeria

**O. M. Egbe**
Department of Plant Breeding and Seed Science, University of Agriculture, P. M. B. 2373, Makurdi, Benue State, Nigeria